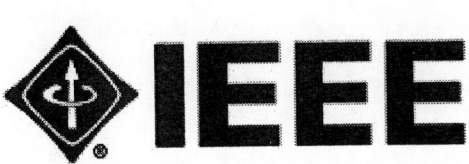 

AA001059

Proceedings of the 15th International Symposium on the Physical & Failure Analysis of Integrated Circuits

IPFA 2008

Edited by: GAN Chee Lip
CHIN Jiann Min
NATARAJAN Mahadeva Iyer
TUNG Chih Hang

Organized by:
IEEE Reliability/CPMT/ED Singapore Chapter

Technically co-sponsored by:

IEEE Electron Devices Society  IEEE Reliability Society

Platinum Sponsor:

Gold Sponsor:
phoenix|x-ray
Part of GE's Sensing &
Inspection Technologies business

Silver Sponsor:

Copyright and Reprint Permission: Abstracting is permitted with credit to the source. Libraries are permitted to photocopy beyond the limit of U.S. copyright law for private use of patrons those articles in this volume that carry a code at the bottom of the first page, provided the per-copy fee indicated in the code is paid through Copyright Clearan e Center, 222 Rosewood Drive, Danvers, MA 01923. For other copying, reprint or republication permission, write to IEEE Copyrights Manager, IEEE Operations Center, 445 Hoes Lane, P.O. Box 1331, Piscataway, NJ 08855-1331. All rights reserved. Copyright ©2008 by the Institute of Electrical and Electronics Engineers

IEEE Catalog Number: CFP08777-PRT
ISBN: 978-1-4244-2039-1
Library of Congress: 2008920146

IPFA 2008 ORGANIZING COMMITTEE

GENERAL CHAIRMAN
TUNG Chih Hang
Institute of Microelectronics

TECHNICAL PROGRAMME CHAIRMAN
GAN Chee Lip
Nanyang Technological University

TECHNICAL PROGRAMME CO-CHAIRMEN
CHIN Jiann Min
Advanced Micro Devices

NATARAJAN Mahadeva Iyer
Chartered Semiconductor Manufacturing

TUTORIAL PROGRAMME CHAIRMAN
John THONG
National University of Singapore

FINANCE CHAIRMAN
TAN Kok Tong
CentiForce Instruments

EXHIBITIONS & SPONSORSHIP CHAIRMEN
Eddie ER
Chartered Semiconductor Manufacturing

Alastair TRIGG
Institute of Microelectronics

WEBMASTER
NATARAJAN Mahadeva Iyer
Chartered Semiconductor Manufacturing

MEMBERS:
Michael CHANG
BASF South East Asia

HO Chaw Sing
Chartered Semiconductor Manufacturing

LIM Yeow Kheng
Chartered Semiconductor Manufacturing

LIM Soon
Systems on Silicon Manufacturing Company

PEY Kin Leong
Nanyang Technological University

M.K. RADHAKRISHNAN
NanoRel, India

TANG Lei Jun
Institute of Microelectronics

SECRETARIAT
Jasmine LEONG
Blk 121 Paya Lebar Way
#03-2801
Singapore 381121
Tel: (65) 6743 2523
Fax: (65) 6746 1095
Email: ipfa@pacific.net.sg
Website: http://www.ieee.org/ipfa

IPFA 2008 TECHNICAL PROGRAMME STEERING COMMITTEE

Christian BOIT
Technical University of Berlin
Germany

Jeffrey GAMBINO
IBM
USA

Guido GROESENEKEN
IMEC
Belgium

Tony OATES
Taiwan Semiconductor Manufacturing Co.
Taiwan

M.K. RADHAKRISHNAN
NanoRel
India

TAO Guoqiao
NXP Semiconductors
The Netherlands

Andre TOUBOUL
Ministry of Research in Bordeaux and Region
Aquitaine France

IPFA BOARD

John THONG (Chairman)
National University of Singapore

PEY Kin Leong
Nanyang Technological University

M.K. RADHAKRISHNAN
NanoRel, India

Alastair TRIGG
Institute of Microelectronics

TUNG Chih Hang
Institute of Microelectronics

IPFA 2008 INTERNATIONAL REVIEWERS:

The IPFA 2008 Organizing Committee would like to thank the following reviewers for their important role in the selection of papers:

D.S. ANG
Nanyang Technological University, Singapore

Ludwig BALK
University of Wuppertal, Germany

Hugo BENDER
IMEC, Belgium

Mike BRUCE
Advanced Micro Devices, USA

Michael CHANG
BASF, Singapore

CHIN Jiann Min
Advanced Micro Devices, Singapore

Alan CRAVEN
University of Glasgow, UK

Hans Juergen ENGELMANN
Advanced Micro Devices, Germany

Eddie ER
Chartered Semiconductor Manufacturing, Singapore

GAN Chee Lip
Nanyang Technological University, Singapore

Y.F. HSIEH
MA Tech, Taiwan

Young-Chang JOO
Seoul National University, Korea

LAM Tim Fai
Spansion, China

Vincent LEE
Institute of Microelectronics, Singapore

Susan LI
Spansion, USA

LI Yuesheng
Fudan University, China

LIM Soon
SSMC, Singapore

LIM Yeow Kheng
Chartered Semiconductor Manufacturing, Singapore

MAHADEVA IYER Natarajan
Chartered Semiconductor Manufacturing, Singapore

C K Maiti
Indian Institute of Technology Kharagpur, India

Alex MENDENILLA
NXP, Philippines

Robert MERTENS
KU Leuven, Belgium

Young-Bae PARK
Andong National University, Korea

PEY Kin Leong
Nanyang Technological University, Singapore

Phillippe PERDU
CNES, France

M.K. RADHAKRISHNAN
NanoRel, India

SIM Kian Sin
Intel, Malaysia

James STATHIS
IBM, USA

Alan STREET
Qualcomm, USA

David SU
Taiwan Semiconductor Manufacturing Co, Taiwan

Sam SUBRAMANIAN
Freescale, USA

TAN Kok Tong
CentiForce, Singapore

Howard T.H. TANG
United Microelectronics Corporation, Taiwan

TANG Leijun
Institute of Microelectronics, Singapore

TEE Tong Yan
Amkor, Singapore

John THONG
National University of Singapore, Singapore

Zsolt TOKEI
IMEC, Belgium

Alastair TRIGG
Institute of Microelectronics, Singapore

TUNG Chih Hang
Institute of Microelectronics, Singapore

J.K. WANG
Broadcom, USA

Mingxiang WANG
Soochow University, China

WU Xiao-Jing
Fudan University, China

ZENG Wei
Fudan University, China

Ehrenfried ZSCHECH
Advanced Micro Devices, Germany

IPFA 2008 TUTORIAL PROGRAMME

Tutorial 1: **TEM AND ASSOCIATED NANOANALYTICAL TECHNIQUES**
Prof Alan J. Craven
University of Glasgow, UK

Tutorial 2: **NBTI AND TDDB METHODOLOGIES AND MODELS: BASICS AND RECENT EVOLUTIONS**
Prof Guido Groeseneken
IMEC, Belgium

Tutorial 3: **IC RELIABILITY ASSESSMENT – WITH A FOCUS ON ELECTROMIGRATION**
Prof Carl V. Thompson
Massachusetts Institute of Technology, USA

Tutorial 4: **IC FAILURE ANALYSIS TECHNIQUES**
Dr Alastair Trigg
Institute of Microelectronics, Singapore
and
Prof Jacob Phang
National University of Singapore

Tutorial 5: **RELIABILITY ISSUES IN LOW-K INTERLEVEL DIELECTRICS**
Dr. Jim Lloyd
IBM, USA

Tutorial 6: **FIB AND SILICON DEBUG & DIAGNOSIS OF ICS IN SUBMICRON TECHNOLOGIES**
Prof Christian Boit
Technical University of Berlin, Germany

CONTENTS

KEYNOTE ADDRESS: Trends and Requirements of Future High-Performance CMOS
A. Khakifirooz and D.A. Antoniadis / *Massachusetts Institute of Technology, USA* *1*

KEYNOTE ADDRESS: Advance Packaging Requirements for Next Generation Microprocessors
R. Master / *Advanced Micro Devices, USA* *7*

SESSION 2: ADVANCED FA I

INVITED PAPER: Scale, Function and Materials: Debug and Diagnosis in Electronic Device Technology Roadmap
C.Boit / *Berlin University of Technology, Germany* *9*

2.1 Scan-Based ATPG Diagnostic and Optical Techniques Combination: A New Approach to Improve Accuracy of Defect Isolation in Functional Logic Failure
A. Machouat[1], G. Haller[1], V. Goubier[1], D. Lewis[2], V. Pouget[2], P. Fouillat[2], F. Essely[2] and P. Perdu[3] / [1]*STMicroelectronics, France;* [2]*IMS Laboratory, France;* [3]*Centre National d'Etudes Spatiales, France* *15*

2.2 Effect of Refractive Solid Immersion Lens Parameters on the Enhancement of Laser Induced Fault Localization Techniques
S.H. Goh[1], ACT Quah[1], CJR Sheppard[1], CM Chua[2], LS Koh[2], JCH Phang[1,2]/ [1]*National University of Singapore;* [2]*SEMICAPS, Singapore* *20*

2.3 Application of FIB Circuit Edit Combined with TIVA in Advanced Failure Analysis
D. Zhu, S.P. Neo, S.K Loh and E. Er / *Chartered Semiconductor Manufacturing, Singapore* *26*

2.4 Failure Analysis Enhancement by Evaluating the Thermal Laser Stimulation Impact on Analog ICs
M. Sienkiewicz[1,2], P. Perdu[1], A. Firiti[2], O.Crepel[2] and D.Lewis[3] / [1]*CNES – French Space Agency, France;* [2]*Freescale Semiconductor, France;* [3]*Université de Bordeaux I, France* *30*

2.5 The Helium Ion Microscope for High Resolution Imaging, Materials Analysis and FA Applications
W.B. Thompson, J. Notte, L. Scipioni and L. Stern / *Carl Zeiss SMT, USA* *36*

SESSION 3: RELIABILITY EVALUATION & ESD

ESREF 2007 BEST PAPER: Time Resolved Determination of Electrical Field Distributions Within Dynamically Biased Power Devices by Spectral EBIC Investigations
A. Pugatschow, R. Heiderhoff and L.J. Balk / *University of Wuppertal, Germany* *43*

3.1 Humidity-Induced Semiconductor Device Electrical Reliability Failures: Mechanism and Wafer-Level Risk Evaluation/ Electrical Screening-Test Proposal
P. Jacob[1,2] and G. Nicoletti[1] / [1]*EMPA Swiss Federal Laboratories for Materials Testing and Research, Switzerland;* [2]*EM Microelectronic Marin SA, Switzerland* *45*

3.2 A Case Study on Different Test Screening Techniques for ICs with High Resistance Vias Interconnects Issues
D. Sim, Y.C. Tan, F. Low, E.G. Foo / *Systems on Silicon Manufacturing Co., Singapore* *49*

3.3 A Novel ESD Device Structure with Fully Silicide Process for Mixed High/Low Voltage Operation
J.H. Lee[1], J.R. Shih[1], D.H. Yang[2], J.F. Chen[2] and K. Wu[1] / [1]*Taiwan Semiconductor Manufacturing Company, Taiwan;* [2]*National Cheng Kung University, Taiwan* *54*

3.4   Optimization on SCR Device with Low Capacitance for On-Chip ESD Protection in UWB
        RF Circuits
        C.Y. Lin and M.D. Ker / *National Chiao-Tung University, Taiwan* 58

SESSION 4: BEOL I - METALLIZATION RELIABILITY

INVITED PAPER: Using Line-Length Effects to Optimize Circuit-Level Reliability
                  C.V. Thompson / *Massachusetts Institute of Technology, USA* 63

4.1   Statistical Modeling of Via Redundancy Effects on Interconnect Reliability
        N. Raghavan[1] and C.M. Tan[2] / *[1]National University of Singapore; [2]Nanyang Technological
        University, Singapore* 67

4.2   Effects of Pulsed Current on Electromigration Lifetime
        M.K. Lim[1,2], C.L. Gan[1], T.L. Tan[2], Y.C. Ee[2], B.C. Zhang[2] and J.B. Tan[2] / *[1]Nanyang
        Technological University, Singapore; [2]Chartered Semiconductor Manufacturing,
        Singapore* 72

4.3   TCAD Solutions for Submicron Copper Interconnect
        H.Ceric, R.L.de Orio, J. Cervenka and S. Selberherr / *Institute for Microelectronics,
        Austria* 78

4.4   Investigation on the Mechanism of the Leakage Failure Between Poly Gate and Contact
        in Subnano Technology
        Q.F. Wang, S.L. Toh, Q. Deng, P.K. Tan, K. Li, J. Teong, Z.H. Mai and J. Lam / *Chartered
        Semiconductor Manufacturing, Singapore* 82

SESSION 5: ADVANCED FA II

ISTFA 2007 BEST PAPER: Raman-IR micro-Thermography Tool for Reliability and Failure Analysis
                  of Electronic Devices
                  A. Sarua[1], J. Pomeroy[1], M. Kuball[1], A. Falk[2], G. Albright[2], M. J. Uren[3] and
                  T. Martin[3] / *[1]University of Bristol, United Kingdom; [2]Quantum Focus
                  Instruments, USA; [3]QinetiQ Ltd, United Kingdom* 87

5.1   An Application of C-AFM as a Tool for SRAM Soft Single-Column Failure Analysis in
        Advanced HV Technologies
        H.S. Lin, M.S. Wu and Y.M. Tsou / *United Microelectronics Corporation, Taiwan* 92

5.2   Conductive Atomic Force Microscopy Failure Analysis for SOI Devices
        S.H. Lim, X.H. Zheng, C.W. Teo, V. Narang, B.H. Teo and J.M. Chin / *Advanced Micro
        Devices, Singapore* 96

5.3   Advanced Localization Technique of Failures in Packages / IO-Stages of Chips Using
        Vector Network Analyser
        B. Krueger, H. Pohl, F. Schumann and S. Schoemann / *Infineon Technologies AG,
        Germany* 100

5.4   Device-Level Fault Isolation of Advanced Flip-Chip Devices using Scanning SQUID Microscopy
        C.W. Teo, H.E. Lwin, V. Narang and J.M. Chin / *Advanced Micro Devices,
        Singapore* 104

5.5   Test Structure Failed Node Localization and Analysis From Die Backside
        Y.G. Li, S.H. Tan and W.R. Sun / *Systems on Silicon Manufacturing Company,
        Singapore* 108

SESSION 6: PACKAGE FAILURE ANALYSIS & RELIABILITY

INVITED PAPER: Reliability of Cu pillar bump for Flip Chip and 3-D SiP
B.J. Kim[1], G.T. Lim[2], J. Kim[3], K. Lee[3], Y.B. Park[2] and Y.C. Joo[1] / [1]Seoul National University, Korea; [2]Andong National University, Korea; [3]Amkor Technology, Korea 111

6.1 Case Study of Copper Dendrite Growth Under HAST Test
S.A. Kim, D.S. Ahn, Y.H. Eum, D.H. Kim and Y.B. Kim / QRT Semiconductor, Korea 116

6.2 Effect of Bonding Pressure on the Bond Strengths of Low Temperature Ag-In Bonds
I.M. Riko[1], C.L. Gan[1], C.K Lee[2,3], L.L. Yan[2], A. Yu[2] and S.W. Yoon[2] / [1]Nanyang Technological University, Singapore; [2]Institute of Microelectronics, Singapore; [3]National University of Singapore 119

6.3 Early Whisker Detection through Intermetallic Compound (IMC) Grain Size
Y.Y. Tan, D. On and J. Krishnan / Infineon Technology, Malaysia 124

6.4 IMC Growth and DR4 Open on TSOP Package
C. Wu, T.F. Lam, X. Song, B. Chen, M. Sun and X. Wang/ Spansion, China 128

6.5 Lid Adhesive Failure Study for Flip Chip Packaging
M.C. Ong[1], X.L. Zhao[1], P.P. Joman[1], J. M. Chin[1] and R.N. Master[2] / [1]Advanced Micro Devices, Singapore; [2]Advanced Micro Devices, USA 132

SESSION 7: POSTER PAPERS

7.1 CAFM Detection of Resistive Tungsten Contacts in DRAM Devices
E. Ng[1], D. Lam[1] and X. Zheng[2] / [1]Micron Semiconductor Asia, Singapore; [2]Nanyang Polytechnic, Singapore 137

7.2 Application of Conductive Atomic Force Microscopy to Study the In-line Electrical Defects
S.L. Toh, Q. Deng, W.T. Tang, V. Lim, F.H. Gn, P.K. Tan, H.Tan, Z.H. Mai, J. Lam / Chartered Semiconductor Manufacturing, Singapore 141

7.3 Investigation of Soft Fail Issue in Sub-Nanometer Devices Using Nanoprobing Technique
E. Hendarto, H.B. Lin, S.L. Toh, P.K. Tan, Y.W. Goh, Z.H. Mai and J. Lam / Chartered Semiconductor Manufacturing, Singapore 145

7.4 Failure Analysis of 65nm Technology Node SRAM Soft Failure
C.Q. Chen, E. Er, S.P. Neo, S.K. Loh, Q.X. Wang, and J.Teong / Chartered Semiconductor Manufacturing, Singapore 150

7.5 A Simple Method for TEM Sample Preparation Without Carbon Film Background
M.L. Lee and R.D. Lin / United Microelectronics Corporation, Singapore 154

7.6 Improved Image Processing To Enhance Thermal Laser Stimulation Signal
A. Deyine-Barth, P. Perdu and K. Sanchez / CNES-THALES Laboratory, France 157

7.7 Near-Infrared Spectroscopic Photon Emission Microscopy of 0.13 μm Silicon nMOSFETs and pMOSFETs
S.L. Tan[1], J.K.J. Teo[1], K.H. Toh[1], D. Isakov[1], D.S.H. Chan[1], L.S. Koh[2], C.M. Chua[2] and J.C.H. Phang[1,2] / [1]National University of Singapore; [2]SEMICAPS, Singapore 162

7.8 IC Package Inspection with Nanofocus X-ray Tubes and NanoCT
H. Roth, Z.H. He, T. Paul / phoenix|x-ray Systems + Services, Germany 167

7.9 EMMI Analysis on Silicon Solar Cell
B. Yeh, R. Huang, K. Chung, A. Chang and C.H. Chu / Materials Analysis Technology, Taiwan 170

7.10   Backside GMR Magnetic Microscopy for Flip Chip and Related Microelectronic Devices
M. Hechtl / *Infineon Technologies AG, Germany* 174

7.11   Chemical and Physical Characterization Techniques in Highlighting Intermetallic Compound
(IMC) Formation
J.C.M. Fernandez / *Fairchild Semiconductor, Philippines* 178

7.12   Influence of Interconnect Dimensions on Electromigration for Cu/Low-k Interconnect Structure:
An Analytical Study
B.N. Joshi and A.M. Mahajan / *North Maharashtra University, India* 181

7.13   A Novel Pseudo Tri-Gate Vertical MOSFET with Source/Drain Tie
J.T. Lin, Y.C. Tsai, Y.C. Eng, S.S. Kang, Y.M. Tseng, H.J. Tseng and P.H. Lin /
National Sun Yat-Sen University, Taiwan 185

7.14   Simulation of the Multi-Source/Drain SOI MOSFET
P.H. Lin, S.S. Kang, J.T. Lin and Y.C. Eng / *National Sun Yat-Sen University,
Taiwan* 189

7.15   Stress-Induced Degradation in Strain-Engineered nMOSFETs
T.K. Maiti, S.S. Mahato, M.K. Bera, M. Sengupta, P. Chakraborty, C. Mahata,
A. Chakraborty and C.K. Maiti / *Indian Institute of Technology Kharagpur, India* 193

7.16   DIBL in Short-Channel Strained-Si n-MOSFET
S.S. Mahato[1], P. Chakraborty[1], T.K. Maiti[1], M.K. Bera[1], C. Mahata[1], M. Sengupta[1],
A. Chakraborty[1], S.K. Sarkar[2] and C. K. Maiti[1] / *[1]Indian Institute of Technology Kharagpur,
India; [2]Jadaypur University, India* 196

7.17   Influence of Hydrogen Annealing on NBTI Performance
L.J. Jin, H.P. Kuan, D. Sim and M. Mukhopadhya / *Systems on Silicon Manufacturing Co.,
Singapore* 200

7.18   Reliability of ZrO_2/GeO_xN_y Stacked High-k Dielectrics on Ge under Dynamic and Pulsed
Voltage Stress
C. Mahata[1], M. K. Bera[1], P.K. Bose[2], A. Chakraborty[1], M. SenGupta[1] and C. K. Maiti[1] /
[1]Indian Institute of Technology Kharagpur, India; [2]Jadavpur University, India 204

SESSION 8: NOVEL DEVICES I - SOLAR CELLS & MEMORY

INVITED PAPER: Trends in Solar Cell Research
R. Mertens / *IMEC and K.U.Leuven, Belgium* 209

8.1   Au Nanocrystal Flash Memory Reliability and Failure Analysis
P.K. Singh[1,2], K.K. Singh[1], R. Hofmann[1], K. Armstrong[1], N. Krishna[1] and S. Mahapatra[2] /
[1]Applied Materials, USA; [2]Indian Institute of Technology Bombay, India 214

8.2   Characterization and Modeling of Program/Erase Induced Device Degradation in
2T-FNFN-NOR Flash Memories
G.Q. Tao, H. Chauveau, D. Boter, E. Vegt, D. Dormans and R.Verhaar /
NXP Semiconductors, The Netherlands 219

8.3   The Effect of Band Gap Engineering of the Nitride Storage Node on Performance and
Reliability of Charge Trap Flash
S. Chandrashekhar[2], U. Ganguly[1], K.K. Singh[1], C. Olsen[1], S. Seutter[1], G. Conti[1], K. Ahmed[1],
N. Krishna, J. Vasi[2] and S. Mahapatra[2] / *[1]Applied Materials, USA; [2] Indian Institute of
Technology Bombay, India* 224

8.4   A Novel Dual-BBHH Erasing Scheme to Improve Endurance and Retention Performances
for Localized Charge Trapping Memories
G.J. Shi[1], L.Y. Pan[1,2], R. Ritzenthaler[1], L. Sun[1], Z.G. Zhang[1,2] and J. Xu[1,2] / *[1]Institute of
Microelectronics of the Tsinghua University, China; [2]Tsinghua National Laboratory for
Information Science and Technology, China* 231

8.5   Study on SRAM Soft Failure Using Planar-View Transmission Electron Microscopy Techniques
P. Liu, K. Li, Y. Li, C.Q. Chen, E. Er and J. Teong / *Chartered Semiconductor Manufacturing, Singapore* 234

SESSION 9: FEOL

INVITED PAPER: Review of Reliability Issue in High-k/Metal Gate Stacks
G. Groeseneken / *IMEC, Belgium* 239

9.1   A New TDDB Lifetime Bi-Model for eDRAM MIM Capacitor with ZrO2 High-K Dielectrics
S.W. Chang, C.L Chen, C.J. Wang and K. Wu / *Taiwan Semiconductor Manufacturing Company, Taiwan* 245

9.2   A Rigorous Model for Trapping and Detrapping in Thin Gate Dielectrics
W. Goes, M. Karner, V. Sverdlov and T. Grasser / *Institute for Microelectronics, Austria* 249

9.3   The Impact of Gate Dielectric Nitridation Methodology on NBTI of SiON p-MOSFETs as Studied by UF-OTF Technique
V. D. Maheta[1], C. Olsen[2], K. Ahmed[2] and S. Mahapatra[1] / [1]*Indian Institute of Technology Bombay, India;* [2]*Applied Materials, USA* 255

9.4   Trapping and De-trapping Characteristics in PBTI and Dynamic PBTI between HfO_2 and HfSiON Gate Dielectrics
W.L. Lin[1], Y.J. Lee[2], W.C. Lo[1], K.S. Chen[2], Y.T. Hou[4], K.C. Lin[4] and T.S. Chao[1] / [1]*National Chiao-Tung University, Taiwan;* [2]*National Nano Device Laboratories, Taiwan;* [4]*Taiwan Semiconductor Manufacturing Company, Taiwan* 260

9.5   A Study of NBTI in HfSiON/TiN p-MOSFETs Using Ultra-Fast On-The-Fly (UF-OTF) I_{DLIN} Technique
S. Deora and S. Mahapatra / *Indian Institute of Technology Bombay, India* 264

SESSION 10: PHYSICAL & CHEMICAL CHARACTERIZATION

INVITED PAPER: Nanoanalysis of High-k Dielectrics on Semiconductors
A.J. Craven[1], M. MacKenzie[1] and D.W. McComb[2] / [1]*University of Glasgow, United Kingdom;* [2]*Imperial College London, United Kingdom* 269

10.1   Physical Characterization Challenges in 45nm Technology Node
K. Li, P. Liu, Q. Wang, I. Tee and J. Teong / *Chartered Semiconductor Manufacturing, Singapore* 275

10.2   Understanding Soft Defect Localization Set-Points for Reducing Cause-Not-Founds in Integrated Circuits
V.K.Ravikumar, S.L. Phoa, V. Narang and J.M. Chin / *Advanced Micro Devices, Singapore* 279

10.3   Gate Oxide Integrity Failure Caused by Molybdenum Contamination Introduced in the Ion Implantation
D. Gui, Y.H. Huang, G.B. Ang, Z.X. Xing, Z.Q. Mo, Y.N Hua and J. Teong / *Chartered Semiconductor Manufacturing, Singapore* 283

10.4   Tool Cleanliness Characterization for Improving Productivity and Yields
V.K.F. Chia / *Air Liquide – Balazs Analytical Services, Singapore* 287

10.5   Studies and Applications of Standardless EDX Quantification Method in Failure Analysis Of Wafer Fabrication
Y.N. Hua, B.H. Liu, Z.Q. Mo, J. Teong / *Chartered Semiconductor Manufacturing, Singapore* 290

SESSION 11: BEOL II - LOW-K & ADVANCED INTERCONNECTS

INVITED PAPER: New Models for Interconnect Failure in Advanced IC Technology
J. Lloyd / *IBM, USA* 297

11.1 Interfacial characterization of ultra low-k film (kappa=2.55) with Time of Flight Secondary Ion Mass Spectrometry (TOF-SIMS)
J. Widodo, Z.X. Xing, Z.Q. Mo, T. Ouyang, D. Gui, Y.N Hua, H. Liu and W. Lu / *Chartered Semiconductor Manufacturing, Singapore* 304

11.2 Etching of Copper in Deionized Water Rinse
J. Gambino[1], J. Robbins[1], T. Rutkowski[1], C. Johnson[1], K. DeVries[1], D. Rath[2], P. Vereecken[3], E. Walton[2], B. Porth[1], M. Wenner[1], T. McDevitt[1], J. Chapple-Sokol[1] and S. Luce[1] / [1]*IBM Microelectronics, USA;* [2]*IMEC, Belgium;* [3]*IBM T.J. Watson Research Center, USA* 308

11.3 Current-Induced Breakdown of Carbon Nanofiber Interconnects
H. Kitsuki, T. Saito, T. Yamada, D. Fabris, P. Wilhite, M. Suzuki and C.Y. Yang / *Santa Clara University, USA* 312

11.4 Solderability and Reliability of Printed Electronics
B. Salam and B.L. Lok / *Singapore Institute of Manufacturing Technology, Singapore* 316

SESSION 12: NOVEL DEVICES II

12.1 The Device Characteristics of Partially Undoped Poly-Silicon Gate P-LDMOS Power Transistors
R.Y. Su[1], P.Y. Chiang[1], J. Gong[1], J.L. Tsai[2], T.Y. Huang[2], Mingo Liu[2] and C.C. Choub[2] / [1]*National Tsing Hua University, Taiwan;* [2]*Taiwan Semiconductor Manufacturing Company, Taiwan* 321

12.2 Hot Carrier Degradation in Nanowire (NW) FinFETs
T.K. Maiti, M.K. Bera, S.S. Mahato, P.Chakraborty, C.Mahata, M.Sengupta, A.Chakraborty and C.K. Maiti / *Indian Institute of Technology Kharagpur, India* 325

12.3 Degradation of Metal-Induced Laterally Crystallized n-Type Poly-Si Thin-Film Transistors Under Dynamic Voltage Stress
M. Zhang, M.X. Wang and H.S. Wang / *Soochow University, China* 329

12.4 Failure Analysis Matrix of Light Emitting Diodes for General Lighting Applications
N. Hwang / *Korea Photonics Technology Institute, Korea* 333

12.5 Failure Site Isolation on Passive RFID Tags
B. Sood[1], D. Das[1], M. Azarian[1], M. Pecht[1], B. Bolton[2] and T. Lin[3] / [1]*University of Maryland, USA;* [2]*Motorola Inc., USA;* [3]*Motorola Electronics, Singapore* 337

KEYNOTE ADDRESSES

KEYNOTE ADDRESS

Trends and Requirements of Future High-Performance CMOS

Ali Khakifirooz and Dimitri A. Antoniadis

Microsystems Technology Laboratories, Massachusetts Institute of Technology, Cambridge, MA, USA.
Phone: (617) 253-4693 Fax: (617) 324-5341 Email: daa@mtl.mit.edu

Abstract- **The outlook of performance scaling in high-performance CMOS is explored by using an analytical expression for the intrinsic MOSFET delay. The historical trend of carrier virtual source velocity, as the main driver for continuous performance increase in the past, is presented and prospects of further velocity increase in future technology nodes are discussed. An optimistic scaling scenario with realistic assumptions about device geometry and electrostatics is presented and it is shown that from the 32-nm node onward the intrinsic transistor performance will not improve with device scaling unless parasitic capacitances are significantly reduced.**

I. INTRODUCTION

CMOS scaling for high-performance (HP) applications has proceeded successfully down to the 65-nm node and the first full 45-nm node technologies have already been demonstrated. While the so-called contacted gate pitch has scaled as planned [1]-[3], the most important deviation from ITRS projections has been the slow-down of the gate length scaling [1]-[6], which has been in part due to the delay in the anticipated introduction of gate stacks with high-κ dielectrics and metallic electrode. Meanwhile, transistor performance, either defined as the inverse of the intrinsic MOSFET delay or measured experimentally as the inverse of the ring oscillator delay, has increased commensurate with the technology scaling down to the 65-nm node. However, slow-down of transistor performance scaling is being observed in the 45-nm node MOSFETs.

This paper explores the scaling trends of MOSFET performance in future technology nodes with realistic, yet optimist assumptions about the device geometry, electrostatics, and carrier transport. An analytical expression for the intrinsic MOSFET delay [7]–[9] is used to study the historical trend and to predict the future of transistor performance scaling. Scaling trends of virtual source velocity, as the main driver of performance scaling, are first examined and prospects of further velocity increase in the future are discussed. A scaling scenario for CMOS performance scaling is then outlined and it is shown that counter-scaling of device performance will be observed form the 32-nm node onward, unless parasitic capacitances are significantly reduced.

II. NEW METRIC FOR MOSFET INTRINSIC PERFORMANCE

It has been already demonstrated that the conventional CV/I metric fails to reproduce the dependence of the intrinsic MOSFET delay on the main technology parameters [9]. On one

---

Parts of this paper have appeared in References [9] and [19].

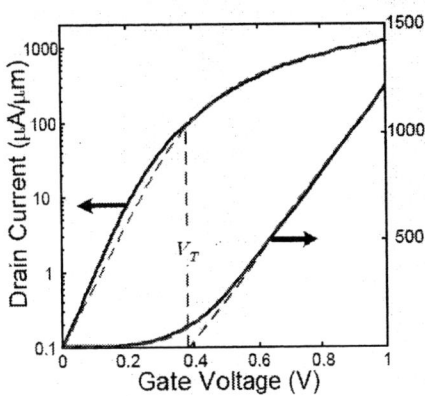

Fig. 1. Demonstration of the fitting the experimental data [2] with simple analytical model. The threshold voltage in saturation is extracted by linear extrapolation of the $I_D - V_{GS}$ characteristics and is significantly higher than the number reported in the literature, which is defined at given current in subthreshold. The continuous model of (6) is also shown for comparison.

hand it uses the saturation drain current, while the transient drain current never reaches this value during the switching [10], and on the other hand, it does not contain the parasitic capacitances inherent to the transistor gate. Nevertheless, CV/I metric provided acceptable results in earlier technology nodes because it also overestimates the inversion charge. However, this will not be the case in the future as the relative importance of the parasitic capacitances increases and as the gate overdrive voltage decreases. In our earlier work [9], an analytical expression was proposed for the intrinsic transistor delay based on the concept of the "effective drain current" [10]:

$$ t = \frac{(1 - \delta)V_{DD} - V_T + C_f^* V_{DD} / C_{inv} L_G}{(3 - \delta)V_{DD} / 4 - V_T} \frac{L_G}{v}, \qquad (1) $$

where V_{DD} is the supply voltage, δ is the DIBL coefficient in V/V, V_T is the saturation threshold voltage, L_G is the gate length, C_f^* is the effective fringing capacitance per unit width that includes inner and outer fringing and overlap capacitances, with Miller effect for the drain side, C_{inv} is the inversion capacitance per unit area, and v is the effective velocity defined by

$$ I_D / W = C_{inv}(V_{GS} - V_T)v . \qquad (2) $$

The effective velocity is related to the actual average velocity of carriers when injected into the channel, called virtual-source velocity, v_{x0}, by:

978-1-4244-2039-1/08/$25.00 ©2008 IEEE

KEYNOTE ADDRESS

$$v = \frac{v_{x0}}{1 + R_S C_{\text{inv}} W (1 + 2d) v_{x0}}, \qquad (3)$$

where R_S is the source series resistance. It should be noted that the threshold voltage in (1) is determined by linear extrapolation of the $I_D - V_{GS}$ characteristics in saturation, as defined in (2), and is usually 200 mV higher than the traditionally defined threshold voltage at a given current in subthreshold. This difference has significant consequences when quantifying the projected requirements for future technology nodes.

The off-current is given by:

$$I_{\text{off}} / W = I_{\text{ref}} \, 10^{-V_T / S^*} \qquad (4)$$

where $I_{\text{ref}} = v_{x0} Q_0$, with Q_0 empirically found to be roughly 8×10^{-8} C/cm^2, and S^* is the effective subthreshold swing.

Eqs. (2) and (5) together describe the $I_D - V_{GS}$ characteristics in saturation as shown in Fig. 1. When a continuous model is desired, the following model can be used [11]:

$$Q_{\text{inv}} = C_{\text{inv}} h f_t \, \log(1 + \exp(\frac{V_{GS}' - V_T'}{h f_t})) \qquad (5)$$

where $\phi_t = \eta k_B T$ is the thermal voltage, $\eta = S^*/0.06$ the ideality factor, k_B Boltzmann's constant, T absolute temperature, q electronic charge, and $V_{GS}' = V_{GS} - R_S I_D$ and $V_T' = V_T + d (V_{DS} - 2 R_S I_D)$ are the internal gate overdrive and threshold voltage, respectively. The drain current is then given by $I_D/W = Q_{\text{inv}} v_{x0}$. The above model requires some numerical iteration to obtain the $I_D - V_{GS}$ characteristics. However, satisfactory results may be obtained by using

$$I_D / W = C_{\text{inv}} h f_t \, \log(1 + \exp(\frac{V_{GS} - V_T}{h f_t})) v, \qquad (6)$$

since it introduces minor deviation from the exact solution of (5) in subthreshold. In fact, since the effective velocity v is slightly smaller than the virtual source velocity v_{x0}, this approximation to some extent takes care of the fact that the virtual source velocity increases slightly with the gate voltage from the weak to strong inversion [8].

Eq. (1) emphasizes the importance of the DIBL in determining the transistor performance: With given saturation threshold voltage, set by the required I_{off} it is important to minimize the DIBL in order to maximize the effective drain current, I_{eff} [12]. This is especially important since the main contributor to I_{eff} is the current at $V_{DS} = V_{DD}/2$ and $V_{GS} = V_{DD}$. Moreover, for stacked transistors, such as those in NAND gate, it has been shown that the transistor current in the linear regime plays a significant role in the overall gate delay and thus it is even more important to minimize the DIBL [13].

III. SCALING TREND OF VIRTUAL SOURCE VELOCITY

Over the past two decades, continuous increase of the transistor performance has been achieved through increasing virtual source velocity with device scaling. Fig. 2 shows the historical trend of the virtual source velocity, extracted from literature data. Down to gate lengths of about 100 nm, v_{x0} has

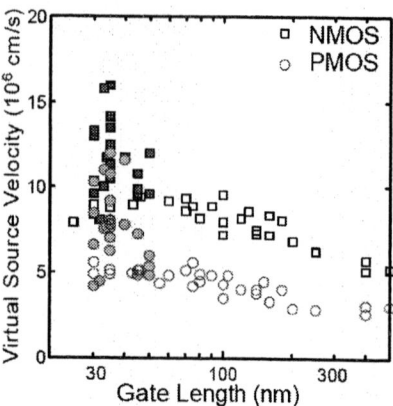

Fig. 2. Scaling trend of the virtual source velocity for benchmark technologies. Filled symbols represent strain-engineered devices..

been steadily increased solely through aggressive scaling of the transistor channel length. According to MOSFET scattering theory [14], the critical length of scattering, i.e. the length of the region near the source where carriers are most likely to backscatter to the source if they encounter scattering event, decreases as the channel length is scaled. Hence, for given ballistic velocity, determined by carrier statistics and transport properties of silicon, virtual source velocity increases as the channel length decreases.

Closer examination of the data in Fig. 2 demonstrates that for unstrained Si transistors with gate length less than about 100 nm, there has been saturation and even slight decrease in v_{x0}. This is most likely due to excessive carrier scattering due to both increased effective electric field normal to the channel, and coulombic scattering from increasingly heavier halos that hindered further increase of the velocity. Further increase of the virtual source velocity was made possible when various strain-engineering methods were first introduced in the 90-nm node. With increased carrier mobility, not only the transistor operates closer to the ballistic velocity, but also since the mobility enhancement in uniaxially strained silicon is mostly due to reduction in the effective mass in the channel direction [15], the ballistic velocity itself increases.

As will be discussed in the next section, further increase in the virtual source velocity is needed in future technology nodes to continue the historical trend of performance scaling. However, the effectiveness of stress liners in applying mechanical strain to the channel decreases as the transistor pitch is reduced [2]. Hence, it becomes challenging to increase or even maintain the current values of the virtual source velocity in future technology nodes. Examination of 45-nm node MOSFETs reveals that in some cases v_{x0} has already stepped back from the corresponding 65-nm technologies as reflected in Fig. 3. This could be either due to higher carrier scattering imposed by high-κ gate dielectrics or to lower strain level due to smaller gate pitch, both inevitable in future technology nodes. A combination of stress liners, stress memorization, stress-engineered substrates, or embedded stressors along with optimized device geometry is

KEYNOTE ADDRESS

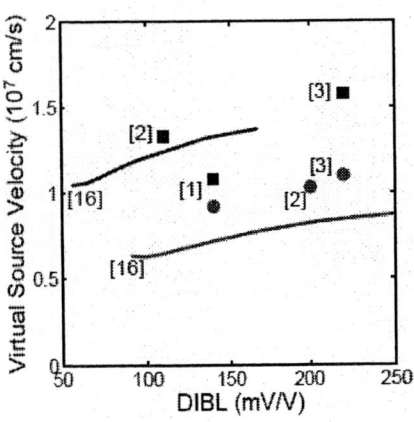

Fig. 3. Comparison of the virtual source velocity in 65 and 45-nm technologies. For a fair comparison, data are plotted against DIBL and only 45nm devices with a contacted gate pitch of 160 nm are included. Solid lines represent electron and hole virtual source velocity in a 65nm technology (L_{pitch} = 220 nm) [16], while symbols show data for 45nm node devices (L_{pitch} = 160 nm) [1]–[3].

thus required to further increase the virtual source velocity in future technology nodes.

Although in principle electron velocity can be increased with higher strain levels, theoretical calculations suggest that hole ballistic velocity in uniaxially strained silicon saturates to about $2-2.5\times$ compared to relaxed silicon. This is due to the fact that the band structure at the top of the valence band does not change with strain once a certain strain level, around 1% for typical carrier densities, is reached [17]. Yet, mobility continues to increase since it depends on the details of the band structure at higher energy levels through reduction in the optical phonon scattering [18].

IV. HIGH-PERFORMANCE CMOS SCALING SCENARIO

Fig. 4 illustrates the scaling trend of the key feature sizes used in this analysis [19]. ITRS projections [20] are also shown in dashed lines for comparison. The main scaling parameter is the contacted gate pitch, L_{pitch}, which historically has scaled by 0.7 per technology node and is desired to continue the same pace in order to enable the doubling of transistor count per each node to reduce the cost and/or increase the chip functionality in accordance with Moore's law. The slow-down of the gate length scaling is taken into account in our projection, whereas ITRS continues to scale the gate length with its "current" pace. We assume gradual decrease in the inversion oxide thickness somewhat similar to ITRS projections. Although some publications on 45nm CMOS continue to use silicon oxynitride as the gate dielectric [2]-[6], we assume high-κ dielectric to allow more aggressive scaling and be consistent with industry announcements [1], [21].

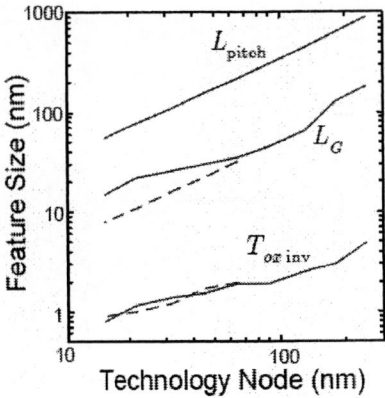

Fig. 4. Scaling trend of key feature sizes for HP CMOS assumed in this work (solid lines) in comparison with ITRS projections (dashed lines). Reprinted from [19].

Starting from the 32 nm technology node, ITRS assumes that the off-current decreases to about 100 nA/ μm, whereas in our projection we let it increase to 300 nA/μm and stay there. Although ITRS does not specify the subthreshold swing and DIBL, from the specified threshold voltage and I_{off} values, the assumed subthreshold swing in ITRS appears to be very optimistic, nearly ideal for double-gate structures. In addition, we assume that the virtual source velocity, v_{x0}, increases somewhat from the 65 to the 45nm node and then stays constant as shown in Fig. 5(a), something that, at least for uniaxially strained silicon, is probably very optimistic as discussed in the previous section.

Models from literature are used to calculate the parasitic components. Parasitic capacitances are calculated using the analytical model introduced in [22]. Fig. 6 shows the scaling trend of the effective fringing capacitance and the intrinsic inversion gate capacitance. The shaded area represents the contribution of the parasitic capacitance between the gate and source/drain contact. Note that the Miller effect is accounted for when calculating the effective fringing capacitances. Also, it should be noted that ITRS requires that the contribution of the parasitic capacitance in the total gate capacitance decreases from about 45% to about 20% as the transistors are scaled. Possibilities to reduce the parasitic capacitance are being sought. However, numbers in the vicinity of those required by ITRS are very optimistic. Another important fact, ignored in ITRS projections, is that unlike parasitic capacitance, which is multiplied by the supply voltage, V_{DD}, to give the parasitic switching charge, the inversion capacitance is multiplied by $V_{DD} - V_{T0}$, where $V_{T0} = V_T + \delta V_{DD}$ is the linear threshold voltage. The ratio V_{T0}/V_{DD} increases as the CMOS scaling proceeds to future technology nodes. Hence, parasitic charge constitutes an ever increasing portion of the total switching charge. An analytical model [23] is used to estimate the source/drain series resistance, R_S. It depends on the sheet resistance of the heavily doped regions, R_{sheet}, which at best is likely to stay constant among generations, the specific contact resistance between silicide and doped silicon, ρ_c, which may be decreasing slightly, and a constant extension resistance, R_{ext}, which is assumed to be

978-1-4244-2039-1/08/$25.00 ©2008 IEEE

KEYNOTE ADDRESS

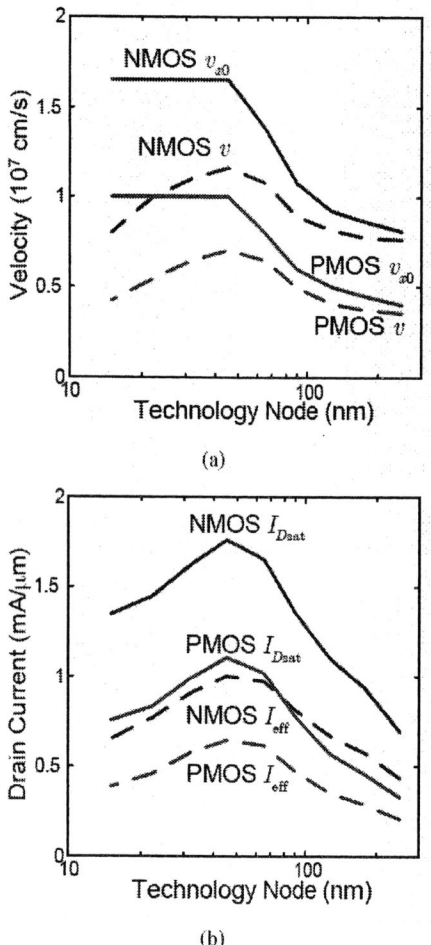

(a)

(b)

Fig. 5. (a) Scaling trend of the virtual source velocity, v_{x0}, and the effective velocity, v for NMOS and PMOS transistors. Despite the fact that the virtual source velocity is kept constant beyond 45-nm technology node and that R_S increases only slightly, the effective velocity drops because of the increased C_{inv} according to (3). (b) With the ratio V_T/V_{DD} increasing with the device scaling, both the effective current, I_{eff}, and the saturation drain current, I_{Dsat}, drop further.

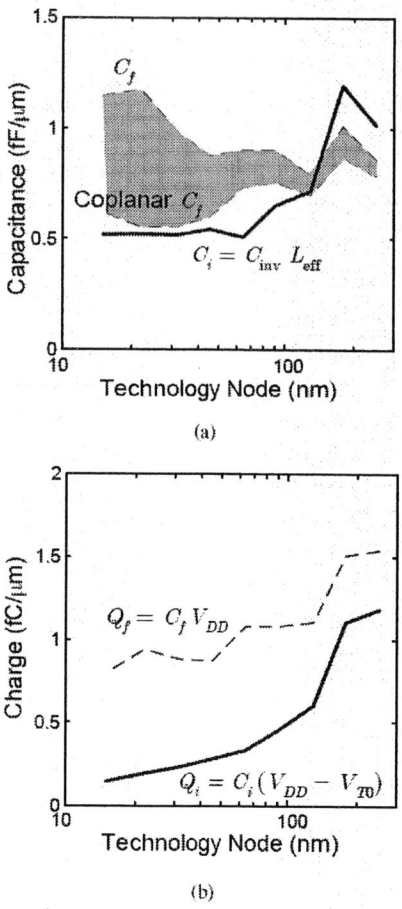

(a)

(b)

Fig. 6. (a) Comparison of the effective parasitic capacitance, C_f and the intrinsic inversion capacitance, $C_{inv}L_{eff}$ as a function of the technology node. The shaded area represents the contribution of the parasitic capacitance between the gate electrode and S/contact studs. Miller effect is taken into account for the drain side when calculating the effective parasitic capacitance. (b) Comparison of the parasitic and intrinsic charge vs. technology node. The effect of the parasitic capacitance is amplified as the ratio V_{T0}/V_{DD} increases with scaling. Reprinted from [19].

roughly 45 and 75 $\Omega.\mu m$ for NMOS and PMOS transistors, respectively. This value comes from empirical observations, but there is certainly fundamental limit to this resistance resulting from the transition from 3D to 2D carrier distribution [24], even though it is not easy to calculate. It is noteworthy that in this study the metal stud resistance or more precisely the Metal 1 to silicide resistance is assumed to be negligible.

Model calculations show that R_S is likely to increase slightly with scaling despite the assumed reduction in the specific contact resistance. This is mainly due to the fact that the area available for silicide formation decreases as the technology is scaled. Again, ITRS requires that the series resistance decreases

with scaling. However, R_S has been almost constant for NMOS transistors over multiple technology nodes in the past and is unlikely to decrease dramatically. The advent of embedded SiGe in S/D has already reduced PMOS series resistance by about factor of two, yet the corresponding numbers are still about twice those of NFETs. According to Fig. 5, although the parasitic resistance increases only slightly based on our assumptions, its impact on the effective velocity increases with increased C_{inv}. Hence the effective velocity shows drop starting from 45-nm technology, despite the fact that the virtual source velocity is kept constant in our projections.

KEYNOTE ADDRESS

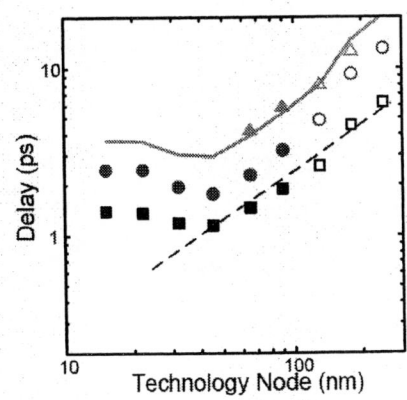

Fig. 7. The calculated intrinsic NMOS (squares) and PMOS (circles) transistor delay as a function of the technology node. Solid line is the calculated ring oscillator delay, using empirical equation $t_{RO} = 1.1(t_{NMOS} + t_{PMOS})$, while triangles represent experimental ring oscillator delay data [16]. Down to 45-nm technology node, the intrinsic transistor delay scales in proportion to the technology scaling. However, from 32-nm node onward the projected delay increases with device scaling. The required "target" delay at each future technology node is given by linear extrapolation of the historical data in log-log scale. Solid symbols represent strain-engineered devices.

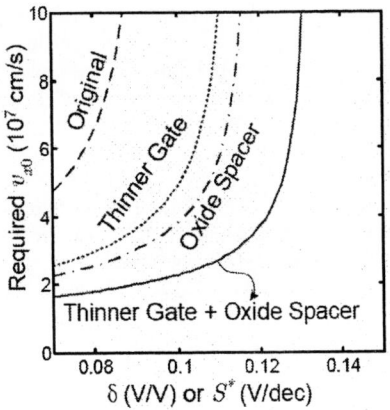

Fig. 8. The impact of reducing the parasitic capacitance on the required virtual source velocity to meet the target delay of 0.6 ps for the 22-nm high-performance NMOS as function of the electrostatic integrity. By reducing the gate height by factor of and replacing the silicon nitride that surrounds the gate by SiO_2, it is possible to significantly reduce the required virtual source velocity. Yet, the required velocity is comparable to the ballistic limit in uniaxially strained silicon with achievable strain levels and stringent control of short channel effects is necessary.

Fig. 7 shows the scaling trend of the intrinsic delay of high-performance transistors, calculated using (1). Experimental ring oscillator data for the past technology nodes [16] are also shown for comparison, along with the calculated ring oscillator delay. It has been shown empirically that the ring oscillator delay can be calculated as the sum of the NMOS and PMOS intrinsic delays multiplied by factor of 1.1 [9], i.e.,

$$t_{RO} = 1.1(t_{NMOS} + t_{PMOS}). \tag{7}$$

The data demonstrate that down to the 65 nm node, the calculated intrinsic delay has been scaled in proportion to the dimensional scaling and the calculated delay using (7) is in agreement with the experimental data. However, starting from the 45nm technology generation, the intrinsic delay is expected to stop decreasing and to exhibit counter-scaling thereafter. One might argue that this behavior is mostly due to the fact that in the scaling scenario presented in Fig. 4, the gate length scaling has been slowed down and this results in less current drive and more switching charge. However, it should be noted that more aggressive gate length scaling does not necessarily translate to higher carrier velocity as long as the electrostatic integrity of the devices is preserved. Also, as shown in Fig. 6, most of the switching charge is associated with the parasitic capacitances that do not scale with gate length scaling. However, with given transistor pitch, a more aggressive scaling of the gate length provides more space between the gate and S/D electrodes and leads to lower parasitic capacitance, which should be helpful.

Since most of the switching charge is due to the parasitic capacitances, the most effective way to meet the required delay in futures technology nodes is to reduce the parasitic capacitances [19]. Fig. 8 shows the impact of reducing the parasitic capacitance on the required virtual source velocity to meet the target delay of 0.6 ps for 22-nm high-performance NMOS as function of the electrostatic integrity. For simplicity we assume that the numerical value of the DIBL (in V/V) is equal to the effective subthreshold swing (in V/dec). Without reducing C_f^*, it is impossible to meet the target delay even with III-V semiconductors as the channel material. However, with simple provisions, such as reducing the gate height by factor of 2 or by replacing the oxynitride material that surrounds the gate electrode with SiO_2, it is possible to reduce the required virtual source velocity significantly. Although the corresponding numbers are still higher than what was assumed in our projections for uniaxially strained Si and also it might not be possible to apply enough strain to the channel once the oxynitride material is removed, Fig. 8 clearly demonstrates the importance of reducing the parasitic gate capacitances in future technology nodes.

V. CONCLUSION

Prospects of performance scaling in high-performance CMOS were explored by using an analytical expression for the intrinsic MOSFET delay. Even with the optimistic scaling scenario used in this analysis, it is difficult to continue the historical trend of the performance scaling from the 32nm node onward. It is demonstrated that the most promising method to lower the required virtual source velocity to what might be feasible with uniaxially strained silicon is to reduce the gate fringing capacitance.

REFERENCES

[1] K. Mistry, et al., "A 45nm logic technology with high-k+metal gate transistors, strained silicon, 9 Cu interconnect layers, 193nm dry patterning, and 100% Pb-free packaging," in *IEDM Tech. Dig.*, 2007, pp. 247-250.

[2] K.L.Cheng, *et al.*, "highly scaled, high performance 45nm bulk logic CMOS technology with 0.242 μm^2 SRAM cell," in *IEDM Tech. Dig.*, 2007, pp. 243-246.

[3] S.K.H. Fung, *et al.*, "45nm SOI CMOS technology with 3X hole mobility enhancement and asymmetric transistor for high performance CPU application," in *IEDM Tech. Dig.*, 2007, pp. 1035-1037.

[4] S. Narasimha, *et al.*, "High performance 45nm SOI technology with enhanced strain, porous low-k BEOL, and immersion lithography," in *IEDM Tech. Dig.*, 2006, pp. 689-692.

[5] H. Nii, *et al.*, "A 45nm high performance bulk logic platform technology (CMOS6) using ultra NA(1.07) immersion lithography with hybrid dual damascene structure and porous low-k BEOL," in *IEDM Tech. Dig.*, 2006, pp. 685-688.

[6] Z. Luo, *et al.*, "High performance transistors featured in an aggressively scaled 45nm bulk CMOS technology," in *Symp. VLSI Tech.*, 2007, pp. 16-17.

[7] D. A. Antoniadis, I. Åberg, C. Ní Chléirigh, O. M. Nayfeh, A. Khakifirooz, and J. L. Hoyt, "Continuous MOSFET performance increase with device scaling: The role of strain and channel material innovations," *IBM J. Research Development*, vol. 50, no. 4/5, pp. 363-376, 2006.

[8] A. Khakifirooz and D. A. Antoniadis, "Transistor performance scaling: the role of virtual source velocity and its mobility dependence," in *IEDM Tech. Dig.*, 2006, pp. 667-670.

[9] A. Khakifirooz and D. A. Antoniadis, "MOSFET performance scaling – Part I: Historical trends," *IEEE Trans. Electron Devices*, vol. 55, no. 6, to be published.

[10] M. H. Na, E. J. Nowak, W. Haensch, and J. Cai, "The effective drive current in CMOS inverters," in *IEDM Tech. Dig.*, 2002, pp. 121-124.

[11] G. T. Wright, "simple and continuous MOSFET model," *IEEE Trans. Electron Devices*, vol. 32, no. 7, pp. 1259-1263, 1985.

[12] E. Yoshida, Y. Momiyama, M. Miyamoto, T. Saiki, M. Kojima, S. Satoh, and T. Sugii, "Performance boost using a new device design methodology based on characteristic current for low-power CMOS," in *IEDM Tech. Dig.*, 2006, pp. 195-198.

[13] K. von Arnim, C. Pacha, K. Hofmann, T. Schulz, K. Schrüfer, and J. Berthold, "An effective switching current methodology to predict the performance of complex digital circuits," in *IEDM Tech. Dig.*, 2007, pp. 483-486.

[14] M. Lundstrom, "Elementary scattering theory of the Si MOSFET," *IEEE Electron Device Lett.*, vol. 18, pp. 361-363, 1997.

[15] K. Uchida, T. Krishnamohan, K. C. Saraswat, and Y. Nishi, "Physical mechanisms of electron mobility enhancement in uniaxial stressed MOSFETs and impact of uniaxial stress engineering in ballistic regime," in *IEDM Tech. Dig.*, 2005, pp. 135-138.

[16] S. Tyagi, *et al.*, "An advanced low power, high performance, strained channel 65nm technology," in *IEDM Tech. Dig.*, 2005, pp. 1070-1072.

[17] A. Khakifirooz, *Transport Enhancement Techniques for Nanoscale MOSFETs*, Ph.D. Thesis, Massachusetts Institute of Technology, 2007.

[18] S. Suthram, *et al.*, "High performance pMOSFETs using $Si/Si_{1-x}Ge_x/Si$ quantum wells with high-k/metal gate stacks and additive uniaxial strain for 22 nm technology node," in *IEDM Tech. Dig.*, 2007, pp. 727-730.

[19] A. Khakifirooz and D. A. Antoniadis, "MOSFET performance scaling – Part II: Future directions," *IEEE Trans. Electron Devices*, vol. 55, no. 6, to be published.

[20] *International Technology Roadmap for Semiconductors*, [Online]. Available: http://public.itrs.net.

[21] M. Chudzik, *et al.*, "High-performance high-□/metal gates for 45nm CMOS and beyond with gate-first processing, in *Symp. VLSI Tech.*, 2007, pp. 194-195.

[22] N. R. Mohapatra, M. P. Desai, S. G. Narendra, and V. R. Rao, "Modeling of parasitic capacitance in deep submicrometer conventional and high-k dielectric MOS transistors," *IEEE Trans. Electron Devices*, vol. 50, no. 4, pp. 959-966, 2003.

[23] S.D. Kim, C.M. Park, and J. C. S. Woo, "Advanced model and analysis of series resistance for CMOS scaling into nanometer regime – Part I: Theoretical derivation," *IEEE Trans. Electron Devices*, vol. 49. no. 3, pp. 457-466, 2002.

[24] R. Venugopal, S. Goasguen, S. Datta, and M. S. Lundstrom, "Quantum mechanical analysis of channel access geometry and series resistance in nanoscale transistors," *J. Appl. Phys.*, vol. 95, pp. 292-305, 2004.

KEYNOTE ADDRESS

Advance Packaging Requirements for Next Generation Microprocessors

Raj Master
Advanced Micro Devices, USA
Email: raj.master@amd.com

ABSTRACT

As the demand for computing performance and density increases, the demand for packaging microprocessors is getting more complex. Computing speed and increased functionalities are achieved by reducing lithography features and increasing no. of transistors and no. metal layers. The performance is further enhanced by incorporating low K dielectric films in the die. While the semiconductor makes leaps, it creates technological challenges and in some areas approach technology barriers. In addition these evolving technical challenges are increasingly dependent on physical failure analysis to understand

This talk would summarize the challenges of bumping at reduced pitch, packaging thermo mechanical issues in packaging fragile low k films in a TCE mismatched system. Discussions on assembly challenges that result from assembling large die in laminate packages will also be described. In summary, talks would discuss the inter dependency between Silicon Technology, packaging Technology and Reliability.

In addition these evolving technical solutions are increasingly dependent on physical failure analysis to understand the root cause and failure mechanisms. The talk will describe some typical failures and the solutions that could be developed by the use of physical failure analysis.

978-1-4244-2039-1/08/$25.00 ©2008 IEEE

SESSION 2:

ADVANCED FA I

INVITED PAPER

Scale, Function and Materials: Debug and Diagnosis in Electronic Device Technology Roadmap

C. Boit,

TUB Berlin University of Technology, Berlin, Germany
Phone +49-30-314-25520 Fax +49-89-30-314-25526 email: christian.boit@tu-berlin.de

Abstract - **Debug and Diagnosis of Integrated Circuits with physical techniques is necessarily correlated strongly to the innovation of electronic device technology. Miniaturization and scaling were main reasons for the introduction of signal access through chip backside in recent years. The introduction of new materials is adding critical challenges, even on the active device level. Based on the technology roadmap to 32nm and below, this article presents the boundary conditions for functional device analysis and discusses the analysis options. The potential of established techniques is assessed with respect to the mismatch of optical resolution and nanoscale device dimensions, to the resolution gain of scanning probe techniques vs. working distance requirements, to the FIB techniques for circuit edit and ultra thin silicon preparation. Applications of innovative approaches are demonstrated, and critical issues like thermal / mechanical management are discussed.**

1 INTRODUCTION

It is not an easy task to anticipate how the inventive community of Electronic Devcie Failure Analysis (FA) will cope with all the technology, circuit design and test innovation of Integrated Circuits (ICs). We have seen an enormous amount of creativity in preparation techniques, repackaging, combination of analysis techniques, use of circuit simulation and scan testing in order to get along with systems on package, minitaturization, clock speed etc., as could be observed at FA conferences worldwide. There has also been external support of new analysis tools like Focused Ion Beam (FIB) and Scanning Probe Techniques (SPT) to move debug and diagnosis successfully into the nanometer and GHz ranges. So, the pessimistic prognoses that FA would fall short of the recent technology innovation did not come true, adding practical inventiveness as one of the most important FA assets to obey Moore's Law in FA [1,2]. But the effort to squeeze out the full potential of the techniques available has given the FA community a growing demand for deeper understanding the interactions of the analysis techniques through device and circuit simulation and meticulous technical optimization.

Taking all this into account, this article will assess the upcoming IC innovation roadmap and condense it to identify issues that FA needs to solve soon. For the identified challenges, the FA state of the art, possible solutions and the degree of demonstrated feasibility will be presented and discussed.

2 ITRS ROADMAP EXTRACT

The International Technology Roadmap for Semiconductors (ITRS) has proven over the years to successfully identify and quantify the IC innovation. Throughout the ITRS progress, a number of separate roadmaps for different product lines have emerged over the years. The most aggressive one, marking the microprocessors, is selected for this investigation. Table 1 and 2 show, starting 1998 as reference, an evaluation of ITRS data from 1999 to 2007 (see examples [3,4]) to get a table of the past and future innovation pace. A look at the data shows that the prediction of the past has described the innovation pace in recent years very well. So microprocessor innovation will go very likely the predicted way. This presentation is looking to just the next 6 years, so there is no reason to hesitate working on the most urgent challenges identified here. On the other hand, not all product lines follow such an aggressive roadmap. ICs for low power and periphery applications, or for medical or automotive products are using relaxed groundrules. The latter keep conservative tehnologies longer in order to meet the high reliability requirements.

Table 1 is giving a very rough overview how circuit performance and device technology will develop in the years to come. Table 2 is focussing on the device concepts and material innovation in the active volume, the gate and the interconnect levels. As most of the IC debug and diagnosis techniques are interacting with the active volume, the new materials employed in strain concepts and Ultra Thin Body (UTB) as major device innovation are discussed in detail. Interconnect and gate stack development is not subject of this study.

2.1 Circuit Performance Aspects:

The major challenge of circuit performance is the clock frequency expanding further into the GHz regime with all the problematic aspects of testing speed, high sensitivity to internal delays and respective weak trigger signals. The internal delay per device is moving into the femtosecond range. IC supply voltage is slowly moving from 1.1V today to 0.8V at around 2014 and not listed in the table . We will see that for IC debug and diagnosis these days, this parameter is not really looming largely.

INVITED PAPER

Approximate Year of Introduction	Min Feature Size [nm]	Performance: On Chip local Clock [GHz]	Delay [ps]	
			Inverter	Intrinsic NMOS
1998	120	2.2	12	2.5
2000	90	2.5	10	1.8
2002	65	3.0	9	1.3
2004	45	3.5	8	1.0
2006	32	4.2	7	0.8
2008	23	5.0	5	0.6
2010	18	5.8	4	0.4
2012	14	6.8	3.5	0.3
2014	11	7.9	3	0.2

Table 1: ITRS roadmap of IC performace after [3,4] The feature in column 2 is physical gate length.

Approx. Year of Introd'n	Feature Size [nm]	Device Concepts	Materials		
			Active	Gate	Inter-connect
1998	120	Bulk Si	Pocket / Halo Implant	High k Metal Gates	Cu Low k
2000	90				
2002	65	SOI / PD	Strain SiGe Raised S/D Ge S/D		
2004	45				
2006	32				
2008	23	Ultra Thin Body / FD	Ge /III-V channels more of above		Ultra low k
2010	18	Dual Gates FIN FETs			
2012	14				Air?
2014	11	Beyond CMOS			

Table 2: ITRS device and materials roadmap after [3,4]. Abbrviations: See paragraphs 2.2 and 2.3

2.2 Scaling of Active Volume 1: Strain

Technologies were really silicon-based as long as physical gate lengths did not fall below 100nm. Performance gain or increase of drain current by further shortening gate lengths requires increased implant doses in the channel region to avoid drain induced barrier lowering (DIBL) which in turn decreases the carrier mobility in the channel. In addition to the geometry scaling, a mobility scaling is becoming essential. Mobility of electrons is increased with tensile strained silicon, holes require compressive strain. The modification of the lattice constants can be implemented by various options (examples in [5,6]), mostly involving deposition of germanium enriched or pure germanium lattices. Once technologies are established depositing active volume material, advantage can be taken by using materials like germanium or III-V compounds exhibiting higher mobilities than silicon in a moderate electrical field. This implies the presence of several band structures in the active volume modifying optical interactions used for debug and diagnosis techniques.

2.3 Scaling of Active Volume 2: Ultra Thin Body (UTB)

Another critical device parameter is the turn on swing. Parasitic capacitances of source/drain (S/D) junctions can be severely reduced if the bottom of the active volume is terminated with a buried oxide layer (BOX) eliminating the lateral space charge layer. This technology, silicon on insulator (SOI), is then called partially depleted (PD). An even thinner active volume would end up being all over in depletion and weak inversion even in 0Volt operation. This is the fully depleted (FD) condition, optimizing the subthreshold slope. As FD active layers are as thin as 20nm, they are called ultra thin body (UTB).

The device concepts that can be anticipated today include the use of the bottom part of the UTB as one more gate area (dual gate), and 3-dimensional geometries like FINFETs have been demonstrated for several years [7]. After 2014, CMOS as logical concept may be replaced completely by new approaches.

3 THE CIRCUIT PERFORMANCE CHALLENGES TO OPTICAL BACKSIDE ACCESS FUNCTIONAL ANALYSIS TECHNIQUES

IC functional analysis techniques like E Beam Probing (EBP) that access the interconnects directly, usually do not match anymore with modern multi interconnect layer and packaging technologies. The established techniques of today are using optical information of near infrared (NIR) in order to transmit the information gained at the active device through bulk silicon to detectors at the backside of the chip.

The following are the main criteria how successfully these techniques will be able to match ITRS performance and technology prospects as shown in chapter 2.
- Wafer technology: How to keep resolving the decreasing feature sizes and how to maintain the optical interaction in the drastically reduced active volume with all the new semiconductor materials.
- Circuit performance: How to get along with low supply voltage of 1V or less, how to keep identifying circuit malfunction in GHz regime (signal pattern, delay).

As the optical resolution is affecting all NIR techniques in a simlar manner, it will be discussed in a separate chapter 4.

3.1 Photon Emission

Photon emission (PE) is the preferred interactive medium of IC functional analysis. It comes as byproduct of electrical operation. Many malfunctions are accompanied with PE, in most cases detectable with Si CCD detectors. For the part of the spectrum that passes bulk silicon, InGaAs has been established as most sensitive detector material [8]. An increasing number of timing issues requires time resolved emission (TRE). In this

INVITED PAPER

case, the light that is emitted during the rise and fall time of inverters, is detected with a time resolution in ps range.

Although this dynamic light signal is faint, PE and TRE are very powerful, reliable and easy to interpret techniques with low risk of artifacts. Exponentially decreasing PE intensity with supply voltage has for a long time been regarded as main risk for the future application of this technique. But, recent studies at low voltage technology devices [9] indicate that the resulting SNR challenge may not be overly severe. The main challenges of PE and TRE are:
- The internal delay per inverter will go below the TRE time resolution in the coming years which makes it more complicted to associate a definite device to a switching event. The switching process itself remains only a subject of TRE investigation if it can be slowed down.
- Only rise and fall phases of the inverter are detected. The photon yield per event at TRE is mainly on the NMOS side and in ppm regime. Signal stability is essential as jitter is decreasing signal to noise ratio (SNR) drastically.
- The mechanisms of PE might be subject of change when all the new active materials with varied band structures will be brought in. Spectral investigations may become important to get footprints of active material composition.
- UTB will be transparent to higher photon energies than bulk silicon, so the visible part of the PE spectrum may become interesting again.

3.2: Laser Stimulation / Laser Voltage Probing

Alteration of device and circuit properties induced by Laser Stimulation (LS), be it by photoelectric (PLS) or thermal (TLS) conversion, can be used in various techniques to localize areas of weak circuit performance [10,11]. In IC debug and diagnosis it is mostly influencing the internal timing, resulting in modified signal patterns of test vectors either for variation of internal delay or pass/fail transitions [12,13,14,15]. The signal can be extracted in various ways, so time resolution and required trigger quality of LS depend strongly upon the extraction procedure. The overall potential of this approach is by far not exhausted yet. More specific applications are expected that increase precision, like pulsed lasers [16] and phase sensitive approaches.

When a laser is interacting with ICs in operation, another option is the evaluation of the reflected signal, modulated by the dynamics of active devices. As this modulation proved to be linear with the voltage level of the investigated node [17], the technique became known under the name Laser Voltage Probing (LVP) as backside access alternative to EBP. Time resolution is in the same range as TRE. The potential of LVP debug and diagnosis is very high as it is detecing the full waveform of the signal pattern thus allowing waveform information still to be obtained when trigger stability is problematic. But the signal modulation is in ppm range, implying risks of SNR and local interferences at unfortunate probing points. These days, LVP seems to improve due to
- better understanding of the interaction mechanisms [18],
- better probe point placement with modulation maps and
- better signal processing using polarized light.

Main future challenges of the laser interaction with the active devices in an IC are:
- The laser wavelength needs to be adapted to the band structure of the active material (composition).
- In SOI or UTB technologies, bulk silicon is usually still present below BOX for thermal management. In such a case, PLS is problematic because the light requires a low absorption rate in bulk silicon in order to transmit some power into the active volume, then a high absorption rate in the active volume to efficiently generate photocurrent.

Despite of the risks and challenges, the optical techniques for physical IC debug and diagnosis presented in chapter 3 prove to be very powerful to analyze (mal)functionality of ICs up to very high performance levels. The challenges resulting of new active materials will restrict application only marginally and it seems possible to develop viable solutions, especially when device and circuit simulation get involved. But, as all these techniques are designed to obtain optical information after passing several hunderd μm of bulk silicon, the transmission needs to be in NIR range, limiting the optical resolution.

4 THE FEATURE SIZE CHALLENGE OF OPTICAL BACKSIDE ACCESS FUNCTIONAL ANALYSIS TECHNIQUES

In many cases, circuit simulation can provide information to identify the device of interest even if the optical techniques can only localize the area of interest but not anymore the exact device. Phase analysis of TLS stimulus to circuit reponse, in conjunction with heat propagation models, may give further information to identify the location of interest.

4.1 Solid Immersion Lens

Maximum microscopic resolution is the light wavelength referring to the length of a repeating structure, the so called pitch. A pitch consists of minimum 2 features, so the minimum resolved structural feature is 2x smaller. Realistic pitch resolution is ca. 800nm, respectively 400nm feature size. Hemispherical solid immersion lenses (SIL) have been introduced, making bulk silicon a part of the microscope and increasing resolution theoretically by the refractive index of silicon n_{Si}=3,5, technically today to about 150nm.

Potentially, a superhemisperical SIL could increase resolution by n^2, to 50nm. This concept, though, has not been demonstrated yet in an IC analysis environment.

4.2 Nanoprobing

As technologies progress further into the nanometer range, scanning probe techniques and derivatives are increasingly inevitable to identify the critical node in nanoscale regime. Nanoprobing, or atomic force probing (AFP), makes use of the lower interconnect levels, preferably the contacts to the active structures under investigation [19,20]. Complex reverse engineering techniques, mostly planar polishing, need to be applied from chip frontside, finally removing several interconnect layers to reveal the interconnect or contact of

978-1-4244-2039-1/08/$25.00 ©2008 IEEE

INVITED PAPER

interest. This process is time consuming and full of risks: If the failing transistor cannot be found in the final measurement, the problem might have been removed with the preparation, leaving no second chance for another analysis approach.

5 CIRCUIT AND FEATURE SIZE CHALLENGES TO CIRCUIT EDIT

FIB-based circuit edit (CE) for ICs operating in GHz regime with ball bonded technology also needs to be processed through chip backside, offering circuit cut and paste operations much closer to the active layers than global preparation techniques can provide. A trenching process using an DCG Systems OptiFIB, equipped with coaxial NIR optical microscope, has been established that removes bulk silicon to the bottom level of shallow trench isolation (STI). This process is performed in three steps, indicated in Fig. 1. After a global polishing technique as applied for the optical

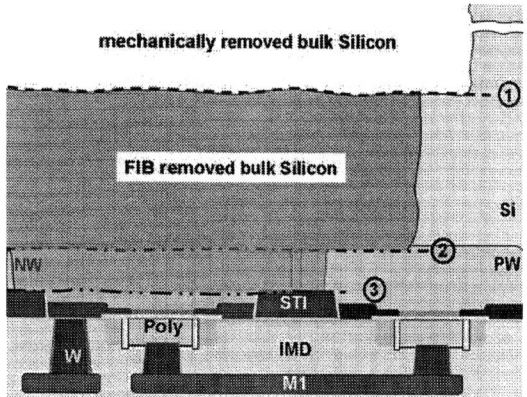

Figure 1: FIB preparation for backside circuit edit

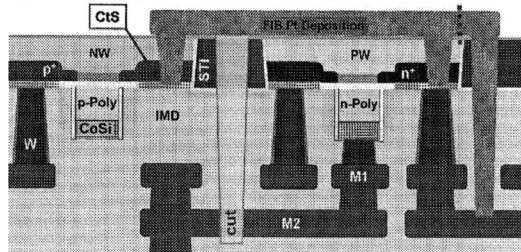

Figure 2: Variety of FIB contacts using UTS

analysis techniques as well, the region of interest can be identified in situ by the coaxial IR microscope (1). After a special FIB cleaning procedure, a max. 200x200µm² wide trench is milled until the endpoint signal of a voltage contrast between the n-wells and the substrate (2). Then, a smaller local trench will be milled until the STI becomes visible as material imaging contrast (3).

By this approach, the FIB trench is producing an ultra thin silicon (UTS) layer of to 300 to 400nm (the depth of the STI trenches) [21]. This structure is offering several more CE

opportunities than the frontside process does: contact to any source/drain area on the chip, even those that have not been contacted on IC interconnect level. All the functional nodes are represented on the active device level. Fig. 2 is illustrating a typical backside CE. Any CE deposition can be implemented using the contact to silicide (CtS) process. The lower aspect ratio for CtS compared to contacts to metal lines (right part of Fig. 2) is very fortunate for the contact resistance, only cut operations still have to be performed by trenching deeper into metal interconnects [22]. With platinum deposition, CtS resistivities are below 60Ω/µm². Typical contacts (1x0.2µm², 300Ω) are acceptable for most CE and probing tasks as FET-channels have resistances of several kΩ. For UTB technologies, the contact height will be reduced by a factor of 10. The utilization of advanced conductor deposition materials may offer further improvement potential.

Taking additionally into account state of the art FIB resolution performance, this backside CE technique can be rated as fully suitable for technologies scaling well into the nanometer regime.

6 BACK SURFACE OF UTS – IDEAL PLATFORM FOR FUNCTIONAL NANO ANALYSIS

An invasiveness study of backside CE processing using intentionally enlarged trench sizes down to STI level (Fig.3), resulted in almost unchanged single device- and circuitry-performance in regular processing regime. No invasive influence of this process to chip functionality could be observed on this ultra thin silicon (UTS) layer [21].

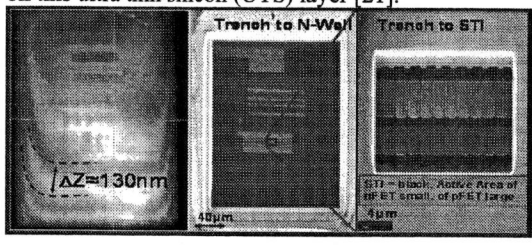

Figure 3: Images of FIB trench location (left: IR-optics, middle: FIB n-well endpoint, right: FIB STI opening)

A wide FIB trench can be performed with remarkable co-planarity to the chip levels. In Fig.3, the coaxial IR microscope image of the trench shows Newton rings indicating the slope of silicon thickness. A planarity mismatch less than 200nm on 250µm trench expansion or 0.08% deviation is state of the art. The back surface of the remaining ultra thin active volume can now be regarded as a new work bench for nano analysis and probing techniques accessing fully functional circuits. The resolution potential of techniques on UTS (Fig.4) is summarized in Table 2.

978-1-4244-2039-1/08/$25.00 ©2008 IEEE   12   *Proceedings of 15th IPFA - 2008, Singapore*

INVITED PAPER

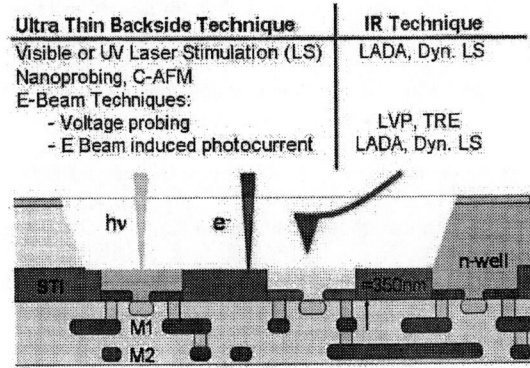

Ultra Thin Backside Technique	IR Technique
Visible or UV Laser Stimulation (LS)	LADA, Dyn. LS
Nanoprobing, C-AFM	
E-Beam Techniques:	
- Voltage probing	LVP, TRE
- E Beam induced photocurrent	LADA, Dyn. LS

Figure 4: Probing on Ultra Thin Silicon

Technique	Resolution	Potential	Status
(IR) Optical bulk Si	400nm	150nm (SIL)	standard technique
(IR) Optical / UTS	200nm	< 100nm (SIL)	future concept
UV / UTS	100nm	< 50nm (SIL)	future concept
E Beam / UTS	100nm	10-20nm	demonstrated
Nanoprobing	50nm	10nm	demonstrated

Table 3: Comparison of Feature resolution potential of Functional Analysis Techniques though moderate and ultra thin silicon. Potential SIL data for hemispherical lens

Nanoprobing can now be performed on a fully functional circuit, avoiding the circuit destruction risks of frontside preparation that are pointed out in chapter 4 [23]. Particle beam techniques like EBP or E Beam Stimulation with the resolution potential of an electron microscope, can be reconsidered for IC analysis on a higher technical level again [24,25]. Optical techniques may be shifted to much smaller wavelengths with the respective resolution improvement as no bulk silicon has to be passed anymore. For SIL, other materials than silicon would have to be utilized. At UTB technologies with an active layer thickness as small as 20-50nm, the full variety of scanning probe techniques may apply from the back suface as well, maintaining full structural resolution power.

Application of the optical functional analysis techniques described in chapter 2 through scanning optical near field microscopy (SNOM) may expand their nanoscale potential even further.if applied to the backside of UTB.

To gain resolution beyond all optical limits though, particle beam techniques seem to be most promising.

In case of substantial power consumption in the IC's area of interest, solutions for proper heat dissipation paths need to be developed.

7 SUMMARY

IC technology innovation will heavily influence active materials. The IR analysis techniques for IC debug are very powerful and offer a huge potential to localize circuit problems which is not fully exhausted yet. The only drastic disadvantage is the increasing optical resolution mismatch in nanoscale technologies, limited even with support of SIL technology to about 100nm. The current practice for nano technologies is 2-level analysis:

Level 1: Optical circuit analysis,

Level 2: Nanoprobing after circuit-destructive polishing.

For circuit edit (CE), a procedure accessing through chip backside is established with a FIB preparation technique processing towards STI level.

The back surface of ultra thin silicon (UTS) may become an ideal platform for nanoscale functional analysis, keeping the Integrated Circuit fully functional. UTS analysis potential matches the requirements for several technology generations to come. The utilization has just begun.

REFERENCES

[1] Can failure analysis keep pace with IC technology development?, (C. Boit), Proc. 7th IEEE IPFA International Symposium on the Physical and Failure Analysis of Integrated Circuits, Singapore, pp. 9 – 14 (1999)

[2] Why Waste Time on Roadmaps When We Don't Have Cars? (D.P. Vallett) IEEE TDMR Transactions on Device and Materials Reliability Vol. 7, No.1, pp. 5-10 (2007)

[3] International Technology Roadmap for Semiconductors, Overall Roadmap Technology Characteristics, Edition 2000, http://www.itrs.net/Links/2000UpdateFinal/ORTC2000final.pdf

[4] International Technology Roadmap for Semiconductors, Executive Summary, Edition 2007, http://www.itrs.net/Links/2007ITRS/ExecSum2007.pdf

[5] Strained silicon MOSFET technology, (J.L. Hoyt et al.), IEEE IEDM International Electron Device Meeting, pp. 23- 26 (2002)

[6] A 90nm high volume manufacturing logic technology featuring novel 45nm gate length strained silicon CMOS transistors, (T. Ghani et al.), IEEE IEDM International Electron Device Meeting, pp. 11.6.1- 11.6.3, (2003)

[7] Silicon CMOS Devices Beyond Scaling ,(W.Haensch et al.), IBM Journal of Research and Development Vol. 50, Nr. 4/5, p.p. 339-362 (2006)

[8] Fundamentals of Photon Emission (PEM) in Silicon-Electroluminescence for Analysis of Electronic Circuit and Device Functionality, (C. Boit), Microelectronics Failure Analysis Desk Reference, 5th Edition, pp356-368, ASM International, Ohio (2004)

[9] Hot Carrier Photon Emission in Scaled CMOS Technologies – A Callenge for Emission Based Testing and Diagnosis (A.Tosi, F. Stellari, A. Pigozzi, G. Marchesi, F. Zappa), Proc. 44th IEEE IRPS International Reliability Physics Symposium San Jose 2006, pp595-601 (2006)

INVITED PAPER

[10] Principles of Thermal Laser Stimulation Techniques, (F. Beaudoin, R. Desplats, P. Perdu, C. Boit), Microelectronics Failure Analysis Desk Reference, 5th Edition, pp417-425, ASM International, Ohio (2004)

[11] Systematic Characterization of Integrated Circuit Standard Components as Stimulated by Scanning Laser Beam, (A. Glowacki, S. Brahma, H. Suzuki, C. Boit), IEEE TDMR Transactions on Device and Materials Reliability, Vol.7, No.1, pp.31-49 (2007)

[12] Soft Defect Localization (SDL) on ICs (M. Bruce et al.), Proc. 28th EDFAS ISTFA International Symposium for Testing and Failure Analysis, Phoenix, AZ, USA, pp.21-28 (2002)

[13] Critical Timing Analysis in Microprocessors Using Near-IR Laser Assisted Device Alteration (LADA), (J. A. Rowlette, T. Eiles), Proc. IEEE ITC International Test Conference, Charlotte NC, pp264-273 (2003)

[14] Delay Variation Mapping Induced by Dynamic Laser Stimulation (K. Sanchez, R. Desplats, F. Beaudoin, P. Perdu, D. Lewis, P. Vedagarbha, G. Woods), Proc. 43th IEEE IRPS International Reliability Physics Symposium San Jose 2005, pp305-311 (2006)

[15] Dynamic Thermal Laser Stimulation Theory and Applications, (K. Sanchez, R. Desplats, F. Beaudoin, P. Perdu, S. Dudit and M. Vallet), Proc. 44th IEEE IRPS International Reliability Physics Symposium San Jose 2006, pp574-584 (2006)

[16] Timing Analysis of Scan Design Integrated Circuits Using Stimulation by an Infrared Diode Laser in Externally Triggered Pulsing Condition (T. Kiyan, C. Brillert, H. Suzuki, T. Nakamura, C. Boit), accepted at 19th ESREF European Symposium on Reliability of Electron Devices, Failure Physics and Analysis, Maastricht, Netherlds, (2008)

[17] Novel optical probing technique for flip-chip packaged microprocessors, (M. Paniccia et al.) Proc. IEEE ITC International Test Conference, Paper 30.2, pp. 740-747 (1998)

[18] Quantitative Investigation of Laser Beam Modulation in Electrically Active Devices as used in Laser Voltage Probing, (U. Kindereit, G. Woods, J. Tian, U. Kerst, R. Leihkauf, C. Boit), IEEE TDMR Transactions on Device and Materials Reliability Vol. 7, No.1, pp19-30, (2007)

[19] Conductive Atomic Force Microscopy Application for Semiconductor Failure Analysis in Advanced Nanometer Process, (K. Lin, H. Zhang, S.-S. Lu), Proc. 32nd EDFAS ISTFA International Symposium for Testing and Failure Analysis, Austin TX, US, pp178-181 (2006)

[20] Case Studies in Atomic Force Probe Analysis, (R. Mulder, S. Subramanian, T. Chrastecky), Proc. 32nd EDFAS ISTFA International Symposium for Testing and Failure Analysis, Austin TX, US, pp153-162 (2006)

[21] Impact of Back Side Circuit Edit on Active Device Performance in Bulk Silicon ICs (C. Boit, U. Kerst, R. Schlangen, A. Kabakow, E. Le Roy, T. Lundquist, S. Pauthner), Proc. IEEE ITC International Test Conference, Austin TX, paper 48.2, pp1-9 (2005)

[22] Contact to Contacts or Silicide by Use of Backside FIB Circuit Edit Allowing to Approach Every Active Circuit Node, (R. Schlangen, P. Sadewater, U. Kerst, C. Boit), Proc. 17th ESREF European Symposium on Reliability of Electron Devices, Failure Physics and Analysis, Wuppertal, Germany, pp.1498-1503, (2006)

[23] FIB Backside Circuit Modification on Device Level Allowing to Access Every Circuit Node with Minimum Impact on Device Performance by Use of Atomic Force Probing (R. Schlangen, U. Kerst, C. Boit, S. Schömann, B. Krüger, R. Jain, T. Malik, T. Lundquist) accepted at 33rd EDFAS ISTFA International Symposium for Testing and Failure Analysis, San Jose, USA (2007)

[24] Functional IC Analysis through Chip Backside with Nano Scale Resolution – E-Beam Probing in FIB Trenches to STI Level (R. Schlangen, R. Leihkauf, U. Kerst, C. Boit), Proc. 32nd EDFAS ISTFA International Symposium for Testing and Failure Analysis, Austin TX, USA, pp376-381 (2006)

[25] Novel Flip-Chip Probing Methodology Using Electron Beam Probing (R. Jain, T. Malik, T. Lundquist, R. Schlangen, R. Leihkauf, U. Kerst, C. Boit), Proc. 14th IEEE IPFA International Symposium on the Physical and Failure Analysis of ICs, Bangalore, India pp39-42 (2007)

Scan-Based ATPG Diagnostic and Optical Techniques Combination: A New Approach to Improve Accuracy of Defect Isolation in Functional Logic Failure

A. Machouat[a], G. Haller[a], V. Goubier[a], D. Lewis[b], V. Pouget[b], P. Fouillat[b], F. Essely[b], P. Perdu[c]

[a] STMicroelectronics, 190 Avenue Célestin Coq, 13106 Rousset, France

[b] IMS laboratory, 351 cours de la libération, 33405 Talence, France

[c] CNES laboratory, 18 avenue Edouard Belin, 31401 Toulouse, France
Phone: +33 442 686 852; Fax: +33 442 685 001; Email: aziz.machouat@st.com

Abstract- **Nowadays, with the increasing complexity of new VLSI circuits, laser stimulation or emission techniques and Scan-based ATPG diagnostic reach their limits in functional logic failure. To overcome these limitations, a new methodology has been established. This methodology, presented in this paper, combines the advantages of both approaches in order to improve accuracy of fault isolation and defect localization.**

I. INTRODUCTION

Nowadays, most integrated circuits manufacturers use "scan path" [1] technique for the digital parts testing. This structural technique uses the internal flip-flops to test digital interconnection nets. This technique replaced the traditional manual approach and allows to generate test patterns rapidly with high test coverage by using ATPG tools. Study led by [2] shows the effectiveness of scan patterns compared to functional patterns. Most of ATPGs integrate a diagnostic tool which enables the prediction of the failing nets in a scan reject. For the most advanced device, the diagnostic can be made on bridge, transition and delay faults models with addition of traditional stuck-at fault model.

The continuous advancement of technological processes in microelectronics industry led to a high density of transistors and interconnections layers in digital parts of Very Large Scale Integration (VLSI) circuits. This density makes functional logic failure analysis more difficult because spatial resolutions of current defect localization techniques do not provide sufficient accuracy in fault site isolation [3].

To improve this accuracy, a new methodology which combines optical techniques and Scan-based Automatic Test Pattern Generation (ATPG) diagnostic has been developed.

The first part of this paper describes the usual Failure Analysis (FA) techniques and their limitations in functional logic failure localization. A new methodology will be described in the second part following by a detailed case study.

II. PRINCIPLE AND LIMITATIONS OF ATPG DIAGNOSTIC

ATPG diagnostic plays an important role in improving yield for functional logics failures by predicting failing nets within a scan design [4]. The standard diagnostic flow is presented in figure 1.

After ATPG process a test patterns are generated, and a datalog is collected from the Automatic Test Equipment (ATE). From this report, logic diagnosis is performed if the device passes the scan chain test but fail for scan test patterns. Some diagnostic tools are able to perform also diagnosis on scan chain failure but the investigations area for physical analysis is often important due to number of flip-flops proposed. For each flip-flop, inputs and output have to be inspected.

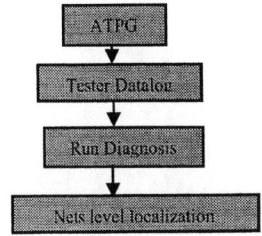

Fig. 1 Standard ATPG diagnostic flow

The advantage of this technique is to speedily locate the failure site without using heavy FA equipments. However, this technique becomes inefficient in some cases. First, the diagnosis result can pinpoints to many candidates as shown in figure 2 (left) and the layout size of localized nets can be widespread on the device. This limitation is illustrated in figure 2 (right); only nets N19 and N20 are reported on the layout. The net N19 have a length of 660μm and the net N20 have a length of 500μm. Length of these nets make physical analysis more complicated and the success rate of analysis is greatly reduced.

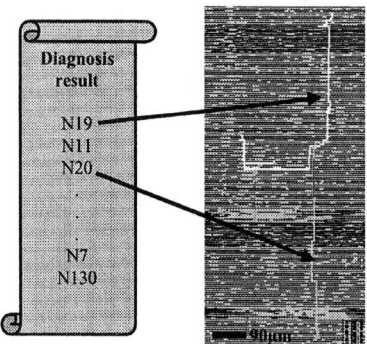

Fig. 2 Limitations of ATPG Diagnostic: number of nets (left) and nets location (right) involved in logical failure

III. PRINCIPLE AND LIMITATIONS OF LASER STIMULATION AND EMISSION TECHNIQUES IN LOGIC FAILURE LOCALIZATION

Functional logic failures are the most challenging to isolate. To analyze a logic failure with mapping techniques such as thermal laser stimulation [5] or emission microscopy [6], the device needs to be conditioned into a particular logic state to generate the failure. Once the vector positioned, mapping techniques can be applied [7].

The spatial resolutions of these techniques are no longer sufficient because optical spot size can cover a lot of connectivity and localizations may contain several nets (see figure 3). In TEM analysis, this limitation becomes a real issue, because a 100nm thin lamella has to be extracted in a 1-2μm² of an OBIRCH or EMMI spot. Thus, the lamella may not contain the defect. Furthermore, the accuracy of physical analysis equipments such as SEMs or FIBs used at highest magnification reaches their practical limits.

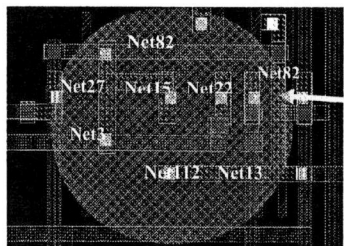

Example of localization obtained with mapping technique (EMMI, OBIRCh, TIVA...)

Fig. 3 Example of mapping technique localization providing 8 nets (130 nm CMOS technology)

Limitations of both optical techniques and physical analysis equipments do not provide sufficient accuracy for defect isolation and observation. Even with improvements in methods and equipments, some of the intrinsic limitations remain and alternatives techniques are needed. One of the attractive approaches is to combine the standard optical localization techniques with software based fault diagnosis. This approach is used in our methodology to improve the

spatial resolution by reducing the number of suspected fault candidates. This methodology is described in the next section.

IV. NEW METHODOLOGY THAT COMBINE SCAN-BASED ATPG DIAGNOSTIC AND MAPPING TECHNIQUES

The proposed methodology takes advantages of both techniques: scan-based ATPG diagnostic and mapping techniques. The principle of this methodology is to combine the list of nets from diagnosis result and the list of nets from mapping technique in order to have a reduced list of candidates. An in-house software has been developed to extract from EMMI/OBIRCh spots coordinates all the nets passing through these spots. These nets are compared with the list of nets given by ATPG diagnosis tool. Figure 4 illustrates the methodology; only nets N112 and N3 are common in the two lists. The physical analysis will be focused only on each layer of these two nets. By doing so, the investigation area for physical analysis is greatly reduced. The efficiency of this methodology is presented in the next section through a case study.

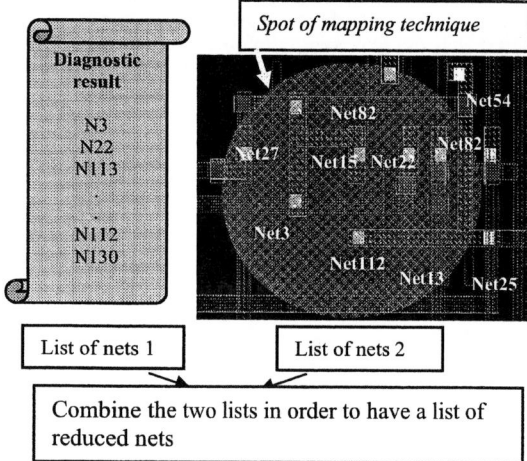

Fig. 4 Principle of a new methodology that combines ATPG diagnostic and mapping techniques

V. CASE STUDY AND RESULT

The device is an IC manufactured in 130nm semiconductor technology with four layers of copper interconnect. The device failed in scan logic test. It has two scan chains; the first scan chain contains 4653 flip flops and the second scan chain, contains 4652 flip flops.

A. Fault site isolation following the methodology

A diagnostic is performed on ATPG Tetramax from the datalog report collected with Personal Ocelot Inovys ATE. The result of this diagnosis is described in figure 5. We noticed that the diagnosis is composed by seven faulty candidates. Another important data to take in account from the diagnosis result is the match score given by the diagnosis tool for the proposed candidates. This match score is based on how well a fault candidate failing behaviour matches the failures seen on the ATE for corresponding locations. In our case, the match score is 100%, thus indicates that we can trust this diagnosis result. However, the faulty candidates are widespread on the layout as shown in figure 6. Due to the length of two nets, the chance of finding the defect during physical analysis is greatly reduced. The first net is "*slim_rom_and_bist_0/I0/U50/A*" which has a length of 470µm and the second net is "*slim_rom_and_bist_0/I0/U51/A*" which has a length of around 500µm.

To correlate the diagnosis result with optical techniques, a failing pattern is extracted from the tester log file. OBIRCh and EMMI analysis is performed using a Hamamatsu confocal microscope (Phemos 1K) by stopping the failing scan pattern at the end of the "shift in" sequence.

EMMI spot is observed as shown in figure 7. The X-Y coordinates of the top-right and bottom-left corners of the box which covers the EMMI spot is extracted and introduced into the in-house software which uses CAD navigation tool Camelot (knight) to extract all the nets cover by the emission spot. This list of nets is presented in figure 8.

```
match=100.00%, #explained patterns:

sa1   DS slim_rom_and_bist_0/I0/U50/Z
sa1   -- slim_rom_and_bist_0/I0/U50/A
sa0   -- slim_rom_and_bist_0/I0/U50/B
sa1   -- slim_rom_and_bist_0/I0/U51/Z
sa0   -- slim_rom_and_bist_0/I0/U51/B
sa1   -- slim_rom_and_bist_0/I0/U51/A
sa1   -- slim_rom_and_bist_0/I0/U422/D
```

Fig. 5 Diagnosis result of failing part

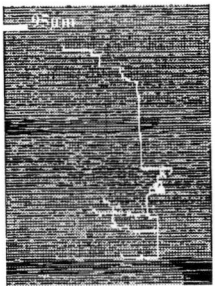

Fig. 6 Layout of faulty candidates

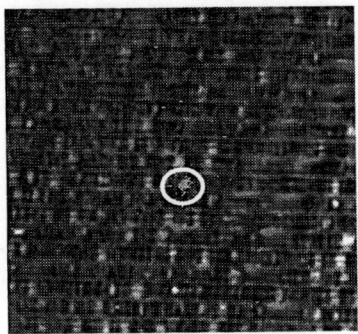

Fig. 7 Emission Microscopy result

```
cab_afibias_ctrl_5_ASTlenNet20302
rbact_rom
cpu_rom_data[7]
slim_rom_and_bist_0/I0/U41/gnd
slim_rom_and_bist_0/I0/U50/gnd
slim_rom_and_bist_0/I0/U50/Z
shscout
slim_rom_and_bist_0/I0/n571
slim_rom_and_bist_0/I0/n570
slim_rom_and_bist_0/I0/U50/A
slim_rom_and_bist_0/I0/U50/B
slim_rom_and_bist_0/I0/U50/net025
```

Fig. 8 List of nets extracted from emission spot

The list of nets obtained from diagnostic ATPG and the list of nets obtained from EMMI analysis are combined to form a reduced list of nets. This list presented in figure 9, pinpoints to only three nets in common. The physical analysis is focused on each layer of these three nets at the emission spot area.

```
slim_rom_and_bist_0/I0/U50/Z
slim_rom_and_bist_0/I0/U50/A
slim_rom_and_bist_0/I0/U50/B
```

Fig. 9 Reduced list of net after combination of ATPG diagnostic and EMMI result

Relying on the ATPG diagnosis, seven nets have been localized with two nets which have a length of around 500µm. On the other hand, Table 1 shows the benefit of methodology by giving the numbers of connectivity to be inspected at the EMMI spot area. The number of connectivity is 11 again 26 using this methodology. Thus, the area for physical analysis to be inspected is greatly reduced.

Number of connectivity			
Without methodology		With methodology	
Metal 4	2	Metal 4	0
Metal 3	3	Metal 3	1
Metal 2	4	Metal 2	2
Metal 1	3	Metal 1	2
Contacts	10	Contacts	4
Polysilicon	4	Polysilicon	2
Total	26	Total	11

Table 1. Number of connectivity to be inspected in physical analysis with and without our methodology

B. Physical analysis result

After deprocessing, SEM passive voltage contrast imaging on via1 level shows two vias with an abnormal contrast (figure 10) at the emission spot localization. These two vias belong to a net present in the reduced list of nets *"slim_rom_and_bist_0/I0/U50/A"*.

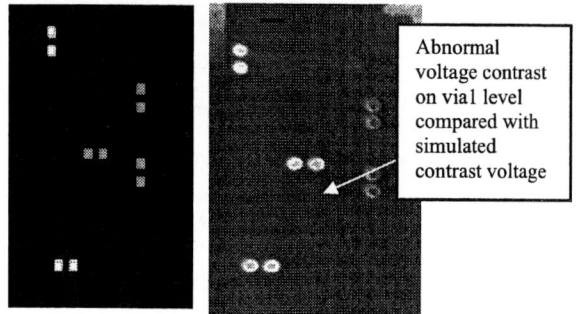

Abnormal voltage contrast on via1 level compared with simulated contrast voltage

Fig. 10 Simulated passive voltage contrast (left), SEM passive voltage contrast (right) at via1 level

After gradual etching of intermediate dielectric IMD1, the defect is observed with SEM image at 25KV. This defect shorts *"slim_rom_and_bist_0/I0/U50/A"* net with two other internal cell nets at metal 1 level (figure 11, left). A cross section (figure 11, right) is performed and a short is confirmed.

The scan logic failure has been physically localized in the chip but the root cause could not be identified from SEM images. For this reason TEM analysis and STEM EDX profile are performed (figure 12). These analyses show that the short is created by an 8nm thick layer of titanium.

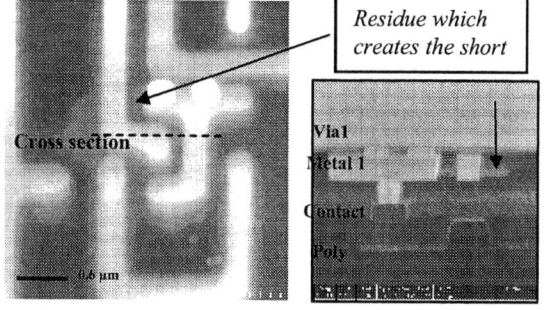

Residue which creates the short

Cross section

Via1
Metal 1
Contact
Poly

Fig. 11 SEM image at via1 level (left), FIB cross section (right)

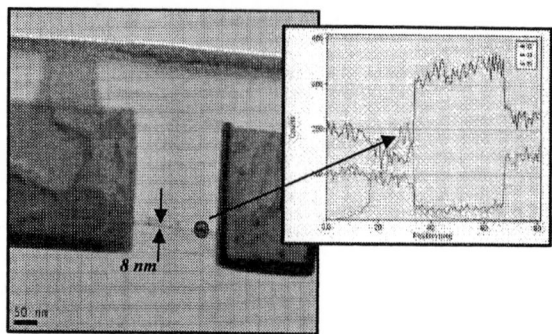

8 nm

Fig. 12 TEM image (letf) and STEM EDX profile (right)

VI. CONCLUSION AND PERSPECTIVES

This paper presents a new methodology by improving the resolution of fault isolation which has become a real issue for advanced technologies and an area of concentration for the FA laboratories. A detailed case study is presented and the result shows an accuracy improvement of defect localization using this new methodology.

Today, our methodology is applied in the functional logic "hard defects". However, with the advancement of technological processes, we observe a new category of defect called "soft defect" in which the device will be operating but not with the performance for which it was designed for. These defects are the most challenging to isolate in failure analysis of integrated circuits. Dynamic Laser Stimulation (DLS) techniques have become a standard failure analysis technique to localize this kind of defect. The spatial resolution provided by this technique is less critical than others optical techniques such as EMMI or OBIRCh techniques, however, in some cases DLS can be inefficient due to the time consuming analysis. To make DLS more effective, we are introducing this technique in our methodology to combine DLS with ATPG diagnosis in order to reduce the area to analyse and the time consumed.

In [8], we are shown that OBIRCh signal depends on materials. Recent simulations show that the geometry and polarization of the defect have also an impact on this signal. Based on these results, we are performing researches to identify the best pattern in which OBIRCh and EMMI signal will be in there highest intensity.

REFERENCES

[1] M. Williams, et al, "Enhancing Testability of LSI Circuits Via Test Points and Additional Logic", IEEE Transactions on Computers, Vol 22, No 1, 1973, pp 46-60.

[2] P. C. Maxwell et Al, "IDDQ and AC scan: the war against unmodelled defects", International Test Conference, 1996, pp 50-258.

[3] M.Lamy et al, "How Effective Are Failure Analysis Methods for the 65nm CMOS Technology Node". International Symposium on the Physical & Failure Analysis of Integrated Circuits 2005 pp 32-37.

[4] C. Burmer et al, "Software Aided Failure Analysis using ATPG tools". International Symposium on the Physical & Failure Analysis of Integrated Circuits 2001, pp. 210-215.

[5] F.Beaudoin et al, Principles of Thermal Laser Stimulation Techniques", Microelectronics Failure Analysis, Desk Reference Fifth Edition, 2004, pp 417-425.

[6] A. Tosi et al, "Hot-Carrier photoemission in scaled CMOS technologies: a challenge for emission based testing and diagnostics", Inetrnational Reliability Physics Symposium, 2006, pp. 595-601.

[7] L. Gao et al, "ATPG Scan Logic Failure Analysis: a case study of logic ICs – fault isolation, defect mechanism identification and yield improvement". Microelectronics Reliability vol. 46, 2006, pp 1458-1463.

[8] A. Firiti et al, "Guideline for interpreting NIR Laser Stimulation signal on semiconductors materials and improving failure analysis flow", International Symposium for Testing and Failure Analysis, 2005, pp 59-65.

Effect of Refractive Solid Immersion Lens Parameters on the Enhancement of Laser Induced Fault Localization Techniques

SH Goh[1], ACT Quah[1], CJR Sheppard[2], CM Chua[3], LS Koh[3], JCH Phang[1,3]

[1]Centre for Integrated Circuit Failure Analysis and Reliability,
National University of Singapore, 10 Kent Ridge Crescent, Singapore 119260
[2] Bioimaging Laboratory, Div of Bioengineering, National University of Singapore, Singapore 117574
[3]SEMICAPS Pte Ltd, 28 Ayer Rajah Crescent #03-01, Singapore 139959
Phone: +65-6516-3890 Fax: +65-6516-7912 Email: gohszuhuat@nus.edu.sg

Abstract - **The effect of Refractive Solid Immersion Lens (RSIL) parameters on the enhancement to laser induced fault localization techniques are investigated. The experimental results of the effect on a common laser induced technique, namely Thermally Induced Voltage Alteration (TIVA), and imaging are presented. A signal enhancement in the peak TIVA signal of close to 12 times has been achieved.**

I. INTRODUCTION

Laser induced techniques are widely used for frontside and backside fault localisation in semiconductor integrated circuit (IC) failure analysis (FA) [1]. For backside thermal stimulation, a 1340 nm wavelength laser is commonly used to obtain high transmittance through the silicon substrate [2]. In far-field systems, the laser spot size at a numerical aperture (NA) of 0.7 is diffraction limited to approximately 1 μm. This resolution limit compromises the performance for both backside optical imaging and the fault localization precision of laser induced techniques especially as ICs scale towards the 65nm and 45nm technology nodes. An effective approach is to apply immersion lens technology. Solid [3] and liquid [4] immersion lens technologies have been used to extend the diffraction limit by increasing NA. The two main types of solid immersion lens (SIL) technologies that are applicable to backside FA are Refractive SIL (RSIL) [3] and Diffractive SIL [5].

RSIL involves placing a truncated spherical silicon (Si) lens on top of the back Si substrate [6]. The resolution enhancement from RSIL leads to a smaller laser spot size which will have a direct impact on Thermally Induced Voltage Alteration (TIVA) signal [7] which is proportional to laser power density. In practice, due to limitations in common mechanical polishers, precise thinning of Si substrate to fit an RSIL design may not be routinely achievable. Furthermore, because of light bending, the aperture angle in air (objective angle) changes when the laser is focused at different focal planes affecting the objective NA. The degree of enhancements from RSIL depends strongly on these factors. In the worst case, enhancements may not be achieved under certain conditions.

Therefore, it is essential to understand these effects prior to the application of RSIL to laser induced techniques.

In this paper, we investigate the effects of varying laser focusing planes and objective NA on the enhancements of TIVA with RSIL. These are the two main RSIL parameters. The focusing plane is varied by changing the RSIL design. The results provide information on the optimum conditions and the effective objective NA required in combination with the RSIL to maximize the enhancements. The latter is termed objective NA matching in this paper.

II. BACKGROUND

A. Refractive Solid Immersion Lens (RSIL)

RSIL is a truncated spherical silicon (Si) lens which is placed on the back surface of the Si substrate [6]. The Centric RSIL involves imaging at the focal plane which is at the centric position (point C) of the Si lens and is aberration-free. This can be achieved by the combination of the Si lens thickness and backside thickness so that the focal plane is at the centric position of the Si lens as shown in Fig. 1(a).

The Centric RSIL reduces the laser spot size by increasing the focusing aperture angle in Si. The aplanatic RSIL further increase this angle through favourable light bending at the spherical interface. The aplanatic focal point (point A) which is also aberration-free is at a distance of equal to Si lens radius divided by the Si refractive index. The effect of refractive index mismatch between the Si lens and doped Si backside is neglected since the real component is close to 3.5 in the light wavelength concerned [8].

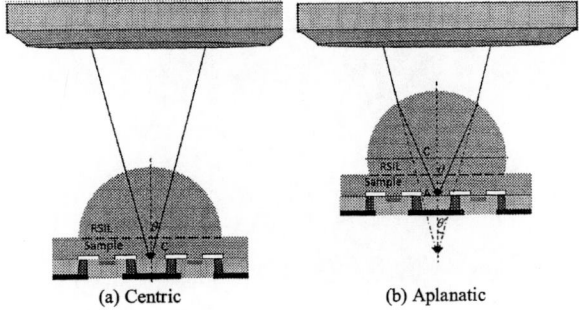

(a) Centric   (b) Aplanatic

Fig. 1. Laser focusing at Centric point (a) and Aplanatic point (b).

The resolution performance achievable by a centric and aplanatic RSIL is comparable. However, the advantage of an aplanatic configuration over the centric is the fact that it requires a low backing objective NA to achieve a good performance. A centric RSIL is usually limited by backing NA. It is reported that a lateral spatial resolution of 0.23 μm has been achieved using a 1.05 μm wavelength laser when imaging close to aplanatic configuration. The objective NA is 0.3 [9].

B. Thermally Induced Voltage Alteration

TIVA is a laser induced technique used for active fault localisation based on thermal stimulation. This technique involves using a scanned 1340 nm laser beam to localize faults which are sensitive to thermal stimulation. It is based on constant current bias and voltage detection configuration. The change in voltage during thermal stimulation is detected by an ac-coupled amplifier, digitized and then mapped into a TIVA image [1]. Subsequently, the TIVA image is overlaid on the reflected laser image for fault localization. The localization precision is dependent on the strength of TIVA signal, signal spot size and spatial resolution of reflected laser image.

The intensity of TIVA signal is proportional to the laser power and laser spot size, i.e. the power density of the incident laser beam. The laser spot size can be significantly reduced by RSIL to enhance the TIVA signal and the TIVA signal spot size. The reduction in beam resolution can also give rise to a better resolved reflected laser image.

III. EXPERIMENTAL SETUP

Fig. 2 shows the schematic diagram of the experimental setup on SEMICAPS SOM 1005 Analytical Scanning Optical Microscope System. Mitutoyo 10X (0.26 NA), 20X (0.4 NA), 50X (0.42 NA) and 100X (0.5 NA) objectives are used. It should be noted that the effective NA for the 10X and 20X objectives are less than the designed values since the objective back pupil is not filled completely by the laser beam. The RSIL setup is in a decoupled configuration as shown in Fig. 2. It is not attached to the objective lens. The add-on components for the decoupled RSIL setup consist of a mechanical aperture structure and a micromanipulator. The truncated Si lens is

placed on top of the device under test and brought close to the area of interest. The mechanical aperture is necessary to apply sufficient pressure between the lens and sample interface for optical coupling [6] during imaging by lowering the micromanipulator. The micromanipulator is also used for fine movement of the RSIL to align the RSIL to the optical axis of the objective lens using a series of alignment steps. A thin layer of optical fluid is used between the lens and sample to avoid scratching the lens during RSIL movement as well as for better contact to eliminate air gaps. In this work, the RSIL is fabricated from a 1 mm diameter Si sphere which is doped in the range of 1×10^{17}-1×10^{18} cm^{-3} and polished to variable thicknesses.

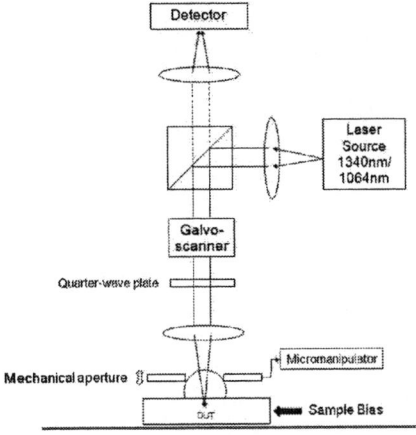

Fig. 2. Decoupled RSIL experimental setup.

The test structure is a 0.5 μm/0.9 μm (line/space) polysilicon serpentine structure on a substrate doped at 5×10^{15} cm^{-3}. The substrate thickness is kept constant at around 60 μm. The mechanical structure has a clear aperture of 90-93% transparency.

IV. RESULTS AND DISCUSSION

A. TIVA Enhancements from RSIL

Fig. 3(a) shows the TIVA signal image using 1340 nm laser with a 20X objective from voiding defects along the serpentine tracks which is biased with a constant current of 10 μA. From the overlay image in Fig. 3(b), it is evident that the defects cannot be precisely localized because the serpentine tracks are not resolvable and the TIVA spots are large. With the same sourcing current, laser power and amplifier settings, the experiment is repeated with RSIL. Fig. 4(a) and 4(b) show the TIVA signal and overlay images respectively.

The serpentine lines are clearly resolved and the defect sites are localized precisely along the lines with stronger TIVA signal. The reflected laser image also reveals the magnification enhancement from RSIL.

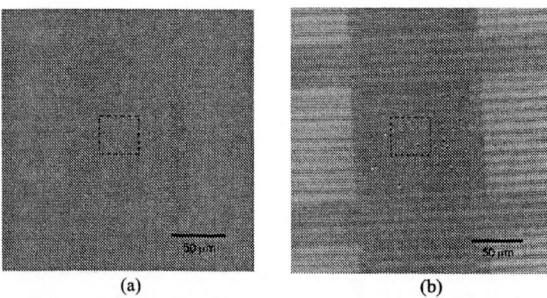

(a)   (b)

Fig. 3. TIVA image (a) and overlay image (b) from a polysilicon structure using Mitutoyo 20X objective.

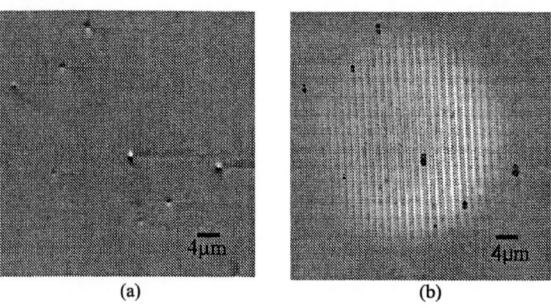

(a)   (b)

Fig. 4. TIVA signal image (a) and overlay image (b) from a polysilicon structure using Mitutoyo 20X objective and RSIL application.

The middle signal is more enhanced than peripheral spots because of increasing aberrations away from the lens optical axis. To quantify the enhancements from RSIL, the line profile AA' is plotted across the middle defect for the case of with and without RSIL as shown in Fig. 5(a). The line profiles are normalized to the same magnification to take into account the increase in magnification due to RSIL.

Fig. 5(b) shows the peak TIVA signal enhancement of close to 4.5 times with RSIL. In this case, the theoretical laser spot size is enhanced by approximately 8 to 9 times. The peak TIVA signal is defined as the signal peak above noise mean. It should be noted that the enhancement in TIVA signal is not directly proportional to the spot size enhancement because of laser power absorption from the additional SIL thickness. The TIVA signal intensity could also be affected by the nature of the defect and the effective interaction area between the defect and laser spot size.

The lines profiles are normalized to signal intensity to compare the Full Width Half Maximum (FWHM) of the TIVA signal. Fig. 5(c) shows that the FWHM of TIVA signal with and without RSIL is around 0.6 μm and 2 μm respectively. The reduction of TIVA FWHM by about 3 times shows that with RSIL, the TIVA signal is reduced close to the defect size of approximately 0.5 μm.

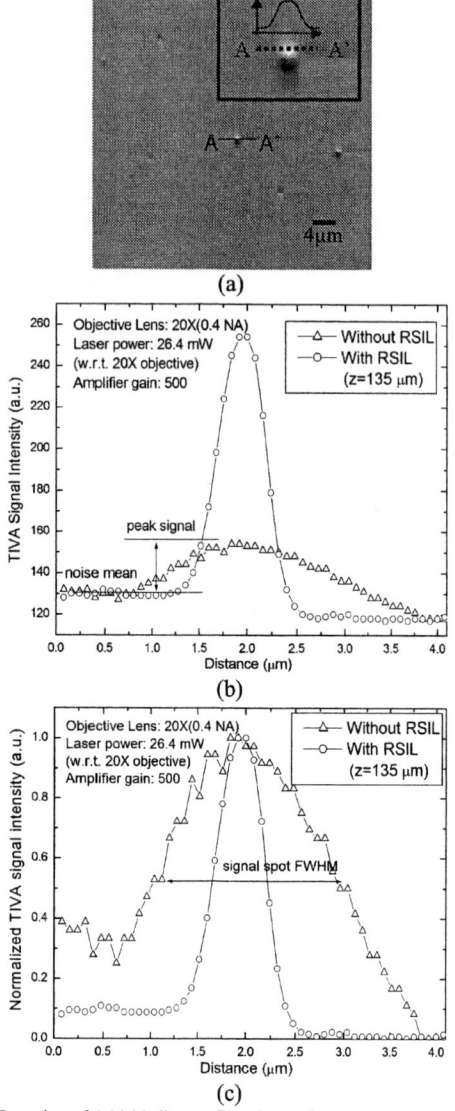

(a)

(b)

(c)

Fig. 5. Location of AA' (a), line profile of AA' (b) and normalized intensity plot (c), across TIVA signal spot.

B. Power Loss from RSIL and Objective Lens

For backside RSIL application, additional laser power losses can arise from reflections at the spherical interface of RSIL, laser power absorption by SIL and power loss due to laser beam truncation by the back pupil of objective lens. Fig. 6(a) shows a comparison between the TIVA line profiles of the cases with selected RSIL designs and without RSIL application. The same laser power of 26.4mW is used and the objective is fixed at 20X. The power loss due to reflections at spherical interface is assumed to be constant for RSILs at different focal planes. Fig. 6(a) shows that TIVA signal enhancement of 4.5 times is achieved at focal planes, Z=13 μm and 135 μm. However at 164 μm, although the laser spot is enhanced, there is no significant enhancement in TIVA signal. This is because the power loss by laser absorption through highly-doped thicker RSIL compromises the increase of power density due to reduction of laser spot size. This is the issue if the RSIL lens is highly doped. The laser spot size enhancement is evident in Fig. 6(b) which shows that the FWHM of the TIVA spot are similar and close to defect size.

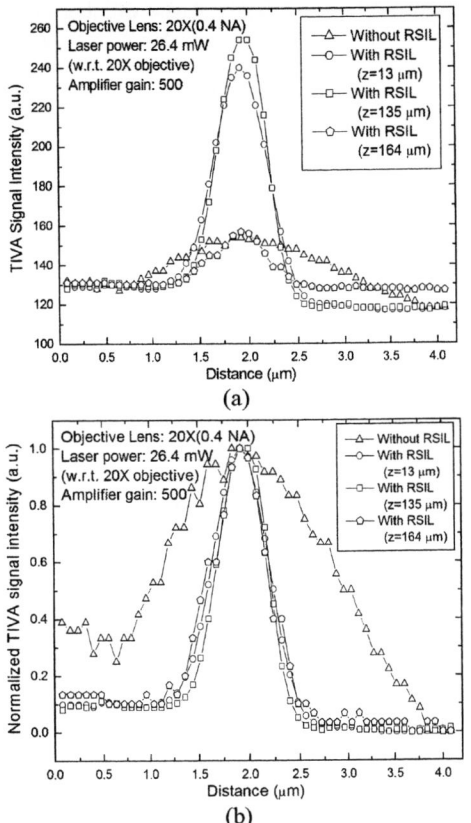

(a)

(b)

Fig. 6. Line profile (a) and normalized intensity plot (b), across TIVA signal spot.

Fig. 7 illustrates the significance of laser beam truncation by the back pupil of the objective lens. In this case, focal plane is fixed at 13 μm below centric point and the same laser power of 36 mW measured before objective lens is used. Fig. 7 shows that switching to higher NA objective lens degrade TIVA signal strength in spite of reduction in laser spot size as NA increases. This is because the power loss by laser beam truncation is more significant than the reduction in laser spot size, giving rise to an overall degradation in power density.

The pupil diameter, PD [10] is defined as:

$$PD = 2*f*NA \qquad (1)$$

As *NA* increases with higher objectives, the objective focal length *f* reduces faster, resulting in an eventual reduction in *PD*. A smaller pupil diameter leads to a larger beam truncation and power loss since the beam diameter is fixed. This loss can be compensated by increasing the laser power.

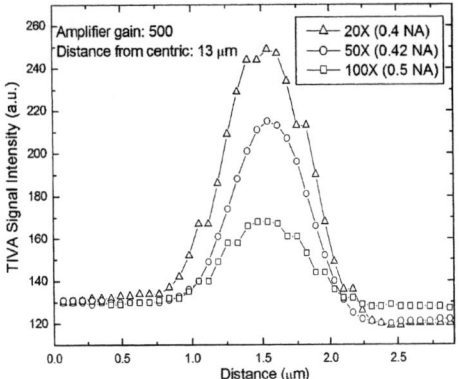

Fig. 7. Degradations from objective lens.

C. Objective NA matching

The mechanical aperture structure plays an important role in RSIL. The collection angle decreases with increase in focusing depths for a fixed clear aperture. Fig.8 shows the corresponding objective NA required based on 90% clear aperture. It is calculated from geometric optics and is an approximation at the paraxial focus. At focusing depths, other than the centric and aplanatic location, focal shifts exist. This can change the objective NA, especially at large angles. Fig. 8 shows that at regions close to the centric location, a high NA is required for optimum resolution. This requirement reduces at greater depths such that an NA as small as 0.23 is sufficient at the aplanatic location.

Although this plot shows the minimum NA required for collection, degradations can occur if excess NA is used for laser delivery. Fig. 9(a) shows the case when the light rays are converged by the objective lens to within the aperture opening. On the contrary, if the NA is in excess of requirement, as shown in Fig. 9(b), there is truncation of the geometrical cone of laser beam which results in power loss and degradation of the TIVA signal and reflected image contrast. Therefore, too much excess NA is undesirable. The NA should preferably be matched close to the minimum requirement for optimum laser delivery.

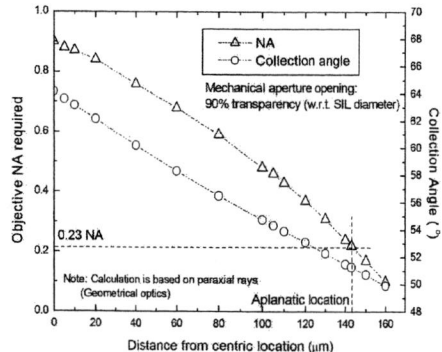

Fig. 8. NA requirement and collection angle at various focal depths.

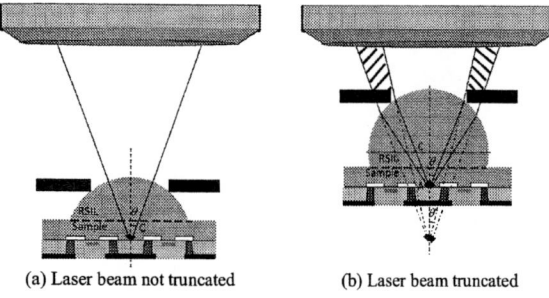

(a) Laser beam not truncated   (b) Laser beam truncated

Fig. 9. Laser beam geometrical cone, within mechanical aperture (a) and larger than mechanical aperture (b).

D. Effect of RSIL parameters on imaging and TIVA enhancement

The key enhancement and degradation factors on TIVA signals have been identified. We investigate the tradeoffs between the laser spot size variation, SIL absorption and objective NA matching at different designed focusing depths. Degradation due to laser beam truncation by the objective pupil is normalized by using variable power for different objective lens to make sure that the resultant laser power measured after objective lens is approximately 10-11 mW. Fig. 10 shows the peak TIVA signal at increasing focal depths, from centric to aplanatic, for selected objective lens.

In general, power absorption from RSIL increases with increasing focusing depths as thicker Si is required. However, the main effects from the variation in TIVA peak signal with focusing depths comes from the spot size variation and objective NA matching. Fig. 11 shows the experimental spatial resolution in imaging. It should be emphasized that the resolution values were quantified based on FWHM resolution. It can be experimentally proven that the experimental observable resolution can vary from as much as 30-38%. Although there are some differences, the reflected laser image resolution is expected to be close to that of laser spot size.

In Fig. 10, at region close to centric location, the highest TIVA signal is achieved by 50X and 100X due to their small laser spot size compared to 10X and 20X. The signal for 10X

is the worst. The contrast and sharpness of the image for 10X is too poor in the image and not quantified. For optimization in this region, the higher the NA, the better will be the signal gain.

With increase in RSIL thickness to focus at larger depths, TIVA signal degrades from 50X and 100X because the power loss from lens absorption is more significant than the resolution enhancement. On the contrary, the signal from 10X and 20X increases because the resolution enhancement is more dominant. When the focal depth increases further to about 90-100 µm below centric location, the NA required for collection starts to reduce to around 0.5. Laser beam from excess NA becomes an issue from this point onwards. This is seen from the increased gradient in signal degradation for 50X and 100X which is especially worse for 100X. Both 10X and 20X continue to achieve increasingly good signal gain from their enhanced resolutions.

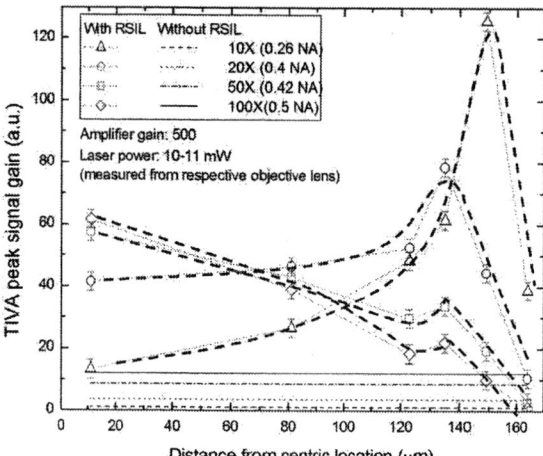

Fig. 10. Change in TIVA signal enhancements with varying laser focusing planes and objective NA.

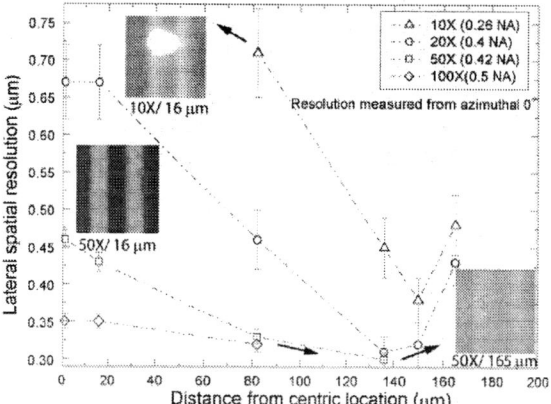

Fig. 11. Experimental spatial resolution at various imaging planes.

At region close to 135 µm, there is a peak in TIVA signal from 20X objective as the resolution is close to the minimum.

This is more significant than the slight laser beam truncation. At this point, the 10X is still not at its optimum resolution. The increase in signal from 50X and 100X is insignificant due to high laser beam truncation. When focusing depths increase further to 150 μm, there is another signal peak from 10X objective because of its enhanced resolution enhancement. This peak is higher than the previous one from 20X although it does not have a minimum resolution better than 20X objective. This is because its beam is not truncated due to its low NA. After this point, TIVA signal degrades drastically for all objectives. Therefore, at regions close to the aplanatic location, it is better to use low objectives of 10X and 20X for NA matching reasons. This is also important for imaging as well. A comparison between the reflected laser images captured at 16 μm and 165 μm as shown by insets in Fig. 11, reveals that beam truncation by mechanical aperture can reduce the laser power to an extent that contrast is significantly affected.

It is clear that TIVA performance with RSIL application is highly dependent on various parameters. The success of TIVA localization technique also depends on the spatial resolution of reflected laser image for overlay. Although it might seem that 10X can provide the highest signal gain, it should not be the design target because of resolution tradeoff. A weak TIVA signal due to power loss can be overcome by increasing laser power delivery. The results in this work are also applicable to the case of RSIL which is formed directly on Si substrate [11].

V. CONCLUSION

This RSIL parameters that affect TIVA and imaging enhancements has been discussed. The results have highlighted the importance of RSIL design to maximize performance both in TIVA and imaging. There is no single optimum condition when applying RSIL. For design close to centric location, a high backing objective NA is necessary. On the contrary, a low backing NA is required at locations close to aplanatic. A too high objective NA will result in degradations from beam truncation by the mechanical aperture of RSIL. Although a low NA will give rise to optimum TIVA signal, it is undesirable because of optical resolution tradeoffs, affecting precise defect localization.

ACKNOWLEDGEMENT

The authors, SH Goh and ACT Quah are supported jointly under the National University of Singapore (NUS) research scholarships and Chartered Semiconductor Manufacturing graduate program grant. We would like to acknowledge Li Lijuan from NUS for helping with sample preparation. We also wish to thank Frank Zachariasse from NXP semiconductors for the valuable discussions.

REFERENCES

[1] Phang JCH, Chan DSH, Palaniappan M, Chin JM, Davis B, Bruce M, Wilcox J, Gilfeather G, Chua CM, Koh LS, Ng HY, Tan SH, "A Review of Laser Induced Techniques for Microelectronic Failure Analysis", Proc Int Symp Physical & Failure Analysis of Integrated Circuits (IPFA 2004), 5-8 Jul 04, Hsinchu, Taiwan, pg 255-261, 2004.

[2] Aw SE, Tan HS, Ong CK, "Optical Adsorption Measurements of Band-gap Shrinkage in Moderately and Heavily Doped Silicon", J Phys: Condens Matter, Vol 3, No 42, pg 8213-8223, 1991.

[3] Mansfield SM, Kino GS, "Solid Immersion Microscope", Appl Phys Lett, Vol 57, No 24, pg 2615-2616, 1990.

[4] Eiles T, Pardy P, "Liquid Immersion Objective for High-Resolution Optical Probing of Advanced Microprocessors", Proc Int Symp Testing & Failure Analysis (ISTFA 2001), 11-15 Nov 01, Santa Clara, California, USA, pg 167-170, 2001.

[5] Zachariasse F, Goossens M, "Diffractive Lenses for High Resolution Laser Based Failure Analysis", Proc Int Symp Testing & Failure Analysis (ISTFA 2005), 6-10 Nov 05, San Jose, California, USA, pg 1-7, 2005.

[6] Ippolito SB, Goldberg BB, Uenlue MS, "Theoretical Analysis of Numerical Aperture Increasing Lens Microscopy", J Appl Phys, Vol 97, pg 053105-1-12, 2005.

[7] de la Bardonnie M, Ross R, Ly K, Lorut F, Lamy M, Wyon C, Kwakman LFTz, "The Effectiveness of OBIRCH Based Fault Isolation for Sub-90nm CMOS Technologies", Proc Int Symp Testing & Failure Analysis (ISTFA 2005), 6-10 Nov 05, San Jose, California, USA, pg 49-58, 2005.

[8] Barta E, "Optical Constants of Various Heavily Doped p- and n-type Silicon Crystals Obtained by Kramers-Krong Analysis", Infrared Physics, Vol 17, pg 319-329, 1977.

[9] Ippolito SB, Goldberg BB, Ünlü MS, "High spatial resolution subsurface microscopy", Appl Phys Lett, Vol 78, No 26, pg 4071-4073, 2001.

[10] Mitutoyo Corporation, "Microscope Units, Objectives, Eyepieces and Accessories", Catalog No E4191-378

[11] Koyama T, Yoshida E, Komori J, Mashiko Y, Nakasuji T, Katoh H, "High Resolution Backside Fault Isolation Technique Using Directly Forming Si Substrate into Solid Immersion Lens", Proc Int Rel Phys Symp (IRPS 2003), 30 Mar 03 – 4 Apr 03, Dallas, Texas, USA, pg 529 - 535, 2003.

Application of FIB Circuit Edit Combined with TIVA in Advanced Failure Analysis

David Zhu*, S.P. Neo, S. K Loh, and Eddie Er

Failure Analysis Group, QRA, Chartered Semiconductor Manufacturing Limited

60 Woodlands, Industrial Park D Street 2, Singapore 738406

Abstract - **This paper presented a failure analysis methodology to overcome the difficulties of fault location encountered by pin leakage and some testing parameter failures in a mixed signal device. These types of failure generally cannot be solved by traditional electrical failure analysis methods.**

I. INTRODUCTION

During failure analysis, a key emphasis is the efficiency of the transition from the failure signature on a wafer map to the identification of the exact process root cause. However, as technology progresses and the feature size shrinks to deep-sub micron technology node, the chip becomes more and more complex. The increased complexity correspondingly leads to increase in difficulties in failure analysis (FA). For example, no passive voltage contrast (PVC) on Silicon on Insulator (SOI) devices and the defective sites of some pins leakage or testing parameters failure cannot be isolated by simply biasing the Vdd and Vss and use the traditional electrical fault isolation methods such as OBIRCH/TIVA. However, if the most suspected failure block in those pins leakage or testing parameter failure units can be identified and Focus Ion Beam circuit edit employed to establish electrical behavior analysis (EFA) in those blocks, coupled with fault isolation technique, the defective site can be quickly and accurately identify and thus the defect and failure mechanism established.

Focused Ion Beam (FIB) plays a crucial role in modern semiconductor failure analysis laboratories. The basic FIB application can be divided into electrical and physical analysis. Physical analysis application is widely varied, includes several cross-sectioning techniques for defect determination, process monitoring, TEM sample preparation, grain size and orientation analysis and material composition analysis via EDX [1][2]. Electrical analysis application includes passive voltage contrast, recently FIB equipped with gas injection system is also widely employed to do circuit edit for failure analysis, because it possesses the ability to remove and deposit material in the circuit with nanometer-scale precision. Circuit edit mainly includes probe-point access, probe-pad formation, connect and disconnect some circuit [3][4]. For failure analysis, the ability to edit a circuit in devices is extremely valuable as it permits us to establish electrical failure analysis in a localized area as needed, such as FIB circuit edit was employed to deposit metal line and micro-pads for accessing to the related internal nodes of the suspected defective transistors in memory failure analysis [5]. In this paper FIB circuit edit was used to make the access point and to form the micro-pads for further electrical analysis.

Thermal Induced Voltage Altering (TIVA) is a powerful electrical fault isolation tool. It has been widely used in conductivity related failure analysis, such as short, via/contact failure, void in metal induced by stress migration or electron migration, abnormal resistance, ESD failure and so on. The operation principle of this kind of tool is illustrated in Fig.1. A constant current is applied to the device or circuit of interest and monitor the voltage without laser scan. As IR laser beam (wavelength 1340nm) is used to scan the sample, the voltage measured will be different from that measured in the absence of the IR laser scan due to the resistance of the circuit being changed by the heating induced by the laser beam scan. Since the resistivity as well as the resistance temperature coefficient (TCR) is different from one material to another, the TIVA detected voltage change will be largely depend on the relative contribution to the current path of each material heated by the IR laser beam [6][7]. When the laser reaches an abnormality in the current path, the voltage change will be significant due to the big difference of TCR between the abnormal point and normal area, thus TIVA system will see it as a hotspot. TIVA analysis can be performed from frontside or backside of the sample since silicon becomes transparent at wavelength of 1070 nm. In this paper TIVA was employed to successfully isolate the fault site after FIB circuit edit.

II. EXPERIMENT AND DISCUSSION

In this paper, a SOC (system on chip) device that

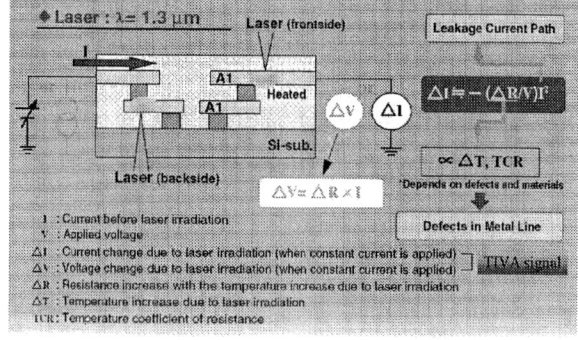

Fig. 1. Operation principle of TIVA

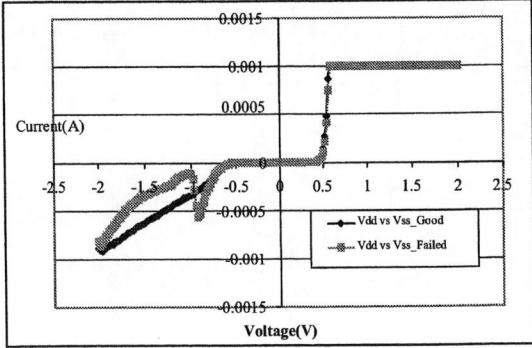

Fig. 2. IV curve of Vdd with respect with Vss

manufactured with 90 nm technology in bulk wafer suffered from power short failure issue was presented. The electrical verification was performed by probing between general power Vdd with respect to ground Vss as shown in Fig. 2. A difference was observed in the reverse region in the failed die as compare with that of a good unit. The current initially increased with the force voltage and then decreased at a certain voltage. Conventional TIVA and liquid crystal (LC) analysis with biasing at the abnormal IV curve region were performed to identify the possible defective location. However, no hotspot was detected regardless of the biasing condition or equipment parameter settings. Generally, for this kind of case, if electrical fault isolation could not locate the possible defective area, random SEM/optical inspection will be performed at all layers, which is a time-consuming and aimless process. Normally, no visible defect will be observed if the defect is not very obvious. So the success rate is very low. Subsequently, we could not identify the failure mechanism and root cause for corrective action.

In order to have an effective and efficient method to overcome the difficulty encountered in this case, information

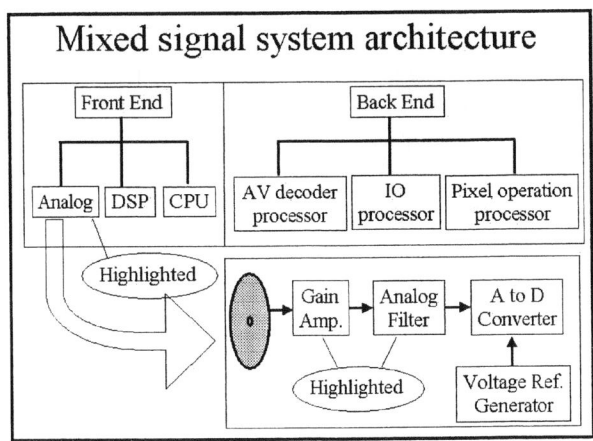

Fig. 3. System architecture and the suspected blocks

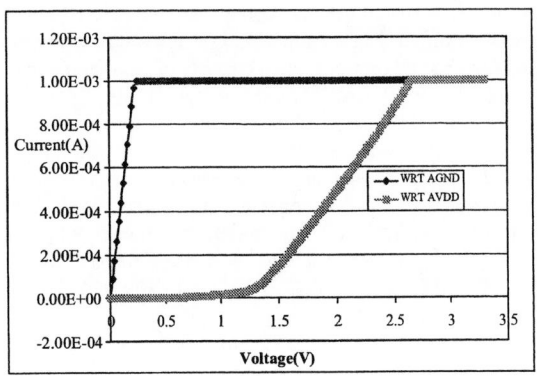

Fig. 4. Micro-probe analysis of the suspected current path revealed resistive short with respect to Vss

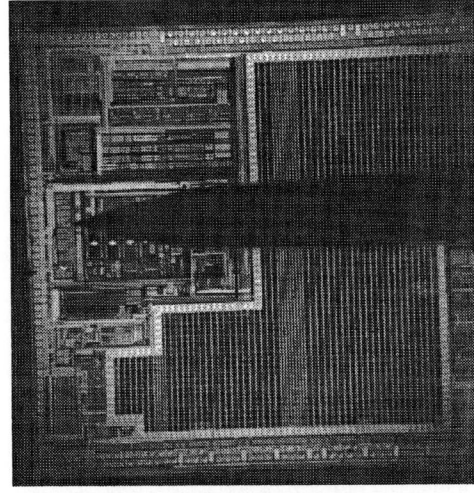

Fig. 5. TIVA image showed the hotspot

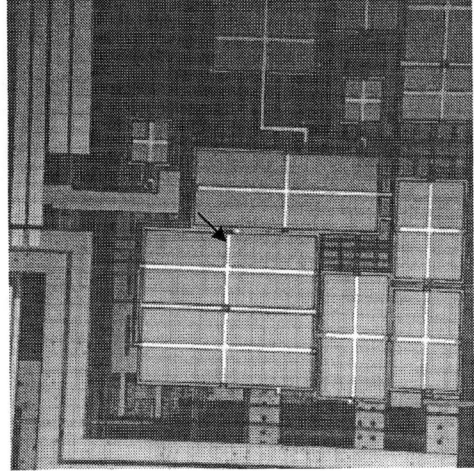

Fig. 6. Higher magnification TIVA image showed the accurate hotspot location

978-1-4244-2039-1/08/$25.00 ©2008 IEEE

about device architecture and testing methods and testing data log were obtained. Fig.3 showed the system architecture of the mixed signal SOC device, which consists of front end and back end, and the front end includes analog, DSP(digital signal processor) and CPU (central processor unit). Due to the nature of the testing, the wafer level testing in mixed signal integrated circuits is mostly based on the measuring and comparing of direct current tests, which means that most mixed signal failures can be sorted by these measuring parameters [8]. After analyzing the testing data and the relationship between different blocks, basically the analog block was suspected to be the defective block. Further analysis was performed in the analog block and the gain amplifier and analog filter was considered to be the most suspected area of failure (see Fig. 3). From the analysis and trace of GDS layout in the suspected functional blocks, a special current path was selected for FIB circuit edit for electrical behavior analysis and fault isolation in the suspected blocks. Therefore one failed unit was de-processed to top metal layer followed by FIB circuit edit to make the access point to the suspected current path and micro-pad for electrical probing. Now the unit was ready for micro-probing to characterize the electrical behavior of the circuit in the suspected blocks. Microprobe analysis of the suspected current path showed an abnormal resistive short with respect to Vss as in Fig. 4. It implied that bridging existed somewhere in the suspected block. Electrical fault isolation using TIVA was performed again by biasing the suspected current path with respect to Vss. The hotspot detected by TIVA analysis was shown in Fig 5 and 6. Thus the unit was submitted to physical failure analysis. FIB slice and view was performed around the hotspot location and metal bridging between two adjacent MIM capacitor's bottom plate was shown in Fig 7. A second unit was analyzed to confirm the failure. Hot spot was detected. In this unit, top-down delayering and SEM inspection was performed at the hotspot location, it revealed the similar metal bridging shown in Fig. 8. Further XTEM and EDX analysis as shown in Fig. 9 and Fig. 109 revealed and confirmed Ta residue causing

Fig. 8. SEM top down inspection showed bridging between metal lines

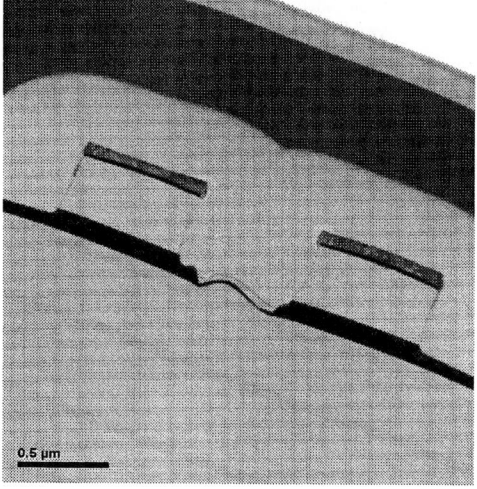

Fig. 9. Cross-section TEM image showed residue between metal lines

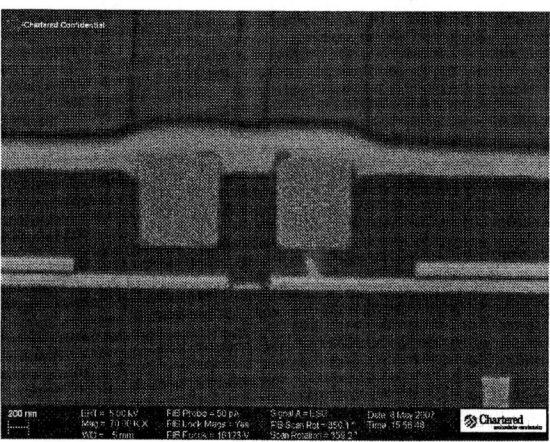

Fig. 7. FIB slice and view showed bridging between metal lines

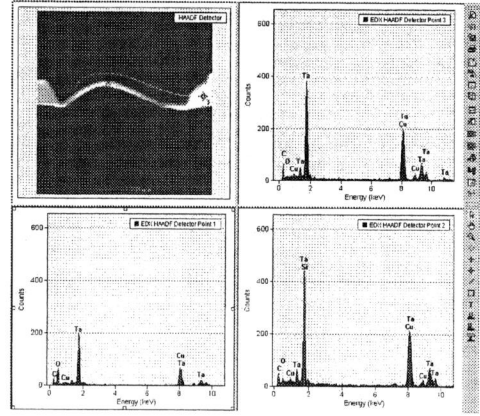

Fig. 10. TEM EDX analysis confirmed Ta residue

the two metal line to be bridged and was responsible for the parameter failure of the device.

Corrective action was taken according to this FA results. A series of split was designed on MIM capacitor etch and the split results showed the Ta residue increase when reduce the Ta over etching. After increasing the Ta over etching time, inline defect scans showed absence of Ta residue and electrical-test data also showed absence of failure on MIM bottom plate bridging.

III. CONCLUSION

This paper presented a FA methodology, which successfully overcomes the difficulties conventional electrical failure analysis method has in isolating the failure location. It bases on the analysis of the system architecture, test methods, test data log and theirs relationship. This leads to excluding of some blocks which are not related to the failure and GDS analysis on the layout in the most suspected blocks was done to choose the common current path to perform FIB circuit edit for localized electrical behavior analysis and electrical fault isolation. In this case, fault isolation technique, TIVA analysis was chosen and the failure location was isolated. This approach is applicable for pin leakage and some testing parameter failure too. It helps to expedite the failure analysis process and improve the success rate.

ACKNOWLEDGMENT

The authors would like to gratefully thank our colleagues Koo Hongtak, Li Yan and Yau Moikian for sample de-processing and SEM inspection, and we also thank TEM group for providing FIB slice & view analysis and TEM inspection and EDX analysis for this case.

REFERENCES

[1] S.B. Herschbein, L.S. Fischer and A.D. Shore, Microelectronic Failure Analysis, pp. 517-526.
[2] Kendall Scott wills and Seshu V. Pabblisetty, Microelectronic Failure Analysis, pp. 527-552.
[3] Vladimir V. Makarov and Nicholas Antoniou, ISTFA proceedings, pp.62-65 (2006).
[4] ACT Quah, JCH Phang, LS Koh, SH Tan and CM Chua, ISTFA proceedings, pp.234-238 (2006).
[5] Z. G. Song, S. K. Loh, X.H. Zhang, S.P. Neo and C.K. Oh, ISTFA pp. 204-207 (2006).
[6] R. Ross, K. Ly, M. de la Bardonnie, L.F.tz. Kwakman, F.Lorut, M. Lamy, C. Wyon, ISTFA proceedings, pp.447-450 (2004).
[7] H.W. Yang, B.H.Lee, R.L.Hwang, L.H.Chu, and David Su ISTFA proceedings, pp.229-302 (2004).
[8] Cha-Ming Shen, Ten-Long Chang and Lian-Fon Wen, ISTFA proceedings, pp.327-330 (2007).

Failure Analysis Enhancement by Evaluating the Thermal Laser Stimulation Impact on Analog ICs

Magdalena SIENKIEWICZ[1,2], Philippe PERDU[1], Abdellatif FIRITI[2], Olivier CREPEL[2], Dean LEWIS[3]

1: CNES – French Space Agency, 18, avenue Edouard Belin, 31401 Toulouse, France
2: Freescale Semiconductor France SAS, 134, avenue Général Eisenhower, 31023 Toulouse, France
3: IMS, Université de Bordeaux 1, 33405 Talence, France

Abstract - **Technology advances in mixed-mode and analog ICs, involving multiple functions inside the device, make Failure Analysis (FA) difficult. Variation Mapping techniques based on Thermal Laser Stimulation (TLS [1],[2],[3]) have already proven their efficiency for the localization of a physical defect or design weakness. Their extension will play a key FA role but the impact of IR laser beam on the basic analog structures has to be better understood. In this paper we will make this necessary study based on simulations & experimentations. The results allow a better understanding of device activities under TLS to, finally, improve the failure localization process.**

I. INTRODUCTION

Nowadays, laser stimulation techniques are commonly used in the failure localization domain. As a function of the failure background and diagnostic, two major phenomena will be employed: photoelectric or thermal. For the first, a 1064nm wavelength laser source, by generating electron-hole pairs in the device active area, induces a photocurrent and allows the localization of sensitive devices at the silicon layer (PLS: Photoelectric Laser Stimulation techniques) [4], [5], [6]. For the second, a 1300nm wavelength laser source, by heating the conductive material in the device, allows the localization of resistive issues (TLS techniques).

Initially, based on these phenomena, static localization techniques were developed: PLS (OBIC [4], LIVA [5]...), TLS (IR-OBIRCH [1], TIVA [2], SEI [2]...). In the course of time, the devices have become more complex (miniaturization, increasing number of components, metal layers, interconnections, etc.) and so their analysis. The number of cases increased where the failure appeared only when the devices were stimulated dynamically. Moreover, a kind of fine failure named "soft" was observed, where the device was failing only for a certain range of its electrical or environmental setup parameters and was working correctly for the other parameters. In consequence, dynamic localization techniques were developed [7]: DPLS (LADA [8]), DTLS (SDL [9], RIL [10]).

These advanced techniques are often applied on digital ICs. The digital circuits represent only Boolean logic functions while analog devices have plenty of analog static and dynamic parameters related to their internal functionalities. Laser stimulation can slightly change these parameters inducing measurable effect at the output of the devices. It can be measured with Variation Mapping Techniques previously presented (xVM [11], [12], [13]).

In the literature we find considerably less case studies concerning debug or failure localization on the analog or mixed-mode devices compared to digital ICs. Understanding of the localization signatures on complex analog and mixed-mode ICs is very difficult comparing to the digital circuit. To enhance this weak point for the FA community and industry, we decided to study signatures on elementary analog devices and especially the impact of the thermal laser stimulation on these elementary devices.

Here, we propose the experimental results and first order simulations on a commercialized mixed-mode golden device where we focus on the analog basic structure. Then, we present the advanced simulations on a similar, separately designed structure.

The above basic analog structure, current mirror, was designed with different configurations at Freescale and specific electrical simulations (CADENCE tool) were run with exact process data. The impact of the thermal laser stimulation (1.3μm wavelength laser inducing heating) was analyzed for different electrical and environmental setup conditions (voltage, frequency, global and local temperature). The goal was to better understand the behavior of the complex analog ICs through the basic devices.

II. POWER AMPLIFIER DRIVER

A. Circuit Presentation

The studied golden device is an advanced mixed-mode Freescale IC, commonly used in the RF domain. It contains three metal layers, hence the observation of the active areas from the frontside is partially handicapped.

Among numerous analog blocks of this device, we selected the Power Amplifier (PA) Driver block to analyze. Its simplified electrical schematic is presented in figure 1.

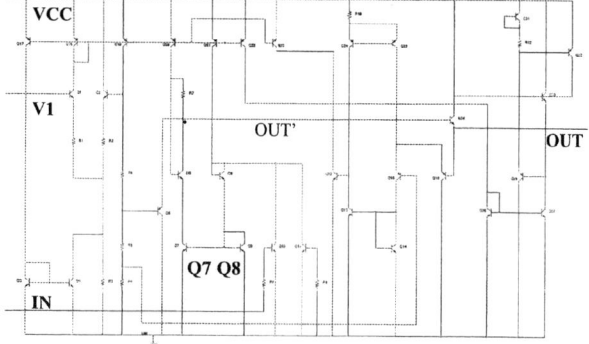

Fig. 1. Simplified electrical schematic of PA driver.

Data is sent on the external IN pin which, further, drives the RF output stage in the PA block. The OUT pin is used to control the device output power. V1 corresponds to the internal voltage level.

B. Experimental Setup & Results

To analyze the device under thermal laser stimulation, timing behaviour was chosen and studied as a function of the laser stimulation through the Delay Variation Mapping technique (DVM [12]). This parametric technique allows us to quantify the sensitivity level and to determine whether the signal is delayed or accelerated. Figure 2 presents the experimental setup used to localize and characterize sensitive areas inside the IC.

The delay between synchronized OUT, and a reference signal, was turned into a voltage level by means of Time to Amplitude Converter (TAC) for each laser beam position during the scanning. In our experiment, only the variation of the rising edge was measured. VCC was fixed at 3V and a square signal of 2.5V at 5kHz was applied at the input IN. The laser beam position was synchronized with the input IN signal. To have a complete view on a studied devices and to avoid any access problem (three metal layers), the device was analyzed from the backside.

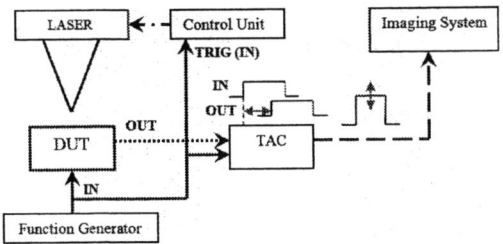

Fig. 2. Delay Variation Mapping (DVM) : Experimental setup.

Figure 3 presents the DVM mapping results: the 1300nm wavelength laser beam was scanning over the DUT and several sensitive areas were detected on the golden device. The white color corresponds to the acceleration of the rising edge, the black areas describe the slowing down of the rising edge.

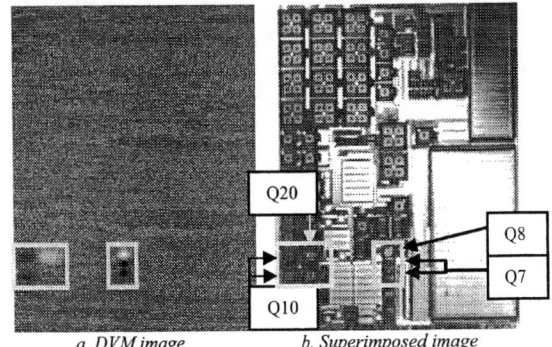

| a. DVM image | b. Superimposed image |

Fig. 3. Sensitive areas localization on a golden device by using DVM technique.

DVM localization highlighted Q7, Q8, Q10 & Q20 bipolar transistors. Q7 and Q10 devices are composed of two parallel bipolar transistors, hence in figure 3 there are two sensitive areas

corresponding to each of these devices. Figure 4 is an extracted schematic of PA driver illustrated in figure 1 including the highlighted devices. The dashed squares correspond to the accelerating devices from figure 3, and the continuous squares describe the slowing down devices from figure 3 when DVM technique is used. The most sensitive devices were NPN bipolar transistors Q8 and then Q7, designed as a current mirror.

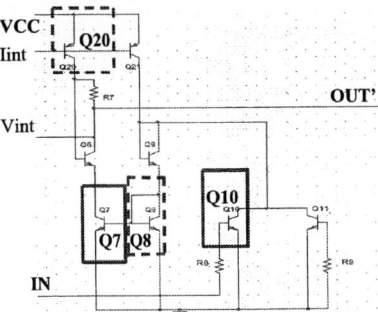

Fig. 4. Extracted electrical schematic of PA driver including devices highlighted by DVM technique.

C. OrCAD Simulation Results

The PA driver presented in figure 1 was simulated with OrCAD tool. The electrical models correspond to the real models without taking account of process data. The electrical setup was exactly the same as in measurements described in the preceding section. To simulate a 1.3μm wavelength laser beam scanning over the device, we increased locally the temperature of each of the device highlighted previously by DVM technique. The time sensitivity was probed on OUT signal. Figure 5 presents obtained results. The REF waveform corresponds to the OUT signal propagation when the device is not submitted to the laser beam. Others waveforms present the OUT signal propagation when the laser scans over Q7 bipolar transistor, then Q8 bipolar transistor, etc. From figure 5 we see the rising edges of Q7 and Q10 transistors are delayed. For Q8 and Q20 transistors the rising edges are accelerated.

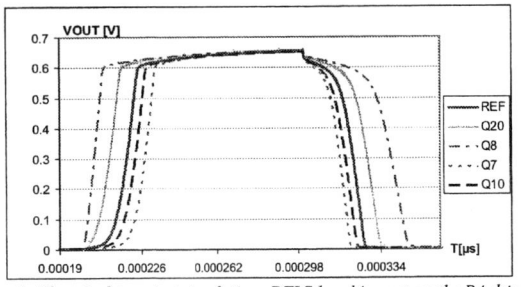

Fig. 5. Electrical transient simulation: DTLS local impact on the PA driver.

Despite using only electrical models (no data process taken into account), these first order simulations with OrCAD tool fit well with the experimental results. Hence, we can use it to study the FA case or to validate the failure localization results. Therefore, the above structures have to be studied with more accuracy. For this reason we designed the test structures: current mirrors based on PNP and NPN bipolar transistors. In the next

978-1-4244-2039-1/08/$25.00 ©2008 IEEE   31   *Proceedings of 15th IPFA - 2008, Singapore*

chapter we describe in detail the announced basic structures and various simulations run on these structures.

III. CURRENT MIRROR

A. Structure Presentation

The current mirrors were simulated in two configurations: with PNP bipolar transistors (Fig. 6a) and with NPN bipolar transistors (Fig. 6b). To note different device sensitivity, different emitter dimensions were chosen. For PNP structures the Q0 and Q1 emitters length were: 2.5μm, 3.5μm and 4.5μm. For NPN structures the Q1 and Q2 emitters width were fixed at 3μ and its length varied as follows: 3μm, 5μm and 7μm. All resistors values were fixed at 1kΩ.

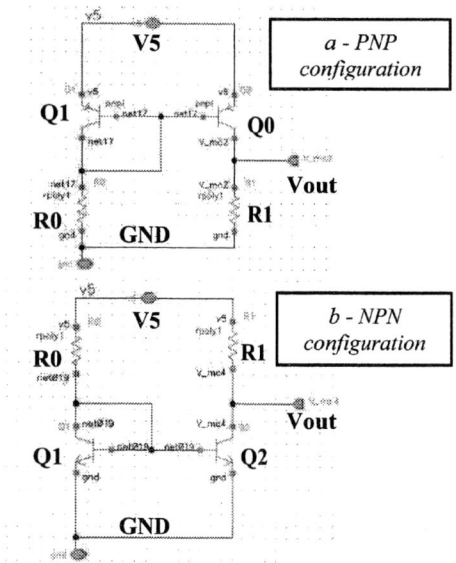

Fig. 6. Electrical schematic of the current mirrors.

The goal of this paper is to look at the temperature sensitivity of all the devices presented in the above structures. In the current mirror structure a key parameter is a current, so we should have simulated and measured this parameter. However, this current strongly depends on the Vout voltage node which is much easier to measure than current. Hence, we will focus on this voltage node. Precisely, the output current Iout = Vout/R1 for PNP structure and Iout = (V5-Vout)/R1 for NPN structures.

Below is presented the nomenclature which will be used in the following text.

Emitter length[μm]	PNP			NPN		
	2.5	3.5	4.5	3	5	7
Structure name	CM1	CM2	CM3	CM4	CM5	CM6

Table I: Structures nomenclature.

B. CADENCE Simulation Setup

All the simulations were run on a Cadence tool with PSpice models libraries. Two kinds of simulations were run: DC and Transient. DC analysis consisted of varying the structure input voltage (V5) from 0 to 5V and observing changes on the output

Vout signal. Transient analysis consisted of applying at the structure input (V5) a square signal with a fixed amplitude and frequency and measuring all variations of time/frequency and voltage level of the output Vout signal.

In the first step, the global temperature was raised 5°C from ambient temperature to observe which of the device output parameters were impacted and to quantify all variations. In the second step, the local temperature of each basic device was increased by 10°C relatively to the global temperature of the structure. The goal was to evaluate the sensitivity of each of the basic devices by observing the variation of the output parameters (voltage level, frequency).

C. Global Temperature Changes Results & Analysis

The purpose of global temperature changes is to check Vout voltage variation versus V5 (DC analysis), signal delay variation and Vout voltage variation (Transient analysis) when the entire device temperature changes. It will confirm the interest of studying analog variations on selected structures.

DC Analysis

The results of DC analysis are presented in figure 7 for PNP structures and in figure 8 for NPN structures. For PNP structures, by increasing the input V5 voltage the output voltage variation (ΔVout) increases. For different size of PNP bipolar transistor active area and fixed input voltage, the output sensitivity is almost the same.

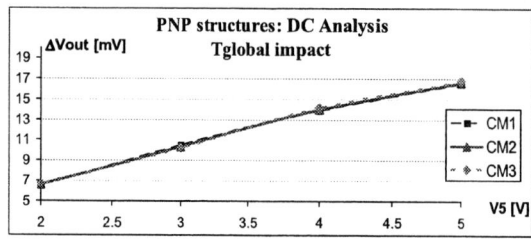

Fig. 7. DC simulations of PNP structures.
ΔVout variation /Vout$_{27°C}$ −Vout$_{100°C}$/ versus V5.

For NPN structures, by increasing the input V5 voltage the output voltage variation increases (Fig.8) but considerably less than in case of PNP structures. For the low input voltage, the output variation is not the same for different size of bipolar transistor active area. For the high input voltage, the output variation converges for different size of bipolar transistor active area.

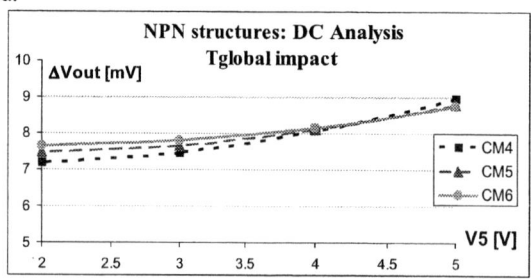

Fig. 8. DC simulations of NPN structures.
ΔVout variation /Vout$_{27°C}$ −Vout$_{100°C}$/ versus V5.

Comparing PNP and NPN structures we observe, that as a function of the input voltage, PNP structures have a higher sensitivity to global temperature changes. Moreover, this sensitivity is independent on the emitter dimensions. When the temperature increases the Vout variation increases. In this case the curve form is kept.

Transient Analysis

The input voltage on V5 was fixed at 5V and the simulations were run for two frequencies: f=5kHz & 250kHz. The results are presented in figure 5 for both kind of structures: PNP and NPN.

When the global temperature changes from 27°C to 32°C, PNP structures have more important signal variation on the Vout (Fig. 9). For all structures, the measurements are not influenced by the size of the active area.

In the time domain, a slight signal delay is observed for all structures: 20ps maximum for PNP structures and 30ps for NPN structures.

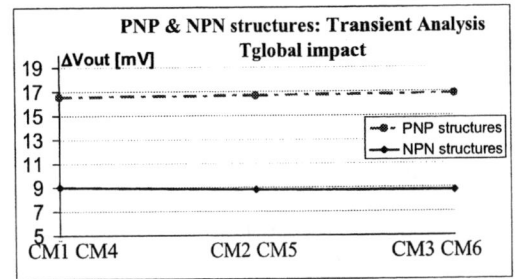

Fig. 9. Transient simulations of PNP & NPN structures (V5=5V).

D. Local Temperature Changes Results & Analysis

First simulations, where the impact of the global temperature is observed, provide interesting indications. For both PNP and NPN structures, DC analysis shows the different output voltage sensitivity when the input voltage varies. Moreover, Transient analysis introduces time delay when the structures are heated. These behaviours encourage us to run advanced simulations where only one device in the structure is heated (local temperature increase).

DC Analysis

DC analysis results are presented in figure 10 and 11 (for PNP structures) and in figure 12 and 13 (for NPN structures). ΔVout defines an absolute value of output voltage difference when the entire structure is at ambient temperature (27°C) and one of the devices is heated (37°C): $\Delta Vout = /Vout_{27°C\ GLOBAL} - Vout_{27°C\ GLOBAL,\ device\ 37°C\ LOCAL}/$.

For PNP structures, the impact of heating is more important (higher variation of the output signal) for a larger sized active area of PNP bipolar transistors. With the increasing input V5 voltage, the output variation increases for three devices: Q0, R0, R1. The R1 resistor is the most temperature sensitive device relatively to the input voltage changes but this variation does not reach 5mV. The local heating of the bipolar transistors results in the highest output signal variation (52mV maximum for Q0 device).

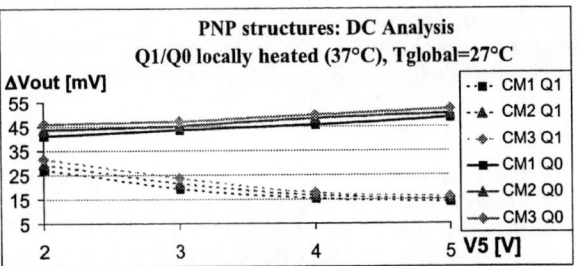

Fig. 10. DC simulations of PNP structures.
Q0,Q1 locally heated vs V5.

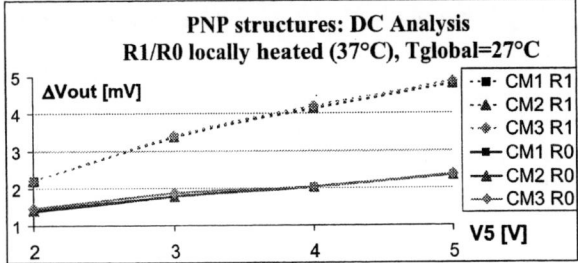

Fig. 11. DC simulations of PNP structures.
R0,R1 locally heated versus V5.

As for the PNP structures, on the NPN configuration the impact of heating is more important (higher variation of the output signal) when the size of NPN bipolar transistors active area is larger. For three devices: Q1, R0, R1, with increasing input voltage V5 the output variation increases. Similarly to PNP structures, the resistors are the most temperature sensitive devices relatively to the input voltage changes but this variation reaches maximum of 14mV. The local heating of the bipolar transistors results in much important output signal variation. The highest variation of the output signal (more than 700mV) is reached for local heating of Q1 transistor.

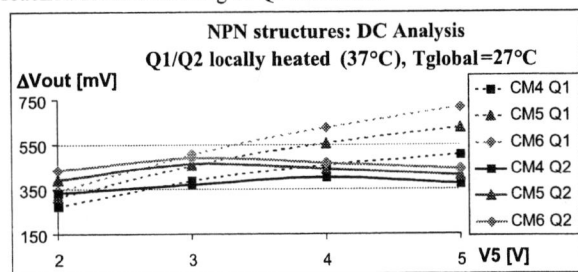

Fig. 12. DC simulations of NPN structures.
Q1,Q2 locally heated versus V5.

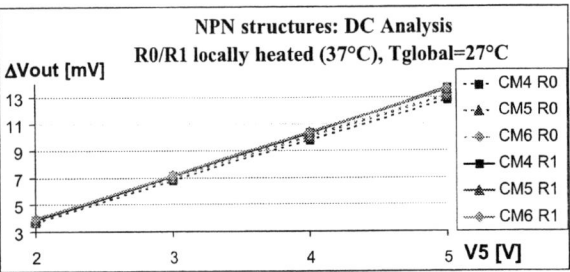

Fig. 13. DC simulations of NPN structures.
R0,R1 locally heated versus V5.

Comparing PNP and NPN structures we observe more important sensitivity to local heating for the second configuration with NPN bipolar transistors (14 times more).

Transient Analysis

For Transient analysis the input voltage on V5 was fixed at 5V and the simulations were run for two frequencies: f=5kHz, 250kHz. The results are presented in figure 14 for PNP structures and in figure 15 for NPN structures. ΔVout defines an absolute value of output voltages difference when the entire structure is at ambient temperature (27°C) and one of the devices is heated (37°C): ΔVout = /Vout$_{27°C\ GLOBAL}$ −Vout$_{27°C\ GLOBAL,\ device\ 37°C\ LOCAL}$/.

For PNP structures the most sensitive device to local heating is the Q0 bipolar transistor. The maximal output variation equals 52 mV for the structure with the larger sized active area (CM3). For NPN structures, Q1 bipolar transistor is the most sensitive device on local heating. The maximal output variation equals 720mV for CM6 structure.

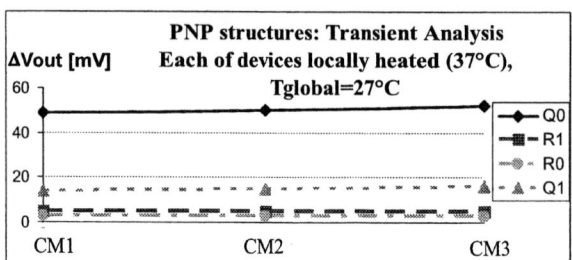

Fig. 14. Transient simulations of PNP structures for V5 = 5V.

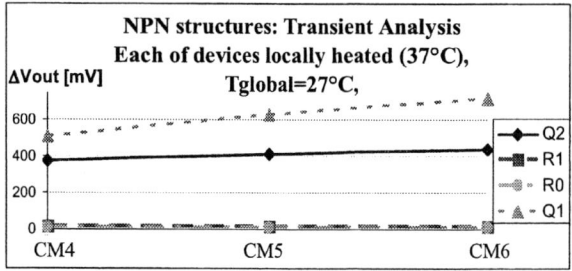

Fig. 15. Transient simulations of NPN structures for V5 = 5V.

In the time domain, a signal delay is observed for both configurations: 2ps maximum for PNP configuration and 250ps maximum for NPN configuration (Tab. II). In both cases, this time variation is measured only for local heating of bipolar transistors. Moreover, after reducing V5 input voltage to 4.7V, this time delay changes for the NPN configuration (Tab. III).

Heated device	ΔTIME [ps]		
	CM4	CM5	CM6
Q1	150	190	230
Q2	70	40	30
R1	No changes		
R0			

Table II: Transient simulations of NPN structures for V5 = 5V. Impact of local heating of different basic devices on the signal time delay.

Heated device	ΔTIME [ps]		
	CM4	CM5	CM6
Q1	100	160	250
Q2	90	110	180
R1	No changes		
R0			

Table III: Transient simulations of NPN structures for V5 = 4.7V. Impact of local heating of different basic devices on the signal time delay.

The above simulation results clearly show that the current mirror structures based on NPN bipolar transistors are more sensitive to local heating compared to the structures based on PNP bipolar transistors. Each of the simple devices (bipolar transistor, resistor) has an impact on the output voltage while the time delay is only observed for bipolar transistor local heating. Hence, the time delay measurements provide limited information compared to the voltage level measurements.

IV. DISCUSSION

Taking account of simulations run on the basic current mirror structures we observed that DVM techniques applied on a complex device showed us sensitive areas in a limited way. For example, in the global simulation on PA driver we could observe a duty cycle variation. Moreover, the simulations on basic current mirror structures showed a high variation of the output signal amplitude. Hence, other techniques, like PVM, may better identify sensitive areas inside devices. The PVM technique was applied on the previously analyzed complex circuit with the same electrical setup. Figure 16 presents the results obtained. As observed before, the PVM mapping gives more information about device sensitivity and for a complete characterization (time delay propagation, duty cycle, voltage amplitude, …) it is generally needed to perform several acquisitions with different parameters. The interpretation of the results can be very difficult for defect localization unless when coupled with a golden device and/or with accurate simulation. In this case, the PVM technique allows to make fully use of its advantages in failing or marginal devices localization.

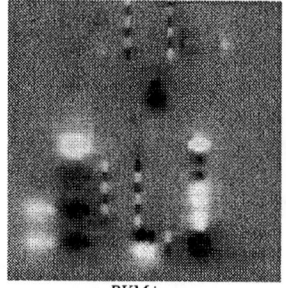

a. PVM image　　　　　*b. Superimposed image*

Fig. 16. Sensitive areas localization on a golden device by using PVM technique.

V. CONCLUSIONS & PERSPECTIVES

In this paper was presented the analysis of the thermal IR laser source impact on the analog structures. By using DVM technique for a complex mixed-mode device, the sensitive areas were highlighted in a selected analog block. Then, global simulations were run to better understand the obtained results. Correlation was then made by precise electrical simulations on separately designed current mirror structures. Two configurations were analyzed. The first, based on PNP bipolar transistors and the second using NPN bipolar transistors. DC and Transient simulations were run to analyze the different output parameters of these structures. In the first step the global temperature was raised 5°C from ambient. The most important output voltage sensitivity was observed for PNP configuration. However, the signal time delay was higher for NPN configuration. In the second step, the local temperature of each basic device was increased by 10°C relatively to the global temperature of the structure. The bipolar transistors were observed to be more sensitive to local heating compared to resistors. This sensitivity increased for a longer sized active area. Moreover, the output voltage changes and the time delay were more important for NPN configuration.

The above analysis of basic current mirror structures allows a better understanding of the complex device functionality. The current mirror based on NPN bipolar transistors under thermal laser stimulation has more influence on the output signal variation than the one based on NPN bipolar transistors (DVM analysis). Moreover, for the NPN configuration, the transistor from where the current is copied (e.g. Fig. 6b Q1) is more sensitive to local temperature changes than the one to where the current is copied (e.g. Fig. 6b Q2). This behavior, seen during measurements of a complex device, was confirmed by simulations on a basic current mirror structure. On the contrary, the simulated PNP configuration revealed the copying current transistor less sensitive.

As we could see, even if the analog structure does not contain a failure, its thermal sensitivity can cause malfunction of the successive blocks when submitted to the thermal laser source. In this case we can omit defect-free but sensitive analog structures during failure localization with TLS. One of these structures, discussed in this paper, is the current mirror. The devices measured the most sensitive to the temperature, do not have to be exactly the same for each of analyzed parameters (voltage level, signal time delay, duty cycle) and also when the setup parameters change (input voltage, temperature, frequency, etc.). Hence, the setup used for the failure localization in the statically or dynamically emulated integrated circuit should be optimized to get information that is easy to interpret. As showed in discussion, the chosen localization technique will play a key role in failure localization.

ACKNOWLEDGMENT

We would like to thank following people at Freescale Semiconductor for their valuable input: P. Sandrez, N. Jarrige, P. Besse and A. Boivin for design support, F. Grave for CAD support, A. Clark for his useful remarks. We would also thank K. Sanchez from CNES for fruitful discussion.

REFERENCES

[1] K. Nikawa and S. Inoue, "New Capabilities of OBIRCH Method for Fault Localization and Defect Detection", ATS, 1997, p214-219.

[2] E.I. Cole Jr. et al., "TIVA and SEI Developments for Enhanced Front and Backside Interconnection Failure Analysis", Microelectronic Reliability, Vol. 39, p991-996, 1999.

[3] F. Beaudoin et al., "Principles of thermal laser stimulation techniques", Microelectronic Failure Analysis Desk Reference, 2004, p417-425.

[4] T. Wilson and E. M. MacCabe, "Theory of optical beam induced current images of defects in Semiconductors", J. Appl. Phys. 61(1), 1987, p191-195.

[5] E.I. Cole Jr, J.M. Soden et al., "Novel Failure Analysis Techniques Using Photon Probing With a Scanning Optical Microsocpe", IRPS 1994, p388-398.

[6] Phang JCH et al., "Single contact beam induced current phenomenon for microelectronic failure analysis", Microelectronics Reliability,Vol.43, 2003, p1595-1602.

[7] F. Beaudoin et al., "Laser Stimulation Applied to Dynamic IC Diagnostics", ISTFA, 2003, p371-377.

[8] J.A. Rowlette et al., "Critical Timing Analysis in Microprocessors Using Near-IR Laser Assisted Device Alteration (LADA)", ITC, 2003, p264-273.

[9] Michael R. Bruce et al., "Soft Defect Localization (SDL) on ICs", ISTFA, 2002, p21-27.

[10] E.I. Cole et al., "Resistive Interconnection Localization", ISTFA, 2001, p43-57.

[11] R.A. Falk et al., "New Application of Thermal Laser Signal Injection Microscopy (T-LSIM)", ISTFA, 2003, p25-35.

[12] K. Sanchez et al., "Delay Variation Mapping induced by dynamic laser stimulation", IEEE International, 2005, p305-311 .

[13] K. Sanchez et al., "Phase Variation Mapping, a Dynamic Laser Stimulation Technique with Picosecond Timing Resolution", IRPS, 2007, p534-541.

The Helium Ion Microscope for High Resolution Imaging, Materials Analysis and FA Applications

William B. Thompson, John Notte, Larry Scipioni and Lewis Stern
Carl Zeiss SMT Inc., ALIS Business Unit
One Corporation Way
Peabody, MA 01960
Phone: (408)-893-9633. FAX: (978)-532-2503. Email: b.thompson@smt.zeiss.com

Abstract -The scanning helium ion microscope (HIM) is capable of sub-nanometer resolution using ion induced secondary electron (SE) mode and back scattered ion (RBI) mode imaging. In addition to its long working distance and sub-nanometer resolution, the helium ion microscope in RBI mode provides contrast that is material Z dependent on a nanometer scale. By the complementary use of charge neutralizing electrons, the helium ion microscope can image dielectric materials at full beam energy and maximum resolution without the need for sample coating. This paper will describe the physics of the subatomic virtual source, the system's ion optics, the types of detectors used, the substrate interactions involved with SE and RBI imaging and the applicability of the helium ion microscope to materials and failure analysis problem solving.

I. INTRODUCTION

The helium ion microscope has its historical origins in the principles of the field ion microscope (FIM). As a consequence the virtual source for the ion column optics is less than an atomic diameter. Although this microscope rasters a sample and collects emitted secondary particles to generate an image in a manner similar to that of an SEM, it differs significantly in its column design, its image information content and its analytical capability. The helium ion microscope and its derivatives that utilized inert gases such as neon and argon were constructed a decade ago to sputter repair phase shift masks [1]. Since that time important ion source stability and geometry issues have been resolved. With the advent of these solutions, the helium ion microscope can now be considered a commercially viable technology for ultra-high resolution microscopy and analysis.

Provided with a stable system, the designers of the helium ion microscope now turned to studying the ion-material interactions of helium ions incident at high resolution on various metallic, dielectric and organic media. These interactions generate electrons, ions and photons – each with their own substrate information content. With a new understanding of the physics of helium ions incident on matter the designers then turned to potential applications. Beyond its high resolution capability a host of other novel system applications arose. Some of the applications of particular importance to failure analysis include device cross section imaging, through dielectric imaging, doping contrast imaging, thin film and small particle materials analysis, capacitive coupled voltage contrast imaging and possibly circuit editing.

This paper will describe the fundamentals of the ion source and its optics. We continue with an introduction of the image generation and image contrast mechanisms. A few examples of non-semiconductor images and applications follow. And we conclude with sections specific to material and failure analysis.

II. THE HELIUM MICROSCOPE ION SOURCE AND ION OPTICS

The ions in the focused probe used for imaging in the helium ion microscope originate in the high field region of a single metal atom at the apex of a source needle that is very similar to the needle of a cold field emitter of an SEM. Fig. 1. shows the atomic lattice plan view before and after tip structuring and the resulting field ion microscope images.

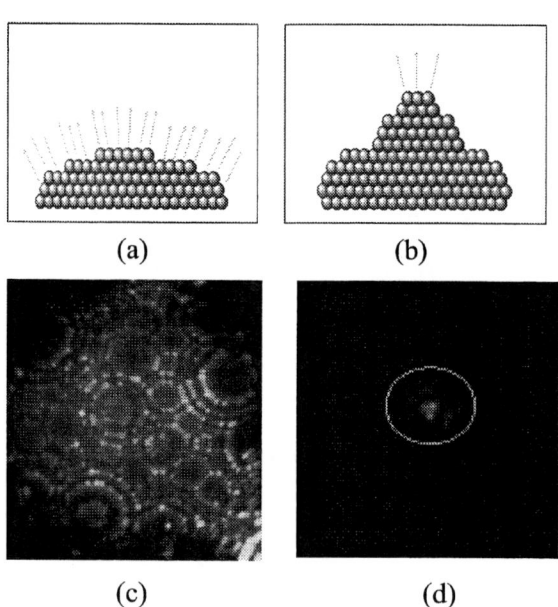

Fig. 1. (a) Illustration (side view) of the end-form of a typical field ion tip, (b) illustration of the HIM emitter, (c) actual FIM image of the geometry illustrated in 1a, (d) actual FIM image of emitter illustrated in 1b and used as the HIM ion source.

The field ion microscope (FIM) was first developed by Muller et al [2] in the 1950's. It consists of a sharpened metal needle chilled to liquid nitrogen temperatures or below and biased to a positive voltage in the vicinity of 10 kV. The applied voltage generates a electric field at the needle tip that is several volts per nanometer. An imaging gas, often an inert gas like helium, neon or argon, is introduced into the vacuum region surrounding the needle. As the gas atoms are cooled by the needle they spend more and more time in the vicinity of the needle high field region. With sufficient cooling a gas atom has time for one of its electrons to tunnel from the gas atom into the metal of the needle. Once ionized at a lattice atom of the metal needle, the helium ion, for example, is launched outward toward, in the case of the FIM, a fluorescent screen generates a pictorial representation of the high field region of the needle lattice atoms. Fig. 1 (c) shows a FIM image of the helium ion microscope needle tip. In the helium ion microscope, a tip restructuring process alters the spherical nature of the emitter to the three metal atom configuration shown in Fig. 1 (d). The helium ions from one of these three metal atoms are then mechanically steered onto the axis of the microscope column. The ions' low energy spread, confined launch location and high current density produce an ion beam brightness of approximately 3×10^9 A/(cm^{2}str). As the ions appear to launch from a location that is behind a single metal atom of the needle, the virtual source size for the microscope is sub-atomic. The energy spread of the ions is around 0.5 eV and their wavelength of the order of 10pm. A consequence of these values is that the final lens aberrations for the focused imaging probe sum to less than 1 nm and sub-nanometer images are possible.

As ions are not easily focused by magnetic fields, the HIM ion optics are electrostatic rather than magnetic. The main optical elements in the HIM column are the ion source, a gun lens, scanning and blanking hardware, a series of apertures and a final lens. The entire column is pumped by turbo pumps and mounted on a 1500 kgm vibration isolation table. Fig. 2 shows a completed helium ion microscope system.

III. ION MATERIAL INTERACTIONS AND THE HIM IMAGING MODES

A. Ion Induced Secondary Electron (SE) Generation and SE Imaging

Any helium ion incident on a sample has a kinetic energy that can be varied from 5 to 45 keV. The helium ion ionization potential energy is 24.6 eV. The incident beam has the capability of ejecting sample electrons and photons both because of its potential energy, potential emission, and because of its kinetic energy, kinetic emission. The maximum kinetic energy, E_e, of the launched electrons is related to the ion beam energy E_0 by kinematics to:

$$E_e = 4E_0(M_eM_{He})/(M_e+M_{He})^2$$
$$\cong 4E_0M_e/M_{He}$$

The maximum secondary electron energy is therefore about two thousand times lower than the ion beam energy. The mean energy of the secondary electrons emitted form the sample has been shown empirically to be about 2eV, more or less independent of the sample composition. This is unlike the kinetic energy of the electrons generated by an SEM where the SE electron energy can range from a few eV to several keV. The range of a 2eV electron in silicon is less than a nanometer. Thus the HIM SE information volume is sub-nanometer. As the probe size of the helium ion microscope is also sub-nanometer, the HIM SE image resolution, in total, is below a nanometer. The 2eV SE energy also means that the helium ion microscope image is extremely surface sensitive.

Fig. 3 shows an aluminum registration mark on a test structure.

Fig. 2. The helium ion microscope source, optics, optical bench and operator console. The system has a beam current range from 1 to 10 pA, a beam energy from 5 to 45 keV and an SE image resolution below 1 nanometer.

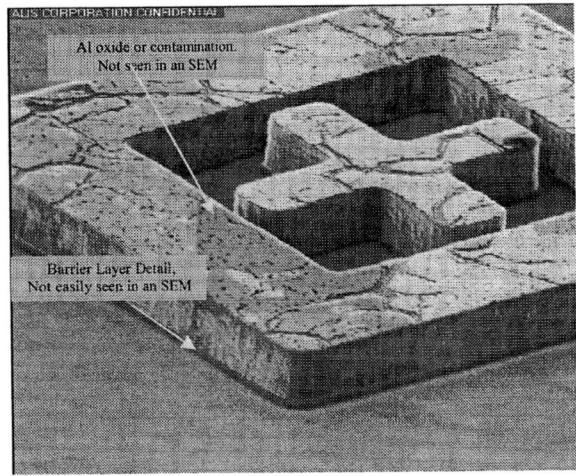

Fig. 3. A HIM SE image of an aluminum fiducial mark showing surface details of contamination and of the underlying barrier layer...information not easily seen in a conventional SEM.

In order to detect electrons in the helium ion microscope a conventional Everhart-Thornley (ET) detector is used. Fig. 3. and all of the SE images shown in subsequent SE mode examples would have been acquired using this ET detector.

The traditional test target for SEM image resolution determination is a gold-on-carbon sample. Fig. 4. shows a HIM SE image of a gold-on-carbon sample imaged at 420,000X magnification demonstrating that the combination of small HIM probe size and sub-nanometer SE range produces HIM resolution that is below a nanometer.

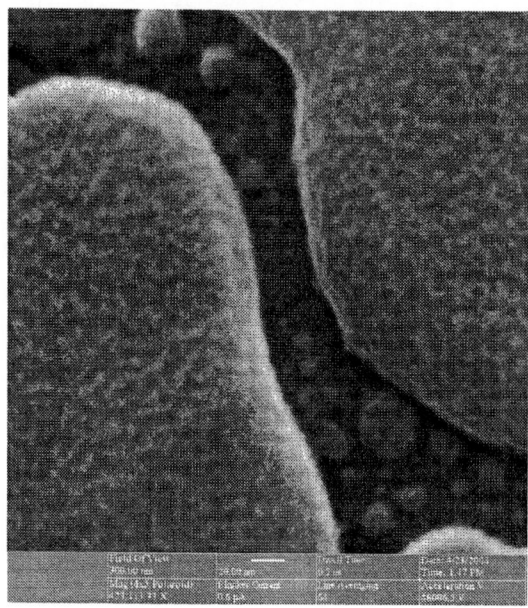

Fig. 4. HIM SE image with a 300nm field of view of a gold on carbon resolution test target.

B. Back Scattered Ion Yields and RBI Imaging

Classical Rutherford backscattering derivations contain two important relationships. The first is that the backscattered ion energy is related to the target mass, the higher the mass of the target, the higher the energy of the backscattered helium ion. The second is that the Rutherford scattering differential cross-section increases with the square of the atomic number of the target, Z_2, and increases inversely with the square of the ion energy, E_0. The differential cross-section is in reality a measure of the amount of helium scattered back by the sample into any detector. For Rutherford backscattered imaging (RBI), the helium ion microscope uses a negatively biased annular microchannel plate (MCP) detector. This MCP rejects the secondary electrons and collects about 80% of all of the backscattered ions in order to form an image. And as a consequence of the second Rutherford relationship, a HIM RBI image has a video gray level and image contrast that is strongly target atomic number dependent. Fig 5. depicts a HIM RBI image and a video level histogram inset of a carbon, nickel, copper and gold containing grid showing the contrast variation among all of the differing atomic number materials.

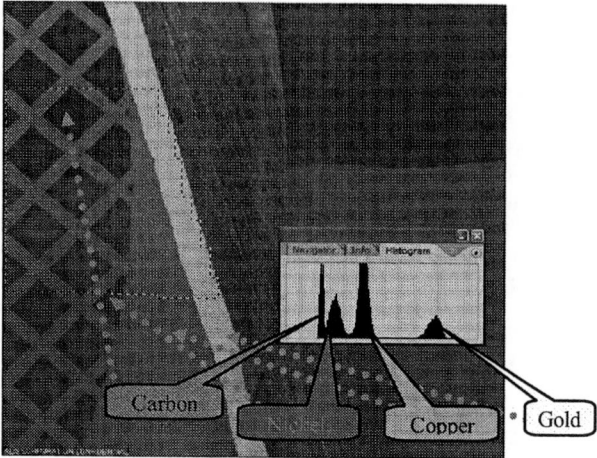

Fig. 5. Carbon, Nickel, Copper and Gold containing grid with associated video histogram depicting the atomic number, Z, dependence of a HIM RBI image gray level. The video gray level is thus an approximate metric for the sample atomic number. The whiter a region, the higher its atomic number.

The present helium ion microscope system configuration, designated the Orion, contains two independent video signal chains that permit the simultaneous acquisition of an SE and an RBI image. Fig 6. shows a comparison of an SE and an RBI image of a tin/lead solder sample.

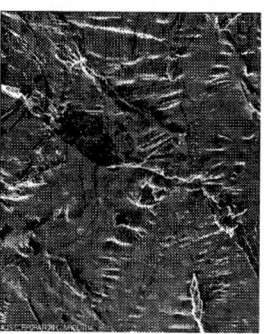 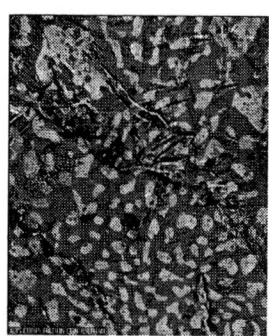

(a) SE image shows topography and surface contaminants (flux residue) in this tin and lead solder.

(b) Backscattered Ion image clearly differentiates the lead from the tin.

Fig. 6 a and b. A simultaneous comparison of SE and RBI images.

An additional strong contrast mechanism exists for RBI generated images, channelling contrast. Fig. 7. gives an example of channelling contrast in a tungsten weld. Well aligned grains permit the ions to penetrate deeply into the sample producing no backscattered signal, while grains oblique to the beam axis back scatter more or less strongly depending on their relative orientation to the beam axis.

Fig. 7. A 150 micron field of view of a tungsten weld imaged in HIM RBI mode.

IV. MATERIALS ANALYSIS USING THE HELIUM ION MICROSCOPE

Although several analytical methods have been suggested for the helium ion microscope, Rutherford backscattering spectroscopy (RBS) sample analysis is the most mature. High energy (~2 MeV) RBS techniques often use silicon surface barrier detectors for conventional RBS methods, while Medium Energy Ion Spectroscopy (MEIS) and Low Energy Ion Spectroscopy (LEIS) may use time of flight or electrostatic analysis methods for RBS ion detection. All of these methods take advantage of the target mass, M_2, dependence of the energy of the backscattered particle for their interpretation of the target composition. With a high energy resolution detector, one might expect mass resolution discrimination of a few percent. It should be noted that the yield as well as the backscattered helium energy contains information about the target. With careful interpretation, film composition and film thickness determination can be made from RBS spectra. Fig. 8. shows two spectra of a thin gold film on silicon sample. The solid curve was produced by the analysis package of the program SIMNRA [3] while the dotted curve is actual early data taken on the microscope. The fixed input to the simulation were the beam current, the data acquisition time, the detector geometries, and the incident beam energy. The output from the simulation, once a satisfactory fit was obtained was the thickness of the gold film, 9.5 nms, being analyzed. The accuracy of any analysis of a sample is strongly sample dependent. For example, thin high Z films on a low Z material produce the best results.

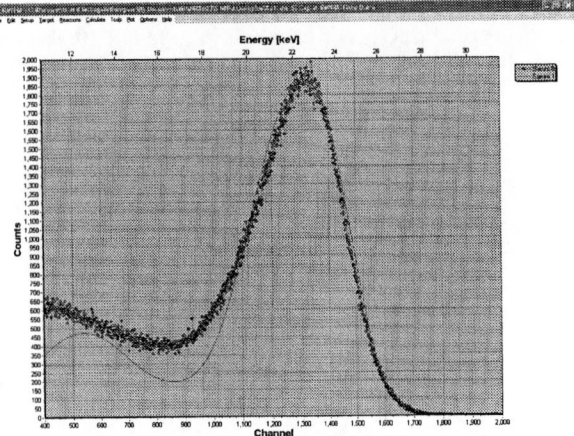

Fig. 8. Gold on silicon film thickness determination using the RBS spectroscopy software SIMNRA and data from the helium ion microscope RBS spectrometer. The gold film was determined from the analysis to be 9.5 nms thick.

In addition to the SEs and RBIs generated by the incident helium ions, photons are emitted as well by a process known as ion beam luminescence (IBL). For transparent and translucent samples IBL spectroscopy may provide the helium ion microscope with even more analytical possibilities. To date, the IBL potential of the microscope has not fully been explored.

V. IMAGING APPLICATION EXAMPLES

A. SE Mode Imaging

The HIM SE mode contrast originates from variations of sample topography, material composition, surface contamination, magnetic and crystallographic properties and, for dielectric materials the underlying voltage and/or resistive and capacitive connection to the substrate. The extremely low energy of the SE's results in very slight modulations of the sample surface potential producing large contrast variation in the SE image. As SE electrons are launched only from the surface, even for low Z materials, carbonaceous and organic sample surface are easily imaged. Fig. 9. shows a 300nm field of view of bundles of carbon nanotubes.

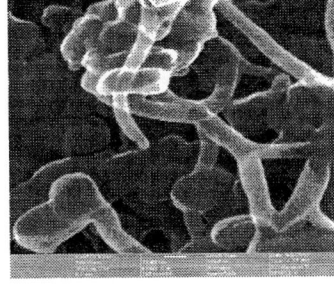

Fig. 9. 300nm SE field of view of carbon nanotube bundles.

Fig. 10. shows a free standing hydrocarbon film of monomolecular thickness with a hole in it. In Fig. 10. one can easily discern the film edge from the hole.

Fig. 10. SE image of free standing film normally electron transparent in an SEM.

Another advantage of SE mode imaging with the helium ion microscope, dielectric samples that would normally charge in an SEM may be charged neutralized with low energy flood electrons. The present microscope permits electron/ion interlacing during SE image acquisition and samples that would normally charge to nearly the incident beam potential can be kept neutral without lowering the beam energy, altering the image resolution, or requiring special sample coating techniques.

Fig. 11 shows a quartz photo mask with regions of isolated chrome imaged in charge neutralized SE mode. No beam energy adjustment is necessary, as it might be in an SEM, to maintain sample neutrality at all ion beam energies.

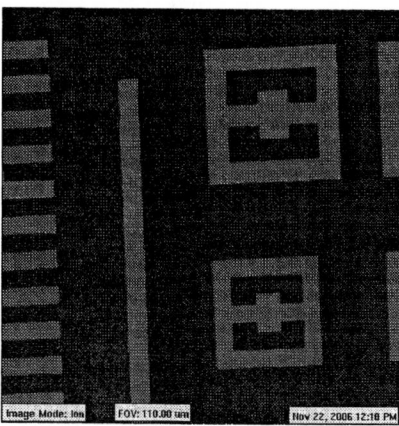

Fig. 11. Charge neutralized SE image of isolated chrome on quartz.

A peculiar property of SE mode imaging of thin dielectrics is that one can actually see through them to materials below. This is believed to be the result of the SE yield of the electrons launched from the sample surface being modulated by capacitive coupling to the materials below. Fig. 12. shows magnesium oxide crystals. A few actually appear transparent in the ion beam and one is able to see the material below the crystal through the crystal. Fig. 12 also shows the results of ion channeling. The largest of the crystals is completely black, implying that the channeled ions are producing electrons so deep into the crystal that no electrons can escape.

Fig. 12. HIM SE mode image of magnesium oxide crystals demonstrating crystal transparency, ion channeling, refraction and the edge enhanced brightness of dielectric samples.

B. RBI Mode Imaging

As mentioned earlier, contrast in RBI mode imaging comes almost exclusively either from crystallographic channeling or from the atomic number, Z, variation of the atoms in the imaged region. This property can often be helpful in identifying composition of an image region. Fig. 13. shows a comparison of two images of silver paint. In the RBI image, on the left, silver grains show up much whiter than the hydrocarbon matrix in which they are embedded.

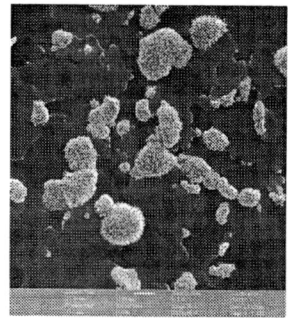

Fig. 13. Silver paint: SE mode image on the left. RBI mode image on the right. High Z materials image brighter.

VI. SEMICONDUCTOR AND FAILURE ANALYSIS APPLICATIONS WITH THE HELIUM ION MICROSCOPE

A. Imaging Examples

Because of the helium ion microscope's unique imaging attributes, semiconductor samples are ideally suited for SE imaging. Contrast in an SE image results for topography, materials, and potential and subsurface composition variations. Cross section samples have even shown contrast in source/drain doped regions. Fig. 14 shows HIM SE images of FIB cross sectioned 45 nm microprocessor vias and wires [4].

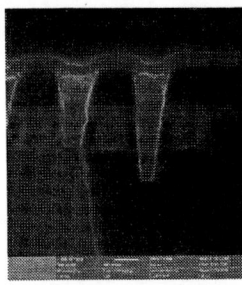

Fig. 14. HIM SE images of 45 nm microprocessor vias and wiring.

Figure 15. shows a thin contaminant film on top of an IC passivation layer lying over top metal wiring. This is another example of HIM SE thin film transparency due perhaps to the influence on the low energy SEs of capacitive coupled voltage contrast (CCVC) effects.

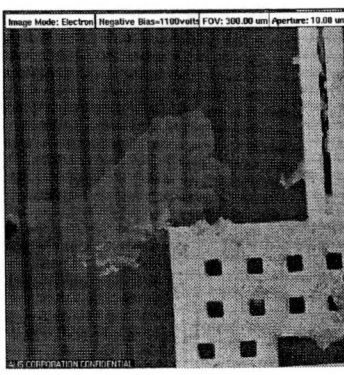

Fig. 15. HIM SE mode imaging through passivation.

B. HIM Failure Analysis Applications and Summary

As a summary of the capabilities of the helium ion microscope we list here some of its possible FA applications.

1.) High resolution imaging of surface contamination and interface films that are transparent to electrons because of their organic composition.
2.) Flood electron, charged neutralized, SE or He+ imaging of uncoated dielectric materials at full He+ energy & resolution
3.) Imaging through thick (100 to 500 nm) dielectrics to underlying defects, opens or shorts.
4.) Thin film and at depth small particle materials analysis with RBS, Elastic Recoil Detections Analysis (ERDA) or IBL techniques.

As the helium ion microscope is just now being introduced, the application space for this technology, especially for FA techniques is virtually unexplored. The HIM FA applications space of the future will be defined by the creativity of its users.

ACKNOWLEDGMENT

The authors would like to thank the R&D, engineering and applications teams of the ALIS business unit of Carl Zeiss SMT Inc. for their untold hours of work invested in the development and commercialization of this technology. Especially deserving of thanks is the founder of the ALIS corporation, Billy Ward, without whose vision, this technology would not have been possible.

REFERENCES

[1] W. Thompson, B. Ward, S. Etchin and A. Saxonis, DARPA Report, 1997, unpublished.

[2] E.W. Muller, *J. Appl. Physics,* Vol. 28 1957 pp 1-9.

[3] SIMNRA, Author: Matej Mayer, The Max-Planck-Institut fur Plasma Physiks.

[4] Sample provided by Chip Works Inc.

978-1-4244-2039-1/08/$25.00 ©2008 IEEE

SESSION 3:

RELIABILITY EVALUATION & ESD

EXCHANGE PAPER (ESREF 2007 BEST PAPER)

Time Resolved Determination of Electrical Field Distributions Within Dynamically Biased Power Devices by Spectral EBIC Investigations

A. Pugatschow, R. Heiderhoff, and L.J. Balk

Department of Electronics, Faculty of Electrical, Information and Media Engineering,
University of Wuppertal, Rainer-Gruenter-Str. 21, D-42119 Wuppertal, Germany
Phone: (+49 202) 439-1572. Fax: (+49 202) 439-1804. Email: heiderho@uni-wuppertal.de

Abstract-**Time resolved quantitative E-field analyses in near-surface regions of dynamically biased power devices can be performed stroboscopically by spectral EBIC investigations. Incorporating filtering and signal separation in time and frequency domains as well as averaging recovery techniques enhance signal to noise ratios and reduce disturbing signals up to 160dB.**

I. INTRODUCTION

Electron Beam Induced Current (EBIC) microscopy is based on separation of excess charge carriers, which are induced in a small volume fragment of the device by a focused electron beam due to internal electrical fields E [1, 2]. This technique has been already successfully applied for microscopic failure analyses and reliability characterization of local electrical device properties, like minority carrier diffusion length, and position of potential barriers [3, 4]. Time-resolved measurements of a pulsed beam or a single particle event supply information on local electric field strengths by extensively acquiring EBIC within the time domain and comprehensive post data processing [5, 6, 7]. At steady state conditions quantitative electric field analyses are only possible if recombination velocity modification [8, 9] or avalanche charge multiplication [10] is present.

Within this contribution inhomogeneities of electrical fields inside advanced dynamically biased power devices and semiconductor materials are detected stroboscopically by spectral EBIC investigations. Local electronic properties are analysed by time resolved quantitative E-field measurements.

II. DETERMINATION OF ELECTRIC FIELD STRENGTHS BY SPECTRAL EBIC INVESTIGATIONS

The density change of generated charge carriers is described by continuity equations coupled with the current density equations for electrons and holes. General analytical solutions for these time dependent partial differential equations are not available in time domain. In order to acquire a system of time independent equations in the frequency domain, generation and recombination terms are assumed to be periodical and can be substituted by their Fourier coefficients. Time dependent terms can be cancelled, because densities of resulting excess charge carriers follow the same periodicity for each individual frequency. At sufficiently high electric fields recom-bination

and diffusion terms are negligible. Finally the Fourier coefficients of the induced current densities outside of the generation volume are given by

$$J_{n,0} = -qg_{n,0}x_g \qquad \text{for } k = 0 \quad (1)$$

$$J_{n,k} = -j\frac{q\mu_n E g_{n,k}}{2\omega_0 k} \cdot e^{j\frac{\omega_0 k x}{\mu_n E}} \qquad \text{for } k \neq 0. \quad (2)$$

with electron charge q, basic frequency ω_0, mobility μ_n, generation $g_{n,k}$, at position x. Considering linear

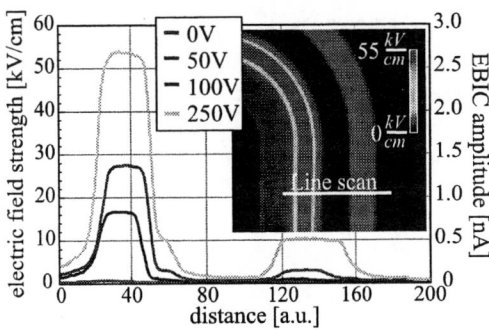

Fig 1:   Electrical field distributions within a statically biased power diode with a guard ring in dependence on applied reverse bias voltages.

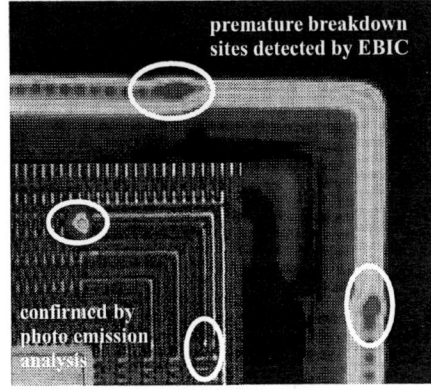

Fig 2:   Detection of premature breakdown locations in an edge termination structure of a MOSFET device by EBIC [11].

EXCHANGE PAPER (ESREF 2007 BEST PAPER)

relationship between EBIC amplitude and electric field, the acquired maps can be correlated to E-field strengths. Therefore the measured signal is integrated spatially along field lines and calibrated to the applied voltage (see figure 1). By means of this technique premature breakdown locations can be found within steady state DUT as illustrated in figure 2.

III. TIME RESOLVED DETERMINATION OF ELECTRICAL FIELD DISTRIBUTIONS WITHIN DYNAMICALLY BIASED POWER DEVICES

For the determination of electric field strengths within a dynamically biased device, transit times of charge carriers and pulse widths of the excitation are assumed to be small in comparison to changes of the electric field. In order to generate EBIC maps from the low signal-to-background for time-resolved and

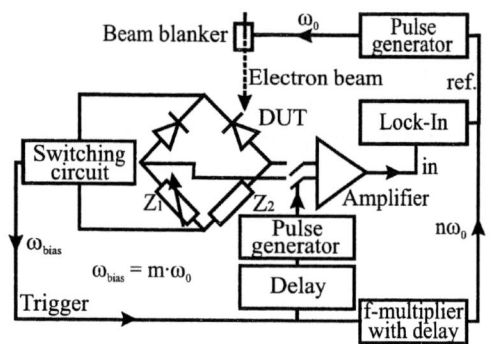

Fig 3: Setup for quantitative determination of electric field strengths within dynamically operated devices using EBIC analysis in the SEM.

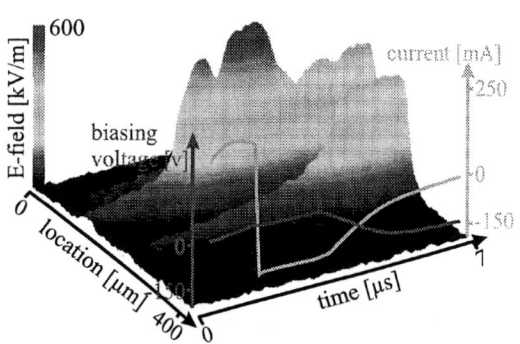

Fig 4: Detected line scan of E-field distribution versus time and transients of IV-characteristics on a power diode with guard ring (generated with electron pulses of 100ns width and 100ns delay steps).

stroboscopic analyses of dynamically biased power devices, characterization methods incorporating filtering in both time and frequency domain, and signal separation respectively, as well as averaging recovery techniques are used. Using a setup illustrated in figure 3 failure analyses and reliability investigations of electrical fields with a total signal recovering up to $160dB$ at $\omega_{bias}=2kHz$, $\omega_0=1kHz$, and time constant of

$1.25ms$ are performed of power diodes under dynamical bias conditions illustrated in figure 4. Charge transport mechanisms lead to electric field enhancements during the switching event which are not detectable under steady state bias conditions.

ACKNOWLEDGEMENTS

We kindly acknowledge sample support from F.-J. Niedernostheide and H.-J. Schulze, Infineon Technologies AG, Munich, Germany.

REFERENCES

[1] Everhart T.E., Matta R.K., Wells O.C., *Proc. of The IEEE*, Vol 52 (12), 1964, pp 1642-1647

[2] Balk L.J., Kubalek E., Menzel E., *Proc. of Scan. Elec. Micr. Symp.*, Part I, 1975, pp 447 – 456

[3] Balk L.J., Menzel E., Kubalek E., *Proc. of International Conference on X-Ray optics and Microanalysis*, Pendell Publish. Company, 1980, pp 613–624

[4] Leamy H.J., *J. Appl. Phys.*, Vol. 53(6), 1982, pp R51-R77

[5] Haynes J.R. and Shockley W., *Physical Review*, Vol. 81(5), 1951, pp 835–846

[6] Fischer P., Bergner H., Stamm U., Süsse K.-E., *Laser und Optoelektronik*, Vol. 24(1), 1992, pp 36–41

[7] Bergner H., Damm T., Stamm U., Stolberg K.-P., *Microelectronic Engineering*, Vol. 12(1-4), 1990, pp 143–148

[8] Donolato C., *Sol. St. Elec., Vol.* 25(11), 1982, pp 1077-1081

[9] Tarento R.J. and Matfaing Y., *J. Appt. Phys.*, Vol. 71(10), 1992, pp 4997–5003

[10] Donolato C., *J. Appl. Phys.*, Vol. 10, 1977, pp 1781-1788

[11] Siemieniec R., Geissler C., Schulze H.-J., Niedernostheide F.-J., Pugatschow A., Balk L.-J., *to be presented at 9th International Seminar on Power Semiconductors*, 2008

Humidity-Induced Semiconductor Device Electrical Reliability Failures: Mechanism and Wafer-Level Risk Evaluation/ Electrical Screening-Test Proposal

P.Jacob[1,2] and G. Nicoletti[1]

1) EMPA Swiss Federal Laboratories for Materials Testing and Reseach
Ueberlandstr. 129, CH-8600 Duebenorf, Switzerland
Tel. +41-44-823-4288, Fax –4054, e-mail: peter.jacob@empa.ch
and
2) EM Microelectronic Marin SA
Rue des Sore 2-3, CH-2074 Marin-Epagnier, Switzerland
Tel. +41-32-755-5327, Fax –5403, e-mail: pjacob@emmicroelectronic.com

Abstract: **Automotive semiconductors are frequently exposed to humidity- if penetrating, a high-ohmic film at the chip surface connects pins to various voltage levels, if they aren't pull up/down protected. Since latest integration combines different technologies/ voltage levels, such films trigger destructive malfunctions. A wafer-level test, high ohmically short-circuiting all existing pins to first maximum, then minimum voltage, should eliminate such design-induced reliability risks.**

I. INTRODUCTION

In recent years we were faced with several field failure cases, most of them from automotive applications: The semiconductor devices were mounted within hybrides, which were coated by a transparent, soft silicone gel. All devices have been already in field operation under harsh environmental conditions, especially considering temperature and humidity. In some cases, the failures were clearly destructive; however, in several cases, the devices showed malfunctions but were not destroyed. If, in the latter cases, the silicone gel had been removed, the devices returned to their regular functionality.
The devices affected were in most cases either mixed signal devices or power semiconductors; sometimes in combined techniques, too.

In other failure analysis cases, we have identified, that EEPROM memories had lost their (on wafer-level) pre-programmed data during wafer sawing. Further examinations have identified surface humidity in combination with electrostatic charging during the sawing and tool-internal post-cleaning step.

These failure analysis cases encouraged us to consider about a special electrical test, which should be added to the regular device functional tests on wafer level. Such test should simulate the introduction of surface humidity in order to test the device's electrical robustness against high-ohmic short-circuits as it could be generated by a surface humidity film.

II. FAILURE ANALYSIS RESULTS:

The main problem was that the silicone gel turned out to be a good humidity transportation medium. It collects humidity like a sponge, resulting in a humidity film at the interface chip-surface-to-silicone gel. Since our former experiments within the LESIT-project in 1994 [1], where we investigated the influence of humidity on different bondwire-alloys on silicone-gel-coated power semiconductor devices, this effect is well-known to us. Fig. 1 shows the result of the combined attack of humidity and high current to an aluminum wirebond.

If, in addition, humidity films at the interface between device surface and silicone coating become ionic-contaminated, a high-ohmic conductive layer (in the Mega- to Giga-ohm-order-of-magnitude) is generated. Unfortunately, not all pads are connected-to outside (some are only used for engineering test purpose or for other-kind-of-use of the device), and some of them had no or insufficient pull-down or pull-up circuitry. In some cases, engineering test pads were more or less directly connected to FET gates and they were already known to be very critical to electrostatic charging in assembly or cleaning processes. Since such pads were not coated by the chip passivation, too, they were accessible to the humidity film.

Fig.1: wire-bond of a power- IGBT,
which suffered humidity through the silicone gel

Depending from conductivity and pad geometry, a voltage divider between GND and Vdd (or maximum potential available on the whole hybride) is provided by such humidity film. In the less severe cases, a more or less reversible, sometimes time-dependent malfunction results from this situation (if the voltage potential at such pads is close to a related switching threshold). In severe cases, such "parasitic" resistive circuitry may lead to a complete destruction. In one of our cases, the semiconductor device included both low-power CMOS and high power bipolar circuitry. Both parts of the circuitry could be connected for engineering test purpose by such a test pad, which was neither protected against ESD nor had defined internal pull-down resistors. The high-ohmic potential provided by the humidity film turned out to be enough to connect high- and low-voltage part within the regular application in a manner, which should never be allowed – causing immediate destruction of the whole chip.

In another, power-semiconductor-related case, the border ring of the high-voltage device had shown cracks within the specific border passivation of this chip. By this, the humidity film connected the high-potential active center area of the chip to local border ring spots (which finally caused a breakdown due to local electrical field strength overload in the border ring.) Fig. 2 shows such a surface distruction of a power semiconductor (gel has been removed) as it can be generated by surface ESD (ESDFOS) during assembly manufacturing [2].

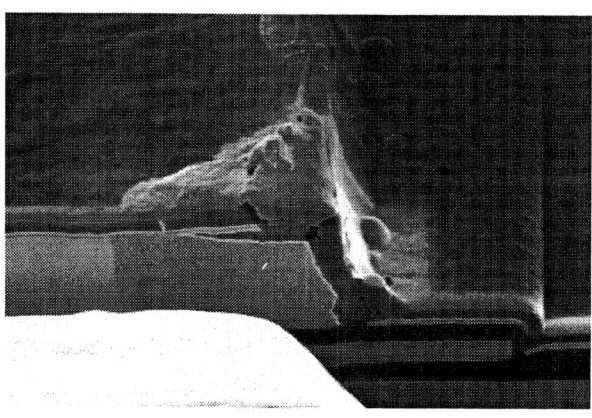

Fig. 2: Surface-ESD on a border ring of a power IGBT device (FIB-cross section). The oxinitride-passivation is cracked and gives access to the metal for penetrating humidity.

III. PROBLEM GENERALIZATION:

To generalize our case studies, we can outline three trends:

- a general miniaturisation of the device structures, combined with space saving by reducing protective and/ or pull up/ down structures. Finally, this makes them more sensitive against unintended static potentials.

- The combination of low- and high-voltage circuitry (in sometimes even different techniques) within one semiconductor chip or as stacked devices; again, such close arrangements are especially sensitive against high-ohmic coupling by humid films.
- The increased circuit complexity requires more often test pads for engineering/ F/A studies. If they are unprotected and still accessible in wafer level final test, later humidity-induced high-ohmic short-circuiting may become a severe reliability risk.

Fig. 3 shows a typical example of a ceramic-based hybride, coated by silicone gel. Similar hybrides are for example used in air flow sensors in the automotive industry.

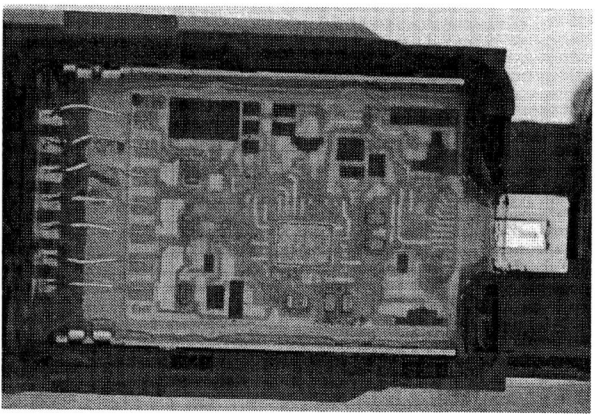

Fig.3: Typical hybride (automotive application) with chips-on-board, coated by soft silicone gel. Humidity impact cused functional problems, which later destructed the device

IV. PROPOSAL FOR ELECTRICAL TESTING AND SCREENING ON WAFER LEVEL

Basic considerations

When passing through the silicone soft gel, air humidity will necessarily gain some ionic contamination if there is not already some from the beginning. The specific electric conductivity σ of water ranges varies over several orders of magnitudes, depending from the degree of ionic contamination. The bandwidth starts with ocean water (specific conductivity σ: 5 S/m), drinking water in the household is around 0.05 S/m and pure water is at 5x10exp-6 S/m.

The specific area resistance R□ is given by R□= 1/ (σ · d), where d is the thickness of the humidity film. Assuming a water humidity film thickness of roughly 0.01μm and an ionic contamination reaching the household water order-of-magnitude, an area resistance R□ of 2000 Mohms/square would result. Assuming a regular device operational voltage between 1 and 10 volts for microelectronic semiconductor devices, the resulting current would be in the nA-order-of-magnitude. For power semiconductor devices with sometimes up tu 7kV operational voltage, however, already the μA-range would be reached.

Reliability testing and screening considerations:
In this topic, we must distinguish at first between screening and reliability. An electrical screening on wafer level might be able to test the principle (resistance-dependent-) <u>sensitivity</u> of the device against potential leakage surface paths generated by humidity films but it cannot simulate the long-term resistance <u>development</u> of such parasitic path under the influence of current, temperature, degree and increase of ionic contamination etc. While the first aspect will indicate the device's robustness against such parasitic leakage, the reliability will be both significantly influenced by this robustness as well as by the time- and environmental condition-dependent in- or decrease of the resistance behaviour of such leakage path. Of course, a short-term screening procedure can only target on the determination of the device robustness, while the environmental influences-based, time-depending development of the surface resistance must remain out of scope (or been evaluated in specific long-term-reliability tests)

Proposal for the screening test:
A humidity film at the passivation surface of the device would more or less short-circuit (high-ohmic) all such parts of the device surface which have an open access to metal structures. In detail, this means the non-passivated pads including wire-bonds as well as other open-access conductive structures, like for example trimming fuses. In the case of hyprides, also open metal parts of the circuitry outside the chip must be considered when determining the electrical maximum and minimum potentials for the screening test.

The basic idea behind this paper is to superimpose a high-ohmic short-circuit network to the regular functional test program, thus electrically simulating the existance of a slightly conductive surface humidity layer. Since the normal functional test should be done at first without such a superimposed resistance network, the difference of both tests will highlight the electrical sensitivity of the device under test (DUT) against parasitic short circuit paths.

Application for CMOS and mixed signal devices:
As a first step, such humidity film could be electrically replaced by a high-ohmic resistor network, which switches all non-used pads (high-ohmically) to at first GND potential and second to high potential (Vdd or highest potential available within the hybride/ module etc) and to observe changes in functionality. If such changes should occur, one gets a fast feedback, whether there might be a missing pull up/ down, latching or missing pad protection problem.

In addition or alternatively, distributed superimposed patterns may be used in order to highlight specifically critical leakage paths. Such patterns could be for example:
- Put (via high ohmic resistors) the voltages of neighboured pads to same/ to contrary/ to neighbour pad signal potentials. An example is shown in Fig. 4.
- Move a "cloud of superimposed high-ohmic electric potentials" through the circuitry by defining maximal/ minimal

voltages and calculating/ simulating the voltage gradient between two neighboured pads under different conductivity conditions of the assumed humidity film.

Of course, also application-specific patterns can be developed, taking into account criticality/ sensitivity of circuit and application, worst-case-scenario-assumptions in mounting/ environment and relevant voltage differences applied to the circuitry. In addition, high-ohmic coupling of pulsed signals from neighboured pads could be included into the design of the superimposed test.

Stacked Devices/ Multi Chip Modules:
In the case of stacked devices, a comparable test, but performed at stacked module level, could give similar information like described in the chapter before. However, the non-accessible test pads and non-connected pads are not included in this kind of evaluation. If the test had been performed already on wafer level of the joint devices, usually no stacked-module test on the issue would be necessary any more – except, if the stacked setup is subject to higher electric potentials than the participating dices on wafer level testing. In such a case, however, it would be recommended (if known and possible) to superimpose such increased electric potentials already on wafer level testing of all participating device types.

Power semiconductors:
Power devices cannot be directly included into such screening recommendations, since no border structures are regularly available with an open metal access. The edge termination is usually designed carefully in order to smoothly distribute the voltage decay between chip border (0V) and active area (sometimes up to kV-range). In some cases, less attention is taken to the specific edge-termination passivation (compared to the field ring designs) and in such cases, a humidity-saturated soft silicone gel can bridge the active area to one of the outer edge termination rings, if such passivation is voidy, cracked or even only too thin. This problem can, of course, not be discovered by the proposed superimposition screening tests. If, however, the robustness against border ESDFOS failures in

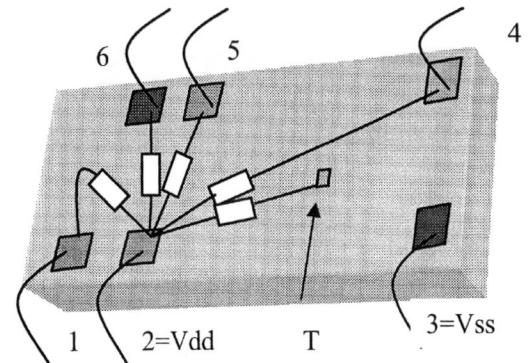

Fig. 4: A normal functional test on wafer level (including test pads), superimposed by a network of high-ohmic resistors, simulating a humidity film, would point out potential circuitry design weakness ([nearly-] floating pads) and thus reduce reliability risks. The resistor center node is set at first to Vdd (as shown in the figure), then to Vss

combination with later humidity filmes should be evaluated, useful PCM structures with border ring test pads would allow such screening, too. Since the failure mechanism has been proven in several cases, we mention it here for paying specific attention to sometimes "neglected" edge termination surface passivation quality. However, this specific aspect should be considered in simulations in case of new designs or design reviews.

Furthermore, we must consider that power semiconductors are frequently mounted into a hybride with neighboured low power devices. In such case, the maximum electric potentials of the power device may exceed the maximum low power device potentials by orders of magnitudes. This must be considered for the definition of the high-ohmic electric potentials to be superimposed in their specific screening tests.

V. OUTLOOK

The implementation of such proposed screening tests could perform valuable feedback to chip designers how to enhance the withstandability against the electrical consequences of humidity on the device surface. This would be beneficial for both manufacturers and clients: Manufacturers would be faced with less customer returns, while for the clients , system robustness – and thus the reliability, too, would increase.

VI. CONCLUSION

New chip-internal integration methods, especially in automotive applications, show often an increased sensitivity for superimposed high-ohmic potentials to accessible pads – not only those which are used in the application, but also for example test pads and other open metals with electric potentials. In the field application, such potential superimposition can be introduced by humidity-saturated plastics, especially soft silicone gels, which are used in manifold hybrides and power semiconductors. A screening test has been presented, which quickly allows to show up related potential design weakness or to screen affected product, such avoidingor at least reducing field reliability problems. This superimposition test, performed in sequence to standard functional tests on wafer level, can be designed in various manners, depending on the electric potentials used in the field application.

REFERENCES

[1] Jacob, P., Held, M. Scacco, P., "IGBT power semiconductor reliability analysis for traction application", Physical and Failure Analysis of Integrated Circuits, 1995 IPFA proceedings, p.169-175
[2] Jacob, P., Thiemann, U.,Reiner, J.C., "Electrostatic discharge directly to the chip surface, caused by automatic post-wafer processing", Transactions on Device and Microelectronics Reliability 2005, 45(7-8), pp. 1174-1180

A Case Study on Different Test Screening Techniques for ICs with High Resistance Vias Interconnects Issues

D.Sim[1], Y.C.Tan[1], F.Low[1], E.G.Foo[1]

[1] Systems on Silicon Manufacturing Company Pte. Ltd., 70, Pasir Ris Drive 1, Singapore 519527
E-mail: derek.sim@nxp.com, Fax: (65) 62487606

Abstract: **Weak ICs that are functional under normal operating conditions are not easily screened out during CP/FT stages. In this paper, different test screening and reliability assessment techniques were applied on a batch of wafers with normal to highly resistive Vias interconnects. The results and effectiveness of these techniques to screen and flag out problematic dies at wafer level testing were presented. . In additional, a fast wafer level interconnect wear-out test (Via electromigration test) and a package level burn-in test were also performed to complete the evaluation.**

I. INTRODUCTION

Weak ICs containing process induced defects or imperfections may not cause failures at normal operating conditions, but will likely degrade quickly over used-operation lifetime and fail to function properly later in the field, thereby causing reliability issues [1]. These kinds of defects that lead to "early failures" include gate oxide resistive shorts or gate oxide pinholes, metal shorts and threshold voltage shifts [2]. Others, like STI, poly and metal extra patterns or residuals [3], and high resistance Vias or metallization interconnects have also been reported. Such ICs with latent defects are not easily screened out during chip probe (CP) or final test (FT) stages. Therefore, to reduce the rate of "early failures" or IC "infant mortality" rate going to their end-customers, many manufacturers have adopted screening techniques during production test stages and/or either a full production or a sample burn-in to stamp out such ICs [4]. Common screening techniques like functional scan tests, IDD_Q (quiescent current) tests, VLV (Very-Low-Voltage) tests, static or dynamic voltage stress tests, delay fault tests [5], when used in conjunction with post-test statistical analysis, distribution testing and maverick screen controls, can effectively contribute towards achieving single digit ppm EFR performance in the field. In this paper, for the purpose of engineering study, some of these screening techniques were applied on a batch of production discarded wafers with clear low yield wafer spatial signature due to high resistive interconnects, to study their effectiveness as a screening method to flag out problematic dies at wafer test level. In additional, a fast wafer level interconnect wear-out test (Via electromigration test) and a package level burn-in test were also performed to complete the evaluation.

II. EXPERIMENT

The devices studied in this work were fabricated using 0.16um logic technology with Shallow Trench Isolation (STI), high performance core transistors, thick gate transistors for I/O, cobalt salicide process, followed by ILD (SACVD & PETEOS oxide), and Fluorine doped Silicate Glass (FSG) for the inter-metal dielectric (IMD), and Aluminium metallization and Tungsten Vias plugs as interconnects.

A single batch of production discarded wafers due to mis-processing caused by equipment EDM (Equipment-down-memo) was specially retrieved to be used in this engineering study. The wafers were found to encounter abnormal localized accelerated etching during Metal2 etch at the 9-to-12 o'clock region of the wafer, due to higher temperature distribution at affected region as a result of equipment electrostatic chuck malfunction. This result in smaller Metal2 CD and top TiN thinning, which then cause subsequent Via2 etch to punch through TiN and etch into the metallization. This leads to the formation of a highly resistive layer at plug-metal interface, due to reaction of Via etchants with Al during Via etch, as shown in figure 1. The elemental mapping of the interface is shown in figure 2.

978-1-4244-2039-1/08/$25.00 ©2008 IEEE

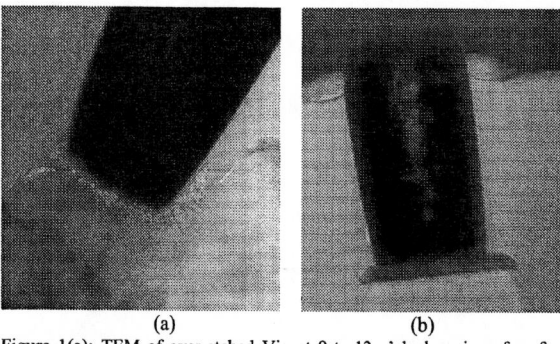

(a)   (b)

Figure 1(a): TEM of over-etched Via at 9-to-12 o'clock region of wafer, forms highly resistive Via-to-Metal interface. Figure 1(b): TEM of a good Via at an unaffected location of the wafer.

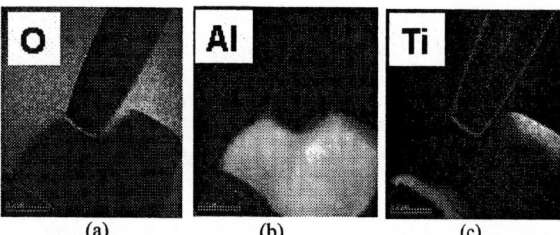

(a)   (b)   (c)

Figure 2: TEM elemental mapping of highly resistive plug-metal interconnect interface: 2(a) Oxygen map, 2(b) Aluminium map, 2(c) Titanium map.

Electrical WAT (wafer acceptance test) measurement was able to detect the high Via2 resistance as shown in figure 3, and subsequent in-house Esort (CP test) also found yield loss (gross fails and scan fails) at same region, (Figure 4).

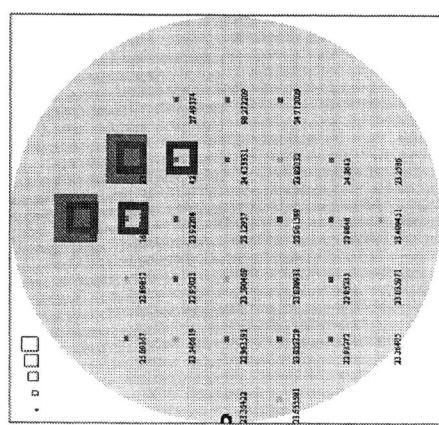

Figure 3: Full wafer map WAT electrical measurement showing high Via2 Rc (resistance) at 9-to-12 o'clock region of wafer (with wafer notch at 6 o' clock)

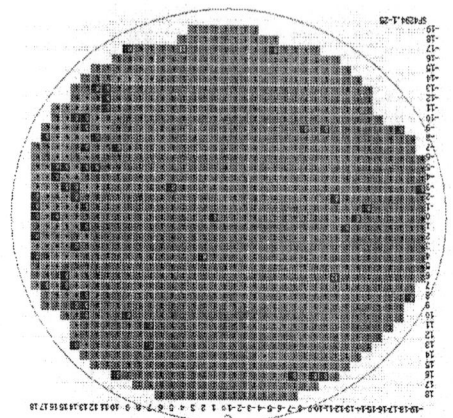

Figure 4: Wafer Esort map showing yield loss (gross and scan fails) region at 9-to-12 o'clock region of wafer (notch 6 o' clock)

Some test screening techniques (VLV tests, frequency tests, DVS tests) and reliability assessments methods (high temp bake, WL temperature cycling) were applied and the corresponding results & effectiveness discussed. Finally, fast wafer level Electromigration (EM) test and package level burn-in test were also performed to complete the evaluation.

III RESULTS AND DISCUSSION

Fast wafer level (fWLR) iso-current Via EM tests using highly accelerated stress conditions (high current density and high temperature) were performed first to study the Via2 interconnect life-times distribution across the wafer map. Although resolution was limited to 1 test structure per field recticle shot, the fWLR EM results and data analysis still managed to be able to yield a distinction between 3 groups of performances, which we labeled as "Poor", "Marginal" and "Normal", as shown in figure 5(a). Sites with "Poor" EM lifetimes coincides with the yield loss region and figure 5(b) show that its EM lifetimes are comparably 10 times worse off than the good or "Normal" sites measured on unaffected regions, elsewhere on the wafer. More interestingly, a "Marginal" EM zone emerges at the transition region (which did not show yield loss) between the "Bad" and the "Normal" sites, with EM lifetimes roughly half of the "Normal" sites. This suggests that the high resistance Via2 issue occurs gradually, turning from bad (at wafer edge), to marginal, to good (towards wafer center) and indicated that passing dies at the transition boundary could have high resistive Vias, but still passes Esort due to device build-in design and test windows. This sets the basis for the further tests and study.

978-1-4244-2039-1/08/$25.00 ©2008 IEEE

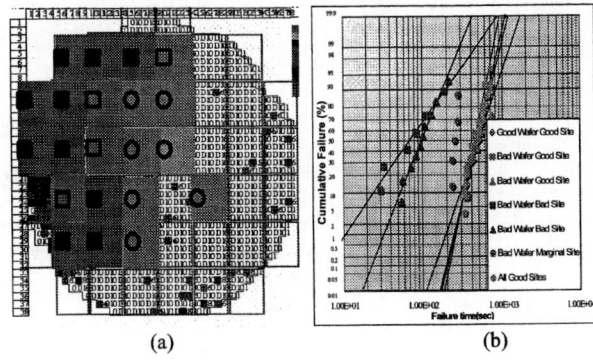

(a)   (b)

Figure 5(a): fWLR EM test results: ■ Solid squares, ▢ Empty squares & ◯ Empty circle represents sites with measured poor, marginal and normal EM lifetimes respectively. Figure 5(b): Cumulative EM lifetimes plot of the 3 groups of results

Based on the mechanism of the Via2 punch-through TiN to form a highly resistive plug-metal interconnect interface, 2 affected wafers were baked at 250°C for 48hrs to check if high temperature acceleration/aging would deteriorate marginally passing dies with borderline high resistive Via2 resistance further so as to result in device catastrophic fails. Post bake Esort test shows the good and bad dies are stable and no die flipping was observed. Similarly, to evaluate the integrity of the high resistive Via2 interconnects under temperature fluctuation and mechanical stress effects, 1 wafer was subjected to 200 cycles of temperature cycling (-65°C to 150°C TMCL). Again, post-TMCL Esort test shows good stability on both the good and bad dies.

Next, Vdd_min stepping and mapping on scan patterns was performed on 2 wafers to study the effect of high resistance interconnects (and therefore higher RC delays) on the minimum Vdd required to enable a die to pass functional tests. Note that standard Esort/CP program is performed at Vdd_nom (1.8V) and does not measure at Vdd_min (below 1.8V) conditions. Also, dies that have already failed during nominal 1.8V CP/Esort tests are omitted from Vdd_min test and these dies are represented in black color. Vdd_min testing is executed by performing CP/Esort testing with increasing Vdd in steps. The Vdd_min voltage recorded is the test voltage in which a die first starts to pass functionally. As shown in figure 6, it was observed that intrinsically good dies in non-impacted region typically registered a scan functional pass in the range of Vdd_min values from 1.40 ~ 1.45V (represented by different shades of blue colors) while some "suspect-weak" good dies (red color) within and directly along the boundary of the low yield region (9-to-12 o'clock region) uncharacteristically logged a higher 1.55V Vdd_min value, indicating the marginal performance of such dies. The abnormally higher Vdd_min values recorded for such dies can also be interpreted as an indication of a smaller reliability lifetime window or margin than normally good passing dies. It is also worthwhile to note also that apart from this observation,

the Vdd_min characterization did not show any gradual performance transition from affected areas (failed dies with no margin) to non-affected areas (pass dies with good margin). The performance transition is sharp and abrupt. This also seems to support the argument that RC delays do not scale when the supply voltage is reduced [6], and VLV testing can hardly improve detecting high resistance interconnects [5]. But in this instance, due to global nature of the increase in Via2 interconnect resistance of dies directly at affected region; VLV sensitivity signature is observable, and only at the very worst-case locations, but not visible elsewhere. This can be explained that although interconnect resistance (RC delays) remains the same at different supply voltages, the device speed and drive current of the transistors will degrade with lower Vdd, allowing dies with higher interconnect delays to be flagged out at higher Vddmin levels than good dies.

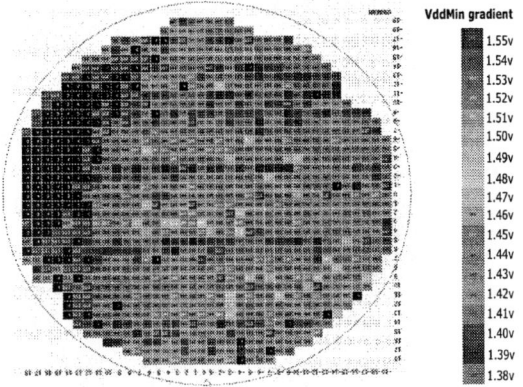

VddMin gradient
1.55v
1.54v
1.53v
1.52v
1.51v
1.50v
1.49v
1.48v
1.47v
1.46v
1.45v
1.44v
1.43v
1.42v
1.41v
1.40v
1.39v
1.38v

Figure 6: Vdd_min mapping found "suspect-weak" good dies (that pass Esort at nominal condition) within and around the boundary of the low yield region (9-to-12 o'clock region) register a higher Vdd_min value, indicating the marginal performance of such dies.

A frequency/speed mapping was done on 1 wafer by performing CP/Esort test with increasing clock period in steps, while holding Vdd at nominal level. The cycle_min (ns) recorded is clock period in which a die first starts to pass functionally, as shown in figure 7. The mapping result shows that good dies around the boundary of the low yield region have slower speed. However, the selectivity and sensitivity of this frequency test module was not high, as good dies from some other unaffected regions on the wafer also show similarly high delay times.

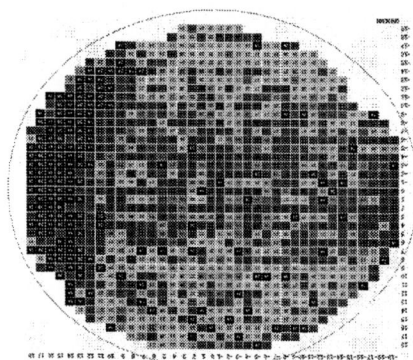

Figure 7: Frequency/speed mapping found good dies around the boundary of the low yield region with indication of higher delay times. However, the selectivity of this frequency test module was not high, as other unaffected regions on the wafer also show similar high delay times.

After that, a dynamic voltage stress test (DVS) was performed on 1 wafer, looping 30x cycles of core functional test at 1.4xVcc stress, (each loop cycle approximately 1.2sec) after which the wafer was re-sorted again. 1.4xVcc is selected as the DVS stress/screen voltage as it corresponds to the AMR (absolute maximum rating: the maximum voltage that can be applied to a minimum gate length device without leading to unrecoverable failure (destructive breakdown) of the device for a particular process node. This stress level is below the intrinsic breakdown voltage of the process. Figure 8 showed that an additional 8 dies flipped (solid squares) from passing (bin1) to failing scan tests after DVS stress cycles. This shows that the short duration test looping at 1.4xVcc exert electrical/current stress on dies with highly marginal Vias, causing the marginal Vias to degrade further, resulting in chip catastrophic fails. It is also conceivable that a less marginal die would take much longer DVS stress cycles to transit from pass to fail status. It is meaningful to take into consideration that the same DVS routine was previously applied on tens of thousands of dies and not a single good die was found to have flipped to fail in that study.

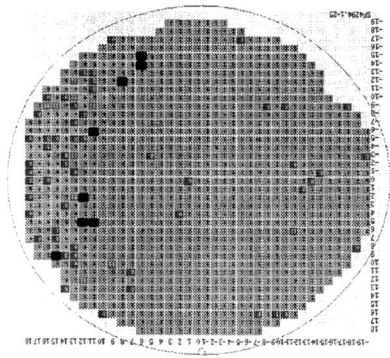

Figure 8: Additional 8 dies flipped (solid square) from passing (bin1) to failing scan tests after 30X loops of dynamic voltage stress (DVS) cycles at 1.4Vcc.

Finally, to explore the DVS method further, 1 good die from unaffected region and a less marginal die located near to the bad region in the x-direction near to wafer center were subjected to extended DVS loop cycling, up to 50,000 cycles. As shown in figure 9, it was observed that both Vdd_min and Cycle_min degradation on the marginal die (solid line) was up to a factor of 2 to 5 times faster than that of the good die (dashed line), which lends further proof that the DVS loop cycles are effective in wearing the device out, and that most likely the different rates of device performance degradation can be attributed to the quicken deterioration of the borderline Vias interconnects in the marginal die.

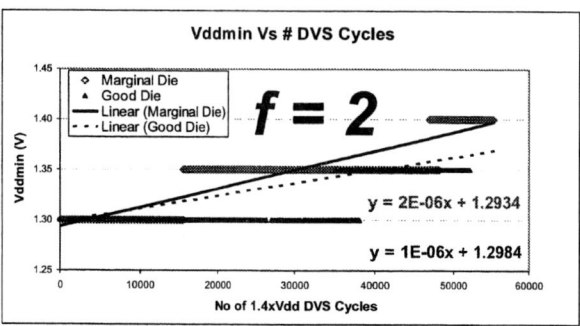

(a)

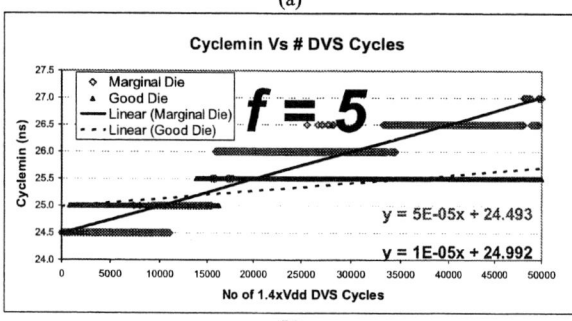

(b)

Figure 9: Rate of degradation in (a) Vdd_min & (b) Cycle_min performance of a marginally passing die (solid line) at the boundary of the low yield region (9-to-12 o'clock) is found to be 2~3 times faster than a good unaffected die (dashed line).

As part of the test setup, a group of specially selected 40 Bin1 passing dies (Group A) in the proximity of the low yield region and another group of 40 Bin1 passing dies (Group B) from the good unaffected region (3 o'clock) of an impacted wafer were send for assembly and packaged for HTOL48hrs testing. 2/40 dies from Group A fallout as scan fails after HTOL168hrs while no fail was registered from Group B.

IV CONCLUSION

In this case study, different test screening techniques and reliability assessment experiments were applied on a batch of wafers with normal to highly resistive Vias interconnects. The results and effectiveness of these techniques to screen and flag out problematic dies at wafer level tests were presented. It is shown that Vdd_min testing and dynamic voltage stress testing could be an effective (quick & low cost) way to detect flaws in weak ICs (pertaining to resistive Vias interconnects) and therefore to improving the quality of the outgoing parts.

ACKNOWLEDGMENTS

The authors would like to thank Joe Hui and Lim Soon for their support during the testing and evaluation stage.

REFERENCES

[1] Hao, H., and E.J. McCluskey, "Very-Low-Voltage Testing for Weak CMOS Logic ICs" *Proc. of ITC*, Oct, pp. 275-284, 1993

[2] Chang J.T.Y. and E.J. McCluskey, "Quantitative Analysis of Very-Low-Voltage Testing" *Proc. of 14th IEEE VLSI Test Symp.*, Apr, 1996

[3] C.Y. Tsao, R.Y. Shiue, C.C. Ting, Y.S. Huang, Y.C. Lin, J.T. Yue "Applying Dynamic Voltage Stressing to Reduce Early Failure Rate" *Proc. of IRPS*, 2001

[4] J.A. van der Pol, E.R. Ooms, T. van Hof, F.G. Kuper "Impact of Screening of Latent Defects at Electrical Test on the Yield-Reliability Relation and Application to Burn-in Elimination" *Proc. of IRPS*, 1998

[5] Chang J.T.Y. and E.J. McCluskey, "Detecting Delay Flaws by Very-Low-Voltage Testing" *Proc. of ITC*, 1996

[6] Wagner, K., and E.J. McCluskey, "Effects of Supply Voltage on Circuit Propagation Delay and Test Application," *Proc. of ICCAD*, Nov, pp. 42-44, 1985.

A Novel ESD Device Structure with Fully Silicide Process for Mixed High/Low Voltage Operation

Jian-Hsing Lee[1], J.R. Shih[1], Dao-Hong Yang[2], Jone F. Chen[2], and Kenneth Wu[1]

Technology Quality and Reliability Division

[1]Taiwan Semiconductor Manufacturing Company, Hsin-Chu, Taiwan, E-mail: jhleea@tsmc.com

[2]Institute of Microelectronics, Department of Electrical Engineering, National Cheng Kung University, Tainan, Taiwan

Abstract-**A novel ESD device structure with non-LDD at drain region has been demonstrated to enhance the ESD immunity of IO circuits with mixed high/low operation voltage. The protection capability of this novel ESD device structure has been proved from 1μm to 65nm technologies with and without fully salicide at the source/drain region. This structure is found to be also very effective to protect the high voltage tolerant (HVT) IO circuits and the drain extended NMOSFET (DEMOS) transistors. The ESD failure thresholds can be improved from HBM < 0.5KV and MM < 50V to HBM 4KV and MM 200V, respectively. In addition, this novel ESD device structure is cost effective because two process modules including RPO and ESD implant can be removed.**

I. INTRODUCTION

It's well known that the conventional ESD protection approach to improve the NMOS ESD performance is with the ESD implant to overwhelm the NLDD implant to build up a pure N+ junction and enlarges the non-silicide region between Poly and Contact at drain region by RPO (Resistor Protection Oxide) process, as shown in Fig. 1 [1]. However, it needs two extra masks (RPO mask and ESD implant mask) and one additional implant (ESD implant) if one would like to improve the device ESD performance by this conventional approach. However, these process steps will increase the process cost and also IO circuit area. In addition, due to analog application requirement, high voltage (HV) transistors integrated to the chip primarily composed of low voltage (LV) devices are inevitable. How to simultaneously achieve good ESD performances of HV device drain extended MOS (DEMOS), LV device and still keep acceptable chip size in mature technologies (Lg > 130nm) and sub-90nm technologies is a big challenge from ESD protection point of view. In this paper, we propose the non-LDD structure to improve the ESD performance for above considerations [2].

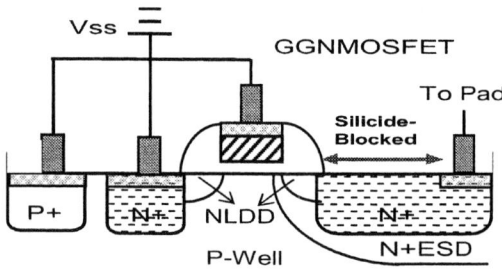

Fig. 1 A ground-gate NMOSFET (GGNMOSFET) with N+ESD implant and salicide-blocked area N+ drain region

II. EXPERIMENTS

To investigate the robustness of the proposal novel ESD protection structure, several technologies including 1.0μm HV 500V/5V process, 0.18μm HV 5V/1.8V process, 0.13μm 3.3V/1.0V process and 65nm 2.5V/1.0V process were verified to evaluate the ESD performance. The key process differences between these technologies include different NLDD implant condition, spacer material and sidewall width, source/drain implant condition, non-silicide scheme (1.0μm), silicide scheme (0.18μm and beyond) and thermal cycles.

III. RESULTS AND DISCUSSION

A. *5V IO ESD Protection for 1.0um 500V/5V Process*

The HV device often needs large dimension rules to sustain the high voltage and high power. There is less benefit to develop the high voltage process in the advanced technology. So, most smart power products still stay at the technologies below 0.18μm process. For these HV products, some IO pads are still designed by 5V device. Except the ESD implant, the drain designed with a ballast resistor by long contact to Poly (Co-to-Po) space, [3] is the conventional scheme to improve the 5V NMOSFET's ESD performance for the non-silicided technologies. Table I and Table II show the layout splits for the 5V device with and without LDD implant at drain region.

Table I shows that the increase of the device total width or Co-to-Po space cannot improve the 5V NMOSFET'S ESD performance for this technology. All devices got the same ESD results of HBM 0.5KV and failed at MM 50V since these devices were damaged after occurred the snapback, which resulted in the leakage current increase (Fig.2a). It can also be observed the ballast resistor formed by Co-to-Po stripe (> 1μm) doesn't work for this technology since the 5V NMOSFET is very difficult to be driven to the snapback region due to long thermal cycle resulting in very gradient doping profile at LDD region. Unlike 5V NMOSFET with LDD implant, the leakage currents of all 5V NMOSFETs without LDD implant in Table II can still keep at very low level ~ 100pA until the devices occur the 2nd breakdown (Fig. 2b) since the abrupt N+ junction can provide a high electrical field to drive the device into the snapback region.

With the same layout, the device without the NLDD can get much better ESD performance, which can increase with device total width even without a ballast resistor (Co-to-Po space = 1μm).

Without the LDD implant, the Non-LDD device only can be used as the ESD protection device and cannot be used as the IO driver. Table III shows the ESD experiment for IO N-driver with Non-LDD ESD protection device. It can be observed the Non-LDD device can protect the IO N-driver if the total width of the Non-LDD device is larger than 300μm. Although the ESD performance of the Non-LDD devices with Co-to-Po space as 2μm and 3μm are the same, using Co-to-Po space 3μm for both Non-LDD device and N-driver can get better ESD performance since the N-driver has higher resistance to the ESD zapping events.

Table I: 5V NMOSFET with LDD, TW: total width

TW	300μm	600μm	1000μm	300μm	300μm	300μm	300μm
Co-to-Po	1μm	1μm	1μm	2μm	3μm	4μm	5μm
HBM	0.5KV	0.5KV	0.5KV	0.5KV	0.5KV	0.5KV	0.5KV
MM	0V	0V	0V	25V	50V	50V	50V
It2	0A	0A	0A	0A	0A	0A	0A

Table II: 5V NMOSFET without LDD, TW: total width

TW	300μm	600μm	1000μm	300μm	300μm	300μm	300μm
Co-to-Po	1μm	1μm	1μm	2um	3μm	4μm	5μm
HBM	1.0KV	2.0KV	3.5KV	4.5KV	4.5KV	4.5KV	4.5KV
MM	50V	100V	150V	250V	250V	250V	250V
It2	0.565A	1.2A	1.6A	2.4A	2.35A	2.32A	2.38A

Table III: N-Driver with 5V NMOSFET without LDD

TW for Non-LDD	TW for N-driver	Co-Po	HBM	MM
200μm	100μm	2μm	1KV	150V
300μm	100μm	2μm	4KV	300V
300μm	100μm	3μm	5.5KV	450V

B. Fully Silicided 5V IO ESD Protection for 0.18um Process

For dual gate oxides process, it is often found the ESD performance of the device with the thin oxide is much better than the device with thick oxide if they are with the same layout. Table IV shows the ESD influences of LDD implant, PW implant and oxide thickness on 1.8V or 5N NMOSFETs with fully silicided source/drain regions. The device is with 720μm total width, 0.25μm Co-to-Po space and surrounded with a P+ guard-ring.

No matter it is a 5V or 1.8V device, the device with or without 1.8V LDD implant can get the robust ESD performance (≥7.5KV HBM and ≥350V MM). However, the performance of device with 5V LDD implant will be degraded to HBM 1KV and MM 100V. It implies that LDD implant dosage has much more impact on the ESD performance than the PW dosage and gate oxide thickness. Figure 3 shows the high current IV characteristics of the devices in Table IV. Although the 5V NMOSFET with 1.8V LDD can get better ESD performance, it will induce too high leakage current (~100nA at 5V) compared with 5V NMOSFET without LDD (60pA at 5V), as shown in Fig. 3. So, the Non-LDD device is still the better choice for fully silicided IO ESD protection circuits.

C. Fully Silicided HVT 5V IO for 0.13μm Process

The cascode NMOSFET structure is a common approach for high voltage tolerant (HVT) IO circuit, which uses two 3.3V IO NMOSFETs to operate at higher voltage (5V). Due to no overlap

Fig. 2(a)

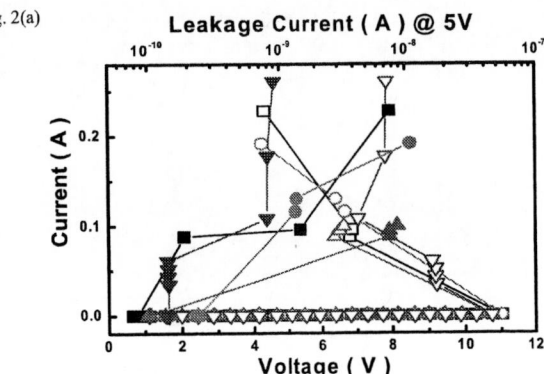

Fig. 2(b)

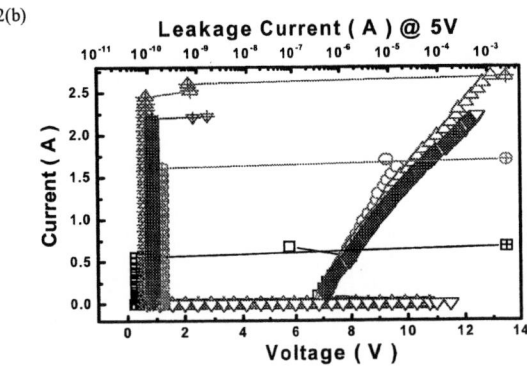

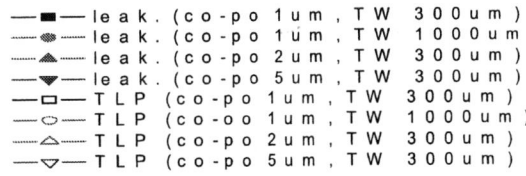

Fig. 2 High voltage IV characteristics of 5V NMOSFETs (a) with and (b) without LDD implant.

Table IV: Co-to-Po: 0.2μm, Fully Silicided, TW: 720μm

PW/Oxide	5V	5V	5V	1.8V	1.8V	1.8V
LDD	5V	No	1.8V	1.8V	No	5V
HBM	0.5KV	7.5KV	>8KV	>8KV	7.5KV	0.5KV
MM	<50V	350V	400V	400V	400V	100V
It2	0.4A	3.79A	4.0A	4.4A	3.85A	1.34A

5V LDD Dosage: ~ 1.5E13 atoms/cm², 1.8V LDD Dosage: ~ 3E14 atoms/cm², PW dosage: 2~6E12 atoms/cm²

region under Poly gate for NMOSFET without LDD implant, there is no gate oxide reliability concern when the device is biased at the voltage higher than 5V, because larger voltage can drop on the non-LDD region under the spacer. However, it was also reported that there existed the gate-voltage-induced channel current crowding (GVICC) effect of the cascode NMOSFET as the ESD protection device for HVT IO circuit [4], which will seriously degrade ESD performance.

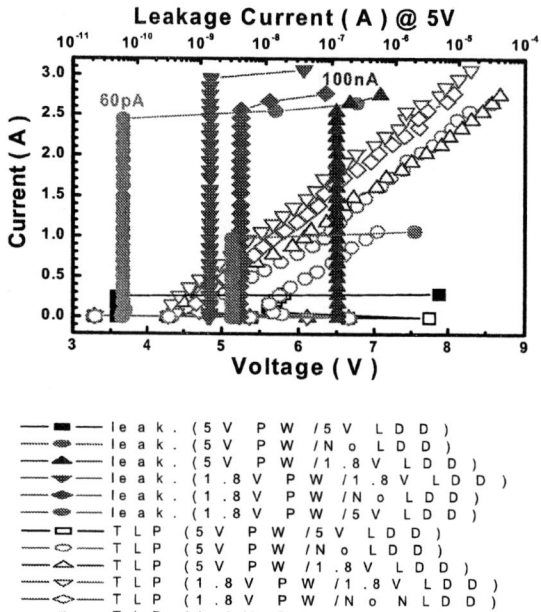

Leakage Current (A) @ 5V

Fig. 3 High voltage IV characteristics of 5V NMOSFETs with different PWell and LDD implant conditions.

Legend:
- leak. (5 V P W / 5 V L D D)
- leak. (5 V P W / N o L D D)
- leak. (5 V P W / 1 . 8 V L D D)
- leak. (1 . 8 V P W / 1 . 8 V L D D)
- leak. (1 . 8 V P W / N o L D D)
- leak. (1 . 8 V P W / 5 V L D D)
- T L P (5 V P W / 5 V L D D)
- T L P (5 V P W / N o L D D)
- T L P (5 V P W / 1 . 8 V L D D)
- T L P (1 . 8 V P W / 1 . 8 V L D D)
- T L P (1 . 8 V P W / N o N L D D)
- T L P (1 . 8 V P W / 5 V L D D)

Table V: 3.3 V NMOSFET with different structure, TW: 720μm

Stru.	NMOS w/ LDD	NMOS w/o LDD	Cascode. NMOS w/ LDD	HVT IO w/ LDD	HVT IO w/o LDD
HBM	4.5KV	6.5KV	4.5KV	<0.5KV	6.5KV
MM	200V	300V	150V	<50V	250V
It2	2.2A	2.94A	2.25A	0.34A	2.86A

Fig. 4(a)

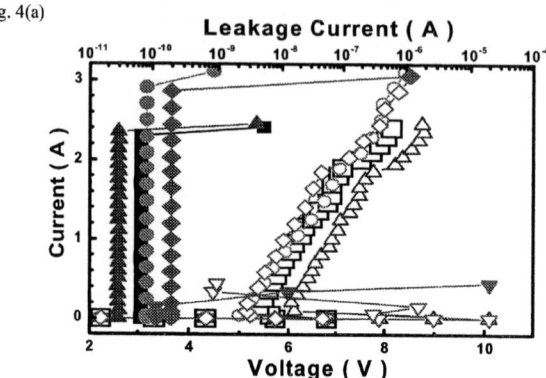

Leakage Current (A)

Fig. 4(b)

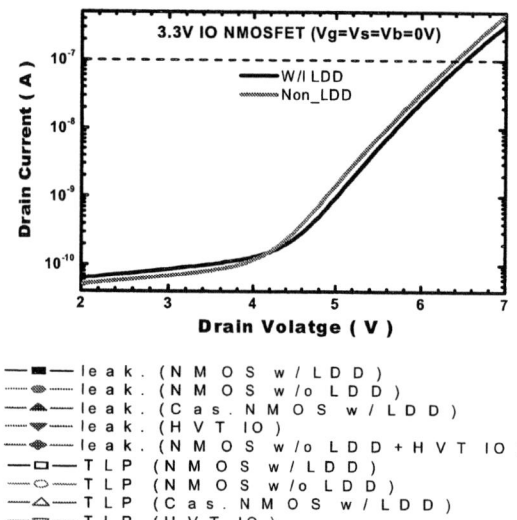

Legend:
- leak. (N M O S w / L D D)
- leak. (N M O S w / o L D D)
- leak. (C a s . N M O S w / L D D)
- leak. (H V T I O)
- leak. (N M O S w / o L D D + H V T I O)
- T L P (N M O S w / L D D)
- T L P (N M O S w / o L D D)
- T L P (C a s . N M O S w / L D D)
- T L P (H V T I O)
- T L P (N M O S w / o L D D + H V T I O)

Fig. 4 (a) Leakage current comparison of 3.3V NMOSFET with and without LDD implantation at drain region, (b) High current IV characteristic comparison of different structures in Table V.

Table V shows the experiment results of 3.3V NMOSFET with and without LDD implant to protect HVT IO circuit. These devices are fabricated by 0.13μm process and have the same layout structure with fully silicided source/drain regions. The total finger width is 720μm and Co-to-Poly space is 0.2μm. It can be observed that NMOSFET without LDD implant has much better ESD performance, but behaves with the same leakage current level of NMOSFET with LDD implant (Fig. 4a). Although the ESD performance of pure cascode NMOSFET is similar to the single NMOSFET's, the ESD performance of the HVT IO circuit composed of the cascode NMOSFET is degraded to below HBM 0.5KV and MM 50V. The mechanism is explained as below.

As shown in Figure 4b, the snapback voltage (Vsp ≈ 8V) of HVT IO circuit is higher than that (6.2V) of the cascode NMOSFET due to the GVICC effect. Attributed to the smaller trigger voltage (Vt1), NMOSFET without LDD implant can turn on before the HVT IO circuit does to protect the HVT during the ESD event. The snapback voltage of the HVT IO circuit composed of NMOSFETs without LDD implant can be increased to the same Vsp value of single NMOSFET without LDD implant. Thus, the GVICC effect can be eliminated, and the ESD performance of HVT IO circuit can be improved significantly. This expectation has been confirmed by the experimental data as shown in Table V, where the ESD failure thresholds of HVT IO circuit without LDD implant are nearly the same as those of single NMOSFET without LDD implant.

978-1-4244-2039-1/08/$25.00 ©2008 IEEE

D. ESD Protection Device for DEMOS in 65nm Process

Drain-extended MOSFET (DEMOS) has widely been used in the mature high voltage (HV) technologies. Now, DEMOS has also become a necessary component of the mix-signal RF circuit for SOC design to integrate the HV function in one chip with sunb-90nm technologies. However, DEMOS does not have any ESD protection capability due to the characteristic of weak snapback [5]. Because the 2.5V NMOSFET without LDD implant is with high trigger voltage (8V, Fig. 5a), and has the acceptable leakage current (~10nA at 5V, Fig. 5b) for 5V operation, it can be a good ESD protection device for DEMOS. Table VI shows the ESD experiment results of DENMOS with and without protection device, that is 2.5V NMOSFET without LDD implant. The Non-LDD device is with a fully silicided source/drain and with 720μm total width. Due to low beta gain of the parasitic bipolar transistor of DEMOS, DENMOS was damaged after went to the snapback, as shown in Fig. 5, which resulted in the device failed at low voltage ESD zapping event. Compared with DEMOS, NMOSFET without LDD implant has smaller Vt1 and excellent ESD protection capability, and it can be turned on before DEMOS does to protect the DEMOS. So, the ESD performance of DEMOS protected by non-LDD NMOSFET can be improved from HBM < 0.5KV and MM < 50V to HBM 4KV and MM 200V, respectively.

Fig. 5(a)

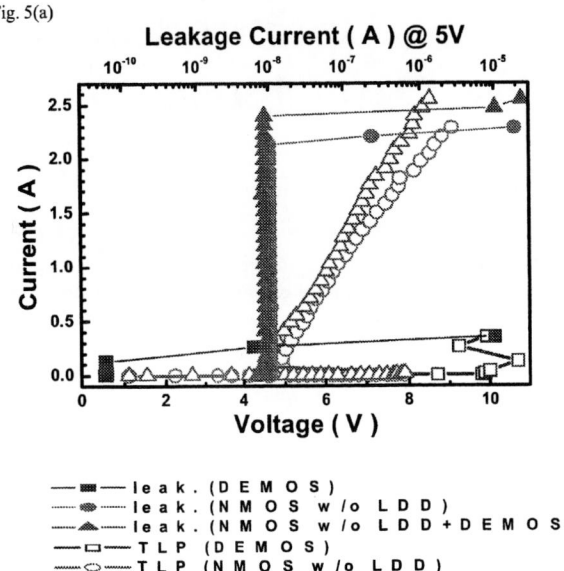

- ■ leak. (D E M O S)
- ● leak. (N M O S w /o L D D)
- ▲ leak. (N M O S w /o L D D + D E M O S)
- □ T L P (D E M O S)
- ○ T L P (N M O S w /o L D D)
- △ T L P (N M O S w /o L D D + D E M O S)

Fig. 5(b)

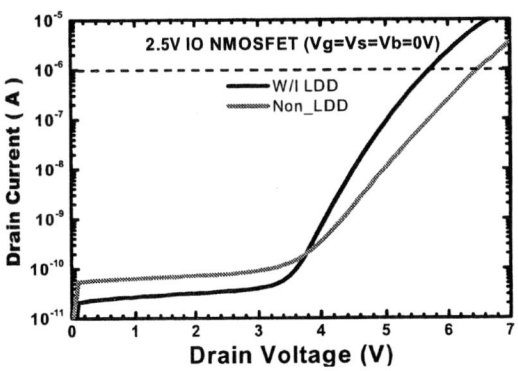

Fig. 5(a) High current IV characteristic comparison of different device structures in Table VI, (b) Leakage current comparison of 2.5VNMOSFETs with and without LDD implantation at drain region.

Table VI: 65nm Technology, Fully Silisided S/D, TW=720um

	HBM	MM	It2
DENMOS	<0.5KV	<50V	0A
2.5V NMOSFET w/o LDD	4.5KV	250V	2.4A
2.5V NMOSFET w/o LDD+DEMOS	4.0KV	200V	2.2A

IV. CONCLUSIONS

A novel ESD device structure composed of NMSFET without LDD implant has been proposed. It has been demonstrated the Non-LDD device is a very powerful ESD protection device for all CMOS technologies. Without any extra process step and additional mask, the Non-LDD device not only can be used to protect the IO circuits with mature CMOS technologies, but also can be used to protect fully silicided IO circuits, HVT IO circuits and DENMOS with sub-90nm CMOS technologies.

REFERENCES

[1] Y. Wei, Y. Loh, C. Wang and C. Hu, "MOSFET Drain Engineering For ESD performance," EOS/ESD Symposium, p. 143, 1992.

[2] JH Lee et al. United States Patent: 5663082

[3] R. N. Rountree, Charles L. Hutchins, "NMOS Protection Circuitry", IEEE ED., No.5, p.910, 1985.

[4] J. H. Lee, J.R. Shih*, Y. H. Wu, T.C. Ong., "The Failure Mechanism of High Voltage Tolerance IO Buffer under ESD," IRPS Proc., p.269, 2003

[5] G. Boselli, V. Vassilev, and C. Duvvury, "Drain Extend nMOS High current Behavior and ESD protection strategy For HV Application in sub-100nm CMOS Technologies ," IRPS, p. 342, 2007

Optimization on SCR Device With Low Capacitance for On-Chip ESD Protection in UWB RF Circuits

Chun-Yu Lin and Ming-Dou Ker

Nanoelectronics and Gigascale Systems Laboratory, Institute of Electronics, National Chiao-Tung University, Taiwan
Phone: 886-3-5712121#54215   Fax: 886-3-5715412   Email: mdker@ieee.org

Abstract-**Low capacitance (low-C) design on ESD protection device is a solution to mitigate the radio-frequency (RF) performance degradation caused by electrostatic discharge (ESD) protection device. Silicon-controlled rectifier (SCR) device has been used as an effective on-chip ESD protection device in RF ICs due to the smaller layout area and small parasitic capacitance under the same ESD robustness. In this paper, the modified lateral SCR (MLSCR) realized in waffle layout structure is studied to minimize the parasitic capacitance and the variation of the parasitic capacitance within ultra-wide band (UWB) frequencies. With the minimized parasitic capacitance, the degradation on RF circuit performance due to ESD protection devices can be reduced. The waffle MLSCR with low parasitic capacitance is suitable for on-chip ESD protection in UWB RF ICs. Besides, the turn-on speed of MLSCR with waffle layout structure is verified to be better than that with conventional stripe structure.**

I. INTRODUCTION

It has been a trend to integrate the whole radio-frequency (RF) circuits into a single chip. With the scaling-down feature size and lower cost, nanoscale CMOS technology is the leading role to integrate RF circuits [1]. However, the thin gate oxide in advanced CMOS processes seriously degrades the electrostatic discharge (ESD) robustness of IC products. Against ESD damages, ESD protection devices must be included in ICs. A general concept of on-chip ESD protection for RF ICs is illustrated in Fig. 1 [2]. The ESD protection devices must be provided for all I/O pads in RF ICs. The parasitic capacitance (C_{ESD}) of ESD protection device is one of the most important design considerations for RF ICs. The parasitic capacitance inevitably contributes capacitive loading to the I/O port, which disturbs the high frequency signals, induces RC delay on the signal path, and causes degradation on RF performance [3]. To mitigate the RF performance degradation caused by ESD protection device, low capacitance (low-C) design on ESD protection device to reduce the parasitic capacitance is a rapid and simple method [4].

With the highest ESD robustness within a smaller layout area and lower parasitic capacitance, the silicon-controlled rectifier (SCR) devices were reported to be useful for RF ESD protection design [5]. The lateral SCR (LSCR) device has been used as the conventional ESD protection device in CMOS technology; however, LSCR has a higher turn-on voltage, which is generally

This work was supported by National Science Council (NSC), Taiwan, under Contract of NSC96-2221-E-009-182.

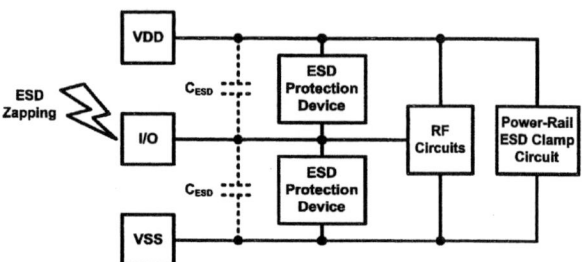

Fig. 1. General concept of on-chip ESD protection for RF ICs.

greater than the gate-oxide breakdown voltage of the MOS transistor in input stage. In order to reduce the turn-on voltage of LSCR to provide more effective ESD protection for the internal circuits, the modified lateral SCR (MLSCR) device has been reported [6].

In this paper, the layout structure of MLSCR device is investigated to minimize the parasitic capacitance. The parasitic capacitance within ultra-wide band (UWB, 3.1~10.6 GHz) frequency band and ESD robustness of MLSCR under different layout structures are presented. The turn-on speed of MLSCR devices under different layout structures are also investigated in this work.

II. MLSCR DEVICES FOR RF ESD PROTECTION

Fig. 2 shows the device cross-sectional view of the stripe-structured MLSCR (SMLSCR) and the waffle-structured MLSCR (WMLSCR). Both devices were designed with the same size of 60.62 x 60.62 μm². MLSCR devices are basically composed of four regions of P+/N-well/P-well/N+. The anode of MLSCR is electrically connected to P+ and N+ which are formed in the N-well. The cathode is electrically connected to N+ and P+ which are formed in the nearby P-well/P-substrate. The trigger P+ diffusions are added across the N-well/P-well junction in MLSCR device to reduce the junction breakdown voltage and the turn-on voltage. When a positive potential is applied between the anode and the cathode, the N-well/P-well junction is reverse-biased, so MLSCR device is kept off under normal circuit operating conditions. When an ESD stress is zapped to the anode with cathode grounded, MLSCR device will become highly conductive to quickly discharge ESD current due to the turn-on of latchup path. In SMLSCR, it discharges ESD current in only two directions, whereas

978-1-4244-2039-1/08/$25.00 ©2008 IEEE

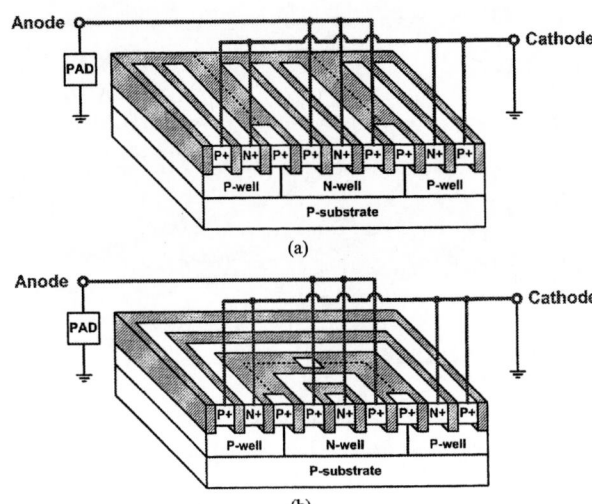

(a)

(b)

Fig. 2. Device cross-sectional view of (a) stripe-structured MLSCR (SMLSCR) and (b) waffle-structured MLSCR (WMLSCR).

WMLSCR discharges ESD current in four directions. Thus, the ESD robustness can be improved under the same parasitic capacitance by using WMLSCR. In other word, the ratio of the parasitic capacitance to ESD robustness can be minimized by realizing MLSCR device with waffle layout structure.

To investigate the relationship between the trigger P+ diffusion area and the parasitic capacitance, MLSCR devices were implemented with different trigger diffusion areas to evaluate the device characteristics and ESD robustness. The trigger diffusion areas of two SMLSCR devices were 123.2 μm^2 and 242.48 μm^2, and those of two WMLSCR devices were 140.48 μm^2 and 264.96 μm^2, which are listed in Table I. These devices have been fabricated in a 0.18-μm CMOS process for experimental investigations.

III. MEASURED DEVICE CHARACTERISTICS

A. Transmission Line Pulsing (TLP) Measurement

The turn-on voltage ($V_{turn-on}$) and secondary breakdown current (I_{t2}) of the fabricated MLSCR devices are characterized by the TLP system. The TLP-measured I-V curves for all MLSCR are shown in Fig. 3, and the extracted device characteristics are listed in Table I.

B. ESD Robustness

The human-body-model (HBM) ESD levels of the fabricated MLSCR devices are evaluated by the ESD simulator. All MLSCR devices pass the HBM ESD test (V_{HBM}) of 8-kV, which is the measurement limitation of HBM ESD tester. In order to distinguish the ESD robustness of SMLSCR from those of WMLSCR, the machine-model (MM) ESD tests are performed. The MM ESD levels (V_{MM}) of all MLSCR devices are within the range of 1.5~1.7 kV, as listed in Table I. Despite the MM ESD robustness of WMLSCR are slightly worse than those of SMLSCR due to the reduction of N-well area in the waffle

Table I
COMPARISONS ON MEASURED DEVICE CHARACTERISTICS OF MLSCR UNDER DIFFERENT LAYOUT STRUCTURES

Symbol	Trigger Diffusion Area (μm^2)	$V_{turn-on}$ (V)	I_{t2} (A)	V_{HBM} (kV)	V_{MM} (kV)	ΔC_{ESD} Within 3.1~10.6 GHz (fF)
SMLSCR$_1$	123.2	12.52	>6	>8	1.63	53.41
SMLSCR$_2$	242.48	12.54	>6	>8	1.68	67.47
WMLSCR$_1$	140.48	11.81	>6	>8	1.59	39.36
WMLSCR$_2$	264.96	12.55	>6	>8	1.56	49.13

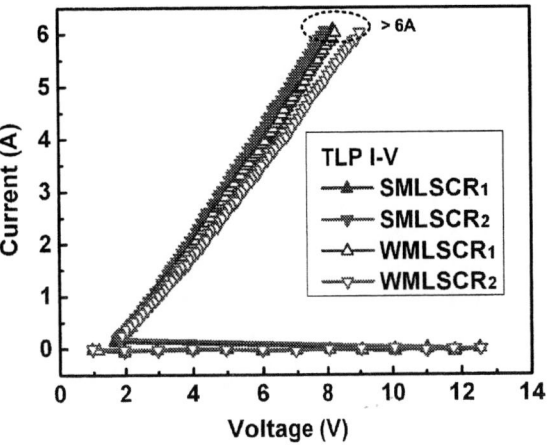

Fig. 3. TLP-measured I-V curves of MLSCR under different layout structures.

layout structure, the parasitic capacitance can be greatly reduced.

C. Parasitic Capacitance

The parasitic capacitance of each MLSCR was obtained from the Y_{11}-parameter, which was transformed from the measured two-port S-parameters. The two-port S-parameters were measured by using the vector network analyzer HP 8510C. To facilitate on-wafer two-port S-parameters measurement, MLSCR devices were implemented with ground-signal-ground (G-S-G) pads. In order to extract the characteristics of the intrinsic device in high frequency, the parasitic effects of the bond pads (PAD in Fig. 2) must be removed. The test patterns, one including the DUT and the other excluding the DUT, as shown in Figs. 4(a) and 4(b), were fabricated in the same experimental test chip. The Y_{11}-parameter can be obtained from the measured two-port S-parameters by using

$$Y_{11} = \frac{(1-S_{11})(1+S_{22})+S_{12}S_{21}}{Z_0\left((1+S_{11})(1+S_{22})-S_{12}S_{21}\right)}, \quad (1)$$

where Z_0 is the termination resistance and equals to 50 Ω. The measured Y-parameter of the including-DUT pattern is labeled as Y_{11_meas}, and the measured Y-parameter of the

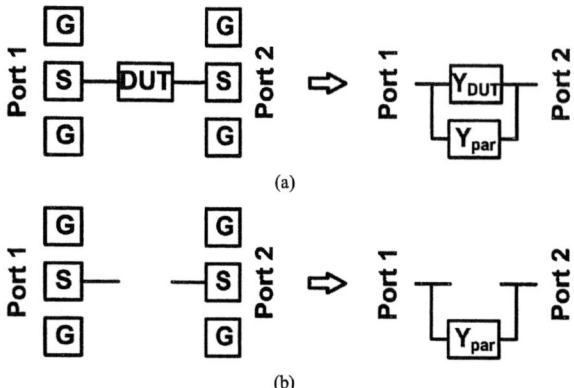

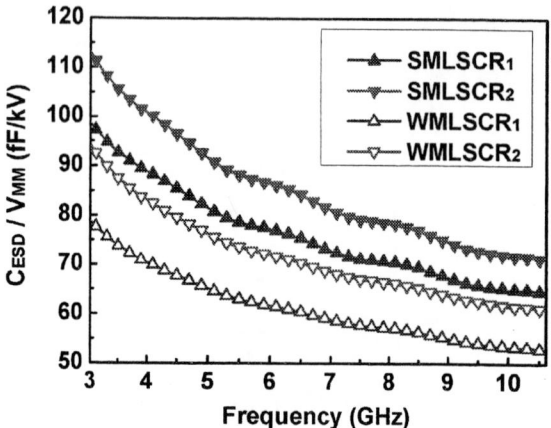

Fig. 4. Layout top view with G-S-G pads and equivalent model of (a) including-DUT pattern and (b) excluding-DUT pattern.

Fig. 6. Ratios of parasitic capacitance to MM ESD robustness within 3.1~10.6 GHz of MLSCR devices under different layout structures.

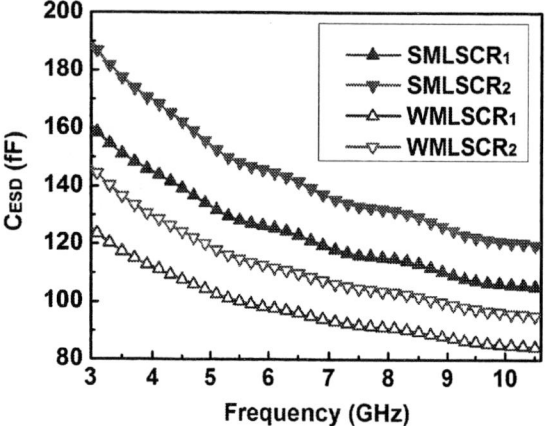

Fig. 5. Extracted parasitic capacitance within 3.1~10.6 GHz of MLSCR devices under different layout structures.

excluding-DUT pattern is labeled as Y_{11_par}. The intrinsic device Y-parameter (Y_{11_DUT}) can be obtained by subtracting Y_{11_par} from Y_{11_meas}. Finally, the intrinsic parasitic capacitance (C_{ESD}) of each MLSCR was extracted from the intrinsic device Y-parameter by using

$$C_{ESD} = \frac{\text{Im}(Y_{11\_DUT})}{2\pi f}, \qquad (2)$$

where f is the operating frequency.

Fig. 5 shows the extracted parasitic capacitance within UWB frequencies of all MLSCR devices. During the S-parameters measurement, the anode of each MLSCR was connected to port 1 and biased at 0.9 V, which is VDD/2 in the given 0.18-μm CMOS process, and the cathode was connected to port 2 and biased at 0 V. The parasitic capacitance of SMLSCR and WMLSCR are about 105~190 fF and 85~145 fF, respectively. For each MLSCR, the parasitic capacitance is decreasing as the frequency increasing. Because the parasitic capacitance was in series with a resistor, which is caused by the parasitic N-well resistance and P-well resistance in each MLSCR, the parasitic capacitance in high frequency is decreasing with the increasing

frequency. The variation of the parasitic capacitance within 3.1~10.6 GHz (ΔC_{ESD}) of SMLSCR and WMLSCR are about 55~65 fF and 40~50 fF, which are summarized in Table I.

D. Comparison on Parasitic Capacitance and ESD Robustness

The ratios of the parasitic capacitance to MM ESD robustness (C_{ESD}/V_{MM}) within UWB frequencies of all MLSCR devices were evaluated and compared in Fig. 6. According to the measured results, the C_{ESD}/V_{MM} ratios of SMLSCR and WMLSCR are about 65~115 fF/kV and 50~95 fF/kV, respectively. The C_{ESD}/V_{MM} ratios are decreased with the decrease of the P+ trigger diffusion area. The C_{ESD}/V_{MM} ratios of WMLSCR have significant decrease of about 30%, as compared with those of SMLSCR. The ratios of ΔC_{ESD} to MM ESD robustness ($\Delta C_{ESD}/V_{MM}$) of SMLSCR and WMLSCR are about 30~40 fF/kV and 20~30 fF/kV, respectively. The $\Delta C_{ESD}/V_{MM}$ ratios of WMLSCR also have significant decrease as compared with those of SMLSCR.

E. Turn-on Speed

To investigate the turn-on speed of MLSCR devices under different layout structures, the experimental setup to measure the required turn-on times of MLSCR devices is illustrated in Fig. 7. The trigger diffusions of MLSCR device were treated as the trigger port, and the trigger pulse was applied to the trigger port to turn on MLSCR device. The pulse with the amplitude of 5 V, rise time of 10 ns, and pulse width of 100 ns was applied to the trigger port. A 5-V voltage bias was connected to the anode of MLSCR through a 10-Ω resistance, which was used to limit the sudden large transient current from power supply when MLSCR device is turned on. The cathode of MLSCR device was grounded. The turn-on time of each MLSCR is defined as the time for MLSCR device to enter its low-voltage holding region and reach 2.5 V. The measured voltage waveforms on the trigger nodes and anodes of MLSCR devices under different layout structures are shown in Fig. 8. The turn-on times of SMLSCR$_1$ and WMLSCR$_1$ are 12.5 ns and 10.9 ns, respectively. The faster turn-on speed is found in WMLSCR.

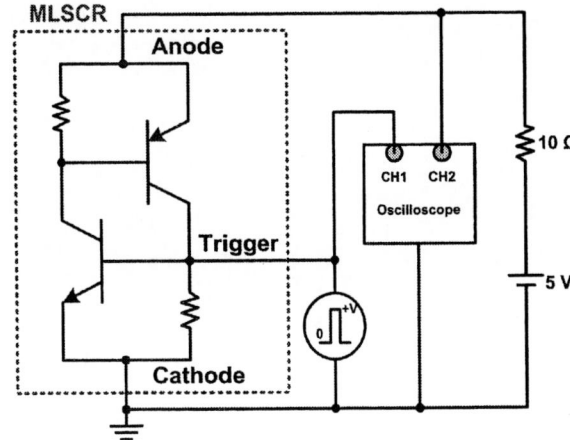

Fig. 7. Measurement setup to find turn-on time of MLSCR.

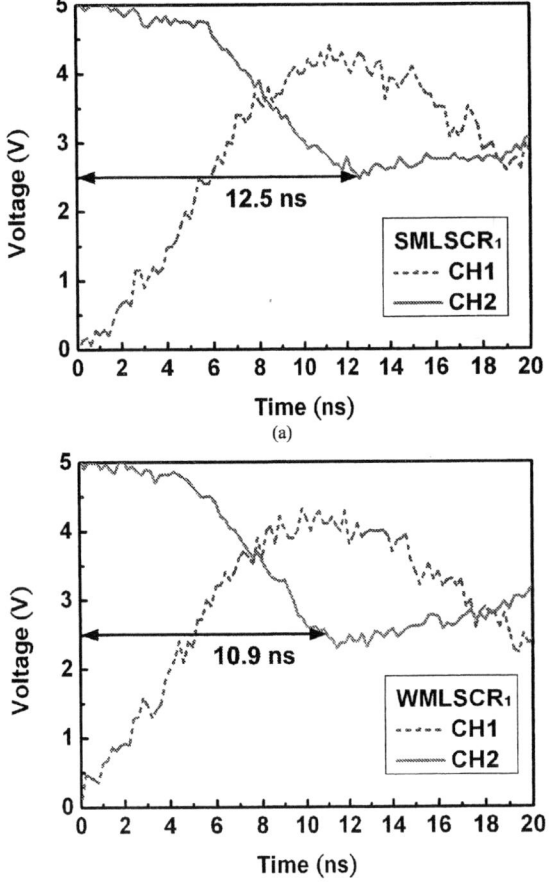

(a)

(b)

Fig. 8. Measured waveforms for turn-on time of (a) SMLSCR$_1$ and (b) WMLSCR$_1$.

The turn-on speed of MLSCR device was related to the base-emitter resistance (P-well resistance) in the NPN transistor of MLSCR device. MLSCR device with the larger P-well

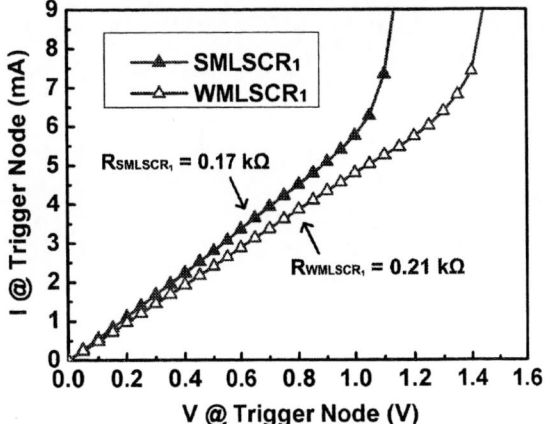

Fig. 9. DC I-V curves at the trigger nodes of MLSCR with cathode grounded.

resistance can be turned on by the smaller trigger current, which leads to the smaller turn-on time. Fig. 9 shows the DC I-V curves at the trigger nodes of MLSCR devices with cathode grounded. The larger resistance is found in WMLSCR (R_{WMLSCR_1}), which agrees with the result that the turn-on time is reduced in WMLSCR.

IV. CONCLUSION

In this work, SCR devices with the waffle layout structure have been verified to have the reduced parasitic capacitance under the same ESD robustness, and they also have been verified to have the reduced variation of the parasitic capacitance within UWB frequencies. Besides, the faster turn-on speed is found in SCR device with the waffle layout structure. Thus, SCR devices realized in the waffle layout structure are more suitable for RF ESD protection than those realized in the traditional stripe layout structure.

REFERENCES

[1] C. Grewing et al., "Fully integrated distributed power amplifier in CMOS technology, optimized for UWB transmitters," in *IEEE Radio Frequency Integrated Circuits Symp. Dig.*, Jun. 2004, pp. 87–90.

[2] M.-D. Ker, T.-Y. Chen, and C.-Y. Chang, "ESD protection design for CMOS RF integrated circuits," in *Proc. EOS/ESD Symp.*, 2001, pp. 346–354.

[3] D. Linten et al., "A 5-GHz fully integrated ESD-protected low-noise amplifier in 90-nm RF CMOS," *IEEE J. Solid-State Circuits*, vol. 40, no. 7, pp. 1434–1442, Jul. 2005.

[4] W. Soldner et al., "RF ESD protection strategies: Codesign vs. low-C protection," *J. Microelectronics Reliability*, vol. 47, no. 7, pp. 1008–1015, Jul. 2007.

[5] J.-H. Lee et al., "The embedded SCR NMOS and low capacitance ESD protection device for self-protection scheme and RF application," in *Proc. IEEE Custom Integrated Circuits Conf.*, 2002, pp. 93–96.

[6] M.-D. Ker and K.-C. Hsu, "Overview of on-chip electrostatic discharge protection design with SCR-based devices in CMOS integrated circuits," *IEEE Trans. Device Materials Reliability*, vol. 5, no. 2, pp. 235–249, Jun. 2005.

978-1-4244-2039-1/08/$25.00 ©2008 IEEE

SESSION 4:

BEOL I - METALLIZATION RELIABILITY

INVITED PAPER

Using Line-Length Effects to Optimize Circuit-Level Reliability

C.V. Thompson
Dept. of Materials Science and Engineering
MIT, Cambridge, MA, USA
Email: cthomp@mit.edu

Abstract - **By taking advantage of short-line reliability improvements in circuit-level reliability analyses and in modified layout strategies, the severe reliability constraints on future interconnect technology can be significantly addressed. To do this though, the effective length of laid-out interconnect trees must be used instead of the lengths of individual segments. Recent results on the reliability of interconnect trees are reviewed, and methods for reliability optimization are suggested.**

I. INTRODUCTION

Copper-based interconnect technology faces very difficult challenges at the 32nm node and beyond. As in the past, needed line widths will continue to decrease with time at an exponential rate, and required current densities will increase at an exponential rate. Also, the total length of interconnect per IC will continue to increase at an exponential rate. As a consequence, reliability requirements increase at an exponential rate (reliability per meter), at the same time that factors which reduce reliability will also increase exponentially. This is why the ITRS [1] indicates that no known solution is available for the requirements of the 40nm node, and difficult challenges remain for the 45nm node.

Electromigration is one of the key reliability concerns for future interconnect reliability. One route to partial relief of concerns arising from electromigration is to take advantage of the benefits of using short lines and reservoirs, implemented in circuit-level reliability assessments and through modified layout strategies [2, 3]. However, a correct understanding of the benefits of these strategies is critical for successful implementation. Incorrect use of short-length effects can not only fail to lead to reliability improvements, but can also lead to false expectations for reliability improvements, and even to reliability degradation. In this paper, recent experimental and modeling research on length and reservoir effects, as well on the reliability of interconnect trees, will be reviewed. This is done in order to provide a background for recommendations for strategies for reliability characterization, circuit-level reliability assessments, and layout strategies for improved and optimized circuit-level electromigration reliability.

II. LENGTH EFFECTS IN ISOLATED INTERCONNECT SEGMENTS

The electromigration reliability of interconnects is typically characterized using individual two-terminal segments of the type shown in Figure 1, or chains of such segments. The refractory metal liners at the base of the cathode and anode vias of these segments block electromigration through the vias, so

that as Cu is transported toward the anode via, a compressive stress develops at the anode, and a tensile stress develops at the cathode (Fig. 2). Compressive stress can lead to extrusion of Cu, leading to a short-circuit failure, while tensile stress can lead to formation and growth of voids that lead to open circuit failure (Fig. 2). A stress gradient develops in such segments due to the action of the electron wind force on the Cu atoms, F_{ew}, that causes their migration. However, as the stress gradient develops, a corresponding gradient in the chemical potential leads to a back stress force, F_{bs}, which opposes F_{ew}. The total force on Cu atoms is therefore given by

$$J_a = \frac{Dc_a}{kT}\rho\, j\, z^*q + \frac{Dc_a}{kT}\Omega\frac{\partial\sigma}{\partial x}, \qquad [1]$$

where J_a is the atomic flux, D is the diffusivity of the Cu, c_a is the concentration of Cu atoms (number/cm^3), ρ is the density of the Cu, j is the current density, z^* is the effective charge of the Cu (a measure of the momentum transfer from electrons to the Cu), q is the fundamental electronic charge ($z^*q < 0$), k is Boltzmanns constant, T is temperature, Ω is the atomic volume of Cu (cm^3/number), σ is the hydrostatic stress, and x is the distance along the interconnect segment, with $x = 0$ at the cathode. The left-hand term on the right side of equation 1 is due to F_{ew}, and the second term is due F_{bs}.

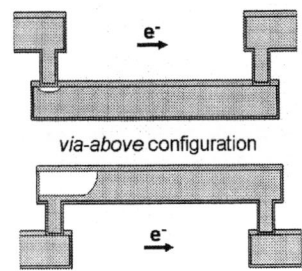

Fig. 1: Two configurations typical of electromigration test segments.

The change in σ, as a function of position x, can be described by

$$\frac{\partial\sigma}{\partial t} = \frac{\partial}{\partial x}\left[\frac{DB}{kT}\left(\Omega\frac{\partial\sigma}{\partial x} + z^*q\,\rho\, j\right)\right], \qquad [2]$$

where B is the effective elastic bulk modulus for the material

978-1-4244-2039-1/08/$25.00 ©2008 IEEE

INVITED PAPER

that surrounds the interconnect [4]. B is a function of the elastic moduli and dimensions of the refractory metal liner, the dielectric diffusion barrier, and the low-k dielectric. B can also vary along the interconnect length, especially near the anode and cathode. It is important to note the elastic bulk moduli of low-k dielectrics are substantially lower than the elastic modulus of SiO_2 (by about a factor of 10) so that B is lower for Cu/low-k than Cu/SiO_2.

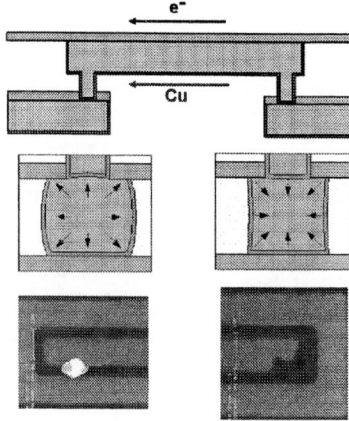

Fig. 2: Stress evolution and failure of test segments.

When the electron wind force and back stress force balance, the atomic flux goes to zero and no further evolution in σ occurs. This condition is achieved when

$$\rho \; j \; z^* q = -\Omega \frac{\partial \sigma}{\partial x} . \qquad [3]$$

In this state, σ is a linear function over x so that

$$(j \; L)_{Blech} = \Omega \frac{\Delta \sigma}{z^* q \rho} , \qquad [4]$$

where L is the length of the segment and $\Delta\sigma$ is the difference in the hydrostatic stress at the anode and cathode. When the product of j and L fall the below the critical value indicated by the left-hand-side of equation [4], electromigration will stop. If this happens before the critical compressive stress required for failure at the anode, σ_a, or the critical tensile stress required for failure at the cathode, σ_c, is reached, the segment will be immune to failure and hence 'immortal'. This effect, and the corresponding immortality condition, is generally referred to as the Blech effect [5] or the Blech immortality condition.

If a void forms it must grow to a critical size, V_{fail}, before it causes failure. It should be noted that V_{fail} depends on the shape and location of the void, and can vary from line to line. However, V_{fail} is always finite. As a consequence, even if $(jL)_{Blech}$ is reached, failure will not occur until V_{fail} is reached. There will be a resistance increase, ΔR, that is associated with void growth. In short lines, a steady state may be reached before ΔR increase reaches an unacceptable value, ΔR_{fail}. In this case, void growth will saturate, and electromigration will cease.

There is therefore a critical jL product below which immortality is achieved due to void growth saturation. This condition is given by [6, 7]

$$(jL)_{sat} < \frac{\rho/A}{\rho_l/A_l} \frac{\Delta R_{fail}}{R} \frac{2\Omega \; B}{q \; \rho \; z^*} , \qquad [5]$$

where ρ/A and ρ_l/A_l, correspond to the ratio of the resistivity to the cross-sectional area of the Cu and the liner, respectively, and R is the initial resistance. It is important to note that $(jL)_{sat} > (jL)_{Blech}$.

There is a significant asymmetry in the reliability of segments with via-above and via-below configurations (Fig. 1) [8]. This arises from the difference in the average V_{fail} for these to structures. In the via-above configuration, the dielectric diffusion barrier does not provide a path for shunting of current around a void, so even a very low-volume void, if placed directly underneath the via, can cause failure (Fig. 1). The value for $(jL)_{sat}$ is therefore lower for via-above segments than for via-below segments. Experimentally determined values of the critical jL product for immortality, $(jL)_{crit}$, include 2100A/cm [9] and 1500A/cm [10] for Cu/SiO_2 via-above segments, 3700A/cm for Cu/SiO_2 via-below segments [11], and 375A/cm for Cu/low-k via-above segments[12]. Note that, as expected, $(jL)_{crit}$ is lower for via-above configurations than via-below configurations, and that it is lower when a low-k (low-stiffness) dielectric is used instead of SiO_2.

A $(jL)_{crit}$ value of 375A/cm indicates a critical length of only 1.5μm at $2.5 \times 10^6 A/cm^2$, typical of test conditions, but 7.5μm at use conditions of 5×10^5 A/cm^2. A large fraction of interconnect segments in an integrated circuit fall below this length, even when stressed at the maximum allowed current density. Under more typical current-stress conditions, the vast majority of segments are, in principle, immune to electromigration failure at service conditions [13]. Allowing for this in circuit-level reliability assessments, and optimizing for this during circuit layout therefore offers significant relief from severe reliability constraints. However, this must be done with knowledge of the effects of linking of segments.

III. THE RELIABILITY OF INTERCONNECT TREES

In assessing circuit-level interconnect reliability it is critical to recognize that the reliability of a segment is strongly dependent on whether it is linked to other segments (without an intervening barrier to electromigration), and on the stress conditions in those linked segments. The appropriate fundamental reliability unit for circuit-level reliability analyses is the interconnect *tree*. An interconnect tree is any collection of interconnect segments that are linked without diffusion barriers between them (therefore typically linked within one layer of metallization). Examples of interconnect trees are shown in Figure 3. It has been shown theoretically [2, 3] that an effective critical jL product, $(jL)_{crit,eff}$, can be defined for an interconnect tree and can be applied to trees, based on values of $(jL)_{crit}$ determined experimentally for isolated two-terminal segments. This has been confirmed in experiments on Al [14] and Cu [15],

INVITED PAPER

though the original approach [2, 3] must be modified for the effects of inactive segments [15]. Therefore, immortal trees can be identified and accounted for in circuit-level reliability analyses [13]. When this is done, it is still found that the vast majority of interconnect trees in an IC are immune to electromigration-induced failure at current service conditions [13]. This offers significant relief from reliability constraints.

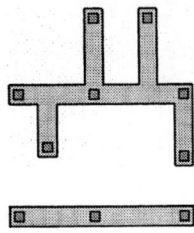

Fig. 3: Examples of interconnect trees.

When interconnect trees have effective jL products, $(jL)_{eff}$, that are near $(jL)_{crit,eff}$, while they are mortal, their reliability is significantly improved over trees with high values of $(jL)_{eff}$. This is illustrated by the experimental data shown for simple three-terminal trees in Figures 4 and 5 [16]. There are two important things to note about these results:

1) The reliability of the left-hand segment can be improved or degraded by the stress conditions in the linked segment, with improvements occurring when the linked segment serves as a reservoir for Cu atoms (Fig. 4) and degradation occurring when the linked segment serves as a sink for Cu atoms (Fig. 5).

2) The positive effects of reservoirs and the negative effect of sinks exist even if there is no current in the linked segment (when the linked segment is 'inactive').

It has also been demonstrated experimentally that the reliability improvements associated with reservoirs are reduced and the negative effects of sinks are increased when going from Cu/SiO$_2$ [17] to Cu/low-k [16] structures. This can be quantitatively understood to be a consequence of the lower stiffness of the low-k dielectric. There will therefore be a significant overall negative effect that will increase as the dielectric constant, and therefore the stiffness (B), of low-k dielectrics is further reduced in the future [16].

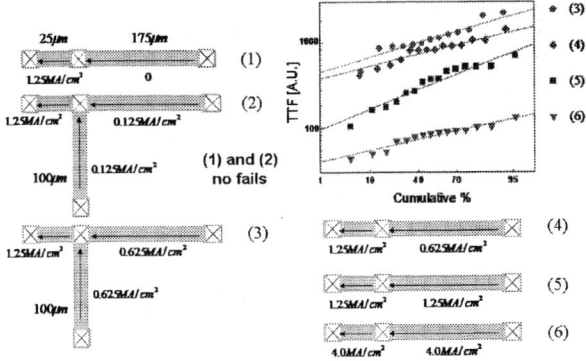

Fig. 4: Cu/low-k trees stressed so that the right-hand segments serve as atomic reservoirs [17].

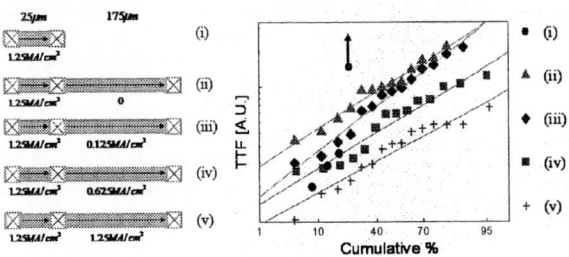

Fig. 5: Cu/low-k trees stressed so that the right-hand segments serve as atomic sinks [17].

Another consequence of the replacement of SiO$_2$ with low-stiffness low-k materials is that compressive failures (extrusions) become more likely [18]. Figure 6 shows such an extrusion. It has been shown that the critical stress at which an extrusion will occur is lower (and extrusion failure more likely) when interconnects are wide and sparsely-packed. Extrusion failures are also favored by interlevel dielectrics with low stiffness and thin liners, both of which are needed in future interconnect systems [18].

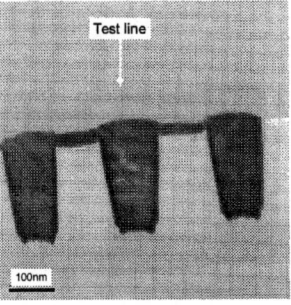

Fig. 6: An extrusion failure in a Cu/low-k tree [18].

IV. CONCLUSIONS: STRATEGIES FOR RELIABILITY OPTIMIZATION

From the results outlined above, several strategies for optimization of circuit-level reliability emerge.

- First, at-risk trees can be identified at the layout stage [2, 3, and 13], so that their $(jL)_{eff}$ can be reduced through layout modifications. This can be done by un-linking large trees to create smaller trees by passing part of the larger ones through a diffusion barrier (e.g. a via).

- Second, linking to active or inactive segments that serve as reservoirs can significantly improve the reliability of at-risk segments. However, this can only be done if the direction of current flow is fixed so that the linked segments do not function as sinks.

- Third, the reliability of at-risk segments will be lower if they can be laid out in via-below configurations.

- Fourth, the risk of extrusion failures can be minimized through the use of narrow lines (when possible) and by placement other Cu features nearby (even if inactive).

978-1-4244-2039-1/08/$25.00 ©2008 IEEE   65   *Proceedings of 15th IPFA - 2008, Singapore*

INVITED PAPER

- Fifth, whenever possible the effective stiffness of the material surrounding an anode or cathode via should be maximized, and the liner thickness minimized.

ACKNOWLEDGEMENTS

This work was sponsored by the Semiconductor Research Corporation. The MIT and NUS-based work described here is the product of hard-working and creative students.

REFERENCES

1. International Technology Roadmap for Semiconductors, (http://www.itrs.net/), 2006 update.
2. S.P. Riege and C.V. Thompson, *A Hierarchical Reliability Analysis for Circuit Design Evaluation*, IEEE Transactions on Electron Devices **45**, 2254 (1998).
3. J.J. Clement, S.P. Riege, R. Cvijetic, and C.V. Thompson, *Methodology for Electromigration Critical Design Rule Evaluation*, IEEE Trans. on CAD of Integrated Circuits and Systems, **18**, 576 (1999).
4. M.A. Korhonen and P. Borgesen, K.N. Tu, and C.-Y. Li, "Stress evolution due to electromigration in confined metal lines," J. Appl. Phys. **73**, 3790 (1993).
5. I.A. Blech, "Eelctromigration in thin aluminum films on titanium nitride," J. Appl. Phys. **47**, 1203 (1976).
6. V.K. Andleigh, V.T. Srikar, Y.T. Park, and C.V. Thompson, *Mechanism Maps for Electrmigration-Induced Failure of Metal and Alloy Interconnects*, J. Appl. Phys. **86**, 6737 (1999).
7. Z. Suo, "Stable state of interconnect under temperature change and electric current," Acta Mater. **46**, 3725 (1998).
8. C.L. Gan, C.V. Thompson, K.L Pey, W.K. Choi, H.L. Tay and M.K. Radhakrishnan, *Effect of Current Direction on the Lifetime of Different Levels of Cu Dual-Damascene Metallization*, Appl. Phys. Letters **79**, 4592 (2001).
9. S. P. Hau-Riege, Probabilistic immortality of Cu damascene interconnects," *J. Appl. Phys.* **91**, 2014 (2002).
10. C. S. Hau-Riege, A. P. Marathe, and V. Pham, "The Effect of Line Length on the Electromigration Reliability of Cu interconnects," in Proc. of Advanced Metallization Conference, p. 169, 2002.
11. K. D. Lee, E. T. Ogawa, H. Matsuhashi, P. R. Justison, K. S. Ko, and P.S. Ho, *Appl. Phys. Lett.* **79**, 3236 (2001).
12. C. S. Hau-Riege, A. P. Marathe, and V. Pham, "The effect of low-k ILD on the electromigration reliability of Cu interconnects with different line lengths," in Proc. of the 41st International Reliability Physics Symposium, pp. 173-177, 2003.
13. S.M. Alam, C.L. Gan, F.L. Wei, C.V. Thompson, and D.E. Troxel "Circuit-Level Reliability Requirements for Cu Metallization", IEEE Transactions on Device and Materials Reliability **5**, 522 (2005).
14. S.P. Hau-Riege and C.V. Thompson, *Electromigration-Saturation in a Simple Interconnect Tree*, J. Applied Physics **88**, 2382-85 (2000).
15. C.W. Chang, Z.S. Choi, C.V. Thompson, C. L. Gan, K. L. Pey W. K. Choi, and N. Hwang, *Electromigration resistance in a short three-contact interconnect tree*, J. Appl. Phys. **99**, 094505 (2006).
16. F.L. Wei, C.S. Hau-Riege, A.P. Marathe, and C.V. Thompson, "Effects of active atomic sinks and reservoirs on the reliability of Cu/low-k interconnects," J. Appl. Phys. **103**, 084513 (2008).
17. C.L. Gan, C.V. Thompson, K.L. Pey and W.K. Choi, "Experimental Characterization and Modeling of the Reliability of 3-Terminal Dual-Damascene Cu Interconnect Trees," J. Appl. Phys. **94**, 1222 (2003).
18. F.L. Wei , C.L. Gan and T.L. Tan, C.S. Hau-Riege and A.P. Marathe, J.J. Vlassak, and C.V. Thompson, "Electromigration-Induced Extrusion Failures in Cu/low-k Interconnects," , to appear in J. Appl. Phys.

Statistical Modeling of Via Redundancy Effects on Interconnect Reliability

Nagarajan Raghavan[♦] and Cher Ming Tan[♣]
[♦] Singapore-MIT Alliance (SMA), National University of Singapore (NUS)
4 Engineering Drive 3, Singapore – 117576.
[♣] School of EEE, Nanyang Technological University (NTU)
Block S2, Nanyang Avenue, Singapore – 639798.
Phone[♣]: (65) 67904567 Fax[♣]: (65) 67920415 Email[♦]: g0702024@nus.edu.sg

Abstract – **Electromigration is an important failure mechanism in the nano-interconnects of modern IC technology. Various approaches have been investigated to prolong the lifetime of an interconnect. One such approach is to have an in-built redundancy in the via structures of the interconnect. The presence of redundant via in a parallel topology helps improve the overall reliability of the via structure. Although reliability improvement due to via redundancy is qualitatively understood, it is necessary to quantify the improvement in reliability through statistical models so that the improvement in lifetime as a result of redundancy can be quantified. A statistical model that incorporates the effects of redundancy is developed in this study and it is used to estimate the reliability of redundant via structures. The Cumulative Damage Model (CDM) is used in conjunction with the Maximum Likelihood Estimate (MLE) method to assess the reliability of load sharing via redundant structures in this study.**

I. INTRODUCTION

As the integrated circuit technology undergoes continued downscaling in accordance to Moore's law to achieve improvements in device and circuit performance and to miniaturize electronic products, many reliability issues have become very critical to the long-term performance of these nanodevices. One of these critical failure mechanisms affecting reliability is electromigration (EM) in the back-end interconnect lines [1].

Electromigration refers to the current-induced atomic flux due to momentum exchange between the electrons and atoms that causes atomic flux from the cathode to the anode terminal. There are various driving forces causing EM some of which include the electron wind force, stress gradients due to hydrostatic stress variations in the interconnect as a result of the thermal coefficient mismatch between various materials in the interconnect structure induced by high temperature process conditions, back flow stresses as well as temperature gradients that arise as a result of the Joule heating effect [2]. Increasing atomic accumulation at the anode causes compressive stresses and depletion of atoms at the cathode leads to vacancies which coalesce to form voids. A steady state is achieved when the atomic fluxes induced by the various driving forces sum up to zero which results in a time-invariant stress profile provided that the peak stresses at steady state are lower than the critical stresses required for cathode void nucleation [2].

The most critical element of an interconnect structure is the via which connects different levels of metallization. The via which has a smaller cross-section than the lines is subject to higher current densities causing it to fail sooner than the lines [3]. In order to improve the overall via reliability, redundancy is incorporated into the via structures by having more than one via connected in parallel in a load sharing configuration so that the multiple via share the current flow. This active load sharing redundancy illustrated in Fig 1 reduces the current density load per via when all the via are operating thereby prolonging their time to failure. After one via fails, subsequent via experience higher current density stresses by sharing the extra current load which the failed via was previously subjected to causing them to then fail sooner. On the whole, via redundancy effects help achieve substantial improvements in interconnect reliability. This is especially the case in Cu dual damascene systems where the via are also made of Cu unlike Al interconnect technology where the via is made of Tungsten (W) which has an intrinsically high electromigration resistance [2].

Having fabricated redundant via structures, it is necessary to quantify the improvement in reliability as a result of the redundancy incorporated. This requires the use of concrete statistical models [4]. Although some literatures in the past have assessed the reliability of redundant via structures [5] – [7], very few have modeled it from a statistical perspective. The statistics describing reliability of redundant via structures is complicated because the nature of degradation cannot be described by a single distribution. As an example, for a two-via system, the failure distribution of both via prior to any one of them failing is different from the failure distribution of one of the via after the other has failed. Therefore, given the different stresses experienced by the via during its lifetime, the distribution of the via elements and the via system need to be modeled in a statistically precise manner.

In this study, we use a robust technique known as the *cumulative damage model* (CDM) [4] in conjunction with the conventional *maximum likelihood estimate* (MLE) [8] method to account for the time-varying stress profile [9] of the via elements during accelerated life test of a via redundant system. Based on these statistical tools the reliability of each "via element" is first estimated. This is followed by an estimation of the reliability of the "via system" based on the "via element" reliabilities accounting for the effect of load sharing in the system.

The structure of this paper is organized as follows. Section II introduces the via load sharing redundant system and analyzes the typical stress profile that the via elements could be subjected to during the accelerated stress tests. Section III presents the cumulative damage model (CDM) along with the life-stress relationship used for accelerated testing of via test structures. Section IV develops the likelihood expression for the via elements based on the CDM model developed. This expression is then optimized based on the maximum likelihood estimate (MLE) method. The values of the statistical distribution parameters which optimize the likelihood function are then used to evaluate the reliability of each via element. In Section V, the reliability of the overall via system is evaluated accounting for the effect of load sharing redundancy. Section VI presents some simulation results obtained based on the theory prescribed above. Finally, the last section concludes with an assessment of the statistical model used and the assumptions it is based upon.

II. STRESS PROFILE IN A LOAD SHARING REDUNDANT SYSTEM

The system in Fig 1 is an *active load sharing redundancy* system wherein the operating via elements share the current density flow equally at time t = 0. The value of current density flowing through each via with time, j(t), depends on the number of unfailed via in the load sharing system, their relative resistance degradation trends and their relative void growth rates due to the EM phenomenon.

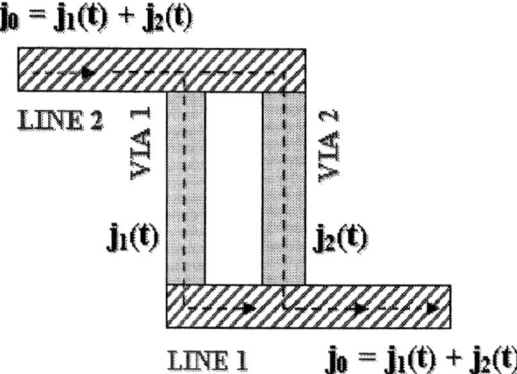

Fig. 1. Load sharing via redundant system for a constant overall current density stress of $j_0 = 2$ MA/cm^2. The current density stresses through the via elements VIA 1 and VIA 2 depend on the relative resistance degradation of the via elements and the relative rates of the EM induced void growth.

Assuming the resistance of via elements VIA 1 and VIA 2 to be $r_1(t)$ and $r_2(t)$ and given the constant overall current density stress of $j_0 = 2$ MA/cm^2, the current splits into $j_1(t)$ and $j_2(t)$ based on the current divider principle as given by (1) and (2). As the resistance of the via elements degrade at different rates and as their voids grow at different rates reducing the effective cross-sectional area for current flow, the current density stress

changes as a function of time in both the via elements. The total current density stress however remains fixed at $j_0 = 2$ MA/cm^2.

$$j_1(t) = \left[\frac{r_2(t)}{r_1(t) + r_2(t)} \right] \cdot j_0 \qquad (1)$$

$$j_2(t) = \left[\frac{r_1(t)}{r_1(t) + r_2(t)} \right] \cdot j_0 \qquad (2)$$

The typical stress profile encountered by a via element in the redundant via system is given by Fig 2. This stress profile indicates a *time varying current density stress* in the via elements and this effect of continuous time varying stress is accounted for in the CDM model. The CDM model considers the cumulative effect of stresses experienced by an element through its lifetime up to failure.

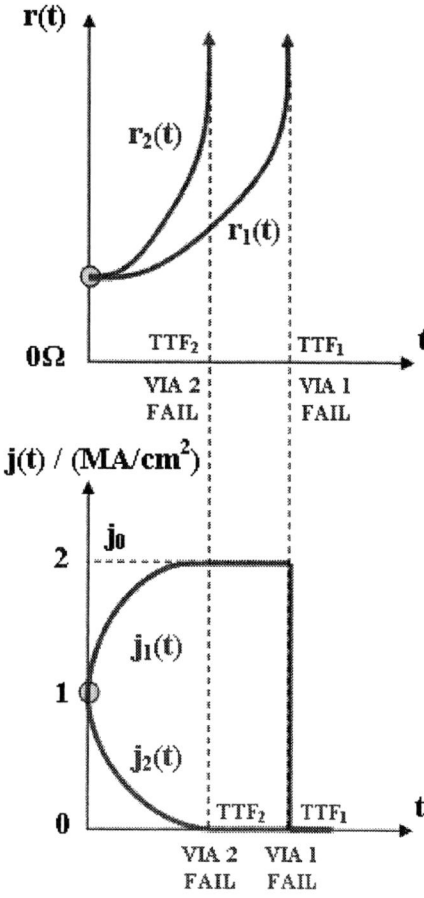

Fig. 2. Typical profiles of resistance degradation and the resulting current density stresses in a two-via redundant load sharing test structure subjected to a constant overall current density stress of $j_0 = 2$ MA/cm^2. The current density profile in the via elements is given by Eq. (1) and Eq. (2). In this case, VIA 2 degrades faster and fails open first at t = TTF$_2$ following which VIA 1 takes in all the current causing it to degrade faster and eventually fail at t = TTF$_1$. The complete via system is considered to have failed at t = TTF$_1$.

III. CUMULATIVE DAMAGE MODEL

For accelerated EM life tests at a given interconnect temperature, we model the scale parameter of an element to depend on the current density stress level experienced by it using the inverse power law life-stress relationship given by (3) where L refers to the scale parameter of the distribution [4]. Eq. (3) suggests that for an instantaneous stress, j(t), at a time interval t to (t + Δt), the failure distribution of the unfailed via elements at time t is represented by a scale parameter, L(t) corresponding to j(t). Therefore, for every instantaneous time interval Δt, the scale parameter, L(t), is different since the stress level and its associated failure distribution at that Δt time interval is different. Based on this scenario, the scale parameter is modeled as a time-varying function that depends on the time-varying stress as given by (3).

Since the process of void nucleation and growth in every via is gradual, the failure of the via element is well represented by the Lognormal distribution [10]. Therefore, the scale parameter (L) in (3) refers to the median life, t_{50}, for the Lognormal distribution..

It is to be noted that although current density is the only stress factor considered, the varying current densities across the redundant vias are likely to result in different local temperature stresses as a result of the Joule heating effect. In this work, we ignore the non-uniformity in the temperature of the redundant vias as a result of Joule heating and assume that current density is the only variable stress factor.

$$L[t, j(t)] = \frac{1}{K \cdot [j(t)]^n} \qquad (3)$$

Based on (3), the via element reliability may be expressed by (4) accounting for the stress dependence and its variation with time. The integral in (4) denotes the cumulative effect of the time-varying current density stress damage experienced by the via element. The parameter σ is the shape parameter of the Lognormal distribution and it is assumed to be constant for all stress conditions since it is indicative of the failure mechanism [11] which is assumed to remain the same for all applied and field stress conditions.

$$R[t, j(t)] = 1 - \Phi\left(\frac{1}{\sigma} \ln\left(\int_0^t \frac{dt}{L[t, j(t)]}\right)\right) \qquad (4)$$

IV. MAXIMUM LIKELIHOOD ESTIMATE

Having modeled the reliability, R(j, t), as a function of time and the time-varying current density stress level, the log-likelihood function of every via element may be expressed by (5) where $T_{F,i}$ represents the time to failure of the ith via element, $T_{S,j}$ is the censor or suspension time of the jth via element, κ is a constant and LKL is the log-likelihood function [8]. In (5), f(t) is the probability density function which is given

by $-dR(t)/dt$. F denotes the number of via element failures observed while S refers to the number of censored units. Note that we have so far been analyzing individual via elements in the redundant via structure. The quantity (F + S) is the total number of redundant via structure units tested. The failure and suspension times are measured for every via element in the (F + S) via structures that are subjected to the accelerated test.

$$LKL = \sum_{i=1}^{F} \log\left(f\left(T_{F,i}\right)\right) + \sum_{j=1}^{S} \log\left(R\left(T_{S,j}\right)\right) + \kappa \qquad (5)$$

Given the log-likelihood function, the set of parameters $\{\sigma, K, n\}$ that optimize the log-likelihood function may be found using various optimization techniques such as the Quasi-Newton method or the global optimization simulated annealing algorithm. The equations representing the optimization problem are given by (6).

$$\frac{\partial LKL}{\partial \sigma} = \frac{\partial LKL}{\partial K} = \frac{\partial LKL}{\partial n} = 0 \qquad (6)$$

Based on Eq. (6), the parameters $\{\sigma, K, n\}$ for every via element may be obtained and the individual via element reliability functions are fully described. Having quantified the reliability functions of the individual via elements, the reliability of the overall redundant via structure needs to be determined.

V. LOAD SHARING SYSTEM RELIABILITY

The "system reliability" of the load sharing redundant via structure for a simple two-via system may be expressed as in (7) where the expressions for $R_{1\&2}$, $R_{1/2}$ and $R_{2/1}$ are given by (8), (9) and (10) respectively. The expression in (7) implies that a two-via system could be operating under three conditions. Either both the vias are functioning or one of them has failed while the other is functional. Equation (8) denotes the system reliability when both VIA 1 and VIA 2 are functional. Equations (9) and (10) refer to the system reliability under cases of VIA 2 functioning while VIA 1 failure and VIA 1 functioning while VIA 2 failure respectively.

$$R_{system}(t, S) = R_{1\&2}(t, S) + R_{1/2}(t, S) + R_{2/1}(t, S) \qquad (7)$$

$$R_{1\&2}(t, S) = R_1(t, S_1) \cdot R_2(t, S_2) \qquad (8)$$

$$R_{1/2}(t, S) = \int_0^t f_1(x, S_1) \cdot R_2(x, S_2) \cdot \frac{R_2(t_{1e} + (t-x), S)}{R_2(t_{1e}, S)} dx \qquad (9)$$

$$R_{2/1}(t, S) = \int_0^t f_2(x, S_2) \cdot R_1(x, S_1) \cdot \frac{R_1(t_{2e} + (t-x), S)}{R_1(t_{2e}, S)} dx \qquad (10)$$

In (5) – (8), S is the total current density stress the via system is subjected to and S_1 and S_2 are the corresponding fractions of the total stress that VIA 1 and VIA 2 experience respectively.

The parameter t_e denotes the equivalent operating time of an element if it had been operating at a different stress level.

VI. STATISTICAL ANALYSIS OF REDUNDANT VIA SYSTEM

To illustrate the application of the above theory, a sample of test data from an electronic device with built-in redundancy was obtained [4]. This set of data is assumed to hold true for the two-via EM test structure examined here. The overall current density stress is taken to be 2 MA/cm^2 while the temperature during the stress test = 300^{0}C. Although current density and temperature are both acceleration factors in general for any EM test, the temperature stress of the line is kept fixed and joule heating induced temperature changes are also ignored making current density the only acceleration factor of focus. The failure data and the predicted stress profile, shown in Fig 2, for both the via elements in the via system are used and statistical analysis is then performed by optimizing the log-likelihood function in (5).

As a first attempt to model the impact of via redundancy statistically, for the sake of illustration and simplicity, we assume that the reliability functions for the two via elements are similar and therefore the failure data for these two via are treated collectively. However, in actual test conditions, the two vias will not be identical since they have different boundary conditions. Optimization of the log-likelihood function results in the values for σ = 0.7744 and n = 0.8072. The Lognormal probability plot for the via element in the presence of a load sharing redundancy is shown in Fig 3. The plot reveals a good lognormal fit to the tested failure data.

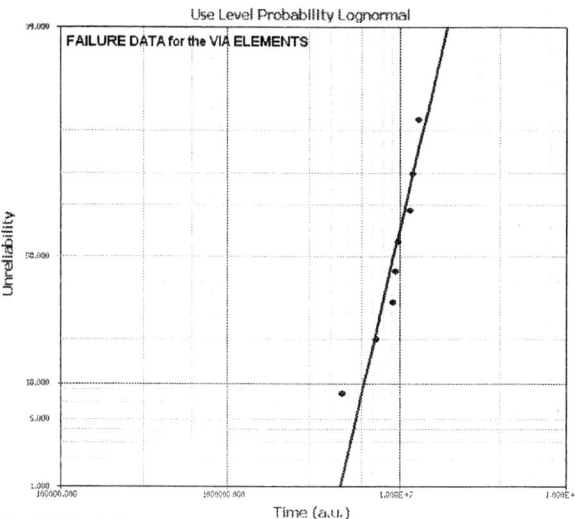

Fig. 3. Lognormal probability plot for the via elements at the use stress level of 0.5 MA/cm^2.

Having characterized the reliability of the individual via elements, we may determine the reliability of the "via system" for any given stress condition by using (7) – (10) where the stresses S_1 and S_2 are expected to be changing with time depending on the resistance degradation profiles of the

individual via elements. Assuming for illustrative purpose that the two via elements each carry 50% of the total current load, the system reliability curve for the overall load sharing system at the field operation stress of 0.5 MA/cm^2 is given in Fig 4. Since the reliability of the via elements degrade gradually, the overall resistance degradation behavior of the via system is also expected to be gradual and hence via redundant system may also be well represented by the Lognormal statistics.

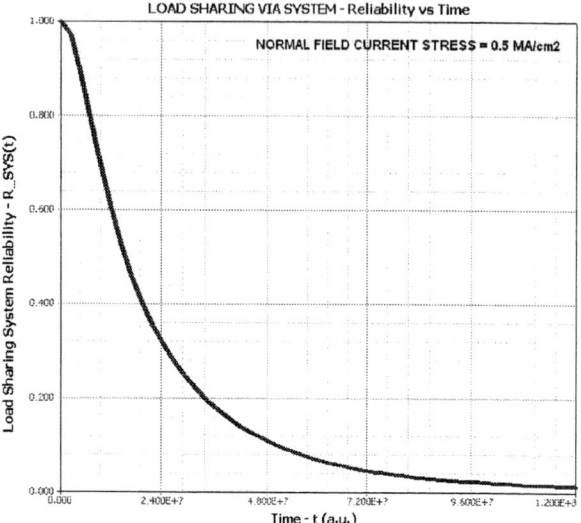

Fig. 4. Reliability function of the overall load sharing via redundancy system for a field stress of 0.5 MA/cm^2 and assuming that both the via elements share 50% of the load for all times up to failure.

VII. CONCLUSION

The novelty and usefulness of the CDM model has been illustrated in this work highlighting the theory involved and the way it accounts for the accumulated damage as a result of time varying stresses. Although the CDM model may not be necessary in the case of single via or single EM line tests where the current density stress is bound to be constant throughout, it is a very useful approach to model redundant via systems wherein the current densities through the individual via elements is bound to change depending on the relative resistance degradation rates of the load sharing via elements. The approach presented in this work may be further extended to analyze and quantify the improvement in reliability that may be observed as the number of via in the EM structure is increased. This is the first work of its kind that explicitly models the impact of via redundancy using a reliability block diagram (RBD) approach.

Although the CDM model appears robust and convenient for use, there are a few inherent assumptions [12] that it is based upon and it is important to take note of these. The theory we have used thus far only applies to the case if there is a single failure mechanism present. Moreover, the model does not account for failures that could occur during sudden instantaneous changes in stress levels, if any, as is the case for step-stress tests. Lastly, the presented version of the CDM model is not capable of accounting for small cyclic changes in the stress levels about a given mean stress which is often the case during fatigue.

It is hoped that this work serves as motivation for further statistical modeling and analysis into the effects of via redundancy. Further research work is under way to consider the effect of multiple failure mechanisms on the CDM model since most EM structures are subjected to bimodal failure distributions and also account for the effect of joule heating while assessing the statistical reliability of redundant via systems.

ACKNOWLEDGMENT

The authors would like to thank the *Nanoelectronics Design Lab, School of EEE, Nanyang Technological University (NTU), Singapore* for access to *Reliasoft® Inc.* software tools that were used to perform the reliability data analysis simulations in this work.

REFERENCES

[1] International Technology Roadmap for Semiconductors (ITRS), Interconnects, (2007).

[2] Tan, C.M. and Roy, A., "Electromigration in ULSI Interconnects", *Materials Science and Engineering R: Reports*, Vol. 58, Issues 1-2, pp.1-75, (2007).

[3] Gill, J., Sullivan, T., Yankee, S., Barth, H. and von Glasow, A., "Investigation of via-dominated multi-modal electromigration failure distributions in dual damascene Cu interconnects with a discussion of the statistical implications", *IEEE International Reliability Physics Symposium*, pp.298-304, (2002).

[4] Mettas, A. and Vassiliou, P., "Application of Quantitative Accelerated Life Models on Load Sharing Redundancy", *Annual Reliability & Maintainability Symposium*, pp.293-296, (2004).

[5] Christiansen, C.J., Li, B., Gill, J., Filippi, R. and Angyal, M., "Via-depletion electromigration in copper interconnects", *IEEE Transactions on Device and Materials Reliability*, Vol. 6, No. 2, pp.163-168, (2006).

[6] Gan, C.L., Thompson, C.V., Pey, K.L., Choi, W.K., Tay, H.L., Yu, B. and Radhakrishnan, M.K., "Effect of current direction on the lifetime of different levels of Cu dual-damascene metallization", *Applied Physics Letters*, Vol. 79, No. 27, pp.4592-4594, (2001).

[7] Lin, M.H., Lin, Y.L., Chang, K.P., Su, K.C. and Wang, T., "Copper interconnect electromigration behavior in various structures and precise bimodal fitting", *Japanese Journal of Applied Physics*, Vol. 45, No. 2A, pp.700-709, (2006).

[8] Jiang, S. and Kececioglu, D., "Maximum Likelihood Estimates, from Censored Data, for Mixed-Weibull Distributions", *IEEE Transactions on Reliability*, Vol. 41, No. 2, (1992).

[9] Mettas, A. and Vassiliou, P., "Modeling and Analysis of Time-Dependent Stress Accelerated Life Data", *Annual Reliability & Maintainability Symposium*, pp.343-348, (2002).

[10] Tan, C.M., Raghavan, N. and Roy, A., "Application of Gamma Distribution in Electromigration for Submicron Interconnects", *Journal of Applied Physics*, Vol. 102, 103703, (2007).

[11] Tan, C.M., Roy, A., Vairagar, A.V., Krishnamoorthy, A. and Mhaisalkar, S.G., "Current crowding effect on copper dual damascene via bottom failure for ULSI applications", *IEEE Transactions on Device & Materials Reliability*, Vol. 5, No. 2, pp.198-205, (2005).

[12] Nelson, W., "Accelerated Testing : Statistical Models, Test Plans and Data Analyses, *John Wiley & Sons.*, (1990).

Effects of Pulsed Current on Electromigration Lifetime

M. K. Lim[1,2,#], C. L. Gan[1], T. L. Tan[2], Y. C. Ee[2], C. M. Ng[2], B. C. Zhang[2] and J. B. Tan[2]

[1]School of Materials Science and Engineering, Nanyang Technological University, 50 Nanyang Avenue, Singapore 639798
[2]Chartered Semiconductor Manufacturing Ltd., 60 Woodlands Industrial Park D, Street 2, Singapore 738406
[#]Phone: (65) 6790 4142   Fax: (65) 6790 9081   Email: X060016@ntu.edu.sg

Abstract – **Asymmetrical Cu interconnect structure, where one end of the metal-2 (M2) test line is connected to M1 while the opposite end is connected to M3, was subjected to very long periods of bipolar pulsed current (i.e. 2, 16 and 48 hours) in this study. The median-time-to-failure (t_{50}) of the samples was found to depend on the direction of electron current in the first half-period, and t_{50} of samples that were subjected to direct current (D.C.) that flow upstream or downstream. Bipolar pulsed current stressed samples showed improvement in lifetimes as compared to that of D.C. stressed samples only when the half-period of bipolar pulsed current was shorter than the t_{50} of D.C. stressed samples.**

I. INTRODUCTION

Integrated circuit chips containing copper (Cu) interconnects were introduced in 1998 [1] to alleviate the trend of increasing wiring resistance due to interconnect scaling. Besides having lower resistivity than aluminum (Al), Cu was chosen as the material of choice for future interconnects because of its higher resistance to electromigration [2]. As scaling proceed on to achieve higher interconnect density, the current density carried by Cu interconnects will also increase. Although higher current density will consequentially lead to shorter electromigration lifetime, as predicted by Black's equation [3] for similar process, the ITRS targets a lower failure-in-time (i.e. better reliability) for future interconnect technology [1]. In addition, the goal of achieving better reliability for future interconnect technology is expected to be even more challenging with the incorporation of low-κ and ultra low-κ materials as intermetal dielectrics, since these mechanically weaker materials aggravate electromigration reliability [4].

While shorter electromigration lifetime is expected under conditions of higher current density and low-κ intermetal dielectrics, reliability studies on the electromigration lifetimes of interconnects under these conditions might report erroneous shorter lifetimes. This is because interconnects are commonly subjected to the most severe form of current stress, i.e. direct current (D.C.), during reliability studies, while most interconnects carry pulsed current, which is less severe, during field operations. Reliability studies [5,6] have shown that Al interconnects exhibit longer lifetime when carrying pulsed current, as compared to those carrying direct current (D.C.). Electromigration lifetime was observed to increase with frequency when Al interconnects were carrying bipolar pulsed current [6]. Although longer lifetimes are also expected from Cu

interconnects when they carry pulsed current, however, substantial information on the electromigration lifetime of Cu interconnects subjected to pulsed current stress is either lacking or limited currently. In this paper, we discuss on the lifetime enhancement when bipolar pulsed current was applied to asymmetrical Cu/low-κ interconnect structure which spans three metal levels (Fig. 1).

II. EXPERIMENTAL DETAILS

The Cu/SiCOH interconnect test structure used in this study was fabricated using a 65 nm CMOS process, which employs single damascene process at metal-1 (M1) and dual damascene process at metal-2 (M2) and metal-3 (M3). The Cu test line, which is in M2, is 200 μm long and 0.3 μm wide. One end of the Cu test line is connected to M1 while the other end is connected to M3, where both M1 and M3 are broad lines whose widths are a few microns wide. The asymmetrical Cu interconnect test structure is illustrated schematically in Fig. 1.

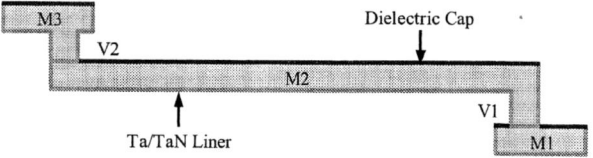

Fig. 1. Schematic diagram of Cu interconnect test structure.

Electromigration test was carried out on ceramics packaged samples using a Xpeqt electromigration test system. Bipolar pulsed current with a duty cycle of 50% and different periods of 2, 16 and 48 hours were applied to the samples in this study. Accelerated test was carried out by subjecting the samples to a temperature of 350°C and bipolar pulsed current whose minimum and maximum current density were -2.0 MA/cm² and 2.0 MA/cm², respectively. A group of 13 to 15 samples were tested for each period. The electromigration induced resistance changes in the samples were measured using Kelvin connection. Baseline direct current (D.C.) electromigration tests were also carried out by subjecting similar samples to a current density of 2.0 MA/cm² and a temperature of 350°C. Failure is defined as a 10% increase in resistance.

III. ELECTROMIGRATION INDUCED RESISTANCE EVOLUTION

A. Bipolar Pulsed Current with 48 Hours Period

The typical resistance evolution obtained from a test sample that was subjected to bipolar pulsed current with 48 hours period is shown in Fig. 2. The sample was effectively subjected to a D.C. electromigration test where electron current was flowing upstream (i.e. M1→M2→M3) during the first half-period (i.e. $0^{th} - 24^{th}$ hour) and downstream (i.e. M3→ M2→M1) during the second half-period (i.e. $24^{th} - 48^{th}$ hour). The direction of electron current was switched after every 24 hours. Due to the long duration of a half-period, the first half-period of the test resulted in the three stages of resistance change that is typical of a D.C. electromigration test [7]. The three stages of resistance change are described as follows: (i) an incubation period where there is insignificant resistance change, thus signifying a void nucleation and growth process; (ii) a sharp increase in resistance which arises at the onset when electron current has to shunt through the higher resistivity Ta/TaN liner, thus indicating that the void has grown to a size that is large enough to obstruct the flow of current, and (iii) gradual increase in resistance where current has to shunt through a progressively longer length of Ta/TaN liner, thus implying that the size of the void is increasing progressively. (We will term a void that is large enough to obstruct the flow of electron current and cause it to shunt through the Ta/TaN liner as a "blocking void".) Note that the sample was considered to have failed during the first half-period given that the increase in resistance within the first 24 hours is more than 10%. The electromigration induced void in M2 Cu interconnect was located at or near to the end that is connected to via-1 (V1) since that was the cathode when electron current was flowing upstream.

Right after proceeding into the second half-period, the sample exhibited a gradual decrease in resistance. This implies that the void, that was located at or near to the end of M2 that is connected to V1, was shrinking. The void shrank because its location has become the anode when electron current is flowing downstream. Therefore the void was gradually filled up by Cu atoms that had diffused to the anode. A sharp decrease in resistance, which marked the end of gradual decrease in resistance, was observed after sometime. This observation indicates that the void has shrunk to a size that is small enough such that the flow of the electron current was no longer obstructed. Hence the electron current no longer needs to shunt through the Ta/TaN liner which has higher resistivity than the Cu metallization. Following the sharp decrease in resistance was a phase of insignificant resistance change, where the resistance of the sample was very close to the resistance at the start of the test. This phase signifies that the electron current was not shunting through the Ta/TaN liner, but it is not an indication that full recovery of electromigration induced damage (i.e. total annihilation of void) had occurred. The sample exhibited a sharp and large increase in resistance (> 100 Ω) followed by gradual resistance increment after the phase of insignificant resistance change. These changes in resistance indicate that a blocking void had developed in M2 and it was growing larger gradually.

The void was located at or near to the end that is connected to via-2 (V2) since that was the cathode when electron current was flowing downstream. We observed that the asymmetrical Cu interconnect structure in this paper exhibited a characteristic large and sharp increase in resistance (> 100 Ω) only when electron current was flowing downstream. Hence, a large difference between the resistance at the start of a test and that at a particular instance is therefore indicative of the presence of a blocking void that was located at or near to the end of M2 that is connected to V2.

Similar serial of void shrinkage at one end of M2 and void growth at the other end occurred as the direction of electron current was changed after every half-period. Beyond the fourth half-period, the resistance of the sample was always found to be much larger than that at the start of the test. This indicates that the blocking void, which was located at or near to the end of M2 that is connected to V2, did not shrink or recover to a size that was small enough such that the flow of electron current was not obstructed.

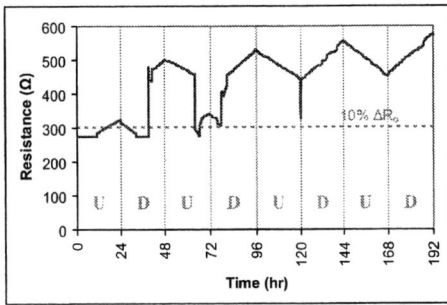

Fig. 2. Typical resistance evolution exhibited by an asymmetrical Cu interconnect structure that was subjected to bipolar pulsed current with 48 hours period. Electron current was flowing upstream during the first half-period.

B. Bipolar Pulsed Current with 16 Hours Period

The resistance evolution shown in Fig. 3 was obtained from a test sample that was subjected to bipolar pulsed current with a period of 16 hours. This test began with electron current flowing upstream, following with a change in the direction of electron current after every half-period of 8 hours. While a phase of insignificant resistance change lasted throughout the entire first half-period when electron current was flowing upstream, there was a sharp and large increase in resistance during the second half-period when electron current was flowing downstream. These changes in resistance indicate that a void that is large enough to obstruct the flow of electron current has not evolved during the first half-period, while a blocking void had evolved during the second half-period. The blocking void was expected to be located at or near to the end of M2 that is connected to V2, since that end was the cathode during the second half-period. The sample was considered to have failed during the second half-period since the failure criterion was fulfilled when the resistance increased sharply.

Sequential sharp decrease and increase in resistance were observed in the subsequent half-periods when electron current was flowing upstream and downstream, respectively (Fig. 3). Sharp decrease in resistance was however absent beyond the eighth half-period. Thereafter, the resistance of the sample remained significantly higher than that at the start of the test, indicating the presence of a blocking void that was located at or near to the end of M2 that is connected to V2. Failure analysis was performed after the sample underwent 750 hours of electromigration test. The FIB/SEM cross-section images that were taken from both ends of the sample are shown in Fig. 4. A void that spanned the entire thickness of M2 was located at the end that is connected to V2, while there is no void at the opposite end that is connected to V1. Although no void was found at or near to the end of M2 that is connected to V1, the cross-section image is not indicative that no void has formed throughout the test. A void could have grown and subsequently healed completely when the direction of electron current was reversed.

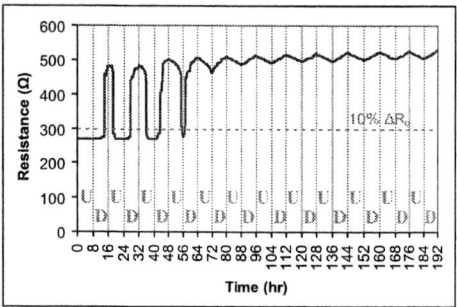

Fig. 3. Typical resistance evolution exhibited by an asymmetrical Cu interconnect structure that was subjected to bipolar pulsed current with 16 hours period. Electron current was flowing upstream during the first half-period.

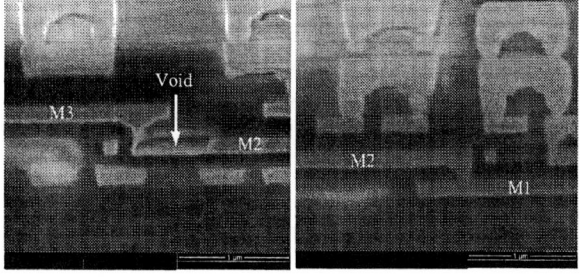

Fig. 4. FIB/SEM cross-section images of a sample that underwent 750 hours of electromigration test. The sample was subjected to bipolar pulsed current with 16 hours period, where electron current was flowing upstream during the first half-period.

Contrasting failure characteristics were observed when similar samples were subjected to bipolar pulsed current with 48 hours period and 16 hours period. When subjected to bipolar pulsed current of 48 hours and 16 hours period, the samples were considered to have failed when electron current was flowing upstream and downstream, respectively. The location of the electromigration induced voids that caused more than 10%

increase in resistance were located at or near to the M2 end that is connected to V1 for the former and V2 for the latter. It was also observed from Fig. 3 that while resistance increased sharply during the second, fourth, sixth and eighth half-period when electron current was flowing downstream, almost full resistance recovery occurred during the third, fifth and seventh half-period when electron current was flowing upstream. We further noticed that sharp increase in resistance was not observed when electron current was flowing upstream. These observations suggest that it is highly probable for a blocking void to evolve at or near to the M2 end that is connected to V2 within an 8 hours half-period, while the opposite end that is connected to V1 might require more than 8 hours for a blocking void to evolve, although the critical size that defines a blocking void from a non-blocking void may differ depending on the location of the void. Based on previous report that via-above-line configuration (i.e. downstream electron current flow) has shorter lifetime than via-below-line configuration (i.e. upstream electron current flow) [8], when electron current was flowing from via to line, we hypothesized that the asymmetrical Cu interconnect structure has a median-time-to-failure (t_{50}) of less than 8 hours when electron current was flowing downstream, and a t_{50} of more than 8 hours but less than 24 hours when electron current was flowing upstream.

In order to verify our hypothesis, two D.C. electromigration tests with electron current flowing upstream and downstream were carried out using similar samples. The t_{50} obtained from both D.C. tests are tabulated in Table 1. Samples that were subjected to D.C. that flow upstream have a t_{50} of 15.8 hours while samples that were subjected to D.C. that flow downstream have a t_{50} of 4.3 hours. The D.C. electromigration experimental results substantiate our hypothesis and thus infer that the contrasting failure characteristics observed in samples that were subjected to bipolar pulsed current with 48 hours period and 16 hours period arose due to the asymmetrical structure of the Cu interconnect. The time-to-failure (TTF) of the first sample that failed (i.e. shortest lived sample) in each D.C. test are also shown in Table 1. The shortest lived samples that were subjected to upstream D.C. and downstream D.C. had a lifetime of 10.8 hours and 1.5 hours, respectively. As the lifetime of the shortest lived sample that was subjected to upstream D.C. test was longer than 8 hours, no failure was found during the first half-period for samples that were subjected to bipolar pulsed current with 16 hours period, when electron current was flowing upstream during the first-half period.

Similar electromigration test subjecting bipolar pulsed current with 16 hours period, but with electron current flowing downstream in the first half-period, were conducted using the asymmetrical Cu interconnect structure. The resistance evolution of one sample is shown in Fig. 5. Our experimental results showed that most of the tested samples (13 out of 14) failed within the first half-period. This outcome is expected since the experimentally obtained t_{50} of downstream D.C. test, which is 4.3 hours, is relatively much shorter than the 8 hours half-period.

Table 1. Time-to-failure (TTF) of the shortest lived sample and median-time-to-failure (t_{50}) of samples that were subjected to bipolar pulsed current and direct current (D.C.) electromigration tests.

Electromigration Test (Direction of electron current in first half-period)	Number of Samples Tested	TTF (hr) [First Sample]	t_{50} (hr)
48 hr Period (upstream)	15	9.5	13.3
16 hr Period (upstream)	15	12.2	21.6
16 hr Period (downstream)	14	1.5	3.7
2 hr Period (upstream)	13	29.2	84.8
D.C. (upstream)	16	10.8	15.8
D.C. (downstream)	16	1.5	4.3

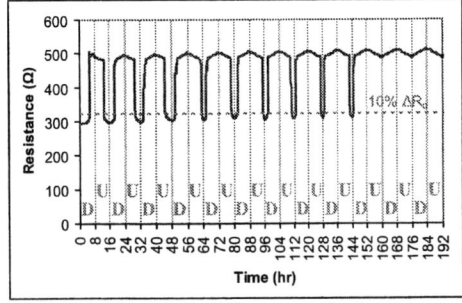

Fig. 5. Typical resistance evolution exhibited by an asymmetrical Cu interconnect structure that was subjected to bipolar pulsed current with 16 hours period. Electron current was flowing downstream during the first half-period.

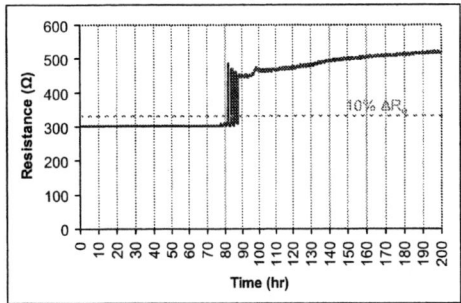

Fig. 6. Typical resistance evolution exhibited by an asymmetrical Cu interconnect structure that was subjected to bipolar pulsed current with 2 hours period. Electron current was flowing upstream during the first half-period.

C. Bipolar Pulsed Current with 2 Hours Period

Samples with asymmetrical Cu interconnect structure were subjected to bipolar pulsed current with 2 hours pulsed period in order to investigate whether the interconnect structure will achieve immortality when the half-period of the bipolar pulsed current is less than the TTF of the shortest lived sample that was subjected to downstream D.C. stress. In this test, electron current flowed upstream during the first half-period and the direction of flow was changed after every half-period of 1 hour. The typical resistance evolution of these samples (Fig. 6) showed that failure do occur, but at a much later time. Sharp and large increase in resistance, which is characteristic of an evolved blocking void that was located at or near to the end of M2 that is connected to V2, was observed at the point of failure. The location of the failure site corresponds to the fact that via-above-line configuration are less reliable than via-below-line configuration [8]. Our experimental results showed that although complete immortality was not achieved, where 9 out of 13 samples fail after 750 hours of stressing, the lifetime of the samples were largely enhanced. As shown in Table 1, the t_{50} of the 9 failed samples is around five times longer than that of samples that were subjected to upstream D.C. stress.

IV. LIFETIME DISTRIBUTION

The lifetime distributions of samples that were subjected to electromigration test with the following current stresses are shown in a lognormal plot (Fig. 7.),

(i) direct current where electron current was flowing upstream, i.e. D.C. (upstream),

(ii) direct current where electron current was flowing downstream, i.e. D.C. (downstream),

(iii) bipolar pulsed current with 16 hours period and electron current was flowing downstream during the first half-period, i.e. 16 hr period (downstream), and

(iv) bipolar pulsed current with 48 hours period and electron current was flowing upstream during the first half-period, i.e. 48 hr period (upstream).

There is a distinct difference between the lifetimes of samples that were stressed with electron current that flow upstream and downstream. As tabulated in Table 1, the t_{50} of samples that were subjected to D.C. (upstream) and 48 hr period (upstream) were 15.8 hours and 13.3 hours, respectively, while much shorter t_{50} of 4.3 hours and 3.7 hours were obtained from samples that were subjected to D.C. (downstream) and 16 hr period (downstream), respectively. This discrepancy in interconnect lifetime was attributed to the less reliable via-above-line configuration as compared to the via-below-line configuration [8], when electron current was flowing from via to line.

Comparable lifetime distributions and t_{50}s were obtained from samples that were subjected to D.C. (upstream) and 48 hr period (upstream). Samples that were subjected to D.C. (downstream) and 16 hr period (downstream) showed similar observations. In both cases, the half-periods of the bipolar pulsed currents were close to or longer than the t_{50}s of samples that were subjected to D.C. stresses. (i.e. D.C. t_{50} of 15.8 hours as compared to a half-period of 24 hours for the former, and D.C. t_{50} of 4.3 hours as compared to a half-period of 8 hours for the latter.) Since the first half-periods of the bipolar pulsed current tests were effectively D.C. tests, more than half of the samples would have failed before the direction of the electron current changes, thus giving rise to comparable lifetime distributions and t_{50}s.

Conversely, Fig. 8 shows that dissimilar lifetime distributions were observed when samples were subjected to bipolar pulsed current with shorter periods. A bi-modal lifetime distribution was obtained when the period of the bipolar pulsed current was 16 hours, where the half-period was shorter than the t_{50} of upstream D.C. stressed samples but longer than the t_{50} of downstream D.C. stressed samples. Hence, 11 out of 15 samples failed during the second half-period whereas the remaining four samples failed at a later time. The lifetimes of the 15 samples were fitted using a liner curve in order to obtain an apparent t_{50}, which was tabulated in Table 1. Although this simple linear fit is inappropriate for the bimodal lifetime distribution, a rough estimation on the magnitude of lifetime enhancement could be obtained using this apparent t_{50}. Mono-modal lifetime distributions were obtained when the periods of the bipolar pulsed current were 48 hours and 2 hours, where the half-periods were much longer or shorter than both the t_{50}s of upstream and downstream D.C. stressed samples, respectively.

The lognormal cumulative failure probability plot in Fig. 8 also shows that samples exhibited longer t_{50} when they were subjected to bipolar pulsed current whose period was shorter than the t_{50} of D.C. stressed samples. Using the t_{50} of upstream D.C. stressed samples as a basis for comparison, our experimental results showed that samples that were subjected to bipolar pulsed current with 16 hours period showed 1.3 times longer lifetime as compared to D.C. stressed samples. The minor improvement in lifetime was observed since the 8 hours half-period was shorter than the t_{50} of upstream D.C. stressed samples, but longer than the t_{50} of D.C. stressed samples. Therefore, most of the samples failed during the second half-period when electron current was flowing downstream, giving rise to a slight improvement in lifetime only. On the other hand, samples that were subjected to bipolar pulsed current with 2 hours period exhibited 5.3 times longer lifetime since the 1 hour half-period is shorter than both the t_{50}s of upstream and

downstream D.C. stressed samples. We postulate that shorter pulse period give rise to longer lifetime because of the following two factors: (1) the resulting shorter half-period is insufficient to incur and accumulate adequate electromigration induced damage for failure to occur before the direction of electron current changes, and (2) there is an increased in the frequency of damage healing at shorter period, which resulted in longer lifetime. When the direction of electron current is changed, Cu atoms will diffuse towards the former cathode to fill up the void, thus resulting in a shrinking void size, i.e. damage healing.

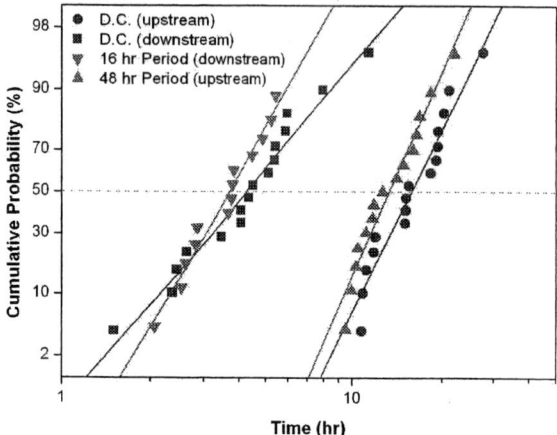

Fig. 7.   Cumulative failure probability versus log(time-to-failure) of asymmetrical Cu interconnects that were subjected to D.C. stresses and bipolar pulsed current with 16 hours period and 48 hours period.

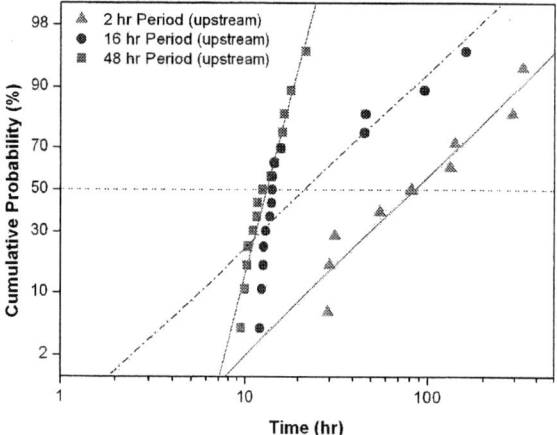

Fig. 8.   Cumulative failure probability versus log(time-to-failure) of asymmetrical Cu interconnects that were subjected to bipolar pulsed current with 2, 16 and 48 hours period.

V. CONCLUSION

Electromigration tests employing very long periods of bipolar pulsed current (i.e. 2, 16 and 48 hours) were conducted. The lifetime of the asymmetrical Cu interconnect structure was found to depend on the direction of electron current in the first half-period when subjected to bipolar pulsed current with 16 hours and 48 hours periods. This is a result of poorer reliability for via-above-line configuration as compared to via-below-line configuration. Our experimental results showed that Cu interconnects can achieve longer lifetime when subjected to bipolar pulsed current, provided that the half-period of the bipolar pulsed current is shorter than the t_{50} of D.C. stressed samples. Minor lifetime enhancement (i.e. 1.3 times longer lifetime) was achieved by samples that were subjected to bipolar pulsed current with 16 hours period, where the half-period is shorter than the t_{50} of upstream D.C. stressed samples but longer than the t_{50} of downstream D.C. stressed samples. Alternatively, 5.3 times longer t_{50}, as compared to the t_{50} of upstream D.C. stressed samples, was achieved by samples that were subjected to bipolar pulsed current with 2 hours period as the half-period is shorter than both the t_{50}s of upstream and downstream D.C. stressed samples. We postulate that shorter pulse period give rise to longer lifetime because insufficient electromigration damage was incurred and accumulated to cause failure before the direction of electron current changes, and shorter period yields a higher frequency of damage healing.

ACKNOWLEDGMENT

M. K. Lim gratefully acknowledges the Joint Industry Postgraduate Program, which is funded jointly by Chartered Semiconductor Manufacturing Limited and Singapore Economic Development Board, for sponsoring his scholarship.

REFERENCES

[1] International Technology Roadmap for Semiconductors (2007): see website http://www.itrs.net/
[2] D. Edelstein *et al.*, *Proc. IEEE IEDM*, pp. 773-776 (1997).
[3] J. R. Black, *Proc. IEEE 6th Ann. Reliability Physics Symp.*, IEEE Cat. 7-15C58 (1967).
[4] P. S. Ho, K.-D. Lee, E. T. Ogawa, S. Yoon, and X. Lu, *Proc. Mat. Res. Soc. Symp.* **766**, E1.6 (2003).
[5] R. E. Hummel, and H. H. Hoang, *J. Appl. Phys.* **65**, pp. 1292-1931 (1989).
[6] J. Tao, J. F. Chen, N. W. Cheung, and C. Hu, *Proc. of 34th Ann. Reliability Physics Symp.*, pp. 180-187 (1996).
[7] P.-C. Wang and R. G. Filippi, *Appl. Phys. Lett.* **78**, pp. 3598-3600 (2001).
[8] C. L. Gan *et al.*, *Appl. Phys. Lett.* **79**, pp. 4592-4594 (2001).

TCAD Solutions for Submicron Copper Interconnect

H. Ceric, R. L. de Orio, J. Cervenka, and S. Selberherr
Institute for Microelectronics, TU Wien, Gußhausstraße 27-29/E360
A-1040 Vienna, Austria
Phone: +43-1-58801-36032 Fax: +43-1-58801-36099 Email: Ceric@iue.tuwien.ac.at

Abstract- **The demanding task of assessing a long range interconnect reliability can only be achieved by combination of experimental and TCAD methods. A basis for TCAD tools is a sophisticated physical model which takes into account the microstructural characteristics of copper. In this work a general electromigration model is presented with a special focus on the influence of grain boundaries and mechanical stress. The possible calibration and usage scenarios of electromigration tools are discussed. The physical soundness of the model is proven by three-dimensional simulations of typical dual-damascene structures used in accelerated electromigration testing.**

I. INTRODUCTION

The main challenge in electromigration modeling and simulation is the diversity of the relevant physical phenomena.

Electromigration induced material transport is also accompanied by material transport driven by the gradients of material concentration, mechanical stress, and temperature distribution. A comprehensive, physics based analysis of electromigration for modern copper interconnect lines serves as the basis for deriving sophisticated design rules which will ensure higher steadfastness of interconnects against electromigration.

An ultimate hope of integrated circuit designers today is to have a computer program at hand which predicts the behavior of thin film metallization under any imaginable condition. Contemporary integrated circuits are often designed using simple and conservative design rules to ensure that the resulting circuits meet reliability goals. This precaution leads to reduced performance for a given circuit and metallization technology.

Relaying of previous work we present our model which reveals an improvement in two mayor points. Firstly a complete integration of mechanical stress phenomena in the connection with microstructural aspects in the classical multi-driving force continuum model was performed, and, secondly, a new finite element based scheme enabling an efficient numerical solution of the three-dimensional formulation of the problem was developed. The satisfying assessment of electromigration reliability can only be achieved through combination of experimental methods and utilization of TCAD tools. Therefore, we also discuss a possible usage scenario of TCAD tools in connection with results of accelerated interconnect tests.

Competitive reliability targets of chip failure rates have been in the order of one per thousand throughout the anticipated life time in the field. The central problem of the interconnect design for reliability is the determination of the long term interconnect behavior.

The TCAD analysis of the electromigration reliability of interconnect structures has to be carried out on at least two levels. The first level is without any doubt a physical one, that means application of most complete and comprehensive models to interconnect portions of moderate size. The restriction in size and complexity arises from the capacity of computers (memory, computational time) but has also a cause in numerical issues.

The first level analysis is based on the simulation of the behavior of characteristic portions of the interconnect, which, known from experiments, represent a high electromigration risk. The goal of the analysis by simulation is to determine the time-to-failure distribution for this specific interconnect part. The second level analysis combines the results of the first level in order to assess electromigration reliability of an entire chip.

II. PHYSICALLY BASED MODELING

Preceding the consideration of the electromigration model equation an electro-thermal problem has to be solved with the goal to obtain an accurate temperature distribution.

All diffusivities in the electromigration model are thermally activated and even a small error in the temperature calculation can lead to a substantial error in the vacancy concentration and stress calculation.

The bulk vacancy transport is given by the following balance equation [1]:

$$\mathbf{J}_v = -\mathbf{D}\left(\nabla C_v + \frac{Z^* e}{k_B T} C_v \nabla \varphi + \frac{f\Omega}{3 k_B T} C_v \nabla \operatorname{tr}(\bar{\bar{\sigma}})\right), \quad \frac{\partial C_v}{\partial t} = -\operatorname{div}\mathbf{J}_v. \tag{1}$$

Here, $k_B T$ is the thermal energy, $Z^* e$ is the effective valence, Ω is the volume of atom, and f is the vacancy to atom volume ratio. $\mathbf{D}$ is the tensorial vacancy diffusivity which is in stress free state set as $D_{ij} = D_{bulk}\delta_{ij}$, where D_{bulk} is the isotropic bulk diffusivity.

Divergences of the vacancy flux produce local strain

$$\frac{\partial \varepsilon_{ij}^v}{\partial t} = \frac{1}{3}\Omega(1-f)\operatorname{div}\mathbf{J}_v\delta_{ij}, \tag{2}$$

which is equilibrated by induced displacements in the copper bulk $\vec{u} = (u_1, u_2, u_3)$ according to the Lamè-Navier equations,

$$\mu\Delta u_i + (\mu + \lambda)\frac{\partial}{\partial x_i}(\nabla \mathbf{u}) = B\frac{\partial \operatorname{tr}(\varepsilon^v)}{\partial x_i}, \quad i = 1,2,3 \tag{3}$$

Here λ and μ are the Lamè coefficients and $B = \lambda + 2\mu/3$. Generally, an elastic deformation of the metal is assumed:

$$\sigma_{ij} = \sum_{ijkl} \mathbf{C}_{kl}\varepsilon_{kl}, \tag{4}$$

with the small displacement approximation:

$$\varepsilon_{ij} = \frac{1}{2}\left(\frac{\partial u_i}{\partial x_j} + \frac{\partial u_j}{\partial x_i}\right), \quad i,j = 1,2,3. \tag{5}$$

Residual process stresses, thermo-mechanical stresses, and electromigration induced stresses cause anisotropy of material transport and therefore in (1) a tensorial diffusivity **D** must be taken into account.

In order to consider the microstructure of copper, the bulk model (1) is extended by fast diffusivity paths models, the most significant of which is the grain boundary model, since it describes also vacancy recombination.

Our modeling approach is based on models of Herring [2] and Fisher [3], which deal with grain boundary mechanics and grain bulk/boundary material exchange, respectively. To connect these models a careful analysis of the chemical potential in the vicinity of a grain boundary and in a grain boundary is necessary.

For the chemical potential in the grain boundary we use Herring's expression [2]:

$$\mu_v^{gb} = \mu_0 + \Omega\sigma_{nn} + k_B T \ln\left(\frac{C_v^{gb}}{C_v^0}\right), \tag{6}$$

where we assume a unique equilibrium vacancy concentration C_v^0 in stress free copper, both, in grain bulks and boundaries. μ_0 is some chemical reference potential, $\sigma_{nn} = \vec{n} \cdot \bar{\bar{\sigma}} \cdot \vec{n}$ and $\vec{n}$ is the normal to the grain surface.

The chemical potential of the vacancies in the bulk is given by [4]:

$$\mu_v = \mu_0 + k_B T \ln\left(\frac{C_v}{C_v^0}\right) + \frac{1}{3}f\Omega\operatorname{tr}(\bar{\bar{\sigma}}). \tag{7}$$

In the continuum modeling approach the grain boundary as a vacancy transport medium is defined by the chemical potential. This chemical potential is constant through the grain boundary thickness and equal to the bulk chemical potential on the interfaces to the bulk regions, eg. $\mu^1_v(-\delta/2) = \mu^1_v(+\delta/2) = \mu_{gb}$, Fig. 1. The vacancy fluxes on both sides of the grain boundary given by:

$$J_v^1 = \frac{D_v C_v}{k_B T}\nabla\mu_v^1,$$

$$J_v^2 = \frac{D_v C_v}{k_B T}\nabla\mu_v^2. \tag{8}$$

The difference $J_v^2 - J_v^2$ is an actual loss (gain) of vacancies which is localized at the thin slice which represents the grain boundary (in continuum modeling).

The recombination rate can now be approximated as:

$$G = \frac{\partial C_v}{\partial t} = -\operatorname{div}J_v \approx -\frac{J_v^2 - J_v^1}{\delta}. \tag{9}$$

In order to include the effect of the grain boundary as a vacancy sink (source) into the bulk vacancy transport model the recombination term G has to be included in equations (1) and (2):

$$\frac{\partial C_v}{\partial t} = -\operatorname{div}\mathbf{J}_v + G,$$

$$\frac{\partial \varepsilon_{ij}^v}{\partial t} = \frac{1}{3}\Omega((1-f)\operatorname{div}\mathbf{J}_v + fG)\delta_{ij}. \tag{10}$$

The full description of the atomic mechanisms of vacancy generation and annihilation in grain boundaries goes beyond the capability of continuum modeling and can only be obtained by molecular dynamics methods.

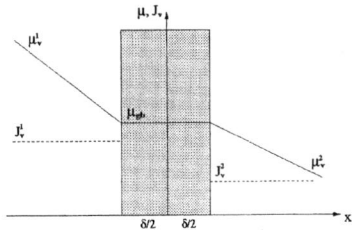

Fig. 1. The grain boundary according to Fisher's model and Herring's relationship.

III. USAGE SCENARIO FOR TCAD TOOLS

Simulation can be used for extrapolation of long time interconnect behavior on the basis of results of accelerated electromigration tests. This capability is clearly superior to an extrapolation by standard statistical methods which rely on Black's equation and extrapolate a time-to-failure (TTF) for a single interconnect structure. The usage of TCAD tools enables a prediction of the behavior for structures which are obtained by variation of geometrical properties and operating conditions of a previously used initial test structure.

The assumed scenario for application of an electromigration reliability TCAD tool is:

1. *Model Calibration:*
For this purpose we use one layout and many test units. At the end of calibration all parameters of the model are fixed.

During this process, different microstructures are considered and simulation parameters are varied with the goal to reproduce experimental failure time statistics, cf. Fig. 2.

2. Model Application:.

The calibrated model is used for simulation. The simulation extrapolates the behavior of the interconnect under real life conditions.

For a given interconnect layout and monocrystalline material, simulation will provide a unique time-to-failure. All impact factors, e.g. geometry of the layout, bulk diffusivity, interface diffusivity, and mechanical properties are deterministic and so TTF is deterministic. However, the situation changes when the interconnect possesses a microstructure. The microstructure has a significant impact on electromigration, since it introduces a diversity of possible electromigration paths and local mechanical properties (the Young modulus and Poisson factor depend on the crystal orientation in each grain). However, the microstructure itself cannot be completely controlled by a process technology. In other words, the position of grain boundaries, angles in which they meet the interfaces, etc. cannot be designed, the process itself determines only statistics of grain sizes and textures.

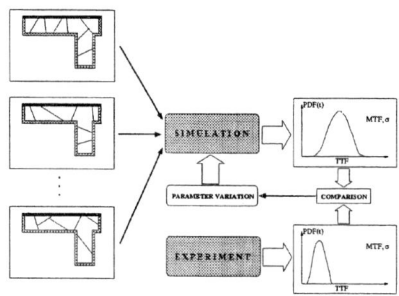

Fig. 2. Electromigration model calibration using a multitude of microstructural inputs.

IV. SIMULATION RESULTS

We have applied our model to the interconnect layout which has been extensively used for accelerated electromigration tests [6]. This layout is typical for dual-damascene 0.18 µm technologies. The Copper microstructure (Fig. 3) is set according to results of EBSD (Electron Backscatter Diffraction) measurements [5]. The peak values of stress and the vacancy concentration values are extracted from three extraction cylinders (C1, C2, and C3), which are presented in Fig. 4a).
The solution of the electro-thermal problem sets the operating conditions for electromigration simulation. The applied mechanical, thermal, and electrical boundary conditions are presented in Fig. 4b). The temperature T is set on constant 673 K which is the temperature level used in accelerated tests [6] and the voltage is V=10 mV. The obtained average current density is 10 MA/cm². Due to the geometry of the problem and

the applied boundary conditions the temperature is only slightly changed by Joule heating.

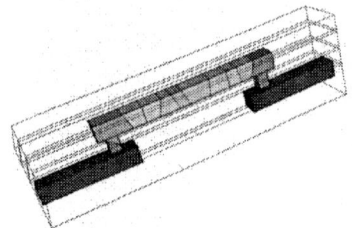

Fig. 3. Three-dimensional dual-damascene structure with polycrystalline copper metallization.

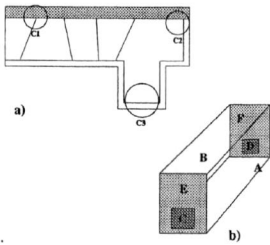

Fig. 4. a) Peak values for hydrostatic stress and vacancy concentration values are extracted from cylinders C1, C2, and C3. b) Applied boundary conditions: A, B: fixed temperature T=673K; C, D: voltage V=10 mV, upstream electromigration from C3 to C1; E, F: mechanically fixed.

In dual-damascene copper technologies interfaces to capping (etch-stop) layers are recognized as the fastest material transport paths for standard capping dielectrics SiNx [6]. Experimental investigations have shown that for such cappings critical voids are formed at the top corner of the cathode edge of the metal line [7].
In order to properly include the effect of fast diffusivity paths, grain boundary, barrier, and capping layer diffusivities are set as $10^2 D_{bulk}$, $10^2 D_{bulk}$, and $10^5 D_{bulk}$, respectively.
For a possible void nucleation high tensile stress and a local interface defect are necessary. Therefore, the primary goal of our investigation is to determine the sites, where high stresses arise.

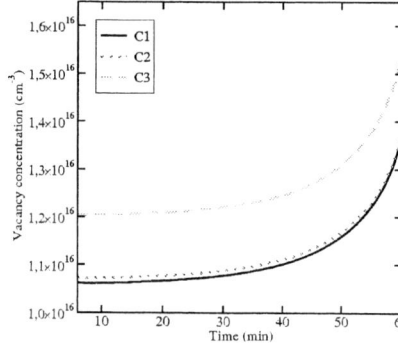

Fig. 5. Time evolution of vacancy concentration at three characteristic spots of the via.

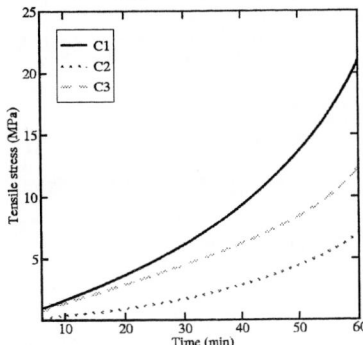

Fig. 6. Time evolution of tensile hydrostatic stress at three characteristic spots of the via.

For this purpose we monitor the hydrostatic stress development in three cylinders placed at the triple point next to cathode end of the via (C1), at the portion of the capping layer which lays directly above the via bottom (C2), and at the via bottom itself (C3), cf. Fig. 4a).

Due to the geometry layout, the highest vacancy concentration always develops at the bottom of the cathode end of the via (C3), cf. Fig. 5.

The peak vacancy concentration on the sites C1 and C2 develops approximately in the same way, but the site of highest stress (C1) does not coincide with the site of highest vacancy concentration (C3) (Fig. 6).

The connection between tensile stress and local vacancy concentration is defined through the relationships (10) and overall mechanical equilibrium (3).

The local stress behavior does not depend only on the local vacancy concentration, but also on mechanical conditions in the neighboring areas. It is obvious that a defect site is necessary for void nucleation and the coincidence of high vacancy concentration and high tensile stress regions with triple points (Fig. 7) indicates that triple points are natural locations of weak adhesion. This assumption was also expressed in the discussion of results of accelerated tests published in [5, 6].

The scenario of weak triple points in combination with a stress threshold would allow multiple void nucleations in the short time interval as it has actually already been observed [6].

These voids can migrate along the copper/capping layer interface, from one triple point to another, and stop at the corners above the via. Here, further void growth takes place eventually resulting in a critical decrease in the effective interconnect cross section leading to failure.

V. CONCLUSION

A comprehensive electromigration model is used as a basis for development of a three-dimensional simulation tool. An influence of residual and electromigration induced strains on material transport is discussed and an earlier electromigration model is extended by introduction of tensorial self-diffusivity.

A discussion of grain boundary physics is provided, whereas a special focus was put on the enlightenment of the stress influence on the grain boundary dynamics.

The physical soundness of the extended electromigration model is verified with several simulation examples. The simulated dynamics of early failure development is in good agreement with experimental observations. The role of triple points regarding void nucleation is discussed on the basis of simulation and corresponding experimental results. A concept for usage of TCAD tools in combination with experimental tests is presented.

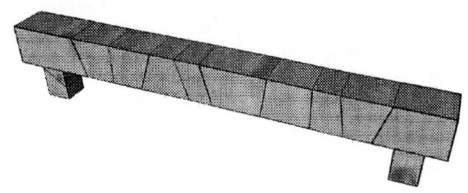

Fig. 7. Peak tensile stress tensor component distribution. Red (dark) color areas marks peak tensile stress.

ACKNOWLEDGMENT

Support by the Austrian Science Fund with the project P18825-N14 is gratefully acknowledged.

REFERENCES

[1] W. W. Mullins, "Mass Transport at Interfaces in Single Component Systems," *Metall. Mater. Trans. A (USA)*, vol. 26, no. 8, pp. 1917-1929, 1995.

[2] C. Herring, "Diffusional Viscosity of Polycrystalline Solid," *J. Appl. Phys.*, vol. 21, no. 5, pp. 437-445, 1950.

[3] J. C. Fisher, "Calculation of Diffusion Penetration Curves for Surface and Grain Boundary Diffusion," *J. Appl. Phys.*, vol. 22, no. 1, pp. 74-77, 1951.

[4] F. C. Larche and J.W. Cahn, "The Interactions of Composition and Stress in Cystalline Solids," *Acta metal.*, vol. 33, no. 3, pp. 331-357, 1985.

[5] E. Zschech and V. Sukharev, "A Model for Electromigration-Induced Degradation Mechanisms in Dual-Inlaid Copper Interconnects: Effect of Interface Bonding Strength," *J. Appl. Phys.*, vol. 96, no. 11, pp. 6337-6343, 2004.

[6] A. V. Vairagar, S. G. Mhaisalkar, A. Krishnamoorthy, and K. N. Tu, "In Situ Observation of Electromigration-Induced Void Migration in Dual-Damascene Cu Interconnect Structures," *J. Appl. Phys.*, vol. 85, no. 13, pp. 2502-2504, 2004.

[7] V. Sukharev, E. Zschech, and W. D. Nix, "A Model for Electromigration-Induced Degradation Mechanisms in Dual-Inlaid Copper Interconnects: Effect of Microstructure," *J. Appl. Phys.*, vol. 102, no. 5, pp. 530501-530514, 2007.

Investigation on the Mechanism of the Leakage Failure Between Poly Gate and Contact in Subnano Technology

Q.F. Wang, S.L. Toh, Q. Deng, P.K. Tan, K. Li, J. Teong, Z.H. Mai and J. Lam
Chartered Semiconductor Mfg Ltd
60 Woodlands Street 2, Industrial Park D, Singapore 738406
Phone: (65) 63604387 Fax: (65) 63622936 Email: wangqf@charteredsemi.com

Abstract - **With the shrinkage of the transistor dimensions, the spacing between the structures become smaller and smaller. However due to the intrinsic characteristic of the CMOS device, the reduction of the operating voltage is limited. The electrical field between different structures keeps on increasing with the shrinkage of the transistor dimensions. Furthermore, many new failure modes were observed with the scaling of semiconductor device. One of them is poly gate to contact leakage. In this paper, the mechanism of the leakage failure between poly gate and the contact in subnano CMOS technology was discussed.**

I. INTRODUCTION

Technology scaling has enabled us to integrate a large number of circuits on a single chip. However, with a decrease of device dimensions and an increase of circuit complexity, many new failure phenomena occur. One of the new failure modes in our sub-nanometer prototypes is poly gate to contact leakage, which may cause column, pair-bit and single bit failure. However, the mechanism of the leakage failure between poly gate and contact is not clear.

In this paper, two hypotheses that result in the poly-gate to contact leakage were proposed based on the Physical Failure Analysis (PFA) results. Planar-TEM and cross-TEM were employed to investigate the mechanism of poly gate to contact leakage failure.

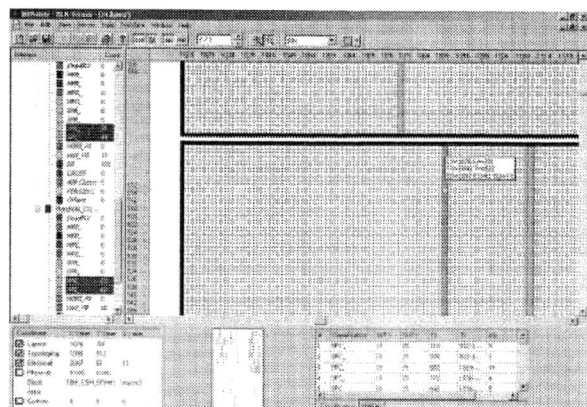

Fig. 1 Bitmap of full column failure except for one bit.

II. RESULTS AND DISCUSSIONS

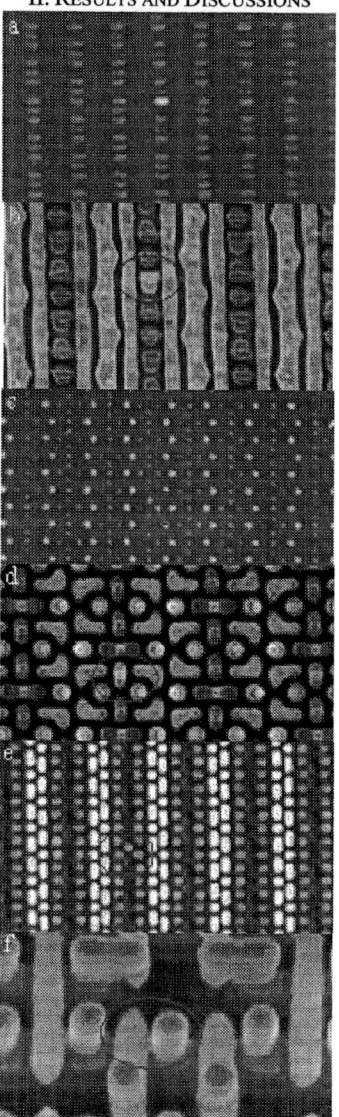

Fig.2 FA results of column failure. Bright voltage contrast was observed at (a) via 2, (b) metal 2, (c) via 1, (d) metal 1 and (e) contact. After poly exposure and spacer removal, poly gate to contact bridging was found at the passing bit in (f).

Figure 1 shows the bitmap location of a column failure in one of the subnano SRAM test chips. In this test chip, one column consists of 512 bits. In the failing column of this chip, all the 511 bits failed except one bit. The PFA results were shown in Figure 2. Bright voltage contrast (VC) was observed at via 2, metal 2, via 1, metal 1 and contact at the pass-gate (WL) location of the passing bit. After the poly was exposed by buffered oxide etching (BOE) at contact level and the spacer was removed with Reactive Ion Etch (RIE), bridging material between poly gate (WL) and contact (BL) was found. In order to investigate the elemental content of the bridging materials, TEM was employed and the images were shown in Fig. 3. EDX spectra clearly showed that the bridging materials were Nickel and titanium, not tungsten. The TEM results also indicated that the height of bridging location was close to the top of spacer. Presence of the bridging materials also helped to protect the underlying Si substrate from directional RIE etching.

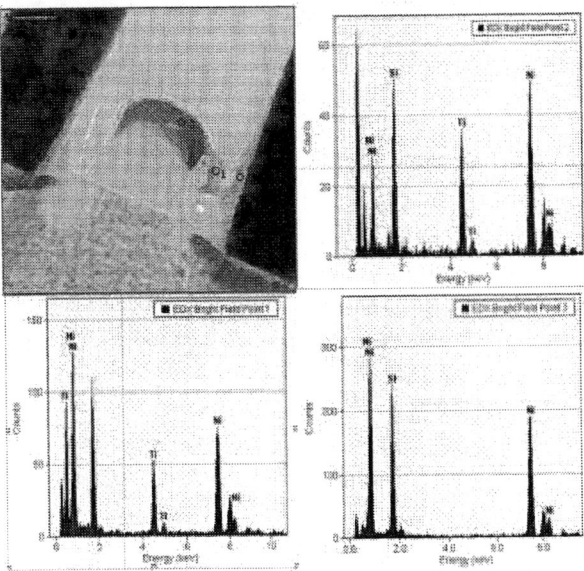

Fig.3 Cross-sectional TEM images of poly gate to contact bridging as shown in Fig. 2(f).

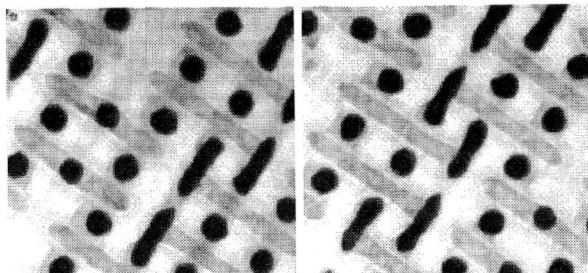

Fig. 4 Planar TEM images on random locations on SRAM.

Based on above PFA results, two hypotheses were proposed. One was attributed to poly gate/contact critical dimension (CD), poly gate and contact overlay, and contact profile consistency.

As IC device scales into nanometer range, the distance between poly gate and contact is only several tens of nanometer. If poly gate/contact CD increases, or poly gate and contact overlay is not good, or contact profile is not consistent, it is easy for poly gate to bridge to contact. Random planar TEM was performed at SRAM area of another chip on the same wafer, as shown in Figure 4. No abnormality in poly gate/contact CD, poly gate and contact overlay or contact profile was found.

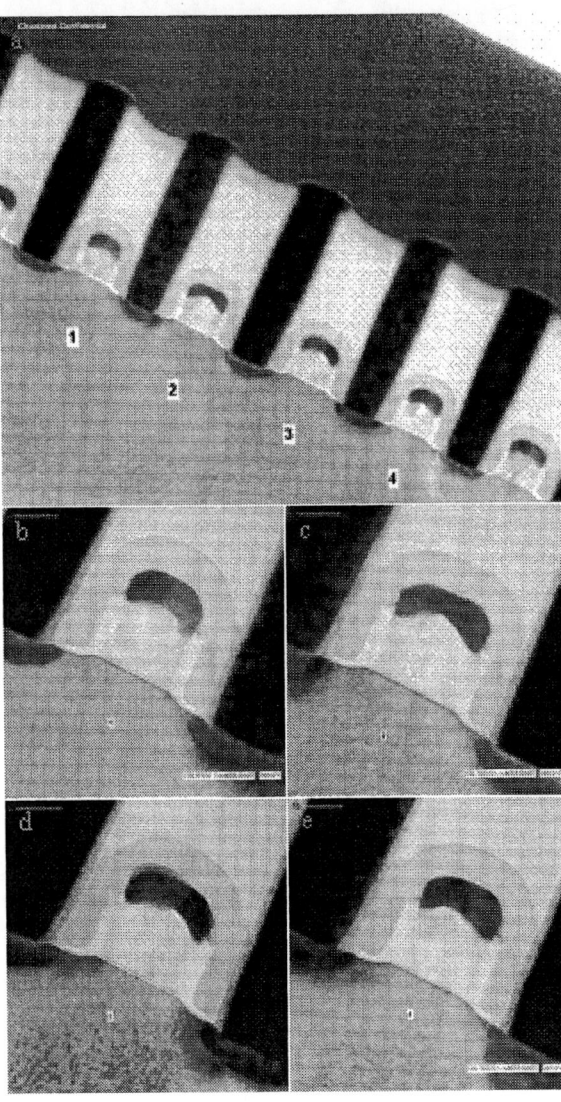

Fig. 5. Cross-sectional TEM images on random SRAM passing bits: (a) low magnification and high magnification of location (b) 1, (c) 2, (d) 3, and (e) 4.

Another hypothesis is that there may exist weak point/path at the interface of spacer and nitride cap. Self-aligned Nickel salicide on active/poly gate may tunnel to contact through the weak point/path. Random cross-sectional TEM was carried out on the SRAM passing bits of third chip on the same wafer. The chip was delayered to contact. No abnormal VC was observed.

Random TEM was done perpendicular to poly gate, as shown in Fig. 5. It clearly showed that there was residual material at the interface of spacer and nitride cap layer of good bits, which verified above weak point/path hypothesis. The residue at the interface of spacer and nitride cap layer may form during process or during testing.

One of the subnano ET structures in both tested and untested wafer was selected to investigate how the residue at the interface of spacer and nitride cap layer was formed. Figure 6 shows its layout structure. When the tested ET chip was delayered to contact, bright VC was observed, as shown in Fig. 7 (a). The poly-gate of the ET structures was situated on top of the shallow trench isolation (STI), hence its VC should appear dark under SEM illumination. Cross-sectional TEM was performed at the bright VC contact and the results were shown in Figs. 7 (b) to (f). Dark-field imaging mode in Fig. 7(f) clearly showed the existence of Nickel salicide between bright VC contact and poly gate. Figure 7 (c) and (d) also indicated that the spacing between the poly gate and bright VC contact was smaller than that of normal contact.

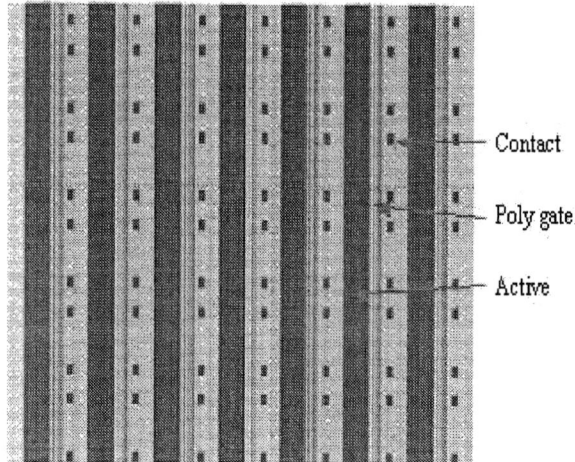

Fig. 6 Layout of the ET structure.

For untested ET chip, no abnormal VC was observed when the chip was delayered to contact. Random cross-sectional TEM was carried out and the results were shown in Fig. 8. It showed that Nickel salicide residue also exists at the interface between spacer and nitride cap layer near the poly gate. This indicated that it was formed during process, not during testing. The formation of Nickel salicide residue at the interface between spacer and nitride cap layer near poly gate may be due to diffusion or incomplete salicide strip.

As IC device continues to scale into sub-nanometer range, the distance between poly gate and contact is only several tens of nanometer. The electrical field between poly gate and contact is very strong. If there is Nickel salicide residue between poly gate and contact at the interface of space and nitride cap layer, it reduces the distance from poly gate to contact and greatly

increases the electrical field during testing. This may promote Nickel electro-migration. The leakage path of Nickel salicide from poly gate to contact may finally form during testing at the narrowest spacing between contact and poly gate because of increased electrical field.

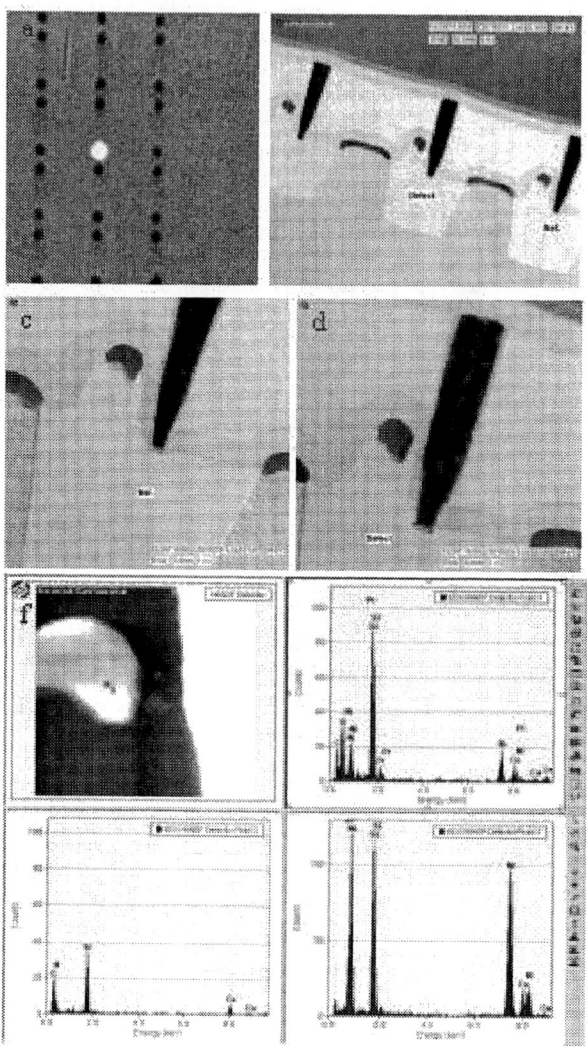

Fig. 7 FA results of the tested ET chip. (a) SEM picture showing bright VC at contact. (b) TEM low magnification image at the bright VC contact perpendicular to poly. (c) high magnification image of reference bit. (d) high magnification image of bad bit (f) high angle angular dark field image and EDX spectra of bad bit.

III. CONCLUSION

One of the new failure modes in the subnano prototypes, poly gate to contact leakage, is identified. The PFA and TEM results showed that there was weak point/path at the interface between spacer and nitride cap layer. Nickel salicide residue formed near poly gate at the interface between spacer and nitride cap layer during process, is enhanced during testing via electro-migration. The leakage path of Nickel salicide from poly gate to contact may form at the smallest spacing between poly gate and contact where the electrical field is the largest during testing.

ACKNOWLEDGMENT

The author would like to thank Chartered TEM team for their great support.

978-1-4244-2039-1/08/$25.00 ©2008 IEEE

SESSION 5:

ADVANCED FA II

EXCHANGE PAPER (ISTFA 2007 BEST PAPER)

Raman-IR micro-Thermography Tool for Reliability and Failure Analysis of Electronic Devices

[1]A. Sarua, [1]J. Pomeroy, and [1]M. Kuball, [2]A. Falk and [2]G. Albright, [3]M. J. Uren and [3]T. Martin

1: H.H. Wills Physics Laboratory, University of Bristol, Bristol BS8 1TL, United Kingdom

1: Phone: +44 (117) 954 6886.   Fax: +44 (117) 925 2564.   Email: a.sarua@bristol.ac.uk

2: Quantum Focus Instruments, Vista CA 92081, USA

3: QinetiQ Ltd, Malvern WR14 3PS, United Kingdom

Abstract-**We report on the development of a integrated Raman – IR thermography technique to probe self-heating in active devices. We compare and discuss advantages of both techniques in terms of spatial resolution on the example of AlGaN/GaN HFET devices. While traditional infra-red (IR) thermography can provide fast overviews of self-heating in the devices over large scales, its use for extraction of channel temperatures is limited by the sub-micron size of the active area in modern devices. Integration with micro-Raman thermography provides not only improvement in spatial resolution down to 0.5 µm on the surface but also unprecedented micron scale depth resolution for true 3D thermography. This enables unique thermal analysis of semiconductor devices on a detailed level not possible before. As it is a generic technique its application can be extended to Si, GaAs and other devices. This opens new opportunities for device performance and reliability optimization, and failure analysis in research and development of modern semiconductor technology, as well as for quality control/ manufacturing environments.**

I. INTRODUCTION

Increased demand in high power and high frequency devices stimulated the emergence of new technologies over the past decade based both on novel semiconductor materials, such as GaN or SiC, but also in the more traditional Si and GaAs device systems. While performance of such devices are reaching new record breaking levels in operating frequency and power density [1,2], often long term reliability is an issue [3], which prevents large scale use and commercialization of these devices. In high power devices reliability is strongly affected by device self-heating, i.e., the temperature rise in a device induced by ohmic heat generation. Since device failure rates can vary with device temperature up to exponentially,[4] dependent on failure mechanism, excessive device peak temperatures confined to the small device dimensions can often result in accelerated device aging and ultimately in device failure. Significant heat generation in high power devices [5] also pose challenges for the device package design, as heat must be efficiently extracted from a device, to prevent degradation and performance loss of packaged devices. Therefore, the importance of thermal management in the device and package design can not be underestimated.

Accurate measurement of device temperature is crucial for device performance and device reliability. Although there are some well-established traditional techniques to assess the temperature rise in operating devices like infrared thermography [6,7], there are key challenges related to the ever decreasing dimension of modern devices with these traditional thermal analysis approaches. For example, in AlGaN/GaN power transistors heat is generated in an area near the gate of less than 0.5 µm width. Such length scales are far beyond the diffraction limit of IR light of about 2-5 µm. As a result IR thermography typically averages device temperature over a several micron size area, leading to an underestimation of peak temperatures reached by the device during operation. Last but not least, an accurate knowledge of the device peak temperature is essential for accelerated lifetime testing [8], where information on the actual junction temperature is needed to extrapolate device lifetimes from electrical test data. Alternatively thermal simulations can be used to obtain an insight into device temperature distribution, however they require a number of input parameters, such as thermal conductivities and/or thermal boundary resistances (TBR), which are often not accurately known. Unless validated experimentally, predictions of these models and their accuracy can bear a large degree of uncertainty. Therefore, there is a clear need for a new thermography technique, beyond IR thermography, which can reliably access temperature rise in devices with sub-micron/micron spatial resolution.

Recently, we demonstrated that Raman spectroscopy together with the high spatial resolution of optical microscopy can be applied to measure non-invasively temperature rises during operation of electronic and opto-electronic devices on the sub-micron/micron length scale (in two and three dimensions) and with nano-second time resolution, which can be used for electrical and thermal model verification [9-18]. This method, however, can not access temperatures of metal contacts in device materials or package materials, which do not have suitable phonon lines in their spectrum. Thus in this work we demonstrate a combined Raman and IR thermal imaging technique. This new approach benefits from combining the strengths of both techniques, i.e., the ability of IR to provide quick large area temperature overviews of devices, and metal contact temperatures, to identify thermal areas of interest, and the high spatial resolution for two and three dimensional thermal analysis of Raman. While the presented results focus on the example of AlGaN/GaN HFETs, which have attracted great interest due to their high frequency RF applications and crucial

978-1-4244-2039-1/08/$25.00 ©2008 IEEE

Proceedings of 15th IPFA - 2008, Singapore

EXCHANGE PAPER (ISTFA 2007 BEST PAPER)

reliability challenges, this technique is essentially generic. We have already demonstrated its applications to other device systems, such as Si and GaAs [13], where it also offers powerful and effective pathways to gain understanding of device functionality and failure analysis on a new qualitative level.

II. EXPERIMENTAL DETAILS AND METHODS

An integrated micro-Raman and IR thermography setup was built using custom-modified commercial equipment. Infrared imaging measurements were carried out with a custom build Quantum Focus Instrument (QFI) Infrascope system, equipped with an LN_2-cooled 256x256 InSb detector array. IR light was collected by a SiGe 15× (NA=0.5) objective lens, with working distance of 16 mm. Theoretical spatial resolution of the system is diffraction limited to about 3-6 μm on the sample surface and pixel resolution of the obtained infrared image is about 1.6 μm. Infrared thermography is based on measuring the amount of radiated thermal IR emission of a body (radiance), which is proportional to its emissivity and the forth power of its absolute temperature. To acquire the temperature a sample radiance calibration was performed [7], allowing pixel to pixel correction of the target emissivity. Once emissivity is obtained, temperature of the selected area can be retrieved almost instantly. Details of the IR thermography method are provided elsewhere [7,11].

This infrared system was integrated into the Leica microscope of a Renishaw InVia system, which was used to perform micro-Raman measurements. As both infrared and Raman objective lenses are mated to the same microscope turret, the switching between Raman and IR mode can be easily achieved allowing quasi-simultaneous measurements over the same devices area. Device temperature using Raman thermography is obtained by measuring lattice vibrations of the device material, so-called phonons, which are dependent on temperature. If temperature of semiconductor material is increased, the distance between its atoms is increased and the vibration frequency, i.e., the phonon energy, is decreasing as a result. Temperature is then extracted using a calibrated phonon dependency on temperature using a two-stage-measurement between off and on state of the device. The phonon energy is measured in Raman by shining a monochromatic laser light on the sample and analyzing a Raman scattered light from the sample using a spectrometer [11].

In our case, the 488 nm line of an Ar+ ion laser was used as optical excitation source. A 50× (NA=0.5) objective lens was used to focus the laser beam onto the sample surface with diameter of about 0.5-0.7 μm. In confocal mode the depth of focus can be estimated as ~2 μm. Spectral shift resolution of the Renishaw InVia system is 0.02 cm^{-1} giving a temperature resolution of better than ±5-10 °C. For the conditions used here on the example system of AlGaN/GaN HFETs, the measured device temperature corresponds to the device temperatures averaged over the Raman probing depth of typically ~2 μm in the confocal microscope used, or over the thickness of the semiconductor layer probed, whichever is smaller. As each semiconductor material has its characteristic phonon modes, in a layered device structure each individual layer and hence its layer

temperature can be probed selectively [11,18]. It is important to underline that the here used laser wavelength is below the band-gap of AlGaN and GaN to prevent local heating and photo-carrier generation due to light absorption. Laser wavelengths above the material band-gap can also be utilized for Raman measurements of the devices, providing that laser absorption effects are kept to a minimum by using a low laser power, which enables surface probing depth on the order of just tens of nanometers, which we illustrated for GaAs pHEMTs [13].

Use of a laser excitation wavelength below the band-gap of the device materials furthermore enables the possibility to extract 3D temperatures by focusing the laser beam with a microscope lens at a defined depth into the device. We note as Raman thermography can only analyze areas in a device where the semiconductor is not covered by metal contacts, temperature underneath metal contact areas can in this case be probed by focusing the laser beam from the back of the die through the substrate onto the areas under the metal contacts. An XY- and Z-mapping stage was used to obtain 2D scans of the device temperature with lateral resolution of 0.1 μm and of 1.0 μm for depth scans for 3D temperature analysis. Temperature of the device and its heat sink during measurements was controlled by Peltier heater stage, and was kept at 25°C, unless stated otherwise. For 3D thermal simulations a commercial TAS finite difference package was used (for details see Ref. 11,16).

The application of the here developed Raman-IR system is illustrated on the example of AlGaN/GaN HFETs. The devices used were 2×50 μm HFETs and consisted of 30 nm AlGaN, 1.2 μm of GaN on 300 μm of 4H-SiC substrate. Standard Ti/Al/Pt/Au Ohmic contacts and Ni/Au Schottky gate contacts were used. Source-drain gap was 4.8 μm and gate length was 1.2 μm. The wafer was cut into 25 mm^2 die which were mounted on a well thermally heat-sunk package. All measurements were carried out under DC electrical bias.

III. RESULTS AND DISCUSSION

A temperature map of a 20 W/mm operated AlGaN/GaN HFET is shown in Fig. 1. The IR temperature image reveals a hot spot in the vicinity of the source-drain gap of the device, depicting an IR measured peak temperature of about 114 °C, measured through the back of the chip. Only one of the two device fingers was operated in this measurement. The temperature decreases away from the hot-spot. Areas underneath the metal contacts away from the channel show a temperature of about 60 °C, while in areas where no metal is on the device surface a temperature of 27 °C is measured by IR thermography, which as will be discussed later is an artifact.

978-1-4244-2039-1/08/$25.00 ©2008 IEEE

EXCHANGE PAPER (ISTFA 2007 BEST PAPER)

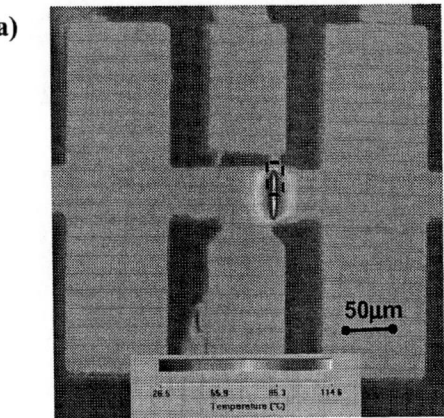

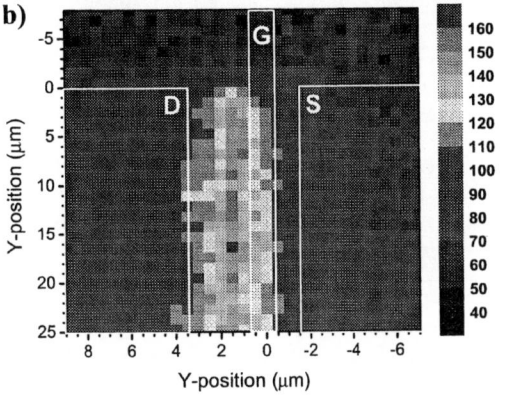

Fig. 1. a) IR temperature map of AlGaN/GaN HFET device at Vds = 40V and Ids = 25 mA (only one finger was operated). b) Micro-Raman map of GaN temperature of the same device over selected area shown by rectangle in a). All measurements were performed through the back of the chip.

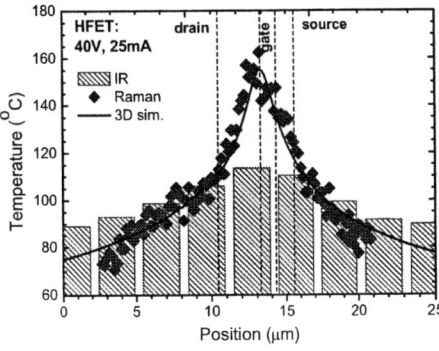

Fig. 2. Temperature profile measured by IR and Raman thermography between drain and source of AlGaN/GaN HFET in the centre of the operating finger under identical operating conditions as in Fig 1.

Micro-Raman thermography was performed over the active finger area of the same HFET, through the back side of chip

through the substrate. This allows accessing all device areas, including those covered by metal contacts, and enables an easy comparison with the IR imaging data. Temperature of GaN was determined from the temperature dependence of the E_2(high) phonon mode of GaN, shown in Figure 1(b). Since depth of the focus of the Raman system is on the order of 2 μm, Raman averages the temperature over the thickness of the GaN layer (1.2 μm). The pinched-off state of the device was used here as a reference for the device off-state to compensate for potential piezoelectric effects on GaN phonons and increase accuracy. Raman measurements show device peak temperatures in the source-drain gap of about 160 °C, significantly higher than the 114 °C measured by IR. This is illustrated in more detail in Figure 2. While both IR and Raman thermography allow detection of the hot-spot location and its basic shape, IR thermography considerably underestimates peak temperatures of the hot-spot area. This underestimation is largely related to the limited spatial resolution of optical systems in the IR thermography spectral region (2.5-5 μm) [11]. Detailed analysis showed that the experimental spatial resolution of the IR system was about 5-7 μm. As a result IR temperature is averaged over several micrometers. IR imaging therefore cannot resolve the temperature rise accurately, since heat generation in HFET occurs on the length of just 0.5 μm near the gate contact [12]. Use of much shorter optical wavelengths in the Raman measurements provides spatial resolutions of 0.5-0.7 μm (Table I). This allows more accurate determination of the hot-spot location and its temperatures, illustrated in Fig. 2 also by the good agreement with 3D thermal device simulations. We note that temperatures away from the source-drain region are similar in Raman and IR (Fig 2), as expected.

Table I
Spatial resolutions of optical systems

		IR	Raman (488nm)
Lateral resolution			
	Estimated (μm)	3-6	0.5
	Measured (μm)	5-7	0.5-0.7
Depth resolution			
	Estimated (μm)	10-15	1.6
	Measured (μm)	-	2-4.5

While Raman thermography proves to be a very powerful technology in terms of spatial resolution and accuracy, it should be noted that each point in the Raman measurement can take up to 10s of seconds to acquire, dependent on material system. As a Raman line-scan or a 2D map is typically recorded by moving the probing laser beam point-by-point sequentially over the device surface, this thermal probing approach is therefore typically restricted to the analysis of limited device areas, and recording temperature over very wide areas similar to the IR map shown in Figure 1a, although possible, would be impractical. This poses challenges for the analysis of large area devices, like multistage HPAs, or a large series of devices during routine testing or reliability optimization stages of a technology development. Thus we believe that the developed

EXCHANGE PAPER (ISTFA 2007 BEST PAPER)

combination of IR and Raman thermography can effectively address many challenges in device thermal characterization and provides a wide range of benefits over the use of a single technique alone. Initially IR thermography is used to pre-screen devices to obtain a quick overview of the device temperature distribution, similar to as shown in Figure 1a, to identify thermal areas of interest, e.g. temperature anomalies related to hot-spots due to a leakage pathway, or cold-spots, i.e., failed parts with less or no current flow in the devices (see Figure 3). Subsequently Raman thermal analysis can be used to focus onto selected device areas in order to obtain high resolution data of accurate device peak temperatures. For instance, Figure 4 shows the effect of a defect in an GaN/SiC device epilayer on temperature, recorded with Raman thermography. A significant temperature rise occurs in the vicinity of this defect, with its size well below the spatial resolution limit of IR optics. Micro-pipe defects and/or micro-pipe clusters are present in the SiC substrates and can propagate into the grown device epilayer. Such defects enhance the risk of device failure and produce variations in device temperature and device lifetime when considering devices from different parts of a wafer [17].

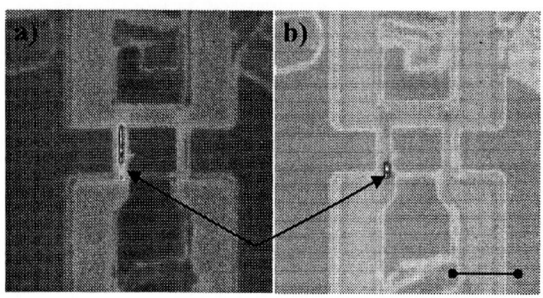

Fig. 3. Radiance map (unprocessed IR intensity image) from a) an 20 W/mm operated AlGaN/GaN HFET device, with one of the two finger operated, and b) the same device when pinched-off. Arrows show an IR anomaly related to a failed part of the device in a), and an associated leakage path when pinched-off in b). Colors are representing IR radiance from the device (red more, blue less IR emission).

temperature is close to room ambient temperature. This is artificially low and reflects more the heat sink temperature rather than the device surface temperature. In device areas covered by metal, which are not transparent to IR light, IR shows a temperature close to the ones obtained in Raman thermography and simulations (Fig 2). This additional depth temperature averaging provides further accuracy limitations for IR temperature analysis, in particular when a vertical temperature gradient is present in a device structure [11]. While IR surface sensitivity can be improved by covering the target surface with a non-transparent paint layer, this method however suffers from an uncertainty in the thermal contact at the paint/device interface and risks surface contamination, i.e., is potentially destructive.

In contrast, Raman thermography provides the additional benefit of full 3D temperature analysis of devices using confocal optical microscopy. In addition to demonstrated 2D maps of device temperature, depth scanning into a device is possible with micron scale resolution, if a laser wavelength below the band-gap of the device layers is employed in Raman thermography [11,16,18]. Figure 5 shows the example of a Raman temperature depth scan in the center of an ungated AlGaN/GaN HFET with contact separation of 20 µm (see Fig. 5 inset). There is a significant gradient of the temperature from the GaN layer, where the heat is generated, towards heat sinking SiC substrate. Remarkably, the measured temperature profile shows a sizable temperature discontinuity ΔT at the GaN/SiC interface. This temperature discontinuity is a direct manifestation of a thermal resistance at the GaN/SiC boundary (TBR). In addition to material bulk thermal conductivities or resistances, thermal properties of interfaces need to be considered in device thermal management. The presence of a TBR at the GaN/substrate interface hinders efficient heat extraction from the device active area and can play an important role in high power devices. As Raman data and thermal simulations show, the associated GaN/SiC TBR can contribute to up to 20-30% of the total temperature rise in the channel during self-heating [16].

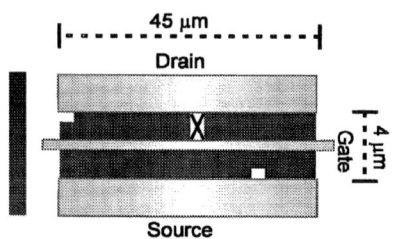

Fig. 4. High resolution Raman temperature map of operated AlGaN/GaN HFET finger, containing a defect (micro-pipe related void in SiC) marked as "x".

Additionally, since most of semiconductor materials are transparent to IR light, IR thermography can suffer from limitations in depth selectivity, when temperature is acquired from a semiconductor material. In the device areas in Fig. 1a where metal contacts do not cover the surface the measured

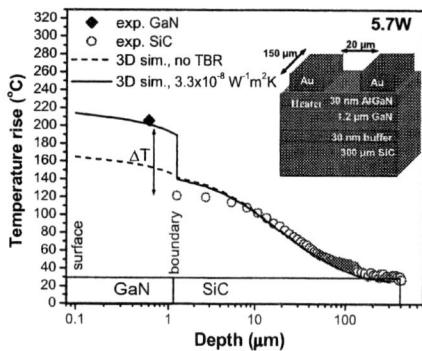

Fig. 5. Raman temperature depth scan in the centre of an ungated AlGaN/GaN HFET structure and simulated temperature profile with and without TBR at the interface (see Ref. 16). ΔT indicates the temperature discontinuity. Inset shows the schematic structure of the device.

EXCHANGE PAPER (ISTFA 2007 BEST PAPER)

IV. CONCLUSION

An integrated Raman-IR micro-thermography system was developed and applied to study the self-heating in electronic devices. Using the example of AlGaN/GAN HFET device structures, both techniques, Raman and IR, were compared side by side in terms of lateral and depth spatial resolution. It was demonstrated that IR imaging is an effective tool for screening of device self-heating to quickly locate and select thermal areas of interest, however has limitations in temperature accuracy due to spatial temperature averaging. The Raman thermal probe has the advantage of high spatial resolution temperature analysis, which allows more accurate determination of sub-micron/micron sized temperature hot-spots in the device active region, however, use of this method for analysis of very large areas is limited. Raman thermography can also be used to reveal depth temperature evolution in a device, i.e., full 3D temperature analysis. This allows better understanding of device thermal properties and enables reliable experimental verification for thermal simulation models. Combining both techniques into one system brings unparalleled advantages for thermal analysis of electronic devices, which brings device performance optimisation and failure analysis onto a new level.

ACKNOWLEDGMENT

The work in Bristol was supported by EPSRC and GWR. QinetiQ was supported by the ES domain of the UK MoD and the KORRIGAN program.

REFERENCES

[1] M. Kuzuhara, H. Miyamoto, Y. Ando, T. Inoue, Y. Okamoto, and T. Nakayama, "High-voltage rf operation of AlGaN/GaN heterojunction FETs", *Phys. Status Solidi A*, vol. 200, pp. 161-167, March 2003.

[2] Y. F. Wu, A. Saxler, M. Moore, R. P. Smith, S. Sheppard, P. M. Chavarkar, T. Wisleder, U. K. Mishra, P. Parikh, "30-W/mm GaN HEMTs by Field Plate Optimization", *IEEE Electron Device Lett.*, vol. 25, pp. 117-119, Sept. 2004.

[3] C. Lee, L. Witkowski, H. Q. Tseng, P. Saunier, R. Birkhahn, D. Olson, G. Munns, S. Guo, B. Albert, "Effects of AlGaN/GaN HEMT structure on RF reliability", *Electronics Lett.*, Vol. 41, No. 3 (2005), pp. 155-157.

[4] W. J. Roesch ,"Reliability – Beyond the basics", Mantech Workshop, Mantech 2006, Vancouver, Canada; W. Roesch, "Lifetesting GaAs MMICs under RF stimulus", *IEEE Trans*, Microwave Theory and Techniques, vol. 40, 1992 Dec, pp 2452 - 2460.

[5] R. Gaska, A. Osinsky, J. W. Yang, and M. S. Shur, "Self-heating in high-power AlGaN-GaN HFETs", *IEEE Electron Device Lett.*, vol. 19, pp. 89–91, March 1998.

[6] D. L. Blackburn, "Temperature measurements of semiconductor devices - A review", *20th Annual IEEE Semiconductor Thermal Measurement Symposium Proceedings* (2004), pp. 70-80.

[7] G. C. Albright, J. A. Stump, J. D. McDonald, and H. Kaplan, "True temperature measurements on microscopic semiconductor targets", in *Proc. SPIE Thermosense XXI*, vol. 3700, 1999, pp. 245–50.

[8] W. B. Nelson, Accelerated Testing. New York: John Wiley & Sons, 1990.

[9] M. Kuball, J. M. Hayes, M. J. Uren, T. Martin, J. C. H. Birbeck, R. S. Balmer, B. T. Hughes, "Measurement of temperature in high-power AlGaN/GaN HFETs using Raman scattering", *IEEE Electron Dev. Lett.*, Vol. 23, No. 1 (2002), pp. 7-9.

[10] M. Kuball, S. Rajasingam, A. Sarua, M. J. Uren, T. Martin, B. T. Hughes, K. P. Hilton, R. S. Balmer, "Measurement of temperature distribution in multi-finger AlGaN/GaN HFETs using micro-Raman spectroscopy", *Appl. Phys. Lett.*, Vol. 82, No. 1 (2003), pp. 124-126.

[11] A. Sarua, H. Ji, M. Kuball, M. J. Uren, T. Martin, K. P. Hilton, R. S. Balmer, "Integrated Raman/IR thermography probe for monitoring of self-heating in AlGaN/GaN transistor structures", *IEEE Trans. Electron Dev.*, Vol. 53, No. 10 (2006), pp. 2438-2447.

[12] S. Rajasingam, J. W. Pomeroy, M. Kuball, M. J. Uren, T. Martin, D. C. Herbert, K. P. Hilton, R. S. Balmer, "Micro-Raman temperature measurements for electric field assessment in active AlGaN/GaN HFETs", *IEEE Electron Dev. Lett.*, Vol. 25, No. 7 (2004), pp. 456-458.

[13] A. Sarua, A. Bullen, M. Haynes, M. Kuball, "High-Resolution Raman Temperature Measurements in GaAs p-HEMT Multifinger Devices,", *IEEE* Trans. Electron Dev., vol.54, no.8, pp.1838-1842, Aug. 2007.

[14] M. Kuball, G. J. Riedel, J. W. Pomeroy, A. Sarua, M. J. Uren, T. Martin, K. P. Hilton, J. O. Maclean, D. J. Wallis, "Time-Resolved Temperature Measurement of AlGaN/GaN Electronic Devices Using Micro-Raman Spectroscopy," *IEEE Electron Dev. Lett.*, vol.28, no.2, pp.86-89, Feb. 2007.

[15] R. J. T. Simms, J. W. Pomeroy, M. J. Uren, T. Martin, M. Kuball, "Channel Temperature Determination in High-Power AlGaN/GaN HFETs Using Electrical Methods and Raman Spectroscopy,"*IEEE Trans. on Electron Dev.*, vol.55, no.2, pp.478-482, Feb. 2008.

[16] A. Sarua, Hangfeng Ji, K. P. Hilton, D. J. Wallis, M. J. Uren, T. Martin, M. Kuball, "Thermal Boundary Resistance Between GaN and Substrate in AlGaN/GaN Electronic Devices," *IEEE Trans. Electron Dev.*, vol.54, no.12, pp.3152-3158, Dec. 2007.

[17] J. W. Pomeroy, M. Kuball, D. J. Wallis, A. M. Keir, K. P. Hilton, R. S. Balmer, M. J. Uren, T. Martin, P. J. Heard, "Thermal mapping of defects in AlGaN/GaN heterostructure field-effect transistors using micro-Raman spectroscopy", *Appl. Phys. Lett.*, Vol. 87, No. 10 (2005), pp. 103508 1-3.

[18] Hangfeng Ji, M. Kuball, A. Sarua, Jo Das, W. Ruythooren, M. Germain, G. Borghs, "Three-dimensional thermal analysis of a flip-chip mounted AlGaN/GaN HFET using confocal micro-Raman spectroscopy", *IEEE Trans Electron Dev.*, Vol.53, Iss.10, Oct. 2006, pp. 2658- 2661.

An Application of C-AFM As A Tool for SRAM Soft Single-Column Failure Analysis in Advanced HV Technologies

Hung-Sung Lin, Mong-Sheng Wu, Yun-Ming Tsou
United Microelectronics Corporation, Ltd.
No. 3, Li-Hsin Rd. II, Hsinchu Science Park, Taiwan 300, R.O.C.
Phone: 886-3-5782258 ext 33231 Fax:886-3-563-6722 Email:giant_lin@umc.com

Abstract - **It has been long recognized that SRAM memory is an ideal vehicle for defect monitoring and yield improvement during process development because of its highly structured architecture and simplified approach using memory bitmapping. However, the success rate of defect detection, especially for soft single-column failures, which is one of the most complex failure modes of an SRAM failure, is decreasing when using traditional physical failure analysis (PFA) with only the bitmap available for guidance due to a variety of invisible or undetectable defects that cause a leakage behavior in the device. In order to understand the leakage behavior in advanced high voltage (HV) processes, Conductive Atomic Force Microscope (C-AFM) [1-5] is introduced to perform junction-level fault isolation prior to attempting PFA. In this study, the quantified data extracted using the C-AFM can also be used to establish a connection between the failure mechanism discovered and the soft single column failure mode.**

I. INTRODUCTION

A typical SRAM architecture is shown in Fig. 1. It contains memory cells, row decoders (X-decoder), column decoders (Y-decoder), column selectors (Y-MUX), sense amplifiers, input/output circuits (I/O), buffer circuits, conditioning circuits, equalizing circuits, data latch circuits, and so on.

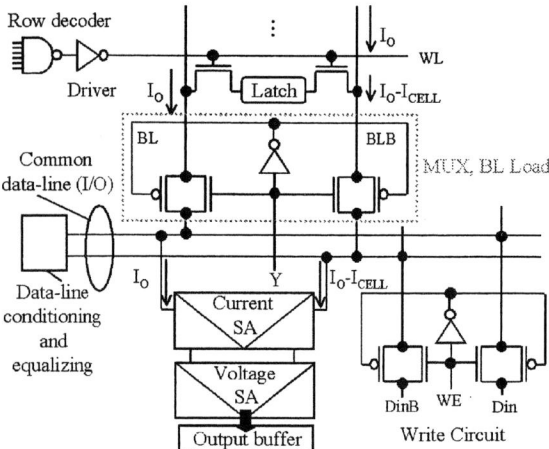

Fig. 1: Simplified SRAM architecture.

Usually the decoding functions in an SRAM are performed by using row and column decoders. The column decoder permits the selection of 1 out of n bits from the accessed row. The selected gates of a column selector controlled by the AND/NAND column decoder permit the transfer of data from the bit-lines to the column data lines. Sense amplifiers are used to detect small variations on the bit-lines (BL) and amplify them to access at the end full-swing signal. During the read cycle, the word-line (WL) is asserted and one word is selected. At this time, part of the constant current I_0, named I_{CELL}, generated by the current source flows into the cell as shown in Fig. 2.

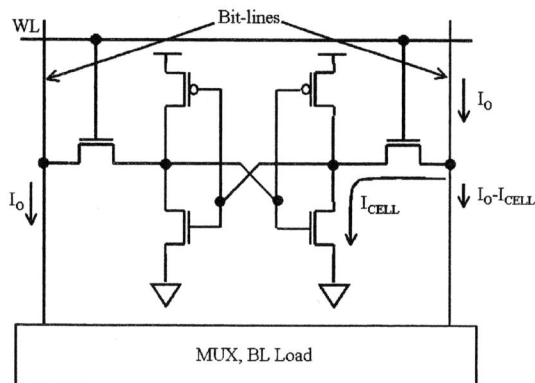

Fig. 2: Simplified SRAM read operation and its equivalent circuit.

A small current change on the bit-line bar (BLB) is desirable in order to achieve a high-speed readout. The Sense Amplifier (SA) amplifies this small current change on the BLB. It should be noted that the SA should provide a wide operating margin over all processes, temperatures and voltage corners. In this case, the failure bit counts were observed to be in a column direction, and were highly sensitive to the voltage supply level. Based on these observations coupled with the principle of SRAM operation, checking the critical path of the read/write circuitry, especially in Y-MUX area, is required in order to quickly isolate the failure location and determine the defect type, which can be used to reasonably account for the soft single-column failure and quickly identify the correct course of action required to achieve rapid yield enhancement.

II. PRELIMINARY BITMAPPING ANALYSIS

Bitmapping analysis enables us to understand failure modes

978-1-4244-2039-1/08/$25.00 ©2008 IEEE

and counts in terms of bit distribution and sensitivity to voltage or testing pattern. A typical failure bitmapping in our case study of a soft single-column failure is shown in Fig. 3. To confirm the failure mode, the SRAM was tested thoroughly at several voltages. The test data showed column failure only at VCC_{MIN}=1.08V and VCC_{NOM}=1.32V. However, it became normal at VCC_{MAX}=1.50V. This provides a valuable insight into the defect properties.

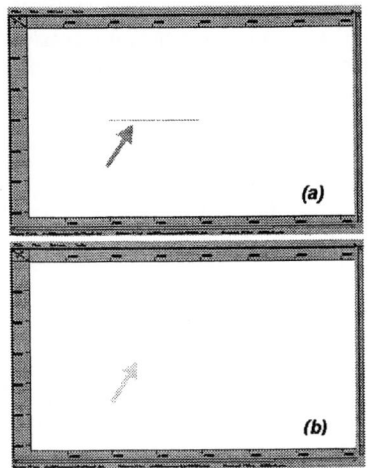

Fig. 3: Illustration of bit-mapping analysis for two bias conditions, 1.32V (a) and 1.50V (b).

III. STATIC THERMAL LASER STIMULATION ANALYSIS

A 1340nm wavelength InfraRed-Optical Beam Induced Resistance Change (IR-OBIRCH) was chosen in order to detect current change signals derived from localized resistive changes from the reverse side of the chip. The results are shown in Fig. 4.

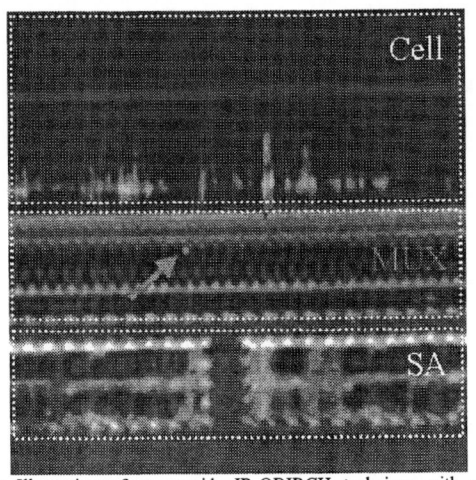

Fig. 4: Illustration of reverse-side IR-OBIRCH technique with a 100X objective.

OBIRCH spots were observed in the corresponding Y-MUX area associated with the failure column identified through the bitmapping of an SRAM instance. The sample was physically de-processed, but no physical defects were observed in the back-end metal layers. To further understand the electrical behavior and accurately pinpoint the defect location, C-AFM was introduced to perform internal probing analysis on contact level.

IV. C-AFM ANALYSIS

Bitmapping analysis showed column failures only occurred at VCC_{MIN}=1.08V and VCC_{NOM}=1.32V. It became normal at VCC_{MAX}=1.50V. The failure bit counts of the columns were highly sensitive to voltage supply level. OBIRCH spots were also observed in the corresponding Y-MUX area associated with the failure column identified through bitmapping. Based on these observations and through an analysis of the schematic circuits of the critical paths of the column decoders, it was suspected that a leakage in the column selectors associated with the failure columns was the electrical root cause. For this analysis, a –2V DC bias was applied to the sample stage during scanning, and the topography and current mapping were recorded simultaneously, as shown in Fig. 5.

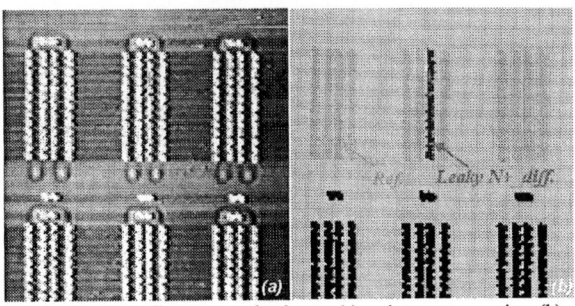

Fig. 5: Corresponding topography image (a) and current mapping (b) of C-AFM on contact level.

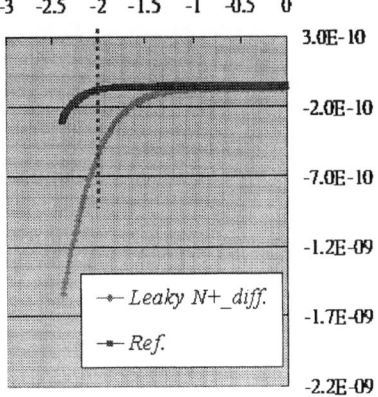

Fig. 6: IV curves measured on contact level by C-AFM.

Current mapping uses different colors to indicate different current levels in the contacts. The darker the color the higher the current level. In this way, the leaky N+_diff. contacts at the Y-MUX of the failing column were successfully isolated. In order to obtain further data, IV characterization of the failed contacts was performed. Fig. 6 shows the IV curves for the both failed and normal contacts used as reference. The failed contacts show a higher leakage current of about 0.5nA@2V, which is higher than normal contact by about 0.4nA@2V. The IV results for the leaky contacts can possibly explain the soft single-column behavior and also give some reliable physical failure hypotheses.

V. PLANE-VIEW AND CROSS-SECTIONAL TEM ANALYSIS

As the failure location was pinpointed precisely, the sample containing leaky contacts was prepared for plane-view (P-V) and cross-sectional (X-S) Transmission Electron Microscope (TEM) inspection. As shown in Figs. 7 and 8, substrate dislocation believed to extend through the junction of the N+/PW diode is responsible for the increased junction leakage.

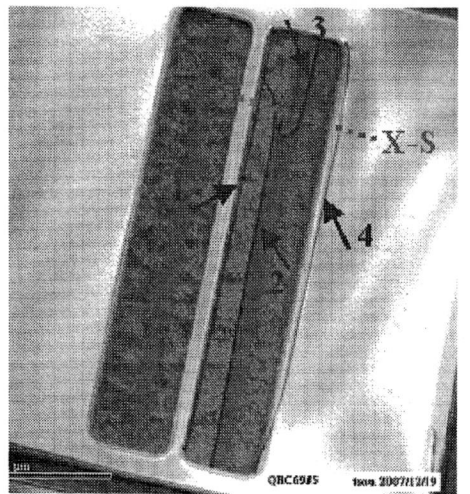

Fig. 7: P-V TEM image on the Y-MUX N diffusion area isolated by C-AFM.

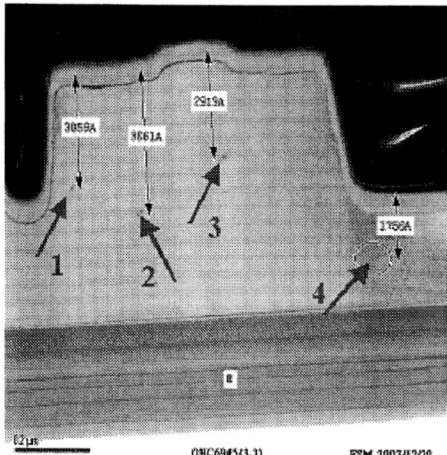

Fig. 8: X-S TEM image on the Y-MUX N diffusion area isolated by C-AFM.

This crystal defect played the role of traps assumed to be distributed energy-wise across the bandgap and spatially uniform, and causing a higher excess carrier generation rate that is responsible for the higher current in reverse-bias direction of the N+/PW junction. As a result, substrate dislocation can explain the IV behaviour of the leaky contacts.

VI. THE INFLUENCE ON SRAM OPERATION

Based on the results from the leaky N+_diff contacts caused by substrate dislocation, a leakage current, I_L, exists in the column selector, as shown in Fig. 9. This current drop also changes the reference current of BL from I_0 to $I_0 - I_L$.

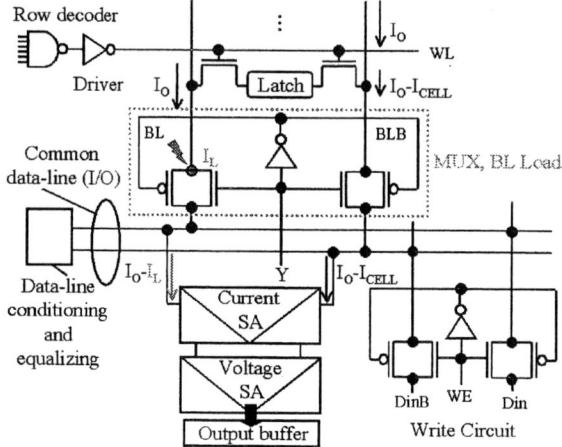

Fig. 9: Simplified SRAM architecture.

Current sense amplifiers are used to detect the small current variation between BL and BLB and amplify it to access the full swing signal at the end.

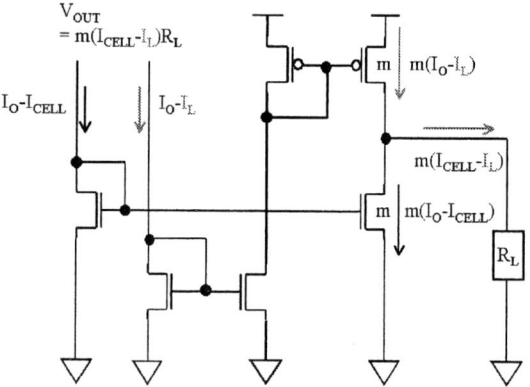

Fig. 10: Simplified current sense amplifier operation and its equivalent circuit.

As shown in Fig. 10, normally, the current sense amplifier amplifies the net current, subtracting the current $I_0 - I_{CELL}$ at the BLB from the reference current I_0 at the BL, in order to determine the full swing m I_{CELL} and $V_{OUT} = mI_{CELL}R_L$. However, the leakage current I_L will erode $V_{OUT} = mI_{CELL}R_L$ and reduce the swing to $V_{OUT}=m(I_{CELL}- I_L)R_L$. The worst case will result in no swing when $I_L=I_{CELL}$, which will in turn result in a solid column failure. As the bit-lines are acutely sensitive to noise, and if the noise margin is severe, dash-like instead of solid columns will occur as in our case. This can also be reasoned to account for why failed BL became normal at $VCC_{MAX}=1.50V$. The explanation is as follows. The ON resistance of the MOS consists of a series combination of drain resistance (R_D), source resistance (R_S) and channel resistance (R_{CH}). Typically, by design, the contribution from R_D and R_S is small enough such that the primary consideration is the (R_{CH}). An expression for the channel resistance can be derived as follows. In the ON state of the switch, the voltage across the switch should be small and V_{GS} should be large. Therefore, the MOS device is assumed to be in the nonsaturation region according to the Linear Bulk Charge Model shown in equation (1). The channel resistance is given as equation (2) where V_{DS} is less than $V_{GS}-V_{tn}$ but greater than zero.

$$I_D = \mu_n C_{OX} \frac{W}{L}\left[\left(V_{GS}-V_{tn}\right)-\frac{1}{2}mV_{DS}\right]V_{DS} \qquad (1)$$

$$m \approx 1+\frac{3t_{OX}}{W_d}$$

$$R_{ON} \approx R_{CH}$$

$$R_{ON} = \frac{1}{\partial I_D/\partial V_{DS}} = \frac{L}{\mu_n C_{OX} W\left(V_{GS}-V_{tn}-mV_{DS}\right)} \qquad (2)$$

I_D : drain current

μ_n : electron mobility

C_{OX} : gate oxide capacitance

    W : channe width

    W L : channel length

V_{GS} : gate to source voltage

V_{tn} : threshold voltage

m : body-effect coefficient

V_{DS} : drain to source voltage

t_{OX} : oxide thickness

W_d : depletion width

It can be seen that a lower value of R_{ON} is achieved for larger values of V_{GS} with a fixed channel width and length when V_{DS} remains at a small value. Therefore, at $VCC_{MAX}=1.5V$, I_{CELL} increased because of the decrease in R_{ON}. In this study, however, the increase in I_L was very limited according to the IV results obtained using the C-AFM, as shown in Fig. 6. As a higher swing for $V_{OUT}=m(I_{CELL}- I_L)R_L$ can be obtained because of the increase in the I_{CELL}, accompanied with a very limited change in I_L, it became normal at $VCC_{MAX}=1.5V$.

VII. CONCLUSION

In this paper, the use of the C-AFM technique is successfully demonstrated in the study of SRAM soft single-column failure and the isolation of nanoscopic defects. After identifying the process step that failed, and determining the defect types, corrective action can be taken to quickly improve yield. In this study, an efficient methodology for overcoming the difficulties in detecting nanoscopic, or even invisible defects, especially for complex failure modes in an SRAM failure, was established.

REFERENCES

[1] Hung-Sung Lin, Wen-Tung Chang, Chia-Hsing Chao, Jesse Wang, Chang-Tan Lin, and Coswin Lin, "Failure Analysis of Soft Single Column Failure in Advanced Nano SRAM Device with Internal Probing Techniques", 31st ISTFA Proceedings, Nov. 2005, pp. 46-48.

[2] Hung-Sung Lin, Wen-Tung Chang, Chun-Lin Chen, Tsui-Hua Huang, Vivian Chiang, Chun-Ming Chen, "A study of Asymmetrical Behaviour in Advanced Nano SRAM Devices", 13th IPFA Proceedings, Jul. 2006, pp. 63-66.

[3] Hung-Sung Lin, Chun-Ming Chen, Kuo-Hsiung Chen, Afung Wang, "A Case Study of Defects due to Process Marginalities in Deep Sub-Micron Technology", 18th ESREF Proceedings, Oct. 2007, pp. 1604-1608.

[4] Hung-Sung Lin, Ying-Chin Hou, Juimei Fu, Mong-Sheng Wu, Vincent Huang, Ching-Heng Chou, " A Case Study of Defects due to Process-Design Interaction in Nano Scale Technology", 33th ISTFA Proceedings, Nov. 2007, pp. 172-175.

[5] T. X. Tong, A. N. Erickson, "Current Image Atomic Force Microscopy (CI-AFM) combined with Atomic Force Probing (AFP) for location and characterization of advanced technology node", 30st ISTFA Proceedings, Nov. 2004, pp. 42-46.

Conductive Atomic Force Microscopy Failure Analysis for SOI Devices

Lim Soon-Huat[†], Zheng Xinhua*, Teo Chea-Wei, Vinod Narang, Teo Beng Hock*, JM Chin

Advanced Micro Devices Pte Ltd, 508, Chai Chee Lane,
AMD Singapore 469032.
Contact: Tel: (65) 6796 9888, Fax: (65) 6233 9080, [†]Email: Soon-Huat.Lim@amd.com
Contact: Tel: (65) 6550 0467, Fax: (65) 6454 9871, *Email: Zheng_Xinhua@nyp.gov.sg

Abstract- **A FIB shorting technique to create a conducting path across the buried oxide to connect active silicon to silicon substrate is demonstrated to allow Conductive Atomic Force Microscopy (CAFM) failure analysis on SOI devices. CAFM is carried out at via and contact levels to provide current images that helped to localize the faulty node and also determine current-voltage characteristics at an area of interest.**

I. INTRODUCTION

Failure analysis is becoming difficult due to the shrinking transistor dimensions and increasing layout complexity. In many cases, physical failure analysis (PFA) fails to identify the anomalies responsible for the failure as scanning electron microscopy (SEM) inspection is not able to provide visual or diagnostic information about the failure. Similarly Passive Voltage Contrast (PVC) technique using electron or ion beam sometimes cannot precisely isolate abnormal contacts and vias due to the lack of sensitivity of the collected electron signals at the region of interest.

One of the methods that are increasingly being used in the failure analysis (FA) process is Conductive Atomic Force Microscopy[1] which is a Scanning Probe Microscope (SPM) based contact probing technique[2]. This technique simultaneously provides topography information with nanometer scale resolution as well as electrical characteristics of the surface. It can perform current mapping of large areas and provide current-voltage (I-V) characteristics of specific contacts, vias and even foreign particles in a non-destructive manner[3]. This current image of the region of interest (ROI) gives an easy to interpret map of the possible defective contacts or vias. This technique is more sensitive than PVC as the contrast mechanism is the direct current measurement of the surface. CAFM technique for bulk silicon devices is easily performed by biasing the silicon substrate while scanning with the conductive tip to measure the current passing through each point of contact.

Increasing number of device fabrication technologies are switching to Silicon-on-Insulator (SOI) technology[4] due to the advantages of low voltage, low power consumption and high speed. However in SOI devices, existing CAFM methodology is not feasible as the conductive path required to bias the active area cannot be not formed. This is due to the presence of a blanket film of silicon dioxide known as buried oxide which isolates active transistors from the silicon substrate.

In this paper, we will describe the focused ion beam (FIB) shorting method that will allow the CAFM technique to work on the SOI devices. We will also elaborate on the information that can be extracted from the CAFM current images and I-V curves. Finally, case studies that show successful applications of the CAFM techniques on SOI defects are presented.

II. EXPERIMENTAL PROCEDURE

In routine PFA process, the die is de-processed using parallel lapping and inspected layer by layer at the possible defective location. Inspection of the ROI is conducted by an optical light microscope or the scanning electron microscope. Preparation of the sample for CAFM scanning can be carried out once the layer of interest is reached or a suspected defect is detected.

In this work, we present a FIB method that allows the CAFM technique to be applicable on SOI substrates. This front-side FIB method involves the use of an FEI Strata Dual Beam System to create a conducting path from the active silicon to the bulk silicon substrate at a suitable location. We first locate the failing location with the continuous active silicon through the electron beam. A FIB trench is milled from the front-side of the die through the buried oxide layer to the bulk silicon near the defect location and platinum metal is then deposited to fill the trench. This results in a conducting path to be formed from the bulk silicon to the active silicon. The voltage bias from the metallic sample holder can then be transmitted through the contacts and vias to the conductive scanning AFM tip at the surface as shown in figure 1. This method also helped to create a FIB marking for easy location during AFM scanning.

978-1-4244-2039-1/08/$25.00 ©2008 IEEE

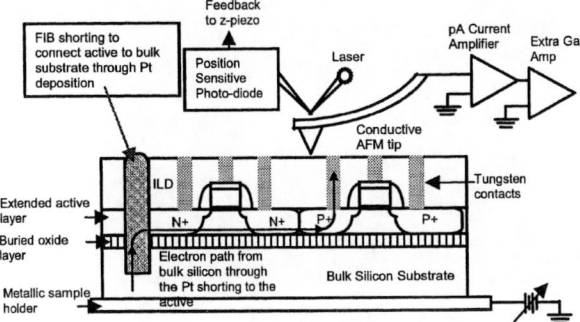

Figure 1: Schematic of FIB shorting technique to facilitate CAFM failure analysis on contact and via level.

A Veeco Dimension V Atomic Force Microscopy was used for imaging purposes, with a TUNA® electrical measurement module which can apply a dc bias of ±12 V and measure current of 60fA to 120pA. Diamond doped coated silicon AFM probes are used for scanning operation due to the increased wear resistance against the metal contacts and vias. These scanning tips have a nominal radius of 35nm and a conductivity of 0.01Ω.cm. The system applies bias voltage on the metallic sample holder while maintaining the tip at ground.

The measured current is in pico-ampere range due to the extended conducting path through the active silicon and resistance of bulk silicon. Using this method, CAFM can be performed at cache memory area with extended active silicon.

III. CASE STUDIES

CAFM failure analysis has been applied at via and contact levels for SOI devices in the following three case studies.

In the 1st case study, the sample was deprocessed to contact level and SEM inspection showed no abnormalities. The FIB shorting method was carried out at a nearby location which connects the extended active silicon to the bulk silicon. Further defect localization was carried out using CAFM and current map was obtained with 0.5V DC bias.

CAFM at contact level with FIB shorting showed the presence of an abnormal poly contact drawing higher current compared to a good poly contact as shown in figure 2. Contacts to active silicon and poly-silicon can be differentiated through the current contrast since poly-silicon is essentially a floating structure due to presence of gate oxide. CAFM analysis clearly showed that the abnormal poly-silicon contact was drawing higher current, which is comparable to those contacts landing directly on active silicon.

This result suggests that the contact to the polysilicon is shorted to the active silicon. Since initial SEM inspection did not show any defects in the area of the contacts, the possible location of the defect would be beneath the contacts, near the

gate oxide and silicide layer. Cross-sectional transmission electron microscopy (TEM) analysis was done on the failing contact location. TEM imaging of the defective area shows nickel silicate extending beneath the spacer towards the poly-silicon in figure 3. The conductive nickel filament could have shorted the floating polysilicon to the active silicon.

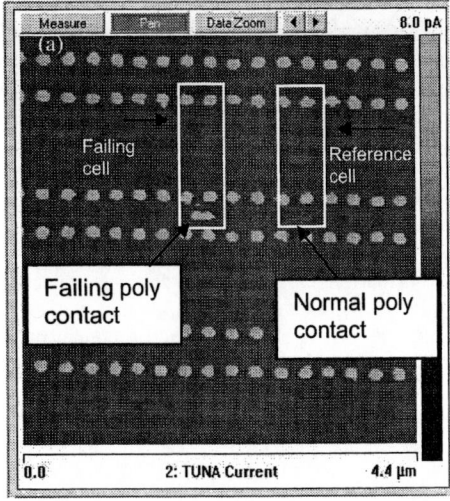

Figure 2: CAFM current image at contact level shows the presence of failing poly contact.

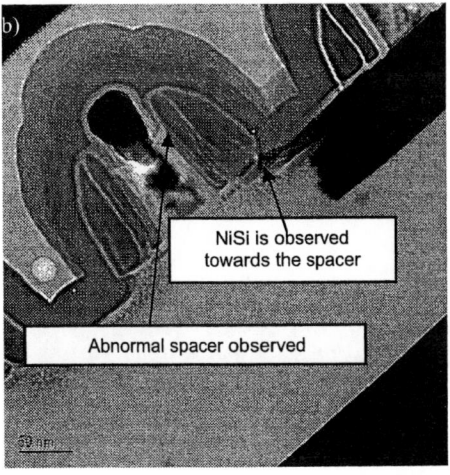

Figure 3: TEM micrograph shows presence of nickel silicate growing beneath the spacer towards the poly-silicon.

In the 2nd case study, the FIB shorting method was applied to higher metal layers.

The sample was deprocessed to via2 level for SEM inspection. Initial SEM inspection at 15keV showed a suspected buried particle bridging the isolated via and the adjacent long line in figure 4(b). However, it was not clear if it was a true defect or a deprocessing artifact. FIB shorting was carried out to connect the long metal line to the bulk silicon for C-AFM analysis. From the layout, the vias on the shorted metal lines were connected to the polysilicon, which were essentially floating structures. Based on this information, the CAFM technique can be used to differentiate the electrical properties of the suspected and normal vias.

CAFM analysis at via level showed an abnormal via as shown in figure 4(a). The suspected via was showing a higher current flowing through as compared to other normal vias. CAFM current map image was able to differentiate electrically the difference between the normal isolated vias that were connected to polysilicon and the faulty via that was shorted to the adjacent metal line. This proved that the particle was a true fabrication process defect.

Using this information, cross sectional analysis was carried out perpendicular to the extended copper lines. Cross-section SEM micrographs at that location identified a copper particle connecting the shorted isolated metal line to the extended metal line as in figure 5.

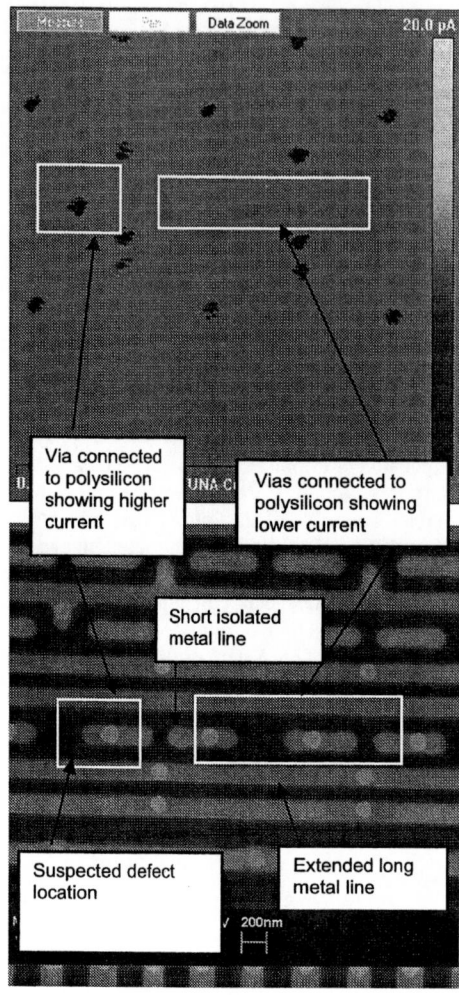

Figure 4: (a) CAFM current image of via level shows presence of conducting poly via (b) SEM images of corresponding location showed suspected particle that caused shorting of isolated via to metal line.

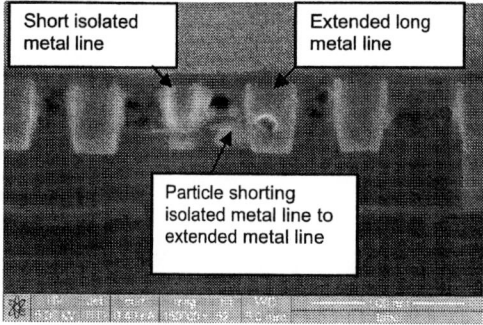

Figure 5: Cross-section SEM micrographs showed particle shorting isolated metal line to extended metal line.

In the 3rd case study, the CAFM technique was used as a non-destructive method for defect isolation and root cause analysis.

The sample was deprocessed to via2 level for SEM inspection. SEM inspection showed a large particle shorting between the isolated metal line and extended long metal line as in figure 6, which was similar to the 2nd case study. SEM imaging could not confirm very clearly if the particle was due to fabrication defect or induced by parallel polishing during deprocessing. FIB shorting was carefully done on the extended metal line to connect the long metal line to the bulk silicon. CAFM analysis was conducted at the defect location.

CAFM current map in figure 7(a) showed that the via with particle short exhibits a more conductive behavior than other normal vias. This confirmed that the particle from fabrication process was causing the isolated via short. Additional information from the CAFM analysis revealed that the particle was buried inside the inter-level dielectric (ILD) material because the current map did not show the presence of the particle. The topography image revealed a small bump at the defect location as the particle was covered by the ILD material. The CAFM technique provides the means to non-destructively scan across the surface of the sample, so that the particle does not get dislodged or scratched off.

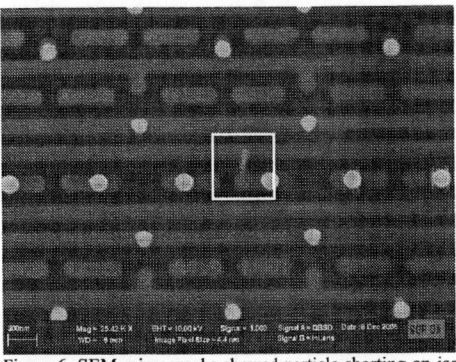

Figure 6: SEM micrographs showed particle shorting an isolated metal line to an extended metal line.

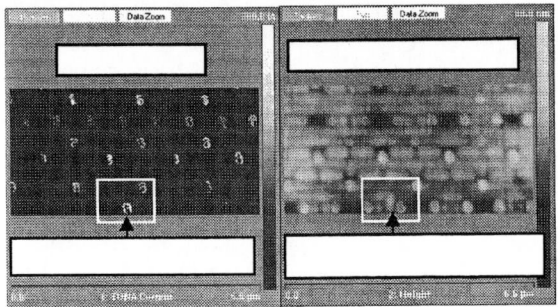

Figure 7: (a) Current map showed that the abnormal via is conductive due to the presence of a shorting particle. (b) Topography image showed a small bump due to the buried particle.

IV. CONCLUSION

The front-side FIB shorting method is able to overcome the inability of performing CAFM analysis on SOI devices due to presence of buried oxide layer. The three case studies illustrate the effectiveness of CAFM fault isolation for SOI devices at both contact and via levels for successful identification of defects. Moreover, this method can be easily integrated into the existing PFA process flow. Electrical information from CAFM analysis has helped to supplement the SEM images to provide more data about the failure mechanism.

REFERENCES

1. K. Krieg, D.J. Thomson, G.E. Bridges, "Electrical probing of deep sub-micron ICs using scanning probes", Microelectronics Reliability, 2001, vol. 41, no. 8, pp 1185-1191

2. M.F. Bailon, P.F. Salinas, J.S. Arboleda, "Application of Conductive AFM on the electrical characterization of single-bit marginal failure", Proc. of 12th IPFA 2005, pp282-284.

3. M. Porti, M. Nafria, X. Aymerich, "Current limited stresses of SiO$_2$ gate oxides with conductive atomic force microscope", IEEE Transactions on Electron Devices, 2003, vol. 50, no. 4, pp933-940

4. S. Herschbein, C. Rue, C. Scrudato, "The Joy of SOI: As Viewed from a Backside FIB Perspective", Proc. of 31th ISTFA 2005, pp78-83.

Advanced Localization Technique of Failures in Packages / IO-Stages of Chips using Vector Network Analyser

Bernd Krueger, Helmut Pohl, Fritz Schumann, Stephan Schoemann

Infineon Technologies AG, Munich, Germany

Phone: (+49) No. 89 234 27088   Fax: (+49) No. 89 234 9553405   Email: Bernd.Krueger@infineon.com

Abstract-**Current Time Domain Reflectometry (TDR) Equipment suffers from certain limitations [1]. These can be resolved by using a Vector Network Analyser (VNA) system [2]. This paper will highlight the advances and drawbacks of using a VNA for the localization of failures within the signal path of packages or input-/output-stages of chips. Various examples will demonstrate the capabilities. Further improvements by using simulation tools may be possible.**

I. INTRODUCTION

There are different methods to visualize current leakages in packages or in the input-/output-stages of dies, but there is only Time Domain Reflectometry (TDR) to localize opens.

Two different instrumental approaches were realized and compared in the scope of this paper. The conventional setup consists of a pulse generator and a high speed oscilloscope that records the reflections in real time. An alternative approach is to apply a Vector Network Analyser (VNA) to receive the device response in the frequency domain. Continuous waves within a specified frequency range are applied to the device under test (DUT) and the fraction of the returned intensity is measured to determine the scattering parameter, usually referred to as S-parameter. The parameter "Direct Reflection (S_{11})" allows the determination of the fault location. This can be done, because the time domain signal can be calculated, by inverse Fourier-Transformation, of the frequency domain data:

$$f(t) = \frac{1}{2\pi} \int_{-\infty}^{\infty} F(j\omega) e^{j\omega t} d\omega$$

(1)

II. COMPARISON TDR- TO VNA-EQUIPMENT

In a first step it could be verified that the information obtained is identical for both methods and can be converted between time- and frequency-domain by Fourier-Transformation without any loss.

However the different instrumental realization causes several differences in performance and accuracy. The most obvious is the bandwidth, which can be as high as 110 GHz for a VNA (the instrument used in our lab had 67 GHz), while the overall bandwidth of the pulse-generator and oscilloscope combination is hard to advance beyond 30 GHz. An example for the dependency of TDR traces on bandwidth is shown for a typical Ball Grid Array (BGA) package in figures 1, 2.

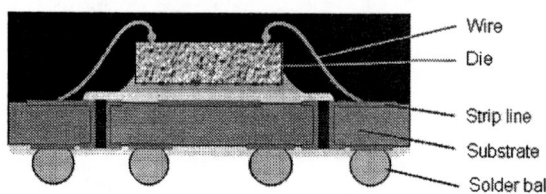

Fig. 1. Schematic cross section of BGA package.

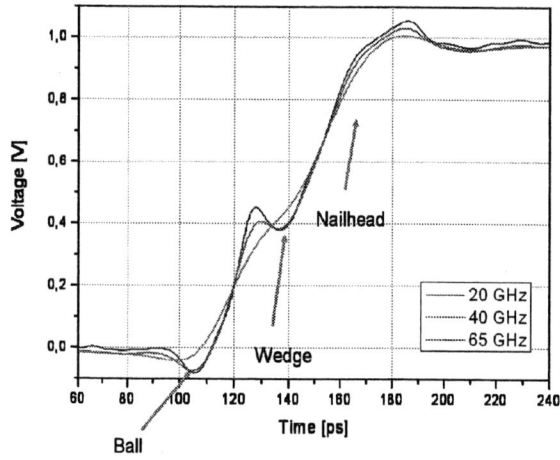

Fig. 2. Examples of TDR traces dependending on band width shown at BGA package.

The effect of improved detail resolution with increasing bandwidth is obvious. With the identification of important features within the signal path, the rough localization of the failure site becomes much easier.

A second major drawback of the conventional setup is the introduction of jitter by the trigger detection system of the oscilloscope. As can be seen in figure 3 the TDR trace from the pulse generator-scope setup is not able to resolve the green and the blue lines because of noise due to random time delay in the order of 10 psec. For the same DUT, the network analyzer shows a perfect overlap of all five measurements.

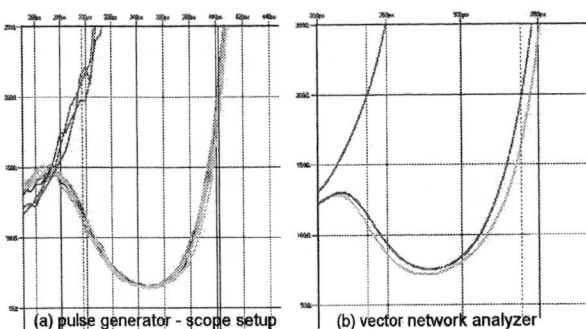

Fig. 3. Examples of TDR traces from different instruments.

Another advantage of the network analyzer is the increased dynamic range of above 90 dB compared to only 40 dB of the conventional TDR. Moreover the VNA signal amplitude can be selected deliberately and even a few mV lead to reliable results. Therefore any non-linear effects and damages can be much better suppressed than for the typical 500 mV pulse generator signals. Finally the VNA is a very versatile High Frequency (HF) tool that can be used additional to characterize a number of reflection-, transmission-, amplification- and cross-talk- properties.

This sensitivity and versatility is at the same time the main disadvantage of the VNA, because it requires extensive HF experience. Especially the correct calibration of the setup and the selection of suited equipment can influence the measurement results significant.

The most critical part for the application is the interface to the device. We tested different probes and achieved best results with the optimized ground-signal-ground architecture shown in figure 4. A set of similar probes for different pitches from 250 to 1250 μm allow measurements on all standard BGA packages and several other package types, too.

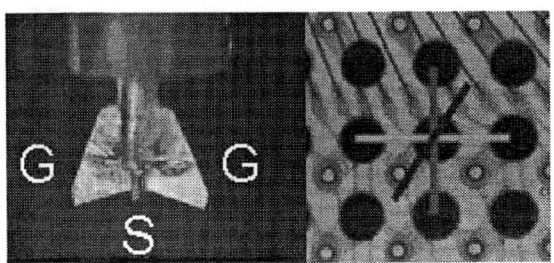

Fig. 4. Instrumental setup components.

The first step of a TDR measurement is always a calibration, to get rid of the delay and any distortions, caused by the cables to the probe station and the probes. Certified standard structures on a ceramic substrate are available (open, short, 50 Ohm).

III. CASE STUDIES

Experience with numerous real-life package problems showed that it was possible to decide in every single case where the failure was located, either at the package or at the silicon die.

The example in figure 5 shows an impedance deviation very close to the ball. XRAY analysis revealed a broken substrate micro via (s. fig. 6).

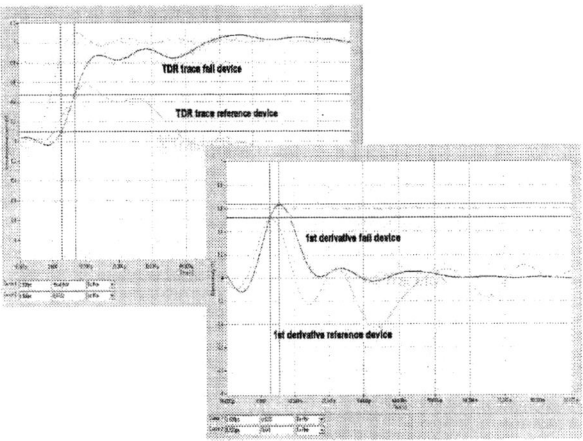

Fig. 5. Example of TDR traces of failure at substrate via.

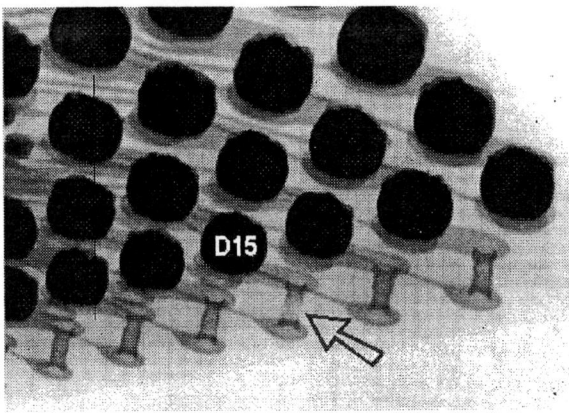

Fig. 6. Example of broken substrate micro via.

The second example in figure 7 shows a perfect match between a failing device and a reference part up to the point marked with a green 1. There is an impedance change just before the signal reaches the ESD protection circuit. The graph contains the original TDR traces in purple/brown and the first derivatives in green/blue, which are used to highlight small differences in slope or position between fail and reference sample. Further failure analysis (FA) work proved that the pad aluminum was corroded (s. fig. 8).

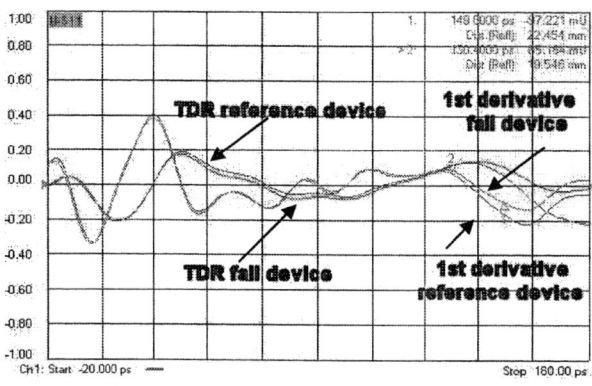

Fig. 7. Example of TDR traces showing failure at pad area.

Fig. 8. Example of pad corrosion.

The third example shows a significant deviation between a failing device and a reference part, which indicates that there is an impedance change at the substrate strip line (s. fig. 9). The diagram contains the original TDR traces in purple/brown and the first derivatives in green/blue.

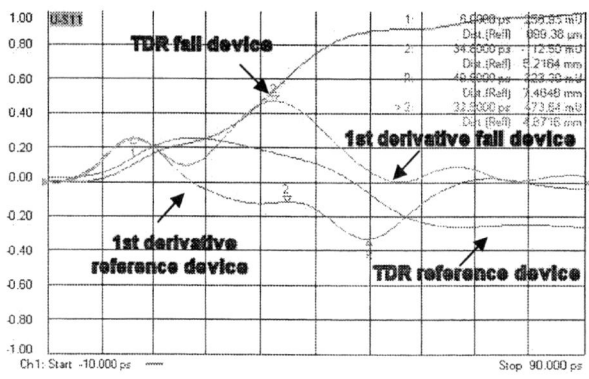

Fig. 9. Example of TDR traces of failure at substrate.

XRAY analysis revealed a crack affecting a substrate strip line (s. fig. 10).

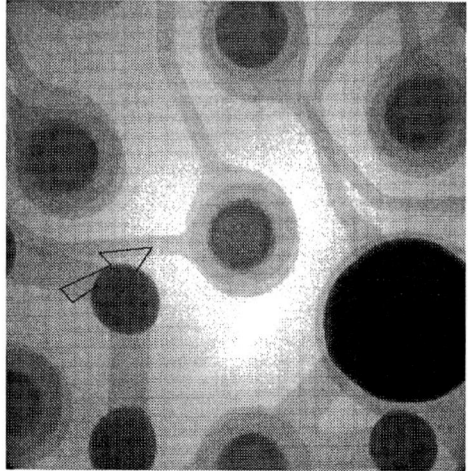

Fig. 10. Example of broken substrate metal line.

IV. BENEFIT FROM SIMULATION TOOLS

It turned out that the overall accuracy is often limited by the unknown local signal velocity and things like overlap of different lines after bifurcation or other ambiguities of the signal interpretation. Therefore the next important step is to develop tools to get a better understanding of the details of the TDR trace shapes – for example by HF simulation. The final example shows a failure analysis case, where a signal starting at a common package ball is bifurcated into 4 different signal lines (s. fig. 11)

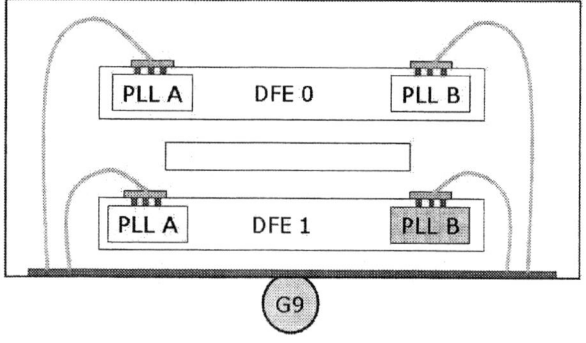

Fig. 11. Schematic of a bifurcated signal trace within a stacked package.

Network analyzer measurement at about 100 MHz, probing ball G9 vs. neighbouring VSS and signal balls, reveals 130 pF for the pass device and only 99 pF for the failing device. Derivative of measurement signal (see figure 12) showed at about 110 ps a pronounced deviation between pass and fail device.

978-1-4244-2039-1/08/$25.00 ©2008 IEEE

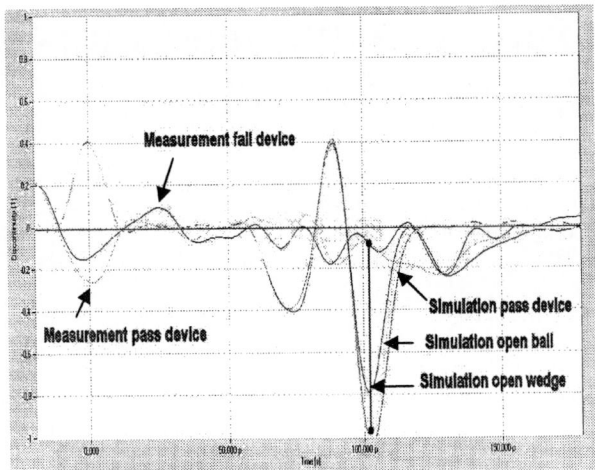

Fig. 12. Examples of TDR traces and simulation results for DFE1/PLL B partial signal path.

Combination of measurement and HF-Simulation of the signal path from ball G9 to PLL_A and PLL_B pads revealed, that the failure signal time estimated the location of the failure around the bonding wire area of the bottom die (s. fig. 13).

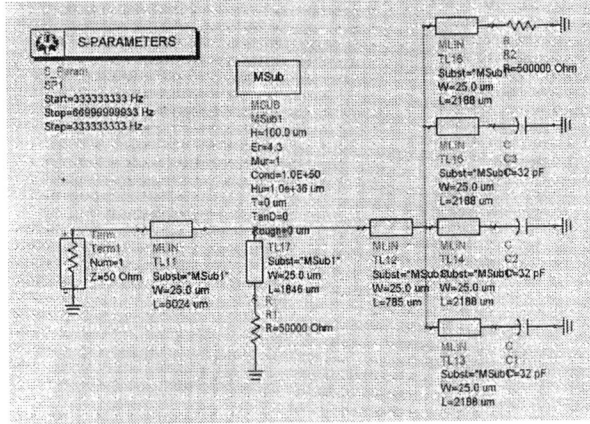

Fig. 13. Examples of simulation setup for open ball failure case.

Further FA work proved that one bond ball was lifted at the bottom die connecting PLL_B (s. fig. 14).

Fig. 14. Examples of lifted bond ball at bottom die of the stacked die package.

V. CONCLUSION

VNA-based TDR – which actually is called FDR because it is frequency-domain – is clearly superior to the combination pulse-generator plus oscilloscope.

The accuracy of about ± 2 ps to determine the failure location in time domain is excellent, but the translation of this number into a length that is needed for failure analysis adds some additional uncertainty, because the signal velocity depends a lot on the dielectric constant of the surrounding material which is not precisely known. With the help of HF-simulation tools, further improvements are achievable.

Nevertheless in several cases the determination of the failure position in the package was better than 500 μm.

ACKNOWLEDGMENT

I would like to thank B. Waidhas and M. Engl for their contributions to this paper.

REFERENCES

[1] Tektronic, Inc., *TDR Impedance Measurements: A Foundation for Signal Integrity,* Application Note, Literature Number 55W-14601-1, 2005.

[2] Agilent Technologies, *Time Domain Analysis Using a Network Analyzer,* Application Note, Literature Number 5989-5723EN, 2007.

Device-Level Fault Isolation of Advanced Flip-Chip Devices using Scanning SQUID Microscopy

C.W. Teo, H.E. Lwin, V. Narang, J.M. Chin
Advanced Micro Devices Singapore Pte Ltd
508, Chai Chee Lane, Singapore 469032
Phone: (65) 67969888 Fax: (65) 62339080 Email: chea-wei.teo@amd.com

Abstract- **This article describes how a scanning SQUID microscope (SSM) enhances the capability of device-level fault isolation on advanced 90 nm and 65 nm flip-chip microprocessor devices. SSM has proved to be very useful in isolating bump shorts and shorts in copper interconnects. For improved resolution and analyzing bumped dies, a front-side SSM technique is developed that has greatly increased success rates and analysis turn-around time. In this paper, we focus on die-level fault isolation on advanced microprocessor devices with numerous metal layers.**

I. INTRODUCTION

As silicon process technology scales down with larger multi-layer interconnects and more complex design, it has become a great challenge to electrically isolate the defects in failure analysis.

This article describes how a scanning SQUID microscope (SSM) enhances the capability of device-level fault isolation for advanced 90 nm and 65 nm flip-chip microprocessor devices. SSM makes use of a highly sensitive magnetic sensor to detect the tiny magnetic field produced by the current flowing in a device. It converts the magnetic field data to current density based on the Biot-Savart law and fast Fourier transform (FFT). With optical and current density overlay capability, SSM helps identify a short location on the device [1].

In previous studies, SSM has proved very useful in isolating bump shorts, shorts between wires in wire-bonded devices, and shorts in copper traces [2-4]. There are also a few die-level fault isolation cases presented in [5], showing the capability of SSM in die-level fault isolation. In this paper, we focus on die-level fault isolation on advanced microprocessor devices with numerous metal layers. We also developed a method involving front-side SSM at the C4 bumps layer with the package removed to isolate the short location. The front-side SSM also proved very useful for analyzing bumped dies having yield issues that required fast analysis turn-around time. Front-side SSM greatly improves resolution because the SQUID sensor is nearer the shorted region and, hence, increases analysis success rate.

In the failure analysis process, infra-red emission microscopy (IREM) and scanning optical microscopy (SOM) tools are widely used for die-level fault isolation.

Unfortunately, these tools are not sensitive to shorts occurring at higher metal layers or to complex routing. Moreover, for flip-chip devices with multiple metal layers, most of the analysis is carried out from the die back-side, thus making the detection of upper metal layer shorts harder. For thermal/laser-based tools, often times, the device must be biased at a higher supply level to detect the short. Higher bias generally worsens the shorted defect and may lead to severe low-k delamination and copper interconnect melting, thus making it hard to find the root cause of failure. SSM sensitivity is not affected by the number of metal layers and complex routing. This gives it an advantage over other fault isolation techniques. Furthermore, low current (<1 mA) is required during SSM.

The separation distance between the sensor and current flow in the die is one of the major contributors to the limited resolution of SQUID. This poor resolution makes the fault isolation for 90 nm and below flip-chip devices with multiple metal layers difficult. However, we combine both current distribution analysis and CAD layout tracing to understand and determine the possible defective location. Selected successful analyses using these methods are presented in the sections below.

II. CASE STUDIES

A. *90 nm Device I/O pin Short*

A 90 nm device that had undergone an ESD stress test showed several I/O pins shorted or having higher leakage. This was a very unusual case as the failing I/O pins were "non-connect" pins, *i.e.*, these pins' bumps are separated from their respective die-level interconnect by a passivation layer. It is not known whether the short occurred in the package or in the die. First, the sample was prepared using die back-side polishing to a thickness of 500 um. The die back-side SSM was then used to determine the short location. With the help of the acquired magnetic and current density data coupled with layout tracing of the interconnection metal layers, it was deduced that the short occurred in the die. As seen in Fig. 1.1, there was an extra current distribution between pin 1 and pin 2. Based on the die layout, we found that the extra current density seen in the SSM image matched the I/O supply power plane. Hence, it gave strong evidence that the short occurred in the die and not in the package. Further physical failure analysis (PFA) showed that the passivation layer had ruptured and severe metal shorting at the top metal layers was observed (see Fig. 1.2). The detailed analysis thus found that the passivation layer for the non-connect pins is vulnerable to existing ESD stress test.

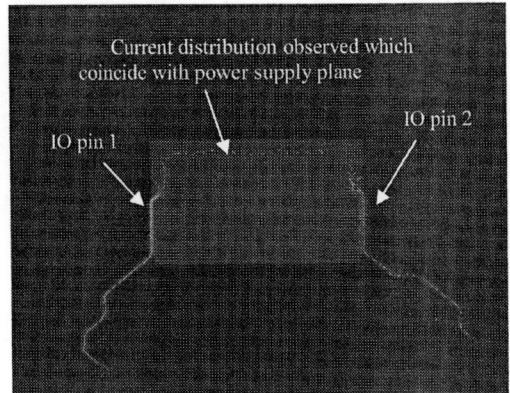

Fig.1.1: Current density overlay of 2 I/O pins shorted through the power supply.

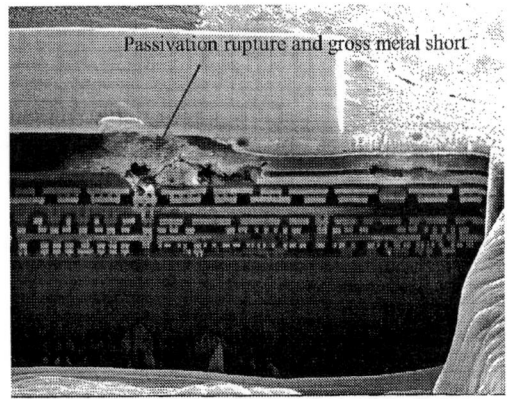

Fig.1.2: Cross-section image showing passivation damage and higher metal layer melting.

B. 90 nm *Device I/O Supply Short*

This is a baseline yield improvement analysis for a 90 nm device. The objective was to determine failure mechanisms that could contribute to the yield loss. We used SSM from die back-side to isolate the short region. Generally, the short between two metal lines is too tiny: it is impossible to detect the spatial resolution of shorted fine metal lines by SSM. As shown in Fig. 2.1, the distribution of the current in SSM suggest that the short could be at a higher metallization layer, based on the longer metal line length and orientation corresponding with die layout. But, for bumps short, the separation between two bumps will be large enough to detect with SSM resolution. PFA at the area of interest showed incomplete copper filling at M3 interconnects (Fig. 2.2). This probably resulted in bridging between metal lines in M4 as chemical mechanical polish (CMP) was unable to completely remove the excess copper.

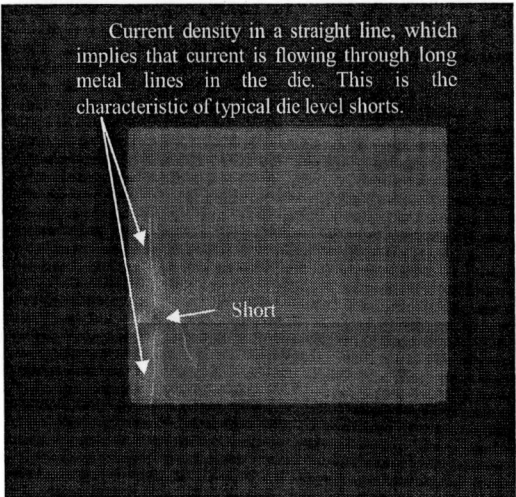

Fig.2.1: Short location identify by SSM.

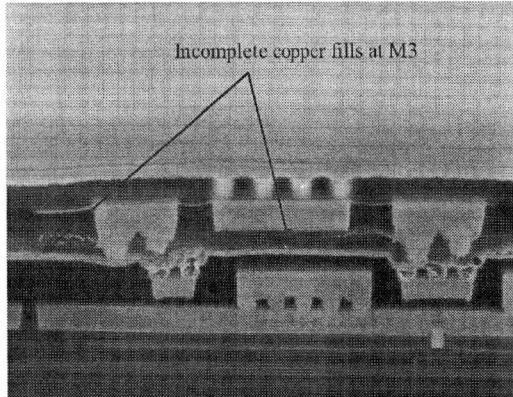

Fig.2.2: Cross-section image showing unfilled M3.

C. 65 nm *Device I/O Supply Short*

This is a 65 nm device failing for I/O power supply short. The initial suspect was bump-to-bump short and PFA was focused on finding defects at the bump. No anomaly was observed at the bumps and it was found that the short might possibly have occurred in the die. This usually would have resulted in a cause-not-found (CNF), since we are unable to use other available fault isolation techniques in such cases. We carried out SSM on the front-side of the die by supplying current to the failing power supply bump and VSS C4 bumps. This technique required extreme care in positioning of the micro-probes as there is a chance of hitting the probe with the SQUID magnetic sensor. SSM scan is carried out from the die front-side, where the bumps are located. As seen in Fig. 3.1, we successfully isolated a current-crowding region using this method. There is also an improvement in the sensitivity and resolution compared to scanning from the back-side of the device. PFA showed metal melting starting from Metal 5 to Metal 2 at the area of interest (Fig. 3.2).

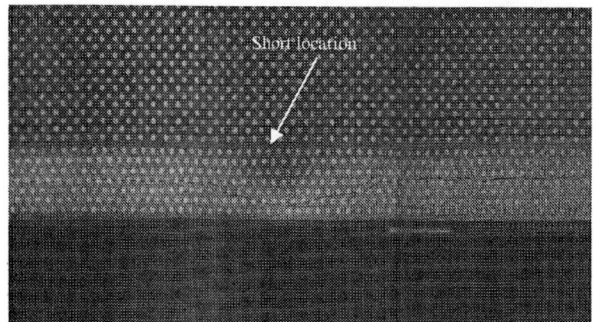

Fig.3.1: Current density overlay at front-side of the die.

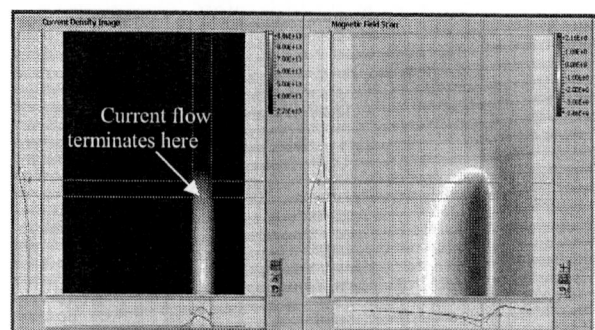

Fig.4.1: a) Current density image and b) magnetic field image.

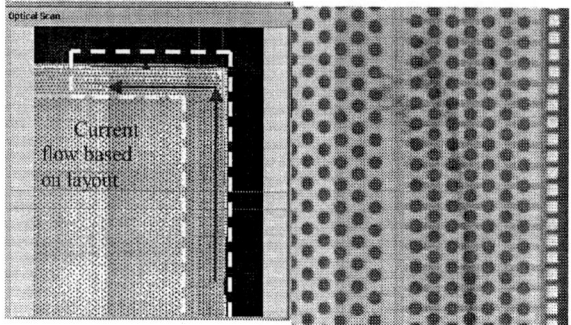

Fig.4.2: a) I/O supply region and b) zoomed-in peak current overlay

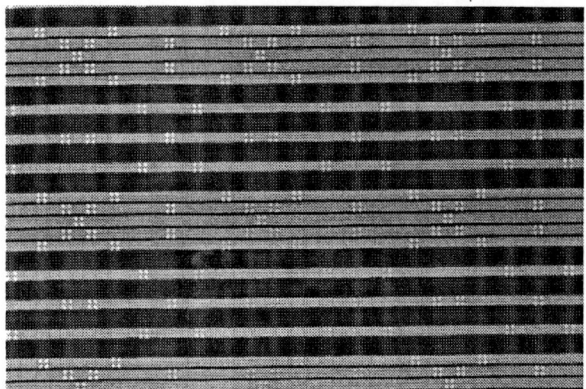

Fig.3.2: Top-down SEM image showing gross metal shorts.

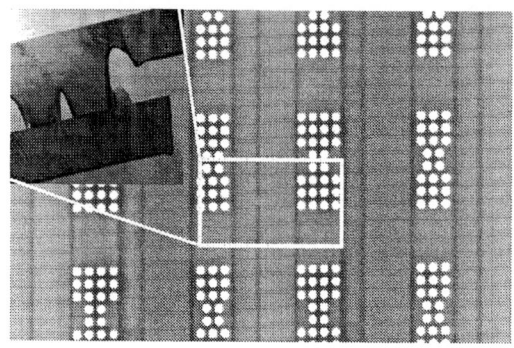

Fig.4.3: PFA based on region of interest identified by SQUID. It showed bridging between two supplies at Metal 8.

D.   *90* nm *device I/O supply short*

A 90 nm device was found to have significant yield loss due to a power supply short. Since packaging the dies involves additional time, resulting in longer fix for yield issues, bumped dies were cut from the wafer for a quick analysis using front-side SSM. In front-side SSM, probes to supply current to the die through the bump pad were used. First, the C4 bumps are etched off to expose the bump pad so the probe can be landed on the sturdier bump pad. Two probes are then placed on failing power supply bump and VSS bump located at one of the corner of the die, so it is easier for the SQUID sensor to scan the other side of the die. It involved judicious use of CAD layout for probe placement. Once the SSM image was acquired, we compared the current flow between the failing unit and how current may possibly flow on a good die based on the layout. As seen in Fig. 4.1a, the current density image showed the current terminating at some point in the die instead of flowing upwards (see Fig. 4.2a.). The short is most likely located at the termination point. Based on the result obtained using SSM, PFA was performed on the area of interest. Metal bridging was observed at metal 8 in top-down SEM analysis and was confirmed with cross-sectional TEM analysis.

III. CONCLUSION

With device technology advancements and ever-increasing numbers of metal layers in the die, it has become a challenge to isolate shorts failures. By making use of layout comparison and current flow analysis, we have successfully isolated die short locations in advanced flip-chip devices with SSM. Front-side SSM methodology has been developed for bumped dies that greatly improves resolution and helps in speedy analysis with higher success rates for yield improvement.

978-1-4244-2039-1/08/$25.00 ©2008 IEEE

ACKNOWLEDGEMENT

We thank Mike Bruce and Cao Li-Hong from AMD Austin for their review and feedback on this paper.

REFERENCES

1. L.A Knauss *et al.*, "Current Imaging using Magnetic Filed Sensors." Microelectronics Failure Analysis Desk Reference 5th Ed, pp. 303-311(2004).

2. L.A Knauss *et al.*, "Detecting Power Shorts from Front and Backside of IC Package Using Scanning SQUID Microscopy." In Proc. of the 25th Int'l Symp. on Testing and Failure analysis, pg. 11, Santa Clara, CA, November 1999.

3. Mai ZH *et al.*, "Short Failure Analysis under Fault Isolation." In Proc. of 8th IPFA, Singapore, 2001, pp. 202-205.

4. Steve K. Hsiung *et al.*, "Failure analysis of Short Faults on Advanced Wire Bond and Flip-chip Packages with scanning SQUID Microscopy." In Proc. of the 30th Int'l Symp. on Testing and Failure analysis, pp. 73-81(2004).

5. D.P Vallet, "Scanning SQUID Microscopy for Die Level Fault Isolation." In Proc. of the 28h Int'l Symp. on Testing and Failure Analysis, pp. 391-396, Phoenix, AZ, November 2002.

Test Structure Failed Node Localization and Analysis From Die Backside

Y. G. Li*, S. H. Tan, W. R. Sun

Systems on Silicon Manufacturing Company Pte. Ltd., 70, Pasir Ris Drive 1, Singapore 519527
* E-mail: yungui.li@nxp.com, Fax: (65) 62487606

Abstract: In this work, backside failure analysis technique on test structure failed node isolation and analysis are presented. Compared to front side failure analysis method, backside failure analysis provides more significant information that is related to the root cause directly. Especially, in the failure situations such as failure related to interface between contact top and metal in the contact chain structure or failure due to high resistance stacked via connected to n+ active P well structure, front side failure analysis isn't effective to localize the failed site or clearly reveal root cause form investigation of defect due to sample preparation shortage. However, Backside failure analysis overcomes the limitation of front side analysis. It was applied successfully on open contact chain observation and passive voltage contract (PVC) failed stacked via localization on the n+ active area. Combined with following cross-sectional TEM analysis, the root cause was firmly concluded.

I. INTRODUCTION

Backside failure analysis is of increasing concern for advanced CMOS integrated circuit (IC) fault location and investigations with technology scaling down; increasing numbers of metal interconnect levels, and packaging technology migration to flip-chip package [1-2]. Livengood et al. has reported focus ion beam (FIB) bulk Si thinning and circuit modification from backside [3]. The pervious works on backside defect localization by using photon emission microscopy (PEM), optical beam induced resistance change (OBIRCH), and backside passive voltage contrast (PVC) [1-2, 4-6] show that backside analysis techniques are significantly important. Especially, for some failures that are related to interface between node and metal or high resistances node chain connected to n+-active, the conventional front side failure analysis technique is impossible to localize the defects due to technical limitations. In this paper, we present some cases details on backside failure analysis using backside PVC or inspection to isolate failed site and identify root cause that can't be confirmed by front side failure analysis method.

II. ACHIEVEMENT OF BACKSIDE FAILURE ANALYSIS

Front side failure analysis usually involves removing insulated layer (passivation or oxide) followed by metal investigation to observe any defects on metal. Sometimes, the metal is mechanically lapped away to expose the node (via or contact) to localize the high resistance (or electrical leakage) via/contact through electron scanning microscopy (SEM)

Via/contact PVC analysis. However, it is impossible for front side failure analysis to localize and analyzes the fault situation shown in figure 1, where the failure happens at the interface between metal 1 and contact. Metal level observation can't localize the failed site as the other end of it was connected to a good node. It is therefore necessary to remove the metal for contact PVC. However, the removal of metal to expose the contact will result in defect being lapped away. Moreover, this is another kind of failed situation shown in figure 2, that happen frequently in the test structure. The via/contact are connected to n+ active, front side PVC approach has shown to be less effective to identify the failed location, because P well can't naturalize positive charge during SEM PVC imaging. To overcome the limitation of front side failure analysis, for those failures as show in figure 1 and figure 2, backside analysis can be the solution. To perform backside investigation and PVC imaging, the Si substrate is removed until contacts are exposed through mechanically lapping. The metal is grounded properly on the sample holder shown as in figure 3. The sample is then analyzed from bottom of contact.

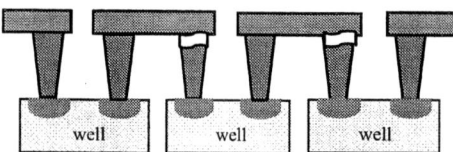

Fig. 1. A typical structure of contact chain, image illustrates the defect are located in the interface of contact and metal 1.

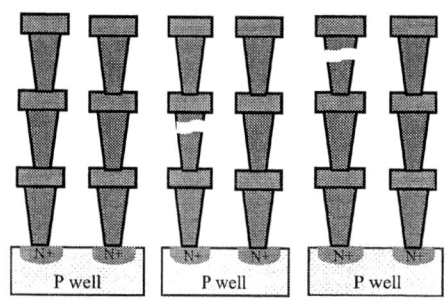

Fig. 2. A typical via chain test structure is connected to n+ active, and two of via are open.

978-1-4244-2039-1/08/$25.00 ©2008 IEEE

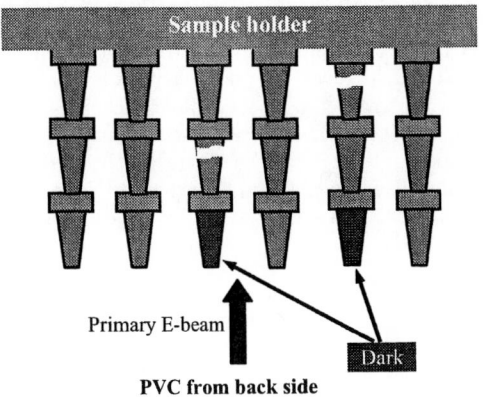

Fig. 3. A typical back die PVC on via chain test structure, and two of open chain will be darker the others.

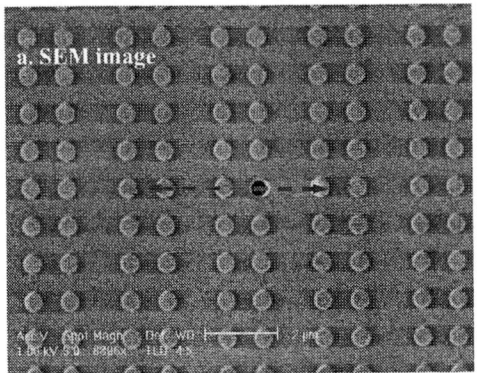

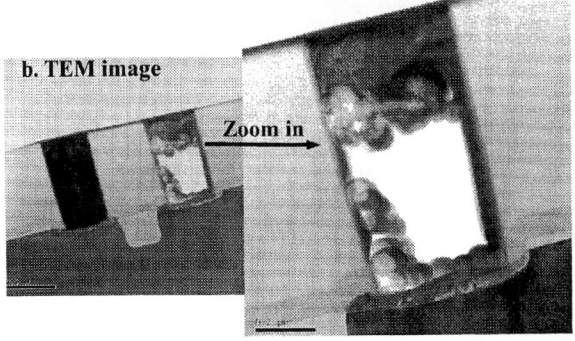

Fig. 4. Front side analysis, a. SEM inspection shows un-filled contact. And cross-sectional TEM sample was processed along the AA' line. b. Cross-sectional TEM image on the un-filled contact.

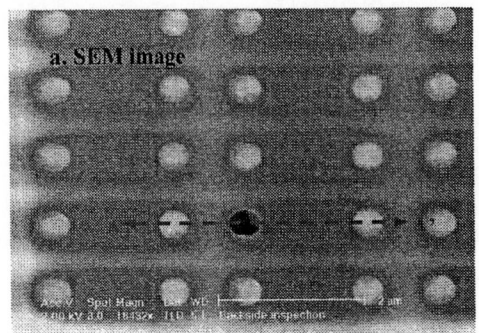

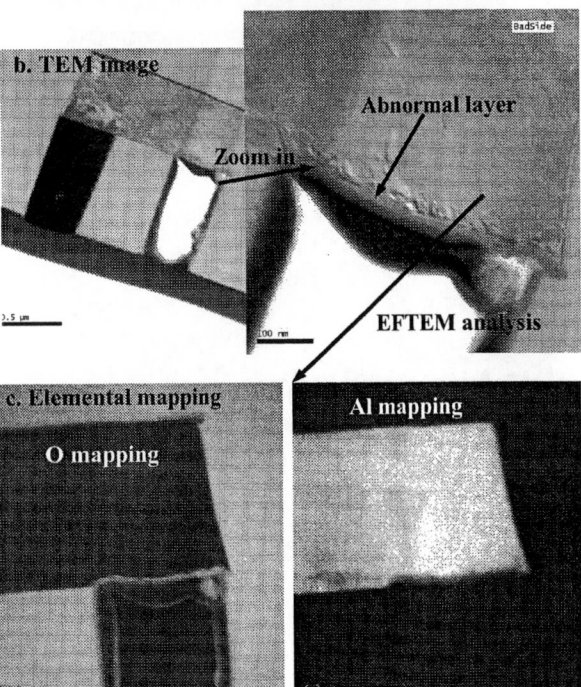

Fig. 5. Back side analysis, a. SEM inspection shows un-filled contact. And cross-sectional TEM sample was processed along the AA' line. b. Cross-sectional TEM image on the un-filled contact.

III. RESULTS AND DISCUSSION

A. Fail Contact Chain Backside Analysis

An open contact chain test structure (similar to figure 1) was found during electrical test. Fault isolation using OBIRCH could not be applied on the failure structure due to no current flowing through the chain. SEM observation on metal layer didn't find any abnormal metal. The metal layer was removed by mechanically polishing to expose the contacts. The unfilled contact, as shown in figure 4a, was observed under SEM inspection. In order to identify the root cause, the cross-sectional TEM sample preparation was performed along A-A' by using FIB. However, the TEM image shown in figure

4b can't reveal any information that caused the unfilled contact.

Backside failure analysis was performed for another sample with similar failure. The Si substrate was removed until the contacts were exposed. SEM observation in figure 5a shows that contact is un-filled at the failed site. The cross-sectional TEM analysis on the unfilled contact, as shown in figure 5b, observed that there is an abnormal layer between contact top and metal. Energy filtered TEM (EFTEM) elemental mapping combined with EDX analysis was performed to identify chemically the abnormal layer. The results in figure 5c show that abnormal layer is silicon oxide. This revealed that top of contact was covered by abnormal oxide layer. Therefore this resulted in the contact can't be filled during w deposition.

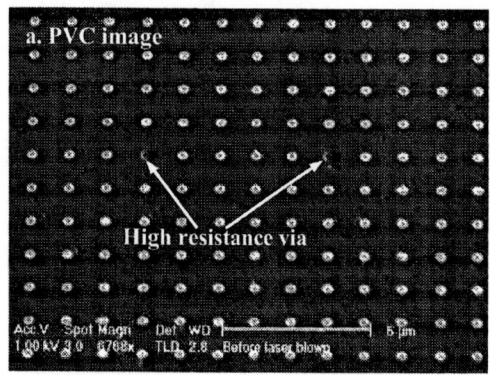

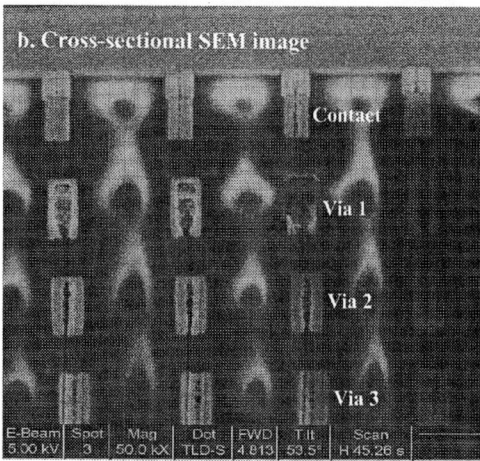

Fig. 6. a. Backside PVC image on the contact; b. Cross-sectional SEM image from the dark PVC contact.

B. High Resistant via Chain Isolation by Using Backside PVC

A stacked via test structure on n+ active was found with high resistance. To isolate the failed site, a typical front side failure analysis was performed for this failed structure. Front side SEM metal layer inspection and PVC imaging could not

localized the failed location. On the contrary, another sample with similar high resistance failure was analyzed by using backside failure analysis method. The failed site was found easily through backside PVC on contact level. Figure 6a shows the SEM 1KV PVC contrast image carried out from the back of die. PVC results reveal that dark contacts are failure path with high resistance. Cross-sectional inspection of the bad connections on the stacked via structure was performed using the FIB cutting. SEM image in figure 6b shows the via is partially un-filled under the dark contact observed from backside PVC image in figure 6a.

IV. CONCLUSION

Backside failure analysis technique provides an alternative novel failure analysis technique to overcome front side failure analysis limitations. Especially, in the failure situations such as failure related to interface between contact top and metal in the contact chain structure or failure due to high resistance stacked via connected to n+ active P well structure, front side failure analysis isn't effective to localize the failed site due to sample preparation shortage, however, backside analysis is powerful to show fail location clearly. The backside failure analysis method was combined with following FIB or TEM analysis, the root cause was identified.

ACKNOWLEDGMENTS

The author would like to thank Joe Hui for supporting the project during our developmental stage of backside sample preparation technique.

REFERENCES

[1] Silke Liebert, "Failure Analysis from the Back Side of a Die", *Future FAB International* issue 12, 2007, p.262.

[2] Loh Ter Hoe, Yee Wai Mun and Chew Yin Yan, "Characterization and Application of Highly Sensitive Infra-Red Emission Microprocessor Backside Failure Analysis" *7th International Symp. Phys.& FA of IC (IPFA'99)*, 1999, p.108.

[3] R. H. Livengood, V. R. Rao "Focused Ion Beam Techniques for Debug of Flip-Chip integrated Circuits", *Semiconductor International*, 1998.

[4] Steven Chen, Brian Shinseki, Cynthia Barutha and Ty Kha, "Infrared Imaging and Backside Failure Analysis Techniques On Multilayer CMOS technology", *6th International Symp. Phys.& FA of IC (IPFA'97)*, 1997, p.17.

[5] F. Beaudoin, G. Imbert, P. Perdu, C. Trocque, "Current Leakage Fault Location Using Backside OBIRCH", *8th International Symp. Phys.& FA of IC (IPFA'2001)*, 2001, p.121.

[6] D. L.barton, K.Bernhard-Hofer and E. I. Coler Jr., "Flip-Chip and Backside Techniques", *Microelectron. Reliab.*, 39, 1999, p.721.

SESSION 6:

PACKAGE FAILURE ANALYSIS & RELIABILITY

INVITED PAPER

Reliability of Cu Pillar Bump for Flip Chip and 3-D SiP

Byoung-Joon Kim[1], Gi-Tae Lim[2], Jaedong Kim[3], Kiwook Lee[3], Young-Bae Park[2] and Young-Chang Joo[1]

[1]School of Materials Science and Engineering, Seoul National University, Seoul, 151-744 Korea
[2]School of Materials Science and Engineering, Andong National University, Andong, 760-749, Korea
[3]Amkor Technology Korea, Seoul, 133-706, Korea
Phone: +82-2-880-8986, Fax: +82-2-883-8197. E-mail: ycjoo@snu.ac.kr

Abstract- **Cu pillar bumps with eutectic SnPb solder were annealed and their microstructures were investigated. Linear relationship was observed between thickness of intermetallic compounds (IMCs: Cu_6Sn_5, Cu_3Sn) and square root of time at 120 and 150 ℃. Kirkendall voids, formed by the diffusivity differences between Cu and Sn, were observed near the interface between Cu and Cu_3Sn. There was a change in slope of the linear relationship between IMCs thickness and square root of time at 165 ℃ when all Sn was consumed. Cu_6Sn_5 growth rate was retarded, while Cu_3Sn growth rate was accelerated. The activation energies for Cu_6Sn_5, Cu_3Sn, and Kirkendall voids growth were estimated to be 1.77, 0.72, and 0.36 eV, respectively. The microstructures of Cu pillar bumps with pure Sn were investigated by *in-situ* scanning electron microscopy under annealing and high current-stressing conditions. It was found that IMC growth rate under annealing condition obeyed parabolic rate law, while that under high current-stressing condition IMC growth rate did not obeyed linear rate law. IMC growth rate under high-current stressing condition was faster than that under annealing condition which is presumed to be caused by the atomic migration enhancement due to the electron wind force.**

I. INTRODUCTION

As the integrated circuit density increases, more input-output (I/O) interconnects are required to connect Si chip and PCB (Printed Circuit Board). Smaller bump size and finer pitch are required to get higher I/O interconnects. However, finer pitch frequently causes a bump bridging [1]. In order to avoid bump bridging, bumps with thick Cu pillar covered with thin solder are recommended to be used instead of conventional solder bumps.

Cu pillar bumps are expected to be used in 3-D SiP (System in a Package) to connect stacked Si chips and PCB. For operating many stacked Si chips, it is unavoidable for Cu pillar bumps to pass high current density which may lead to electromigration (EM) and elevation of the bump temperature owing to joule heating which are known to be detrimental to bump reliability.

EM is an atomic migration due to the electron wind force along the electron flow direction. It has been reported that Cu pillar bumps showed better resistance against EM than conventional bumps because the straight shape of Cu pillar decreases the maximum current density [2]. However, the using of Cu pillar bumps can cause another reliability issues under high current and high temperature condition. Sufficient Cu supply by Cu pillar forms thick intermetallic compounds (IMCs) such as Cu_6Sn_5 and Cu_3Sn through the reaction between Cu and Sn. Thick IMC layers reduce mechanical strength of the

Cu pillar bumps because IMCs are brittle. At the same time, Kirkendall voids formed by the IMCs growth also decreases mechanical strength of the Cu pillar bumps. Therefore, it is important to understand growth behaviour of IMCs and Kirkendall voids in the Cu pillar bumps.

In this paper, we investigated the growth behaviour of IMCs and Kirkendall voids in the Cu pillar bumps under high temperature and high current stressing condition. Activation energies for IMCs and Kirkendall void growth under high temperature condition were obtained and discussed.

II. EXPERIMENT

Cu pillar bumps with eutectic SnPb solder were used for annealing experiment. The schematic of Cu pillar bump used in the annealing experiment is shown in Fig. 1(a). Cu pillars were formed on the Si chip by electroplating and were connected to under bump metallization of PCB by eutectic SnPb solder. The height and diameter of the Cu pillar were $60\mu m$ and $100\mu m$, respectively. The samples were polished by using sand paper and the half-cut samples were annealed at 120, 150, 165°C in oven, respectively. The microstructures of the samples were investigated by scanning electron microscopy (SEM). The maximum annealing time was 300 h and the microstructures of samples were observed at every 20 h.

Cu pillar bumps with pure Sn solder solder were used for high current-stressing experiment. The aim of this experiment was the investigation of EM effect on the IMCs growth behavior. The schematic of Cu pillar bumps with Sn is shown in Fig. 1(b). The diameter and height of the Cu pillar bump with Sn were $70\mu m$ and $50\mu m$, respectively. The thickness of Sn solder was $2\mu m$. The samples were polished by using sand paper and the half-cut samples were exposed to high current-stressing condition. The applied current density was 5×10^4 A/cm^2 and the sample temperature was 150°C. The microstructures of the samples were investigated by *in-situ* SEM as shown in Fig 2. The current density was 5×10^4 A/cm^2 and the temperature was 150°C. Another sample was annealed at 150°C for 360 h and its microstructure was investigated by *in-situ* SEM to compare with the sample undergone high current stress.

978-1-4244-2039-1/08/$25.00 ©2008 IEEE

INVITED PAPER

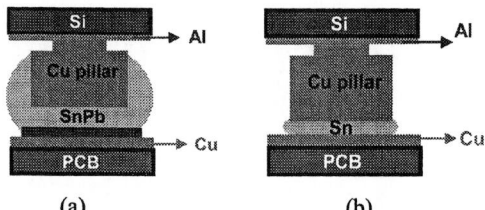

(a)   (b)

Fig. 1. Schematics of Cu pillar bumps for (a) annealing and (b) electromigration test.

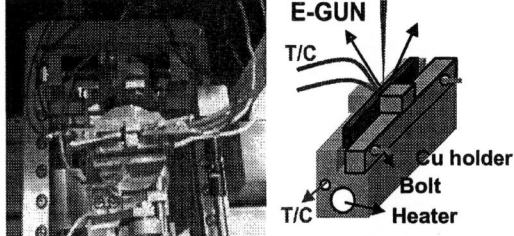

Fig. 2. In-situ electromigration observation system
(a) SEM chamber picture (b) schematic of SEM holder

III. RESULTS AND DISCUSSTION

A. Annealing

Microstructure evolution of Cu pillar bump during annealing at 165 °C is shown in Fig.3. Cu reacted with Sn and formed two kinds of IMCs (Cu_6Sn_5 and Cu_3Sn). Cu_6Sn_5 formed earlier than Cu_3Sn, which implies that Cu diffused into Cu_6Sn_5 continuously considering that Cu to Sn ratio. Pb did not react with Sn, thus it remained in solder region.

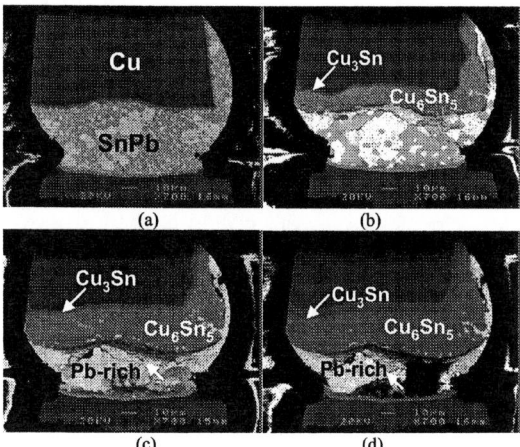

Fig. 3. Microstructures of Cu pillar bumps annealed at 165°C for (a) as-reflowed (b) 80 h, (c) 160 h, (d) 300 h.

To quantify IMCs growth, we measured IMCs area in the SEM image and converted it into IMCs thickness. Cu_6Sn_5 and Cu_3Sn areas at each annealing time were measured by using image analyzer. IMC thickness was calculated by dividing IMC

area with interface length. The obtained IMCs thicknesses were plotted in Fig. 4. It was observed that Cu_6Sn_5 thickness and Cu_3Sn thickness increased linearly with respect to square root of time at 120°C and 150°C, which means both IMCs grow obeying parabolic rate law governed by diffusion rate (diffusion-controlled process) at 120°C and 150°C [3].

IMCs formed at 165°C thicker than those formed at 150°C as well as 120°C. Cu_6Sn_5 grew linearly with square root of time until 160 h. After 160 h, Cu_6Sn_5 growth rate was retarded while Cu_3Sn growth rate was accelerated. Sn rich phase was not observed after 160 hour annealing in the solder region, in other words, only Pb-rich phase was observed in the solder region, which means Sn in the solder region was exhausted entirely. Growth rate of Cu_3Sn increased abruptly after 160 h. Cu_3Sn formed by the reaction between Cu_6Sn_5 and Cu in Cu pillar.

In diffusion controlled process, we can obtain activation energy by using Arrhenius relation.

$$k^2 = k_0^2 \exp(-\frac{E_a}{RT}) \qquad (1)$$

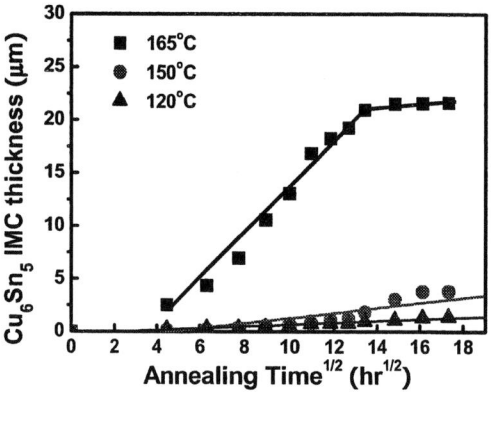

(a)

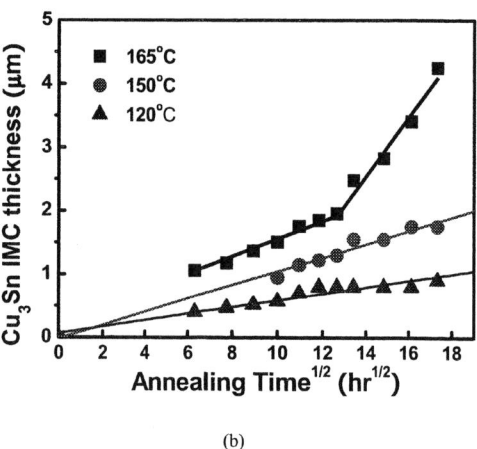

(b)

Fig. 4. IMCs thickness as functions of square root of time: (a) Cu_6Sn_5, (b) Cu_3Sn.

INVITED PAPER

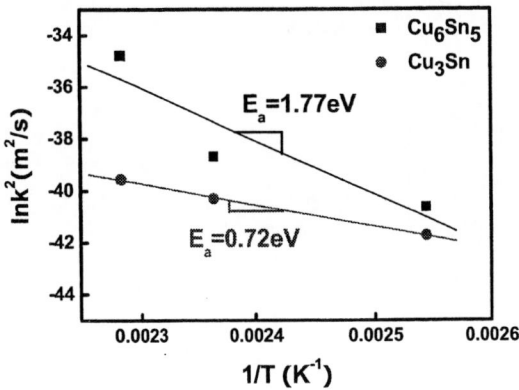

Fig. 5. Activation energies for IMCs growth: (a) Cu_6Sn_5, (b) Cu_3Sn.

To calculate the activation energies for Cu_6Sn_5 and Cu_3Sn growth, the data acquired before IMC growth rate changed were used. Arrhenius plot of each IMC growth is shown in Fig. 5. The activation energies of Cu_6Sn_5 and Cu_3Sn growth are 1.77 and 0.72 eV, respectively. These values are similar to those appeared in the reports [4,5].

Kirkendall voids formation was observed near the interface between Cu pillar and IMCs during the annealing process. Kirkendall voids were known to be formed due to the difference in diffusivities between Cu and Sn. It is known that diffusivity of Cu in IMCs is higher than that of Sn in IMCs [6]. Therefore, vacancy migrates toward Cu pillar and forms Kirkendall voids. Kirkendall voids growth behavior with time at 165°C is shown in Fig. 6.

Kirkendall voids were quantified by the same method used in the quantification of IMCs. Areas occupied by Kirkendall voids in the SEM image were measured by image analyzer, and subsequently, the measured areas divided by interface lengths between Cu pillar and IMC. The void width was plotted in Fig. 7. Activation energy for Kirkendall void growth was obtained

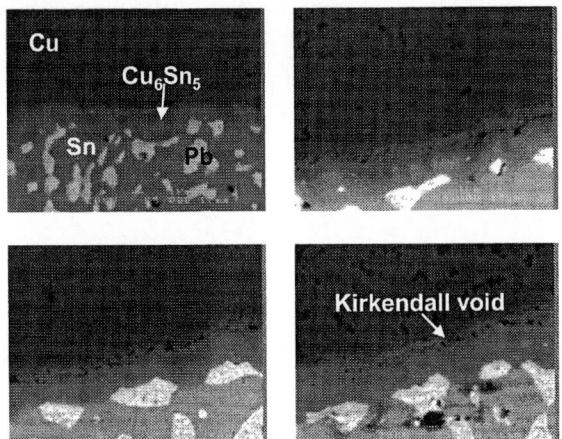

Fig. 6. Observation of Kirkendall voids near the interface between Cu pillar and solder interface: (a) as-reflowed, (b) 60 h, (c) 100 h, and (d) 140 h.

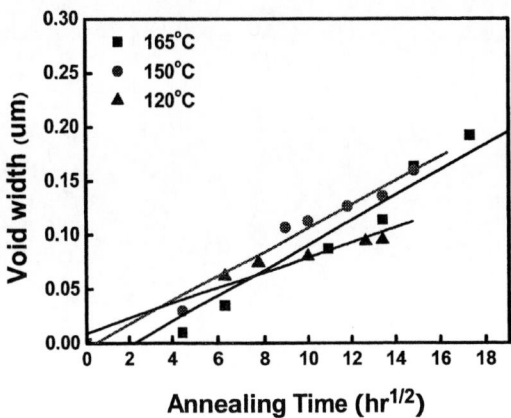

Fig. 7. Kirkendall void width as functions of square root of time.

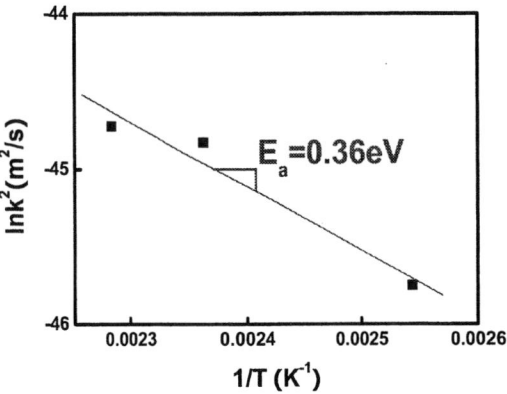

Fig. 8. Activation energies for Kirkendall void growth.

using Arrhenius relation. The activation energy for Kirkendall void growth was 0.36 eV, which is smaller than those for Cu_6Sn_5 and Cu_3Sn growth because Kirkendall void growth is governed by the vacancy migration but IMC growth is governed by the atomic migration.

B. Current stressing test

Electromigration test under 150°C/5 $\times$ 10^4 A/cm^2 current-stressing condition was performed, and annealing at 150°C was also performed to investigate the effect of high current-stressing condition on the microstructure evolution. The cross-sectional images of Cu pillar bumps with pure Sn after 360 h annealing at 150°C and after 200 h current stressing at 150°C are shown in Fig. 9. Cu_6Sn_5 and Cu_3Sn were formed by the reaction between Cu and Sn. Kirkendall voids were observed near the interface between Cu pillar and IMCs.

INVITED PAPER

Fig. 9. Cross-sectional images of Cu pillar bumps with pure Sn: (a) as-reflowed, (b) 150°C, 42 h annealing, (c) 150°C, 360 h annealing, (d) 150°C /5 × 10⁴ A/cm² stress for 200 h.

IMCs thickness with respect to annealing time was obtained by image analysis followed by calculation. Total IMC thickness was the sum of Cu_6Sn_5 thickness and Cu_3Sn thickness. IMCs thickness, as functions of square root of time, is presented in Fig. 10. Total IMC thickness was increased proportionally to square root of time until 240 h and then saturated because Sn was entirely consumed in IMC formation. Cu_6Sn_5 thickness and Cu_3Sn thickness increased until 240 h. After 240 h, Cu_6Sn_5 thickness decreased, while Cu_3Sn thickness increased at a higher rate, which is probably caused by change of Cu_6Sn_5 into Cu_3Sn.

IMCs thickness, as functions of current-stressing time, is shown in Fig. 11. As current stressing time increased, total IMC thickness increased but was not linear to square root of time. This is presumably because both thermal energy and electron wind force gives rise to Cu migration [7]. Cu_6Sn_5 and Cu_3Sn thickened until 60 h. After 60 h, Cu_6Sn_5 thickness decreased,

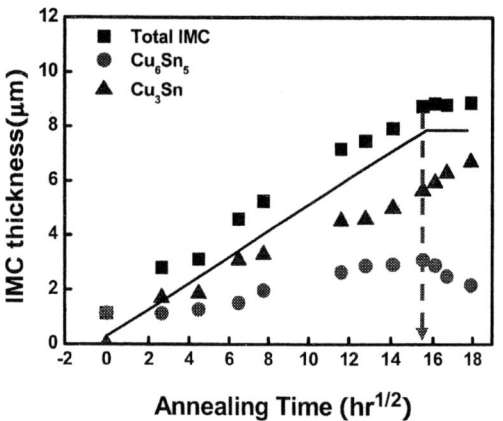

Fig. 10. IMCs thickness as functions of square root of time at 150℃ annealing condition.

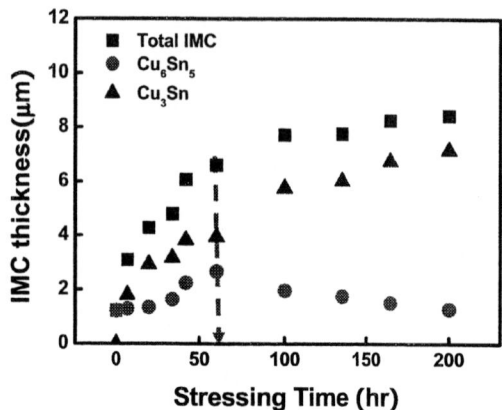

Fig. 11. IMC growth behavior under 150°C /5 × 10⁴A/cm² current stressing condition.

while Cu_3Sn thickness increased at a higher rate, which is probably caused by change of Cu_6Sn_5 into Cu_3Sn.

It is noteworthy that the changing point of total IMC-growth rate appeared earlier under current stressing condition (60 h) than under annealing condition (240 h), indicating that electron wind force enhances Cu migration.

IV. CONCLUSION

IMCs growth in Cu pillar bump with eutectic SnPb solder by annealing was quantified. Cu_6Sn_5, Cu_3Sn, and Kirkendall voids growth rate obeyed parabolic rate law. The activation energies for Cu_6Sn_5, Cu_3Sn, and Kirkendall voids growth were 1.77, 0.72, and 0.36 eV, respectively. In case of annealing at 165°C, Cu_6Sn_5 growth rate began to decrease and Cu_3Sn growth rate started to decrease at 160 h, which agrees well to the time when Sn in solder region was consumed completely.

IMCs growth in Cu pillar bump with pure Sn solder by annealing and high current stressing was investigated by using *in-situ* SEM. In case of annealing, IMC growth obeyed parabolic rate law. In case of current stressing, contrary to our expectation, IMC growth rate did not obeyed linear rate law. The changing point of total IMC-growth rate appeared earlier under current stressing condition (60 h) than under annealing condition (240 h), which reveals that electron wind force enhances Cu migration.

ACKNOWLEDGMENT

This work was supported by next-generation growth engine project of the Korea Ministry of Commerce, Industry and Energy.

978-1-4244-2039-1/08/$25.00 ©2008 IEEE

INVITED PAPER

REFERENCES

[1] H. Y. Son, G. J. Jung, J. K. Lee, J. Y. Choi, K. W. Paik, "Cu/SnAg double bump flip chip assembly as an alternative of solder flip chip on organic substrates for fine pitch applications", *Proc 57th Electronic Components and Technology Conf,* Reno, Nevada, May. 2007, pp. 864-871.

[2] J.W. Nah, J. O. Suh and K. N. Tu,, "Electromigration in flip chip solder joints having a thick Cu column bump and a shallow solder interconnect", *J. Appl. Phy.*, Vol. 100, 123513, (2006).

[3] King-Ning Tu, James W. Mayer, and Leonard C. Feldman, *Electronic thin film science*, Macmillan (New York, 1992), p. 324.

[4] R. Labie, W. Ruythooren, J. A. Humbeeck, "Solid state diffusion in Cu-Sn and Ni-Sn diffusion couples with flip-chip scale dimensions", *Intermetallics*, Vol. 15 (2007), pp. 396-403.

[5] D. R. Flanders *et al.*, "Activation energies of intermetallic growth of Sn-Ag eutectic solder on copper substrates", *J. Electron. Mater.* Vol. 26 (1997), pp. 883-887.

[6] K. Zeng, R. Stierman, T-C. Chiu, D. Edwards, K. Ano, K. N. Tu, "Kirkendall void formation in eutectic Sn-Pb solder joint on bare Cu and its effect on joint reliability", *J. Appl. Phy.*, 97, 024508, (2005).

[7] C. Y. Liu, L. Ke, Y. C. Chuang and S. J. Wang, "Study of electromigration-induced Cu consumption in the flip-chip Sn/Cu solder bumps", *J. Appl. Phy.*, Vol. 100, 083702 (2006).

Case Study of Copper Dendrite Growth Under HAST Test

Sang-Ah Kim*, Do-Seok Ahn, Yong-Hui Eum, Duck-Hyun Kim, Young-Bae Kim
Failure Analysis Team, QRT Semiconductor
San 136-1 Ami-ri Bubal-eub Ichon-si Kyoungki-do 467-701 Korea
Email: ksanga@qrtkr.com

Abstract- **The HAST is performed for the purpose of evaluating the reliability of nonhermetic packaged solid-state devices in humid environments. This paper discusses the issue of copper dendrite growth, electrochemical migration, at memory fuse box and die edge areas under HAST test.**

I. INTRODUCTION

Redundancy scheme in memory device is very important to improve manufacturing yields. The fuse box area is opened, so it is important to protect the fuse box area from humidity. Many chips failed during HAST in FBGA Package. In this paper, failure mechanism of electrochemical migration will be discussed below.

II. TEST AND ANALYSIS

A. Reliability Test Results

To evaluate reliability of the fine pitch ball grid array package type memory devices, we tested several environmental reliability tests. The test items were Highly Accelerated Temperature and Humidity Stress Test, Temperature Cycling, Accelerated Moisture Resistance-Unbiased Autoclave, Steady State Temperature Humidity Bias Life Test, and High Temperature Storage Life Test. The results were summarized in Table1. The devices were tested with the HAST 96 hours, they failed parametric limits were exceeded and functionality could not be demonstrated. The test conditions consisted of a temperature 130 □, 85 % relative humidity, and 1.9 V bias.

The HAST is performed for the purpose of evaluating the reliability of nonhermetic packaged solid state devices in humid environments. It is a highly accelerated test which employs temperature and humidity under noncondensing conditions to accelerate the penetration of moisture through the external protective material or along the interface between the external protective material and the metallic conductors which pass through it.

Test Item	HAST	TC	PCT	THB	HTS
Test Condition	130℃/85%RH /1.9V	-55℃/ 125℃	121℃/100%RH	85℃/85%RH /1.9V	150℃
Duration	96 Hours	1000 Cycle	240 Hours	1000 Hrs	1000 Hrs
Test Result	180000 ppm	0 ppm	0 ppm	0 ppm	0 ppm

Table 1. Environmental reliability test results

B. Electrical & Physical Failure Analysis

The failures were subjected to electrical failure analysis. The devices no longer met the part drawing requirements using parametric and functional testing. Parametric limits were exceeded and all functionality could not be operated. (Fig.1)

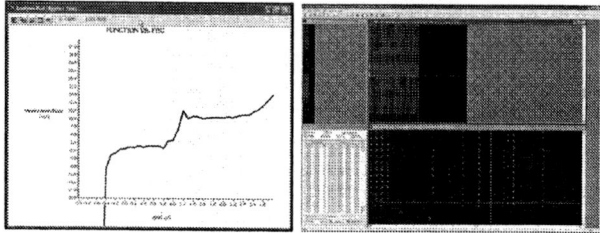

Fig. 1. Parametric and functional test results

In the process of physical failure analysis, copper dendrite growth was observed during failure analysis for the devices with the failure modes of the parametric and functional failures. It was found at the memory fuse box areas and the die edge areas. Figure 2 shows the top view of the dendrite growth and figure 3 shows the cross sectional view of the same point. There is the delamination between the passivation layer and the PIQ layer.

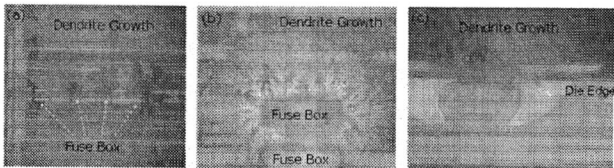

Fig. 2. (a) and (b) show the dendrite growth at the fuse box areas. (c) shows the dendrite growth at the die edge areas

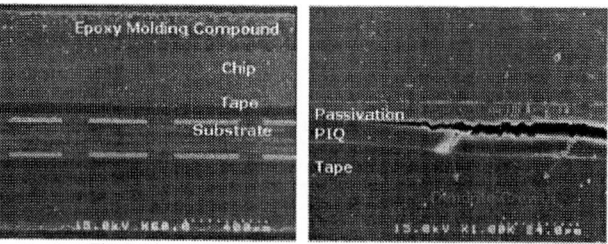

Fig. 3. The cross sectional view of the failure point.

The dendrite growth was subjected to energy dispersive X-ray spectroscopy(EDS) analysis. EDS spectrum picked on the dendrite growth showed carbon, oxygen, and aluminum peak, in addition to copper peak(see Fig. 4). It was believed that the carbon and aluminum were from sample preparation for analysis and oxygen was due to copper surface oxidation. It means the element of the dendrite growth is copper.

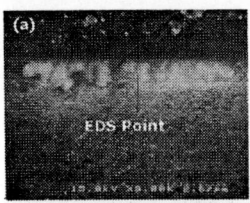
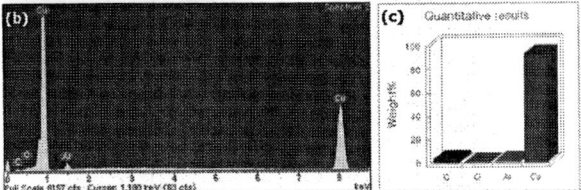

Fig. 4. Energy dispersive X-ray spectroscopy analysis results

C. Root Cause Analysis

Figure 5 shows the dendrite growth is localized under the VDD copper tracers. It means the copper tracer patterns influence the electrochemical migration.

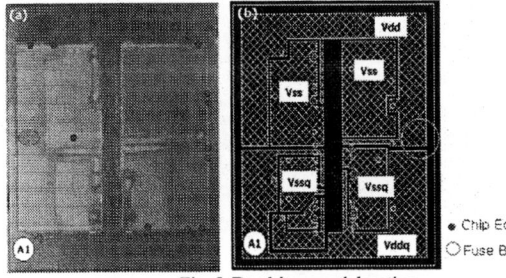

Fig. 5. Dendrite growth location

Electrochemical migration occurs when moisture adheres between electrodes made of material such as copper, solder, or silver.

When the devices are tested with the HAST, the moisture penetrates between the printed circuit board and the die surface. When the bias voltage is applied, metal from the anode is ionized and moves toward the cathode.

Figure 6 shows the electrochemical migration phenomenon. The moisture vaporizes into steam, and the heat of gasification causes damage to the surroundings. So there is a delamination between the passivation layer and the PIQ layer.

- *Copper is dissolved at the anode*

$$Cu \rightarrow Cu^{2+} + 2e^-$$
$$H_2O \rightarrow H^+ + OH^-$$
$$2H + 2e^- \rightarrow H_2$$

- *Copper electrons transfer from the anode to the cathode*

$$Cu^{2+} + 2OH^- \rightarrow Cu(OH)_2$$

- *Copper is deposed at the cathode*

$$Cu(OH)_2 \rightarrow CuO + H_2O$$
$$CuO + H_2O \rightarrow Cu(OH)_2 \rightarrow Cu^{2+} + 2 OH^-$$
$$Cu^{2+} + 2e^- \rightarrow Cu$$

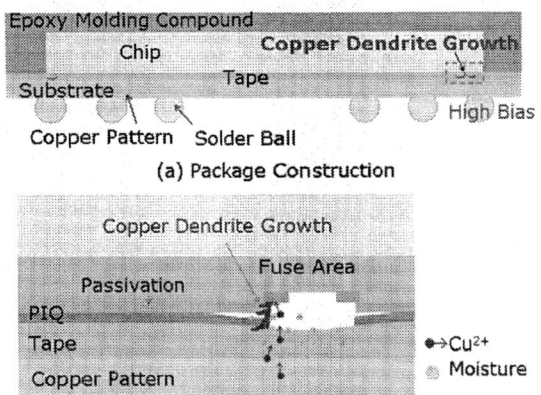

Fig. 6. The electrochemical migration phenomenon.

III. CONCLUSION

The reason of HAST failure was the moisture penetration between the PCB and the die surface. To avoid electrochemical migration phenomenon, we changed the PCB pattern design and the tape material design. We performed several reliability tests again and all devices passed(see Table 2).

Test Item	HAST	TC	PCT	THB
Test Condition	130℃/85%RH /1.9V	-55℃/ 125℃	121℃/100%RH	85℃/85%RH /1.9V
Duration	96 Hours	1000 Cycle	240 Hours	1000 Hrs
Test Result	0 ppm	0 ppm	0 ppm	0 ppm

Table 2. Environmental reliability test results

REFERENCES

[1] D. LAMBERT, R. GANNAMANI, R.C. Blish, II, "Dendrite Fuse Re-growth Kinetics on Organic Substrates for Microprocessors", IRPS 2004
[2] IEC 60749-4, Semiconductor devices-Mechanical and Climatic test method-Part.4:Damp heat, steady state, highly accelerated stress test(HAST)
[3] N.H. Yeung, Victor Lau, Y.C. Chan "Bias-HAST on tape ball grid array(TBGA) test pattern", Micro-electronics Reliability 44 (2004)
[4] Failure Mechanism Models for Conductive Filament Formation"
[5] Yoshinori Kin, "A Consideration of Methods for Evaluating Reliability of Electric Parts", ESPEC Technology Report, 1996 No.2

[6] Hiroko Katayanagi, Hirokazu Tanaka, Yuichi Aoki, Shigeharu Yamamoto, "The affects of adsorbed water on printed circuit boards, and the process of ionic migration", ESPEC Technology Report, 2000 No.9

[7] Hirokazu Tanaka, "Factors leading to ionic migration in lead-free solder", ESPEC Technology Report, 2002 No.14

Effect of Bonding Pressure on the Bond Strengths of Low Temperature Ag-In Bonds

Riko I Made [a] *, Chee Lip Gan[a], Chengkuo Lee[§b,c], LilingYan[b], Aibin Yu[b], Seung Wook Yoon[b]

[a]School of Materials Science and Engineering, Nanyang Technological University, Singapore 639798
[b]Institute of Microelectronics, 11 Science Park Road, Singapore Science Park II, Singapore 117685
[c]Department of Electrical and Computer Engineering, National University of Singapore, Singapore 117576
* imaderiko@ntu.edu.sg; § elelc@nus.edu.sg; leeck@ime.a-star.edu.sg

Abstract- **Bonding of multiple indium-silver intermediate layers facilitates precise control of the formed alloy composition and the joint thickness. The bonding temperature and post-bonding re-melting temperature can thus be easily designed by controlling the multilayer materials and structure thicknesses. However, joining different materials involves the formation of intermetallics, which is known to be brittle. In this paper, In-Ag intermetallic phase formation under different applied pressure is studied.**

I. INTRODUCTION

Solid-liquid interdiffusion bonding uses a multilayer of high-melting and low-melting materials to form a joint. Bonding is carried out at temperatures above the melting point of the low-melting material, during which the low-melting material is gradually consumed, resulting in a micro-joint which contains only intermetallics. Through proper control of the multilayer film thickness and processing conditions, it is possible to form a joint free of the low-melting material [1].

Indium based solders are attractive for low temperature bonding due to their low melting temperatures. Furthermore, by reacting with a higher melting metal such as Cu, Au or Ag, they can easily form a variety of intermetallic compounds with high melting temperatures [2-5].

The concept of low temperature bonding but achieving high re-melting temperature is attractive for applications such as Micro-Electro-Mechanical Systems (MEMS) packaging, stacked-chips processes and wafer-to-wafer bonding. Besides eliminating the undesirable phase transformation that occurs during high temperature bonding, a low temperature bonding process can also drastically reduce the process-induced residual stress of multilayer structures, while achieving a high post-package temperature resistance.

Due to the high rate of inter diffusion within bonded couples formed between the noble metals and indium, intermetallic compound (IMC) formation can even occur at ambient temperature [6]. Solid solubility of indium in silver is much higher (approximately 20 at.% In in Ag) than that of silver in indium (<1 at.% Ag in In). At temperatures between the melting points of indium (156.7°C) and silver (961.9°C), solid silver is dissolved by the molten indium. The interaction between the solid silver and the molten indium results in an eutectic compound and a number of IMCs. The melting point of these compounds increases with the silver content, i.e.

eutectic (144°C) < $AgIn_2$ (166°C) < γ (300°C) < ζ (670°C) < β (695°C) [7].

Through the measurement of electrical resistances, Roy and Sen reported that the formation of $AgIn_2$ phase is a diffusion-controlled process with an activation energy of 0.43 eV [8]. Lee et al. have proposed a two-stage bonding-annealing process consisting of a short duration exposure of the Ag-In thin layers to slightly above 200°C, followed by a lengthy heat treatment under hydrogen environment below 150°C. The high re-melting temperature of the joint was produced by consumption of all the indium in the bonded couple to form an IMC phase of Ag_2In [9].

So far, the application of pressure during bonding has not been considered to be important to the IMC phase formation. Bonding is usually carried out at low pressure or even without pressure at all [7, 10]. When the supply of In is sufficient and the bonding temperature is well above the system melting temperature, pressure is applied just to maintain contact between the mating surface. However, as the bonding temperature becomes lower, the morphology of the matting solder surface becomes more important. Microscopically rough surface is unavoidable and will cause the reduction in the true contact area between the mating couple. Application of load during bonding helps to flatten the surface, and thus may improve the percentage of true contact area [11, 12].

The technique of simultaneous application of pressure and temperature during the bonding process is known as thermocompression bonding. The surfaces can be brought closer together by the application of pressure. Metallic bonds form when the distance between the two substrates is so small that it becomes energetically favorable for surfaces to coalesce in order to eliminate the interfacial energy. In this paper, we report on the intermetallics formation from low temperature thermocompression bonding of Ag-In system. The focus of the study is on the effects of bonding parameters, especially applied pressure, to the formation of IMCs at the bonding interface. The results shown could provide an alternative method to produce high temperature joints other than the conventional heat treatment method.

978-1-4244-2039-1/08/$25.00 ©2008 IEEE

II. EXPERIMENT

The Ag-In multilayer structure for the low temperature bonding study is shown in Fig. 1. Side A consists of a 50 nm-thick SiO_2, a 25 nm-thick Cr (to improve adhesion), a 800 nm-thick In and a 100 nm-thick Ag. The Ag layer was used as an oxidation prevention layer, by reacting with In to form $AgIn_2$ during deposition [8]. Similarly, side B consists of a 50 nm-thick SiO_2 followed by a 30nm-thick Cr, a 2.6 μm-thick Ag and a 30 nm-thick Au. The In layer on side A is thin enough such that it will be fully consumed to form IMCs by Ag on side B during thermocompression bonding. Upon uniform distribution of In at the bonding layer, the overall composition should be approximately 17 at.% In. In other words, the average atomic ratio of a homogeneous phase depends on the layer structure thickness ratio, and should be Ag/In = 83/17 in our Ag-In solder system. It has been suggested that the thickness of In on side A and Ag on side B determines the final phase composition of the bonding couple [9].

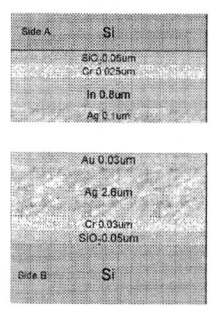

Fig. 1 Multilayer bonding couple

The 50 nm-thick SiO_2 on both sides A and B were deposited by Plasma Enhanced Chemical Vapor Deposition (PECVD) at 300°C on a 200 mm bare silicon wafer. The metal film stack was deposited on separate wafers for side A and side B. This was done in a single evaporation step without breaking the vacuum (10^{-5} torrs) to prevent indium oxidation. Next, pre-prepared wafers of both sides A and B were cut into squares of 5 mm x 5 mm. Thereafter, both sides were aligned and bonded at 180°C for varying time and pressure using a Suss FC150 flip chip bonder. Temperature was only applied when the target pressure had been achieved. The thermocompression cycle was completed by cooling down the bonder to room temperature while maintaining the applied pressure.

Several methods were used to characterize the bonded samples. Cross section elemental composition was characterized by Energy Dispersive X-Ray spectroscopy (EDX). Samples were mounted on epoxy resin and subsequently polished. Arrays of EDX spectrum were taken from side B to side A to measure the elemental composition as a function of various depths. X-Ray Diffraction (XRD) method was used to determine the exact phases that had formed under the various bonding conditions. Bonded samples were pulled apart by mechanical force to expose the bonded interface. X-ray diffraction spectrum was then obtained from a Bruker-General Area Detector Diffraction System (GADDS).

III. RESULT AND DISCUSSION

Fig. 2 (a) to (d) show the Ag and In compositions across the bonded interface for samples bonded at 180°C for 10 minutes with applied load of 1 MPa, 1.2 MPa, 1.4 MPa and 1.6 MPa, respectively. In all cases, the cross sections consist of pure Ag, Ag-rich and In-rich regions. It is also observed that the depth of indium diffusing from side A (In-rich side) to side B (Ag-rich side) is shorter as the applied bonding pressure increases. Shear tests on the bonded samples show that the joint's strength increased with the applied bonding pressure as shown in Fig. 3.

Fig. 4(a) shows that two types of IMC formation had occurred on side A of the bonded couple, where it was initially the In-rich side. $AgIn_2$ is the dominant phase formed from 1.0 to 1.4 MPa bonding pressure. At 1.6 MPa, the main phase is Ag_9In_4. Although pure In was not observed on side A in the EDX analysis, XRD analysis shows that pure In disappeared from side A as the bonding pressure increased beyond 1.4 MPa. On the other hand at side B as shown in Fig. 4 (b), both $AgIn_2$ and Ag_9In_4 phases are observed in the case of 1 MPa bonding pressure, while Ag_9In_4 dominates again at higher bonding pressure.

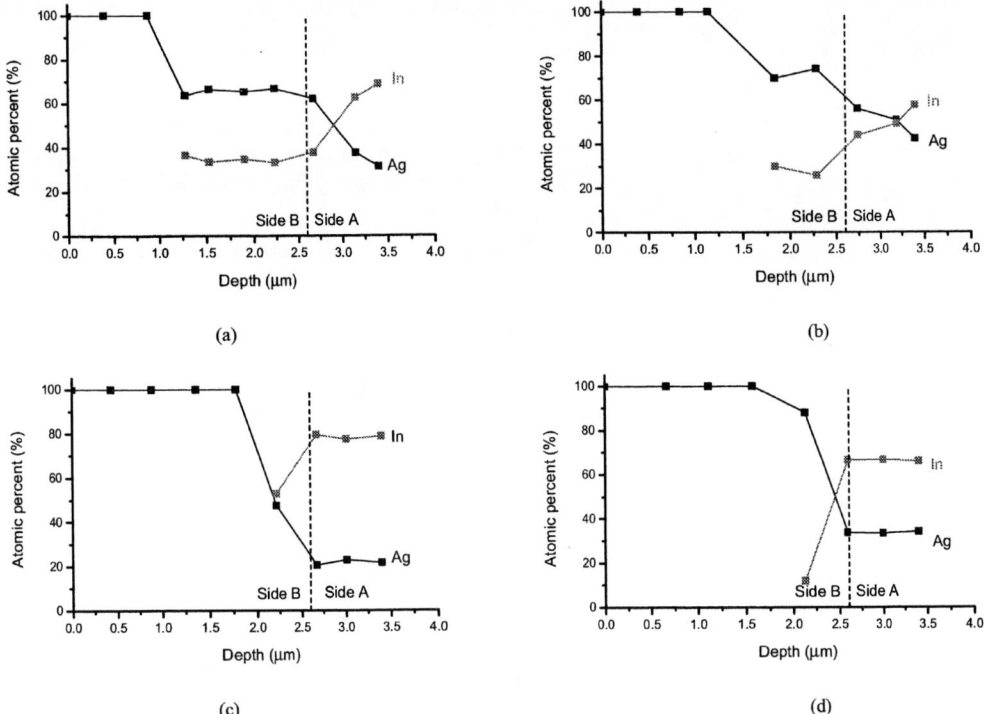

(a)

(b)

(c)

(d)

Fig. 2 EDX cross-section scans from couples bonded at 180°C – 10 minutes and at different bonding pressure: (a) 1.0 MPa, (b) 1.2 MPa, (c) 1.4 MPa, and (d) 1.6 MPa. Zero reference was taken from the Ag-Cr interface.

It is interesting to note the effects of applied pressure during bonding on the IMC phase formation. Ag_9In_4 is observed to form at high bonding pressure while $AgIn_2$ at low bonding pressure. Ag_9In_4, which has higher silver content according to [7], will have a higher melting temperature than $AgIn_2$. Indirectly, we can deduce that a higher bonding pressure not only increases the bond strength of a bonded couple, but also increases the overall joint's re-melting temperature.

From the perspective of contact mechanics, the application of pressure will flatten rough surfaces, and lead to an increase in the contact area and diffusion paths between the two bonded couples. Evaporation deposited film is known to be microscopically rough, and thus both sides of the bonding couples are microscopically rough. At low bonding pressure, there was less force applied in flattening the rough surfaces, which resulted in low contact area and thus low shear strength. Whereas at higher bonding pressure, more area between the couple would be in contact, and thus increase the shear strength. Higher contact area could also increase the diffusion paths for In diffusion to side B or Ag diffusion to side A. Thus, application of pressure will also enhance the interdiffusion of Ag and In.

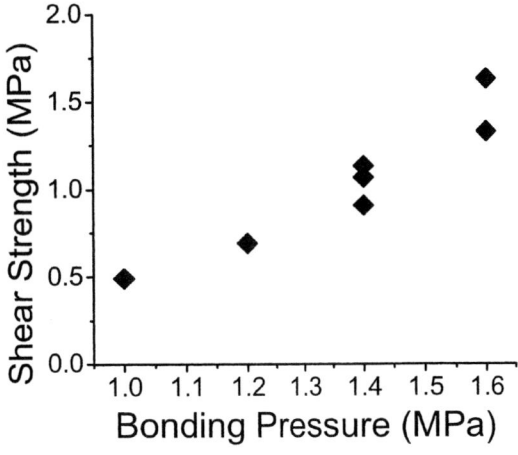

Fig. 3 Plot of bond strength of the joint vs the applied bonding pressure.

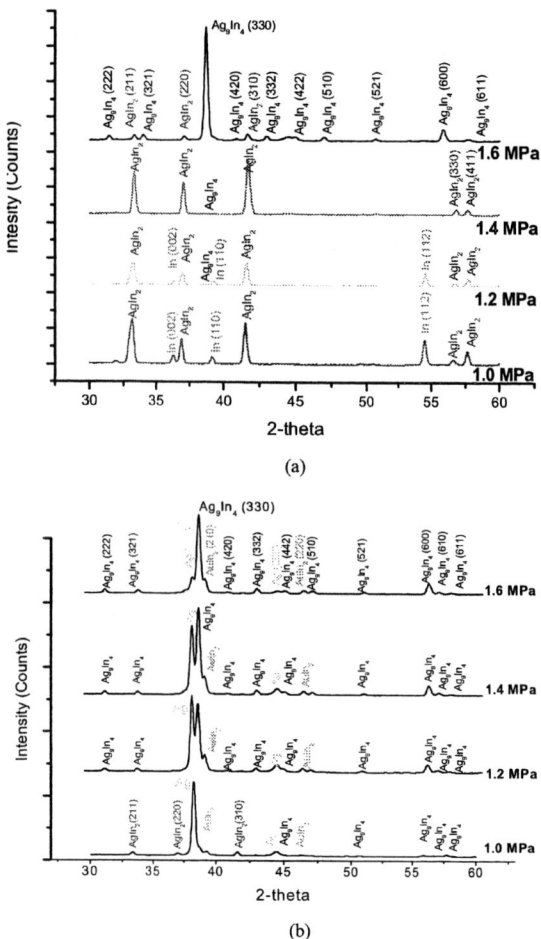

(a)

(b)

Fig. 4 Effect of bonding pressure on the IMC formation for bonding at 180°C - 10 minutes, (a) on side A: Ag_9In_4 is the dominant phase as the bonding pressure increases. (b) On side B, consistent with side A, Ag_9In_4 is the dominant phase as the bonding pressure increases.

On the other hand, the applied pressure during bonding also affects the system's melting temperature, which determines the supply of molten indium for reaction kinetics, such as diffusion and reaction. The pressure dependent melting temperature of materials is thermodynamically stated by Clapyeron equation [13, 18] as shown in Eqn. (1).

$$\frac{dP}{dT} = \frac{\Delta H}{T\Delta V} \qquad (1)$$

where T is the phase changes temperature, P is the pressure, ΔH is the change in enthalpy of fusion in solid-liquid transition and ΔV is the change in system volume. Integration of T and P results in Eqn. (2) which shows the exponential relation

between melting temperature and applied pressure, where T_1 is the phase transition temperature at standard pressure (1 atm).

$$T \propto T_1 e^{\frac{\Delta V}{\Delta H}P} \qquad (2)$$

The exact relation of $\Delta V/\Delta H$ is beyond the scope of the current study, thus quantitative prediction cannot be made. However, the melting temperature typically increases as the pressure increases.

In this study, an increase in In melting temperature would reduce the molten In supply, leading to a reduction of solid-liquid reaction and In diffusion. This effect is clearly observed on the cross section EDX scans in Fig. 2 (a) to (d). As the pressure increases, the In diffusion distance becomes shorter. Furthermore, as the supply of Ag was virtually unlimited, liquid In would quickly become Ag-saturated which may initiate IMC nucleation.

The presence of IMC on the bonding interface may increase the bonding strength since IMC is known to be hard and brittle. However, when in composite with ductile materials, such as Ag, it may enhance the bonding toughness.

Besides the melting temperature, the applied pressure could affect the driving force for IMC formation. In references [14] and [15], it had been mentioned that the driving force is related to the changes in enthalpy of formations (ΔH). Although ΔH is usually considered to be weakly dependent on pressure, it may have a significant effect on the IMC driving force. On the other hand, based on Eqn. (2), ΔV significantly determines the favorable phase at certain pressure. Generally, phase transformation is accompanied by mass density changes or volume changes (ΔV). Positive ΔV could increase the transition temperature as the pressure increases; while negative ΔV could decrease the transition temperature as the pressure increases [18].

In Ag-In system bonding, $AgIn_2$ has lower mass density (8.23 g/cm^3) than Ag_9In_4 (9.90 g/cm^3) which translates into a negative ΔV. Based on Eqn. (2), the phase transition temperature from $AgIn_2$ to Ag_9In_4 will decrease as the bonding pressure increases. Thus, the Ag-rich phase (Ag_9In_4) will be favorable at high pressure. On In-rich side (Fig. 4 (a)), the intensity of Ag_9In_4 peaks becomes higher as the bonding pressure increases. On the other hand, the intensity of $AgIn_2$ peaks decreases as the pressure increases. Similarly, in Fig. 4 (b), where it was initially Ag-rich side, $AgIn_2$ peaks disappear at higher bonding pressure, while Ag_9In_4 peaks become stronger at higher bonding pressure. These results support the prediction based on Eqn. (2) that Ag_9In_4 is the preferred phase at high bonding pressure. Furthermore, Ag-rich phase (Ag_9In_4) could give a higher re-melting temperature of the bond [7, 16, 17].

978-1-4244-2039-1/08/$25.00 ©2008 IEEE

IV. CONCLUSION

In this paper, the low temperature bonding in Ag-In system has been studied. By increasing the bonding pressure, the shear/bond strength of the sample could be increased. It was observed that the Ag-rich intermetallic compound is preferred at higher bonding pressure, which could give a higher re-melting temperature joint. This result is interesting since it provides a method for low temperature bonding to achieve stronger joint with high re-melting temperature.

REFERENCES

[1]. Bartels, F., et al., *Intermetallic phase formation in thin solid-liquid diffusion couples.* Journal of Electronic Materials, 1994. **23**(8): p. 787.

[2]. Shieu, F.S., et al., *Intermetallic phase formation and shear strength of a Au-In microjoint.* Thin Solid Films, 1999. **346**(1-2): p. 125-129.

[3]. Wang, T., et al., *Die bonding with Au/In isothermal solidification technique.* Journal of Electronic Materials, 2000. **29**(4): p. 443-447.

[4]. Litynska, L., et al., *Characterization of Interfacial Reactions in Cu/In/Cu Joints.* Microchimica Acta, 2004. **145**(1): p. 107-110.

[5]. Dae-Gon, K., L. Chang-Youl, and J. Seung-Boo, *Interfacial reactions and intermetallic compound growth between indium and copper.* Journal of Materials Science: Materials in Electronics, 2004. **V15**(2): p. 95-98.

[6]. Simic, V. and Z. Marinkovic, *Room temperature interactions in Ag-metals thin film couples.* Thin Solid Films, 1979. **61**(2): p. 149-160.

[7]. Lin, J.-C., et al., *Solid-liquid interdiffusion bonding between In-coated silver thick films.* Thin Solid Films, 2002. **410**(1-2): p. 212-221.

[8]. Roy, R. and S.K. Sen, *The kinetics of formation of intermetallics in Ag/In thin film couples.* Thin Solid Films, 1991. **197**(1-2): p. 303-318.

[9]. Chuang, R.W. and C.C. Lee, *Silver-indium joints produced at low temperature for high temperature devices.* IEEE Transactions on Components and Packaging Technologies, 2002. **25**(3): p. 453-458.

[10]. Lasky, J.B., *Wafer bonding for silicon-on-insulator technologies.* Applied Physics Letters, 1986. **48**(1): p. 78-80.

[11]. H. Hertz and J.R. Angew, *Translated and reprinted in English in "Hertz's Miscellaneous Paper"* 1986: Macmillan & Co., London.

[12]. Greenwood, J.A. and J.B.P. Williamson, *Contact of nominally flat surfaces.* Proc. Of Roy. Soc. London Series A, 295, 300, 1966.

[13]. Wark, K., *Generelized Thermodynamic Relationship, Thermodynamics .* 5 ed. 1988, New York: McGraw-Hill, Inc.

[14]. Kawamura, S., et al., *Three-dimensional CMOS IC's Fabricated by using beam recrystallization.* Electron Device Letters, IEEE, 1983. **4**(10): p. 366-368.

[15]. Choi, W.K. and H.M. Lee, *Prediction of primary intermetallic compound formation during interfacial reaction between Sn-based solder and Ni substrate.* Scripta Materialia, 2002. **46**(11): p. 777-781.

[16]. Riko, I.M., et al. *Study of Ag-In solder as low temperature wafer bonding intermediate layer.* in *Reliability, Packaging, Testing, and Characterization of MEMS/MOEMS VII* 2008. San Jose, CA, USA SPIE.

[17]. Karpenkopf, L., et al. *Sealing technique for wafer-level integrated cavity using In-Ag multilayers.* in *Reliability, Packaging, Testing, and Characterization of MEMS/MOEMS IV.* 2005. Bellingham, WA: SPIE.

[18]. Porter, D.A and Easterling, K.E *Phase Transformations in Metals and Alloys.* Chapman & Hall, London UK: 1992.

Early Whisker Detection through Intermetallic Compound (IMC) Grain Size

Y. Y. Tan, Dinah On and J. Krishnan

Infineon Technologies (M) Sdn. Bhd.

Batu Berendam FTZ 75710 Melaka

Phone: +60 (6) 2303577   Fax: (+60 (6) 2325069   Email: yik-yee.tan@infineon.com

Abstract—Sn whisker has been a concern in the semiconductor industry. The whiskers will grow within few hours to months with a diameter of 1 micron up to a length of several millimeters. It is difficult to access the whisker impact for Sn plated lead frame. There are no test acceleration factors that can be established to promote a faster growth rate for the whisker. The countermeasure that generally practices is post-bake at 150°C for 1 hour on product. In this paper, an early whisker detection method is introduced. It is observed that the intermetallic compound (IMC) for Sn plated copper lead frame will be developed with different IMC grain size at different annealing hours. The IMC grain size will be developed along the grain boundaries of the Sn deposited and at the Sn-Cu interface layer. It is shown that the larger the IMC grain size will results in reduction of the whisker growth. An IMC grain size grading was developed and it can be used as an early detection method to judge on whisker impact.

I. Introduction

Many semiconductor companies have adopted the "Restriction of certain Hazardous Substances (RoHS)" [1] and converted using the pure Sn plating. Lead-free solder finish on copper is known to form whisker. There are many studies on whisker growth mechanisms [2]-[4]. However, it is noticed that only limited studies focused on whisker test method to judge on the whisker impact onto semiconductor packages. No accelerated test method is known as of today, which can be used to predict whiskers in a reliable way. Therefore, it is vital for the industry to find out other product characterisation methodology that can give an indication to the parts which are free from whisker risk.

Semiconductor pioneers such as Freescale, Infineon, Philips and STMicroelectronics (E4) have proposed the implementation of whisker countermeasure by introducing the annealing step of 150°C for 1 hour on product, within 24 hour of plating [5]-[8]. This step will promote bulk diffusion, forming regular intermetallic layer, which consists of Cu_3Sn and Cu_6Sn_5. Further growth of Cu_6Sn_5 intermetallic with respect to the induced 150°C will acts as a barrier layer that prevents the whisker growth. The grain boundaries of the Sn will then be shifted, resulting in larger grains and less grain boundaries. However, to-date there is no studies to examine critical IMC grain size that prohibits further Cu diffusion that can lead to whisker free.

The annealing proposed by E4 at 150°C for 1 hour on product, is assuming that parts subjected to be less than 1 hour will have whisker risk and more than 1 hour will have whisker free. E4 has suggested that if the annealing countermeasure is implemented, the IMC thickness will grow more than 0.5μm. In order to further investigate this phenomenon, a design of experiment (DoE) is conducted to find out the IMC grain size that is developed as a function of different post-annealing durations and correlate these findings to the IMC thickness and whisker growth.

II. Experimental Procedure

Two types of lead frame base materials (KFC and C70250) were selected in this study. The lead frames are plated using automated production line with Solderon™ ST-300 pure Sn plating process. The plating thickness is controlled at 10μm. Then, the samples are annealed at different hours, ranging from 10 minutes to 2 hours. Each of the samples is further splited into three groups for different analysis approach.

The first sample will then undergo a Sn stripping process using L100 at room temperature for few minutes. The purpose of this process is to remove Sn, leaving only the substrate and the IMC interface. The sample will then be further investigated under Scanning Electron Microscope (SEM). IMC grain size images are taken at 8.0x magnifications and the average measurement of the grain size is further calculated manually by summing the grain size and then divided by number of grain size in an image.

Next, the second sample will be micro-sectioned in a direction perpendicular to the lead frame. The purpose is to investigate the IMC thickness under different annealing hours. To enhance the phase contrast of the IMC, these cross-sectioned samples will be ion polished and inspected under SEM using back-scattered detector (BSE) to observe the thin layer of Cu_6Sn_5 intermetallic.

978-1-4244-2039-1/08/$25.00 ©2008 IEEE

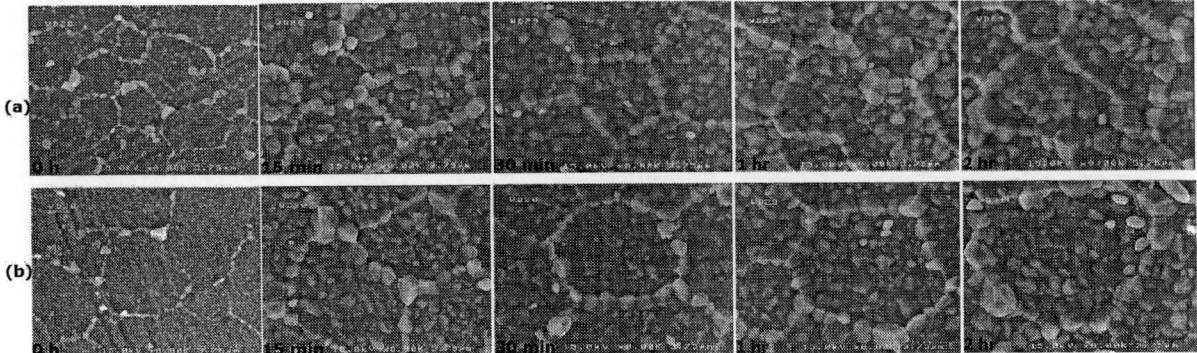

Fig. 1. IMC after stripping off the Tin layer. Image (a) is KFC lead frame at 0 h, 15 min, 30 min, 1 h and 2 h. Image (b) is C70250 lead frame at 0 h, 15 min, 30 min, 1 h and 2 h.

Finally, the third sample will be stored in room ambient environment to monitor the whisker growth. They will be monitored every week for a period of 3 months. The whisker inspection is performed using an optical microscope with a minimum magnification of 100x. If whiskers are detected, it will be further investigated under SEM for whisker length measurement.

III. RESULT AND DISCUSSION

Figure 1 shows the SEM image of the IMC formation after striping-off the Sn at different annealing durations. It was observed that the grain size along the grain boundaries of the Sn deposition and at the base of the Sn grains becoming bigger with respect to increasing annealing duration.

Without annealing, majority of the IMC will grow along the Sn grain boundary and very little grow at the Sn-Cu interface layer. These results in larger IMC grain size along the Sn grain boundary and very little of IMC found at the Sn-Cu layer. This observation had explained the Sn grain boundary 'fill-up' within the IMC will slow down the lateral orientation change along the Sn grain boundary. These results in the Sn grain size boundary at without post-annealing is smaller compare to post-annealing. When the IMC grows only along the grain boundary, the compressive stress in Sn layer will be increased because Cu atoms are dominant diffusing species [9]. The stress that builds over time can shear the surface Sn oxide layer

TABLE I
WHISKER INSPECTION , AVERAGE GRAIN SIZE AND IMC THICKNESS RESULT
(A) KFC LEAD FRAME

Annealing Duration	1 wk (μm)	2 wks	3 wks	4 wks	5 wks	6 wks	7 wks	2 mths	3 mths	Average Grain Size (μm)	IMC thichness (μm)
No annealing	No	Yes Whisker <25μm	Yes Whisker <30μm	Yes Whisker <40μm	Yes Whisker <45μm	Yes Whisker <55μm	Yes Whisker <70μm	Yes Whisker <80μm	Yes Whisker <90μm	0.59	0.15
15 min	No	No	No	No	No	Yes Whisker <20μm	Yes Whisker <20μm	Yes Whisker <25μm	Yes Whisker <30μm	0.79	0.40
30 min	No	No	No	No	No	Yes Whisker <15μm	Yes Whisker <15μm	Yes Whisker <20μm	Yes Whisker <30μm	0.95	0.50
1 h	No	No	No	No	No	No	No	No	No	1.45	0.85
2 h	No	No	No	No	No	No	No	No	No	1.65	1.50

(B) C70250 LEAD FRAME

Annealing Duration	1 wk	2 wks	3 wks	4 wks	5 wks	6 wks	7 wks	2 mths	3 mths	Average Grain Size	IMC thichness
No annealing	No	No	No	Yes Whisker <10μm	Yes Whisker <15μm	Yes Whisker <25μm	Yes Whisker <35μm	Yes Whisker <50μm	Yes Whisker <55μm	0.65	0.13
15 min	No	No	No	No	No	No	No	No	No	0.80	0.45
30 min	No	No	No	No	No	No	No	No	No	1.15	0.55
1 h	No	No	No	No	No	No	No	No	No	1.50	0.90
2 h	No	No	No	No	No	No	No	No	No	1.80	1.70

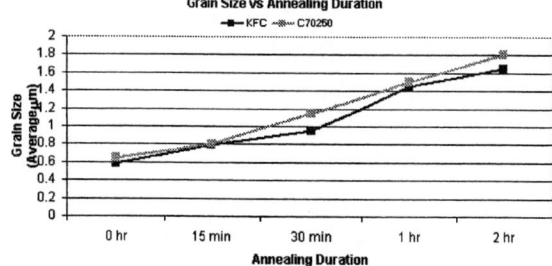

Fig. 2. IMC average grain size at different annealing hour.

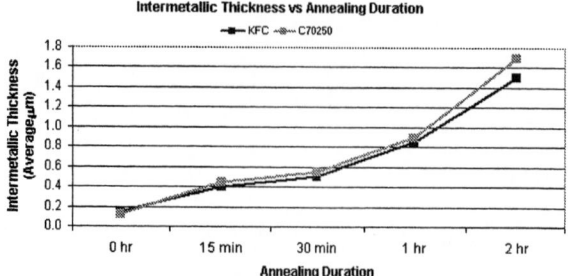

Fig. 4. IMC thickness at different annealing hour.

along the boundaries of grains having orientation different from the majority. This resulted in whiskers grow once the oxide layer is sheared. This correlates with the whisker inspection finding as shown in Table 1. Without annealing, monitoring results clearly shows the formation of whisker growth after a short monitoring of aging duration of two week.

When the annealing was introduced, the IMC will grow along the grain boundary and also the base of Sn-Cu interface. It is observed that the IMC grain size along the grain boundary has minor change only when the annealing increases from 15 min to 2 hours. However, the grain size below the Sn-Cu interface grows bigger from 15 min to 2 hours post-annealing. All the images from Fig. 1 are taken at the same magnification, it is obvious that the Sn grain boundary showing bigger size after heat treatment. When annealing is applied, the lateral orientation change along the grain boundary takes place and resulted in bigger grain boundary. Thus, the compressive stress is built-up by the IMC along grain boundary was released and no whisker detected for post-annealing process.

The grain size was further investigated by calculation of the summation of the grain size for each IMC grain and divided by number of grains within the corresponding image taken at that

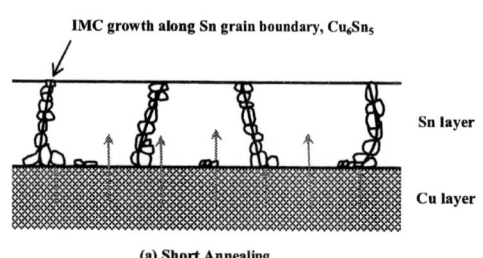

(a) Short Annealing

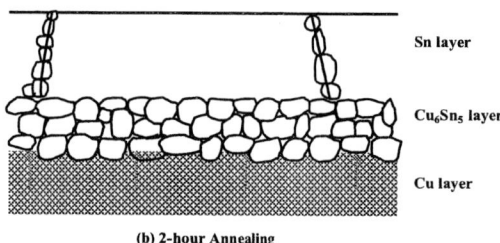

(b) 2-hour Annealing

Fig. 5. Cross section schematic diagram. Image (a) is short annealing process. Image (b) is with 2-hour annealing

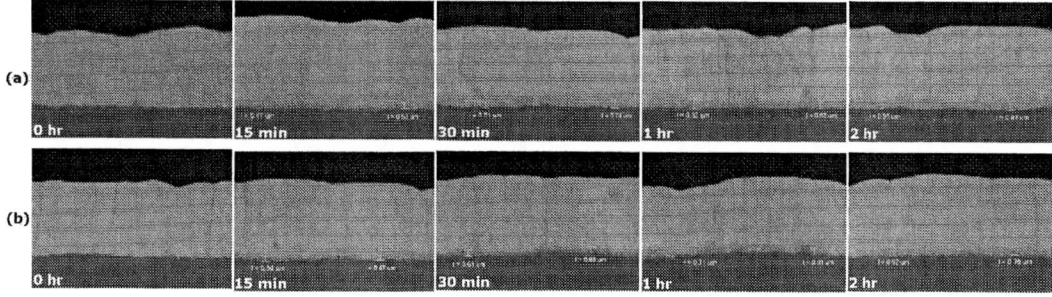

Fig. 3. IMC thickness after cross section. Image (a) is KFC at 0 h, 15 min, 30 min, 1 h and 2 h. Image (b) is C70252 at 0 h, 15 min, 30 min, 1 h and 2 h.

978-1-4244-2039-1/08/$25.00 ©2008 IEEE

particular time. The result is illustrated in Table 1 and is graphically plotted in Fig. 2. Clearly, the average grain size grows with respected to increasing annealing duration. This observation is applicable for both lead frame materials used in this study. The rate of growth of the grains size is independent regardless of the lead frame materials.

Fig. 3 shows the IMC thickness formation from cross-section samples and its corresponding data is plotted in Fig. 4. The results correlate with the IMC grain size findings. Very little evident of IMC is observed without annealing. The IMC thickness grows in parallel with the increasing annealing hour. High densities of large IMC formation contribute to less whisker growth; this observation has been discussed by Egli el at. [10]. In Fig. 5, for short annealing hour, the IMC formation is thin, and the IMC grain size is small. Correspondingly, Sn grain boundary is also small. For long annealing hours, the IMC is thick, IMC grain size is big and Sn grain boundary is large. This explains the purpose of annealing process after Sn plating to form a bulk IMC layer to prevent further diffusion of Cu into Sn layer. When the IMC grain size is big, the rate of diffusion will be slower. Thus, no stress is build-up within the Sn layer and this will prevent further whisker growth. With larger the IMC grain size, the barrier is enhanced further to prevent Cu inter-diffusion. Hence, the objective of the annealing process as countermeasure introduction to create bulk diffusion and form a regular IMC layer is achieved. This will results in less compressive stress, larger IMC grains size and larger Sn grain boundaries.

From the IMC grain size and whisker result in discussed Table 1, we can give an average IMC grain size grading with reference to the whisker inspection result. For IMC grain size greater than 1.5μm is graded as good, followed by a bad grade with IMC grain size is less than 1.5μm. Excellent IMC grain size strongly indicates the part will be free from whisker issues. Bad grade indicates high potential risk of whisker risk. The repeatability of the DoE shows that our new test methodology is highly effective as early detector of any possible whisker issue.

IV. CONCLUSION

Annealing process is well known to be a vital recovery process in the semiconductor industry. The guideline of annealing at 150°C for 1 hour on product is sufficient to create big IMC grain size, with thick IMC thickness and larger Sn grain boundaries. These are the product characteristics to determine any possible whisker risk. The new detection method through IMC average grain size grading is an early detection method for whisker impact. In this study, if the part is having average grain size more than 1.5μm, it is possible to inhibit further whisker growth. This guideline is applicable for lead frame base material KFC and C70250 plated with Solderon™ ST-300 pure Sn. However, the average grain size needs to be characterized if different plating solution used.

ACKNOWLEDGMENT

The authors would like to express their gratitude to all those within Infineon that are involved in the whisker evaluation for their contribution, in particular, Adrian Tong, VK Leong, Kit KT and Dittes Marc.

REFERENCES

[1] Directive 2002/95/EC of the European Parliament and the Council of 27 January 2003. *Official Journal of the European Union*

[2] S.E. Koonce and S.M Arnld, "Growth of Metal Whisker," *J. App;. Phys.* (letter to the editor), 24(3): 1954, pp. 365-366

[3] S.C. Britton," Spontaneous growth of Whiskers on Tin Coatings: 20 Years of Observation," *Trans. of the Institute of Metal Finishing*, 52: pp. 95-102, Apr. 1974.

[4] B.D. Dun, "A Laboratory Study of Tin Whisker Growth," *European Space Agency (ESA) Report STR-223*, pp. 1-51, Sep. 1987.

[5] P. Obernforff, M. Dittes, P. Crema, "Whisker Formation on Sn Plating" *Proc. IPC/JEDEC 5th Inern, Conference on Lead free Electronic Assemblies and Components*, San Jose, USA, Mar. 2004.

[6] M. Ditters, P. Obermdorff, and L. Petit, "Tin Whisker formation – Results, test methods and contermeasures," in *Proc. 53rd Electron. Compon. Technol. Conf.*, New Orleans, LA, 2003, pp. 822-826.

[7] P. Oberndorff, M. Ditters, L. Petit, "Intermetallic Formation in Relation to Tin Whiskers", *Proc. IPC/Soldertec International Conference on Lead free Electronics – Towards Implementation of the RHS Directive*, June 2003, Brussels, Belgium, pp. 170-178.

[8] P. Oberndorff, M.Ditters, L.Petit, C.C Chen, J.Klerk, E.E de Kluizenaar, "Tin Whiskers on Lead free Platings" *Proc. SEMI Technology Symposium*, Advanced Packaging Technologies II, August 2003, Singapore, pp. 51-55, Aug. 2003.

[9] B.Lee and D Lee, "Spotaneous Growth Mechanism of Tin Whiskers", *Acta Mater. 46*, 10.10, pp.3701-3714, 1998.

[10] A. Egli, W. Zhang, f. Schwager, " New approaches to whisker free tin deposits," *Electronics Packaging Technology*, 2003 5th Conference (EPTC 2003) Volume , Issue , 10-12, pp. 55 – 58, Dec. 2003.

[11] P. Oberndorff, M. Ditters, P. Crema, P Su, and E. Yu, "Humidity Effects on Sn Whisker Formation," *IEEE Trans .on Electronics Packaging Manufacturing*, vol. 29 , pp. 239-245, Oct. 2006.

[12] K.N. Tu, J.C.M. Li, "Spontaneous whisker growth on lead-free solder finishes," *Material Science & Engineering: A*, vol. 409, pp. 131-139, Nov. 2005.

IMC Growth and DR4 Open on TSOP Package

Wu Caihong, Lam Tim Fai, Song Xianzhong, Chen Bin, Sun Mingxia, Wang Xiangru

Spansion (China) Limited, No 33 Xinghai Street, Suzhou 215021, China

Phone: 86-512-62523333-36575 Fax: 86-512-62523006 Email: Xianzhong.Song@spansion.com

Abstract: One kind of TSOP package (Device 98M16 encapsulated by mold compound 7351LS) encountered serious open failure after Data Retention 4 test (DR4, 5000 hours baking at 150 °C) and the failure rate is almost 100%. But a similar package (Device 98R16 encapsulated by mold compound G700L) is normal after test. Failure analyses were done for the two packages. The results show that the wire bond IMC are different for failed package and good package. Kirkendall voids and cracks were found on the wire bond area for failed package. FEA simulation results show that the mold compound used in failed package induces high stress status in wire bond area and leads to the bad performance of the wire bond.

I. INTRODUCTION

One kind of TSOP package (Device 98M16 encapsulated by mold compound 7351LS, we will call it "failed package" later for short), which passed Data Retention DR3 reliability test (2000hr baked at 150 °C), encountered serious OPEN failure after Data Retention 4 test (DR4, 5000 hours baked at 150 °C) and the failure rate is almost 100%. But a similar package (Device 98R16 encapsulated by mold compound G700L, we will call it "good package" later for short) is normal after the same test.

Detailed failure isolation experiments were performed. The dimensions and geometries of the die and other components in good and failed package are quite similar. The oxidation status for the I/O leads in two kinds of packages are almost the same. The only differences are:

A. The mold compound used are different.

B. The wire bond situations are different:

1) When the packages were decaped, and the wire bonds were subjected to shearing test, the averaged bond shear strength for the good package is 32.33 gf, while for failed package, it is just 9.13 gf.

2) The optical inspection (Fig. 1) for the bond shear fractures shows that fracture occurs between IMC/Au bond interfaces for failed package (left picture in Fig.1), but for good package, the fracture occurs in the Au wire itself (right picture in Fig 1).

C. After decap, the open failure recovered in failed package.

Form the above analysis, it can be concluded that the bonding area is abnormal for the failed package after 5000-hour baking and cooling down to room temperature.

Fig.1: Bond pad fracture after bond shearing test. (Left : for failed package, Right : for good package)

II. DETAILED STUDY ON IMC IN WIRE BOND

Detailed SEM/EDX study was performed on the cross-sections of the wire bond. The EDX quantitative analysis results on the IMC were compared with Au-Al phase diagram (Fig. 2) for phase identification. The results are as follows.

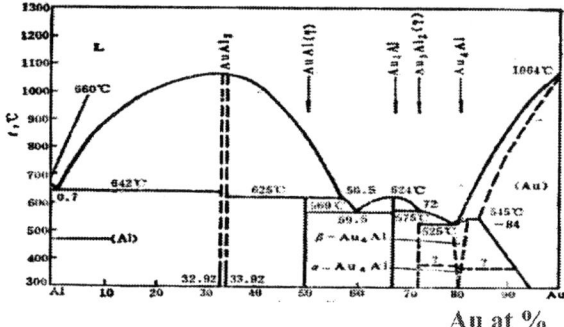

Fig.2: Au-Al phase diagram

A. Original packages before baking (Fig 3): IMC have grown after wire bonding. The IMC are Au5Al2 both for good and failed packages. No void and crack can be seen.

B. 2800-hour baked package: IMC in central part of the bond become thicker. At the periphery of the bond, IMC Au2Al begins to grow. Kirkendall voids were found on the periphery area for failed package (Fig. 4). For good package, no void can be seen.

C. 5000-hour baked package (Fig. 5). The majority of IMC are Au4Al for good package and Au5Al2 for failed package respectively. For failed package, Kirkendall void and serious cracks were found. Obviously, these voids and cracks form a high resistance layer at the Au/IMC interface and leads to an OPEN failure (high voltage is needed to drive a current).

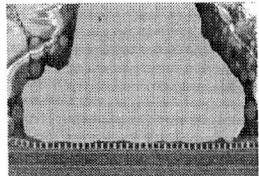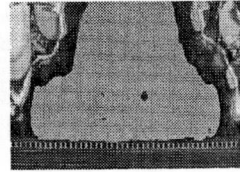

Fig.3: SEM images at wire bond area for original packages (Left for good package, Right for failed package)

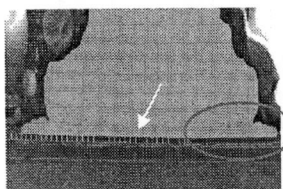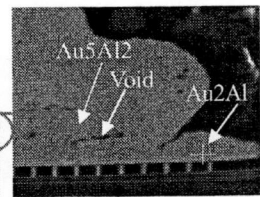

Fig.4. SEM image of wire bond for 2800 hr. baked failed package.

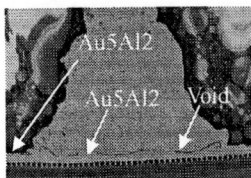

Fig.5: SEM images of wire bonds for 5000 hr baked packages (Left for failed package, Right for good package.)

In summary, SEM/EDX results show that:

1) Majority of the IMC in the bonds is Au4Al for good package and Au5Al2 for failed package respectively.

2) Kirkendall voids were found between Au bond and IMC in failed package as early as 2800 hr baked.

3) Cracks/Separation were found between Au bond and IMC in failed package when it cooling down to room temp after 5000hr baking.

IMC in Au-Al system have been studied in details. Some examples are given in [1-2]. But few of them related the IMC growth and Kirkendall voids to the stress status in the bond area. To get better understanding of our findings mentioned above, FEA simulation was performed.

III. FEA (Finite Element Analysis) SIMULATION

¼ symmetrical models were built for good and failed package with dimensions and material properties given in Table I and II. Au, Cu and Al are treated as bi-linear plastic materials. The models are shown in Fig 6. Wire bond and its surroundings are built in a fine mesh model, as shown in Fig 7 .The fine models are linked to the main models (coarse mesh model) by using constrain equation technique in FEA.

The models were run in 3 steps, as shown in Fig 8:
Step 1: after post-mold cure, the package cools from 175°C to 25°C,
Step 2: baked at 150°C,
Step 3: after bake, the package cools down to 25° C.

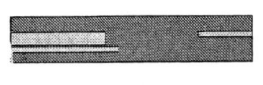

Fig.6:   The ¼ symmetrical model

Table I: The dimensions of the packages

dimensions, all in um	98R16	98M16
	good package	Fail package
PackageX	18400	18400
PackageZ	12000	12000
Diex	7112	7670.8
Diez	5130.8	4368.8
Die thickness	280	280
D/A thickness	30	30
Top MC thickness	330	370
Bottom MC thickness	370	330
Copper thickness	127	127

Table II: The material properties for the components

	E1	E2	CTE1	CTE2	Nu	Tg
	Gpa	Gpa	ppm/K	ppm/K		C
die	131		2.3		0.3	
die attach	1		80		0.3	
Cu	76		16.9		0.3	
Au	77.2		14.2		0.42	
Al	69		23		0.42	
IMC Au5Al2	93		14		0.42	
IMC Au4Al	114		12		0.42	
Failed:MC(98M16)	13.13	0.24	5.23	49.3	0.3	176
Good: MC(98R16)	12.38	0.064	15	45	0.3	137

The component's compositions of the models are based on the EDX results mentioned above. Therefore, as can be seen in Fig 8, for example, in step 3, the IMC underneath   the Au bond for good unit are Au4Al, while for failed units they are Au5Al2. Different IMC have different CTE and E, as can be seen in Table II. Different steps also have different working temp range. Therefore, we used different effective CTE (CTE-eff) and effective E (E-eff) for mold compound at different step and different package, as shown in Table III.

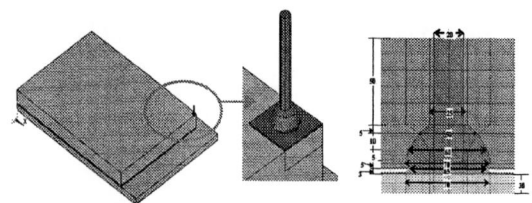

Fig 7: The wire bond on die top

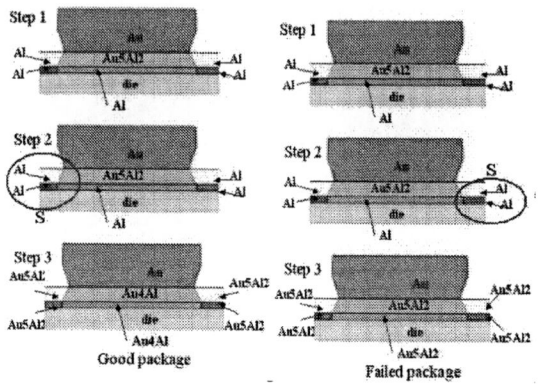

Fig 8: The load steps and the components at bond/IMC interface
(Left for good package, Right for failed package)

Table III: CTE-eff and E-eff used for mold compound at different step

Compound	Eeff (step 2), (Gpa)	CTE eff (step 2), (ppm/k)
Failed: 7351LS	0.654	5.23
Good: G700L	0.0616	34.9
	Eeff (step 1,3), (Gpa)	CTE eff (step 1,3), (ppm/k)
Failed: 7351LS	8.553	5.23
Good: G700L	4.571	15.8

Below are the FEA results for step 2 and step 3. It should be noted that in our models, y-axis is parallel to the wire, and x, z-axis are perpendicular to the wire.

A. *Step 2 Results*

Fig 9 shows the stress contour in x and y directions, Sx (Left) and Sy (Right), at IMC top for good (top) and failed (bottom) package in step 2. Fig 10 shows the Sx (left) and Sy (right) contour at Au-bond-bottom for good (top) and failed (bottom) package. It can be seen that, most of the IMC top area are in tensile stressed (Sx, Sy are positive, see Fig 9). This should help Au atom from Au-bond bottom to diffuse down into IMC top. However, from Fig 10, quite a big area of Au-bond-bottom is in compress-stressed (Sx, Sy are negative). This is no good for Al atom in IMC top to diffuse up into Au-bond-bottom.

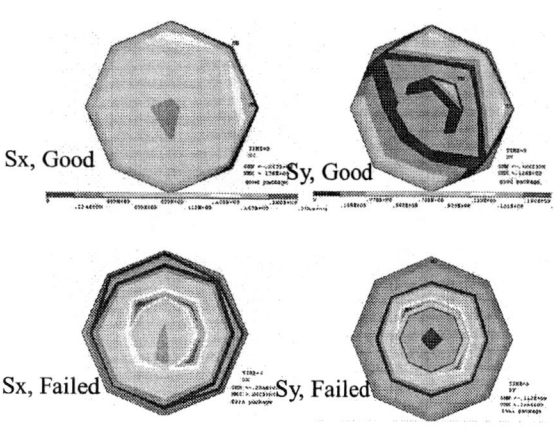

Fig.9: Sx (left) and Sy (right) at IMC-Top for good (top) and failed (bottom) package in Step 2. Only positive part is shown.

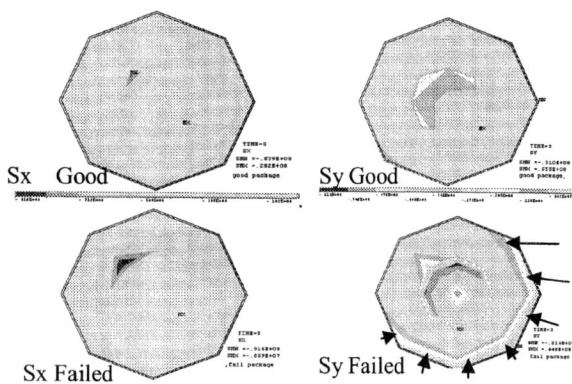

Fig.10: Sx (left) and Sy (right) at step 2 at Au-bond-bottom for good (top) and failed (bottom) package. Only negative part is shown.

But looking at Fig 10, the comparison of the stress between good package (top) and failed package (bottom) (compare Sx-good vs Sx_failed and Sy-good vs Sy-failed in Fig 10) finds that, the Au-bond-bottom in failed package has more area with negative Sx and Sy. This implies that the diffusion process is even more difficult in failed package than in good package. So the IMC Au5Al2 in failed package is difficult to transfer into Au4Al. And this is in agreement with the SEM/EDX finding 1)mentioned above: majority of the IMC in wire bonds are Au4Al for good package and Au5Al2 for failed package respectively.

However, to help the Al in periphery area (marked by "S" in Fig.8) to transfer into Au2Al and then Au5Al2, Au-Al inter-diffusion have to occur between Au-bond-bottom and IMC-top at this area. But for failed package, this area happens to be the area having significantly negative Sy (the area marked by arrows in the right-lower picture of Fig. 10). In other words, this is the area, where Au atom can diffuse into IMC easily, but Al atom is difficult to diffuse into Au-bond-bottom. From the Kirkendall void theory, if inter-diffusion rates in two opposite directions are quite different, Kirkendall voids will create when IMC grows at this area. It is also noted that this happens to be the area where Kirkendall voids were found for failed package as early as at 2800 hr baked, as mentioned above in the SEM/EDX finding 2).

B. Step 3 Results

Step 3 simulate the situation after long time baking and cooled down to 25 °C. Based on EDX results, the IMC in this step is Au4Al in good package and Au5Al2 in failed package respectively. The FEA results for step 3 are shown in Fig 11. It can be seen that, the tensile stress Sy in step 3 are higher than in step 2 .The failed package has more area with higher Sy (compare the top pictures with the corresponding bottom pictures in Fig 11). It means that, after long time baking and cooling down to room temperature, the bond in failed package has more chance to be lift and the separation will occur between Au-bond and IMC. This is in agreement with the SEM/EDX finding 3) mentioned above.

It should be noted that all the high stress status for failed package mentioned above is due to the fact that the wire bond is encapsulated by mold compound and the mold compound and the Au wire bond have different CTE and E. (thermal mismatch). Once the mold compound is removed after decapsulation, the stress will be relaxed, and the resistance between Au bond and IMC will decrease remarkably. This is why the open failure can be recovered after decapsulation.

From the data in Table I, the CTE for Au, mold compound in good package and mold compound in failed package are 14.2, 15 and 5.23 ppm/K respectively. The thermal mismatch between Au and mold compound in failed package is significantly higher than in good package. That is why the stress in wire bond area in failed package is much higher than in good package. And this finally leads to the Kirkendall void growth and bond separation in IMC area for failed package.

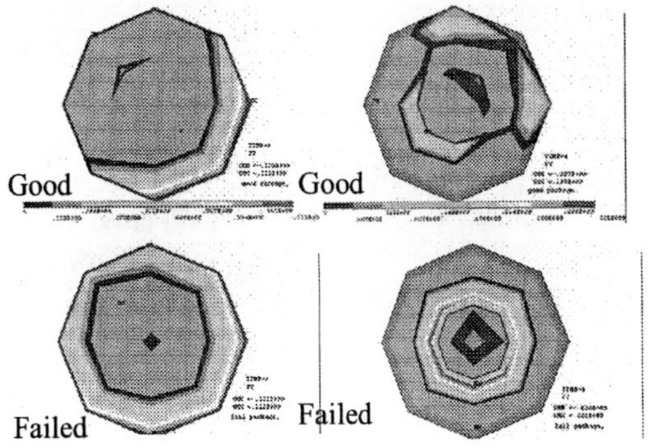

Fig.11: Sy at step 3 at Au bond bottom (left) and IMC-Top (right) for good (top) and failed (bottom) package. Only positive part is shown

IV. CONCLUSIONS

1. The open failure in failed package was caused by Kirkendall voids/ cracks between Au bond and IMC in the wire bond.

2. The FEA stress analysis at different stages for different packages showed that because the different stress status in different package, the IMC in failed package (Au5Al2) is more difficult to transfer into Au4Al, while this can be done in good package. The Au bond periphery in failed package has more chance to create Kirkendall voids. The Au bond in failed package is much easier to separate from IMC than in good package after long time bake and cool down to RT.

3. Since the dimensions and geometries for the good and failed package are almost the same, the main difference between them is the mold compounds used. The mold compound used in failed package has bigger thermal mismatch with Au than the mold compound in good package. And this finally leads to the bad performance of the Au-bond in the failed package.

REFERENCES

[1] S. Sutiono, A. Seah, S. Chew, et al, "Intermetallic Growth Behavior of Gold Ball Bonds Encapsulated with Green Moulding Compounds", Proc. 7th Electronic Packaging Technology Conference, Singapore, December 2005.

[2] E. Galli, G.Majni, C. Nobili, and G. Ottavinil et al,: "Gold-aluminum intermetallic compound formation" Proc 1979, Modena, Italy.

Lid Adhesive Failure Study for Flip Chip Packaging

M. C. Ong, X. L. Zhao, P. P. Joman and J. M. Chin
Device Analysis Lab
Advanced Micro Devices (Singapore) Pte Ltd
508 Chai Chee Lane, Singapore 469032
Phone: (65) 67969888 Fax: (65) 62339080 Email: mei-chyn.ong@amd.com
Raj N. Master
Manufacturing Services Division
Advanced Micro Devices Pte Ltd
Sunnyvale, California, USA 94088-3453

Abstract - **This paper studied the factors of causing adhesion failure between lid and adhesive. Adhesive curing mechanism has been experimentally investigated in combination of surface behaviour analysis on lid by using FTIR and XPS. The results showed residue on lid surface caused by low water rinse flow can affect the curing condition.**

I. INTRODUCTION

Flip chip packages generally consist of an active die mounted down on an organic carrier with the backside available for heat transfer. Attaching a high-conductivity lid directly to the die backside can enhance heat spreading and power dissipation to the environment. The advantages of a lidded package include, among several, (1) increased surface area for better heat transfer to the environment; (2) protection from handling damage and scratching of die; (3) larger surface for marking; (4) easier assembly of heat sink; and, (5) optional mechanical balancing and reduction of package stress. [1]

The lid is attached to the die by thermal interface material (TIM) and attached to the substrate by adhesive material. The lid adhesive is a silicone material, which acts as a very effective sealant because of the high elongation and ability to maintain excellent strength through a cross-linked formulation.

The quality of lid adhesive plays an important role in ensuring the package reliability. There are some factors that would affect the adhesive quality such as curing temperature, curing time, cleanliness of the adhesion surface, etc.

Generally, there are two types of lid adhesive failure mode: cohesive failure and adhesive failure. If the bonding force is higher than the intrinsic strength of the adhesive, the failure mode is said to be **cohesive failure**. It is characterized as an observation after lid shear test with the adhesive presenting more than 70% on the lid flange area. If the bonding force is lower than the intrinsic strength of the adhesive, the failure mode is said to be **adhesive failure**. In this case the characterization is one whereby the adhesive presenting less than 70% on the lid flange area (shown in Fig. 1).

Fig. 1. Adhesive failure mode

The poor lid adhesion problem is commonly observed on lidded package. Therefore, root cause investigation was experimentally carried out by using Fourier transform infrared spectroscopy (FTIR) to study adhesive curing condition and X-ray photoelectron spectroscopy (XPS) to study surface behaviour on lid.

II. EXPERIMENTAL PROCEDURE

The FTIR instrument was a Thermo Nicolet spectrometer equipped with the mid-IR source of radiation. The device recorded FTIR spectra at a spectral resolution of $4cm^{-1}$. The spectra over the wavenumber range $4000cm^{-1}$ to $400cm^{-1}$ with 32 scans were collected through transmission mode. KBr pellet was prepared and a scan of the FTIR background was done before coating adhesive on the pellet. Uncured adhesive was then coated on KBr pellet for FTIR measurement. Automatic baseline correction and automatic smooth were done to minimize baseline shift problem. The same pellet was then placed in the conventional oven at 100°C for pre-cure for one minute and FTIR spectrum was collected again. Subsequently, the same pellet was placed in the conventional oven at 125°C for 2, 5, 10, 30, 60, and 180 minutes and at 200°C for 180 minutes. FTIR spectrum was collected at every interval (see Table I).

XPS measurement was performed using a PHI Quantera SXM Scanning X-ray Microprobe with an Al Kα X-ray source (224eV) with take-off angle of 45° and spot size of 20 um.

III. RESULTS & DISCUSSION

Lid adhesive material is a silicon-based organic polymer called vinyl-terminated poly(dimethyl siloxane) (PDMS). The chemical structure for vinyl-terminated PDMS is shown in Fig. 2, where n is the number of repeating monomer $[SiO(CH_3)_2]$ units.

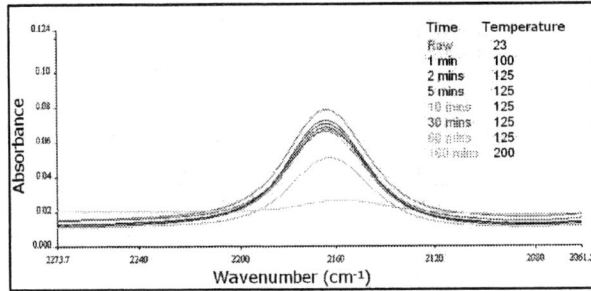

Fig. 2. Chemical structure of vinyl-terminated PDMS polymer

A hydrosilylation cross-linking reaction occurs during curing in which silyl (SiH) groups in a trimethyl-terminated poly(hydrogen methyl siloxane) cross-linker react with vinyl ($-CH=CH_2$) groups on PDMS [2] (shown in Fig. 3). The extent of cross-linking and cure was determined through FTIR analysis on the changes of peak area in the characteristic SiH absorption band which is at spectral region of around $2150 cm^{-1}$ [3].

$$\sim Si - H \; + \; CH_2 = CH - Si \sim \xrightarrow[\text{heat}]{} \; \sim Si - CH_2 - CH_2 - Si \sim$$

Fig. 3. Chemical reaction between silyl groups of cross-linker and vinyl groups of PDMS

It was observed that the cross-linking reaction for the adhesive is the SiH ($\sim 2150 cm^{-1}$) spectral region. There were no significant changes in the spectra elsewhere. Fig. 4 showed the FTIR spectra at each interval of curing condition, with significant changes occurring at the absorption band of $\sim 2150 cm^{-1}$.

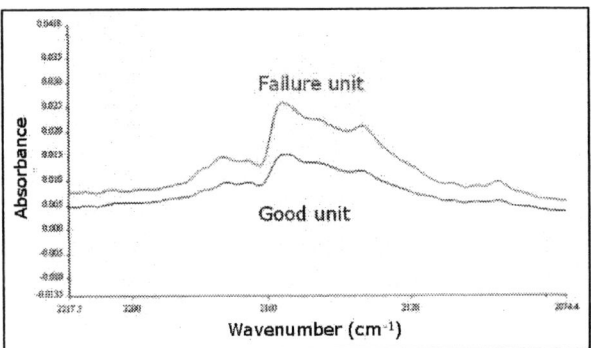

Fig. 4. FTIR spectra of lid adhesive under several curing conditions

The area of the peak at $\sim 2150 cm^{-1}$ was calculated and hence the degree of cure was determined (shown in Table I).

Table I
Degree of cure for lid adhesive

Lid adhesive Curing time	Curing Temperature (°C)	Corrected area	Cure%
Raw	23	2.4689	0.00
1 min	100	2.4425	1.07
2 mins	125	2.3084	6.50
5 mins	125	2.2112	10.44
10 mins	125	2.1236	13.99
30 mins	125	2.0885	15.41
60 mins	125	1.5088	38.89
180 mins	125	0.7315	70.37
180 mins	200	0	100.00

The experimental results showed that the peak with a larger area has a lower degree of curing. Shorter curing time on adhesive leads to insufficient hardness and weaker adhesion, exhibiting adhesive failure mode.

Now with the understanding of the relationship between the lid adhesive curing condition and curing degree, FTIR analysis was also performed on good and failing units in order to check their curing status. Subsequently, FTIR analysis with attenuated total reflection (ATR) mode was performed on 10 lids of a good unit lot (lot with high shear strength) and 10 lids of a failure unit lot (lot with low shear strength). Fig. 5 showed the FTIR spectra around the peak at $2150 cm^{-1}$, while Table II showed the degree of cure.

Fig. 5. FTIR spectra on lid of good unit and failure unit

Table II
Degree of cure on lid of good unit and failure unit

Lid adhesive	Temperature (°C)	Corrected area	Cure%
Raw	23	2.4689	0.00
Good unit	Per spec condition	0.2482	89.95
Failure unit	Per spec condition	1.1227	54.53
180 mins	200	0	100.00

The curing condition for both good and failure unit lots was the same. However, the average degree of cure for the good unit was about 90%, but only about 55% for the failure unit. Compared to the failure unit, the good unit with higher shear strength has a smaller peak area at ~2150cm^{-1}, indicating a higher degree of curing.

From this FTIR observation of non-correlation between the curing condition and the degree of cure, the other factor that could contribute to adhesive failure mode is the cleanliness of the lid surface. Therefore XPS analysis was performed to determine the lid surface condition, and Fig. 6 showed the XPS analysis result.

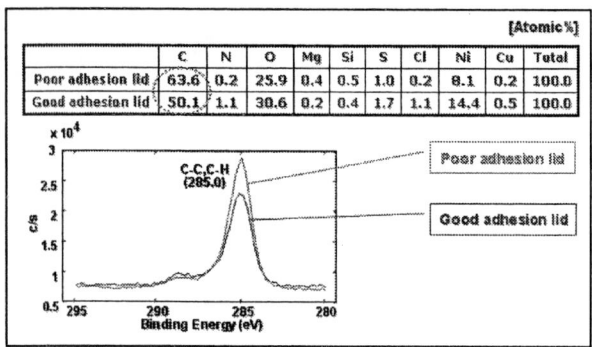

Fig. 6. XPS analysis result of good and poor adhesion lids

XPS analysis of good and poor adhesion lids detected more residue of organic material on the poor adhesion lid. The organic residue was the fatty acid salt (sodium lauryl sulfate $C_{12}H_{25}NaO_4S$ and sodium stearate $C_{17}H_{35}COONa$) (shown in Fig. 7) which was the surface-active agents added in the nickel (Ni) plating bath solution.

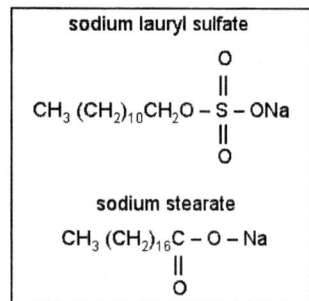

Fig. 7. Chemical structure of fatty acid salt

The Ni plating bath solution, which is used in the Ni lid plating process (shown in Fig. 8), contains a brightening agent composed of an organic solvent and a surface-active agent of either ester or ether. The surface-active agent is added to eliminate Ni plating surface pits. Proper rinsing between process stages is essential to successful pre-treatment; poor rinsing will leave residues on the part surface that will interfere with the remaining processes.

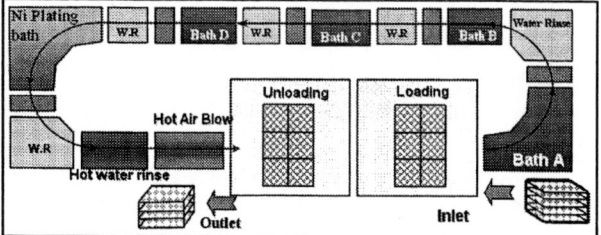

Fig. 8. Ni lid plating process

The organic residues on the lid surface can be eliminated very easily by the pure water rinse, but they cannot be eliminated by low water rinse flow. Therefore, the process step condition was checked thoroughly, revealing abnormal flow conditions for the water rinse flow meter (0.5x LT/min) after the Ni plating bath solution, whereby normal condition should be 1.5x LT/min. This abnormal flow allowed organic residues to remain on the lid surface and affected the lid adhesion.

The fatty acid salt organic residues would attract to hydrogen of silyl (SiH) groups and, thus, were held in suspension (shown in Fig. 9). In such circumstances, there will be less SiH groups to react with the vinyl-terminated PDMS during curing and cause incomplete curing and failure on lid adhesive.

Fig. 9. Attraction between fatty acid groups and silyl groups

For those good units with 90% of curing, re-cure could further improve the shear strength. However, it is difficult to improve the adhesive properties at the interface between adhesive and contaminated lid for those adhesive failure units.

Re-cleaning for lids was performed and higher water rinse flow was carried out after Ni plating process. Fig. 10 evaluated the XPS analysis of the re-cleaned lids.

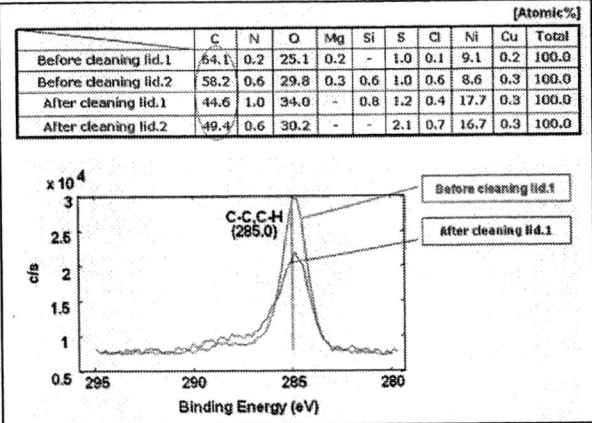

	C	N	O	Mg	Si	S	Cl	Ni	Cu	Total
										[Atomic%]
Before cleaning lid.1	64.1	0.2	25.1	0.2	-	1.0	0.1	9.1	0.2	100.0
Before cleaning lid.2	58.2	0.6	29.8	0.3	0.6	1.0	0.6	8.6	0.3	100.0
After cleaning lid.1	44.6	1.0	34.0	-	0.8	1.2	0.4	17.7	0.3	100.0
After cleaning lid.2	49.4	0.6	30.2	-	-	2.1	0.7	16.7	0.3	100.0

Fig. 10. XPS analysis result of re-cleaned lids

From the XPS result, the re-cleaning process decreased the organic material detected on the lid surface. Additionally, the adhesive shear strength increased (shown in Table III); thus, adhesive failure mode was no longer observed and lid adhesive had significantly improved, as shown in Fig. 11.

Table III
Adhesive shear strength results before and after lid cleaning

Before cleaning		After cleaning	
Unit#	Shear strength (lbs)	Unit#	Shear strength (lbs)
1	234	1	504
2	210	2	554
3	301	3	531
4	255	4	561
5	230	5	498
Average	250	Average	530
Min	210	Min	498
Max	301	Max	561

Fig. 11. Condition of lid adhesive before and after lid cleaning

IV. CORRECTIVE ACTION

As it was proven that the cause of the contaminant was due to the low water rinse flow, the corrective actions included (1) implementing a filter in the water supply line to prevent particle accumulation; (2) increasing the water rinse flow rate by 30%; and, (3) narrowing the control limit of the water rinse flow.

V. CONCLUSION

FTIR database has been successfully established through design of experiment (DOE) for further application. The root cause of lid adhesive failure mechanism has been experimentally studied and successfully remedied. It was due to the organic residues of Ni lid plating process and causing no cross-linking reaction for adhesive during curing. Further, corrective actions have been implemented in the process.

ACKNOWLEDGMENT

The authors would like to thanks the colleagues and the lid supplier for their support.

REFERENCES

[1] J. Wakil, "Thermal performance impacts of heat spreading lids on flip chip packages: With and without heat sinks," Microelectronics Reliability 46, 2006, pp. 380-385.

[2] T.R.E. Simpson, Z. Tabatabaian, C. Jeynes, B. Parbhoo and J.L. Keddie, "The Influence of Interfaces on the Rates of Crosslinking in Poly(Dimethyl Siloxane) Coatings," Journal of Polymer Science Part A: Polymer Chemistry, Vol. 42, Issue 6, 2004, pp. 1421-1431.

[3] D.R. Anderson, A.L. Smith (ed.), *Analysis of Silicones*, John Wiley & Sons, London, 1974, Ch. 10.

978-1-4244-2039-1/08/$25.00 ©2008 IEEE

SESSION 7:

POSTER SESSION

CAFM Detection of Resistive Tungsten Contacts In DRAM Devices

E. Ng*, D. Lam*, X. Zheng#
*Micron Semiconductor Asia Pte Ltd, 990 Bendemeer Road Singapore 339942
Phone: (+65)-6398 4101   Fax: (+65)-6290 3457   Email: ericng@micron.com
#Nanyang Polytechnic, 180 Ang Mo Kio Avenue 8 Singapore 569830

Abstract- **Resistive contacts in multilayer interconnects have been an ongoing challenge for semiconductor industries. The failing mechanisms behind them were difficult to uncover, which made it necessary to initiate deeper research into each issue that we managed to isolate. The traditional approach of using passive voltage contrast (PVC), scanning electron microscope (SEM), or focused ion beam (FIB) [1] to identify resistance gate contacts is getting more difficult, especially for those contacts with resistance that is only marginally higher than normal contacts. In this paper, we discuss the use of CAFM current scanning mode to detect a failing resistive gate contact as well as a failing resistive substrate contact. Finding the root cause for these resistive contacts is vital to solving reliability and yield losses, and that cannot be achieved unless we first find the resistive contacts with great accuracy. This novel approach to electrical fault isolation using conductive atomic force microscopy (CAFM) will enhance the advanced failure analysis techniques which has so far failed to achieve the accurate detection of resistive tungsten contacts.**

I. BACKGROUND

As semiconductor devices have miniaturized, defects are becoming increasingly more difficult to find and more challenging to prove. Resistive contacts, which can result in high yield loss, usually escape our failure mechanism understanding. Due to their subtle resistive nature, they usually fail in electrical tests marginally, but recover under higher temperature or higher stress voltage. Given that there are many possible transistors, package level characterization does not narrow down exactly which transistor gate or drain output is resistive on the die. Unlike open contacts, which are easily seen and can be found by traditional methods such as PVC under SEM or FIB, SEM imaging of cross-sectional, subtle resistive contacts typically do not reveal much information of the actual root cause of the failure mechanism. As a result, when working with resistive contacts, traditional methods have a physical limitation.

An example of a resistance contact was demonstrated between barrier metal layer and $TiSi_2$ which was identified as TiOx causing the resistive layer [2]. PVC under SEM and FIB usually do not reveal any abnormalities in terms of contrast. So we submitted our samples for CAFM analysis using the Dimension V scanning probe mMicroscope system from Veeco where they were mechanically grinded down to W contacts level.

II. RESULTS AND DISCUSSION

An area of suspected resistive contacts in our reliability tests' fall-out was isolated electrically. Electrical characterization using CAFM were performed on these contacts to prove they have higher resistance. Current image scanning was subsequently developed to enhance the accuracy of resistive contacts detection.

A. CAFM IV-curve results for transistor gate contacts

The topography image of the W contacts to the gates is shown in *Figure 1a*. From the I-V plots of the forward-biased turn-on region shown in *Figure 1b*, it was observed that there is a lateral shift to the right for the resistive gate contact compared with the good gate contacts. Similarly, for the I-V plots of the reversed-biased breakdown region shown in *Figure 1c*, there is a lateral shift to the left. This indicates that the suspected failing gate did have a higher resistance compared with the passing gates.

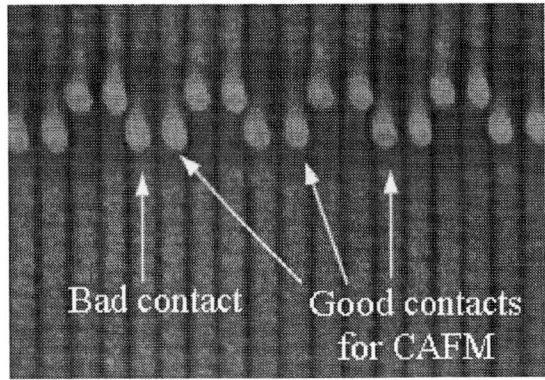

Figure 1a. Topography image of the W contacts to the gates

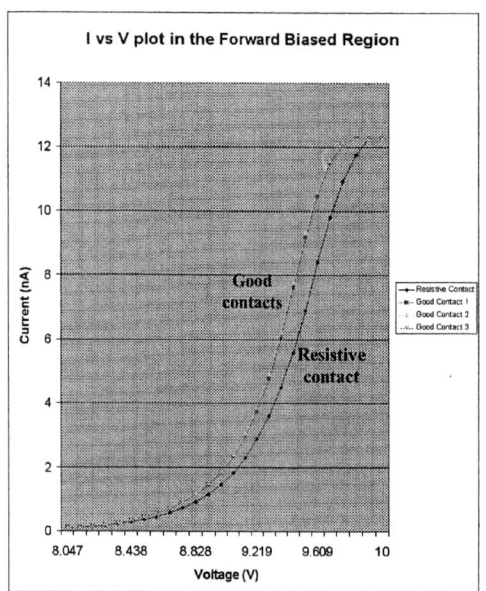

Figure 1b. Current vs. voltage plot of the bad contact and good contacts in the forward-biased region

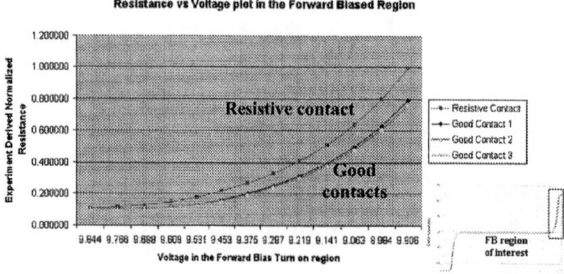

Figure 1d. Normalized resistance vs. voltage plot of the bad contact and good contacts in the forward-biased region

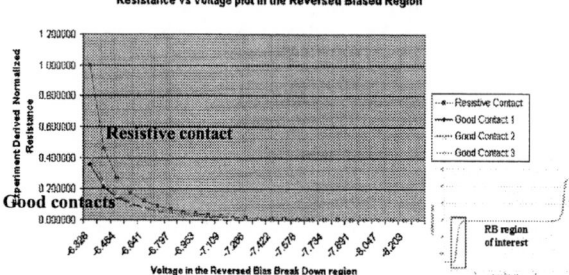

Figure 1e. Normalized resistance vs. voltage plot of the bad contact and good contacts in the reversed-biased region

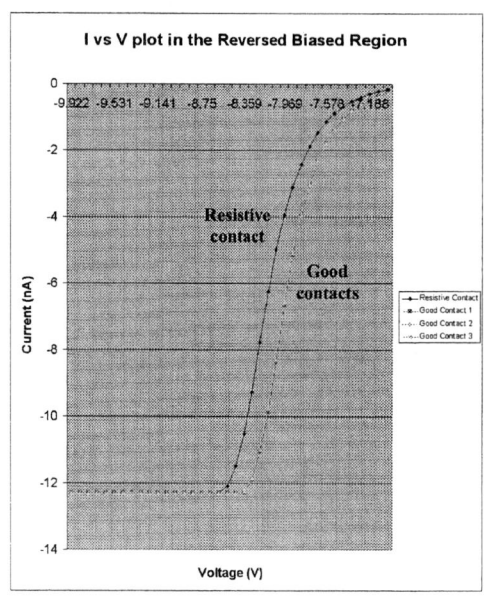

Figure 1c. Current vs. voltage plot of the bad contact and good contacts in the reversed-biased region

The normalized resistance vs. voltage (R-V) plots of the forward-biased turn-on region and reversed-biased breakdown region are shown in *Figure 1d* and *Figure 1e*, respectively. It was observed that at the start of forward-biased region and at the start of the reversed-biased region, the failing gate contact exhibits a higher resistance compared with the good gate contacts.

B. CAFM current image scanning for transistor gate contacts

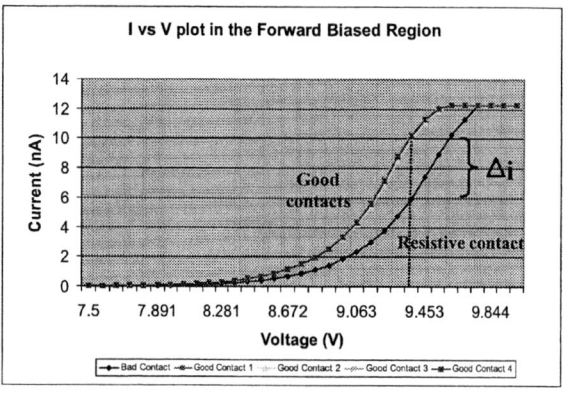

Figure 2a. Choosing current image scanning voltage

As the IV curves obtained require the CAFM tip to come into contact with each suspected contact, it will not be efficient in terms of electrical fault isolation. Hence the CAFM current image scanning mode was brought into play to enhance the fault isolation technique. By running the IV curve previously, we know that for resistive gate contacts in particular, the current difference is greatest at approximately 9.4V as shown in *Figure 2a*. Therefore a current scanning was chosen to perform on the CAFM using 9.4V as bias voltage to the sample.

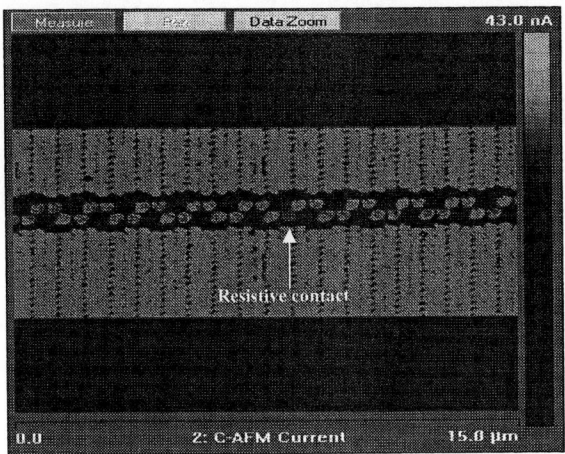

Figure 2b. Scanning out the resistive gate contact using
CAFM current image mode

From *Figure 2b*, the resistive gate contact is successfully detected when we use the correct bias voltage. The approximately range of good detection for transistor gate contact is 9.3-9.5V. With the knowledge of this importance of bias voltage, efficiency is greatly maximized when we need to detect the failing resistive contact from a pool of possible contacts.

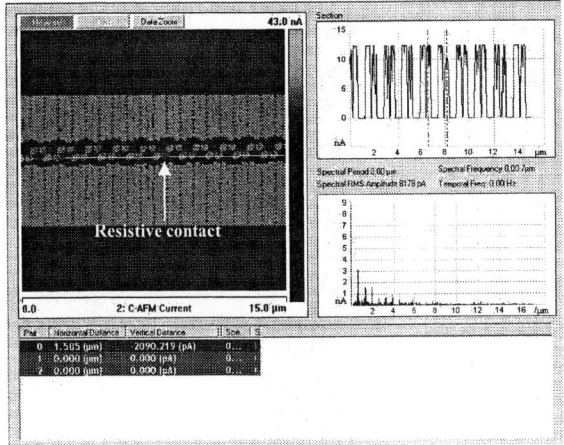

Figure 2c. Current image line scan for resistive gate contact

By performing a line scan (*Figure 2c*) across the gate contacts, the resistive gate contact shows that it conducts 2nA less current than the rest of the good contacts.

C. CAFM current image scanning for resistive substrate contacts

When the technique is applied to resistive substrate contacts, a new bias needs to be re-established. Similarly by running an IV curve on a substrate contact in Figure 3a, we observed that current saturates at bias greater than 1.5V and conducts in the

opposite direction when lesser than -8V. Hence a bias voltage of -8.5V was selected to perform the current image scan.

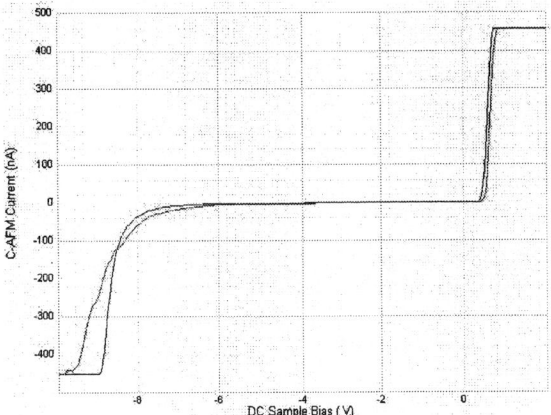

Figure 3a. Current vs. voltage plot of a substrate contact

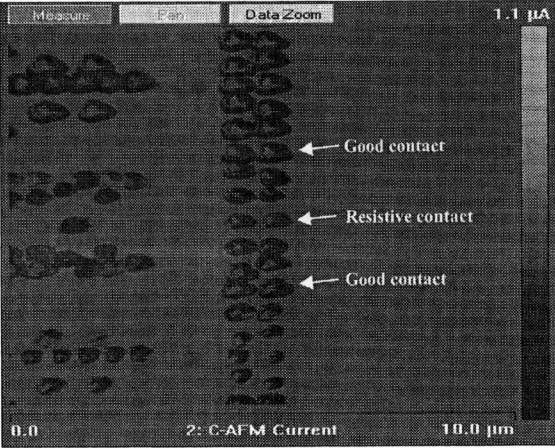

Figure 3b. Scanning out the resistive substrate contact using
CAFM current image mode

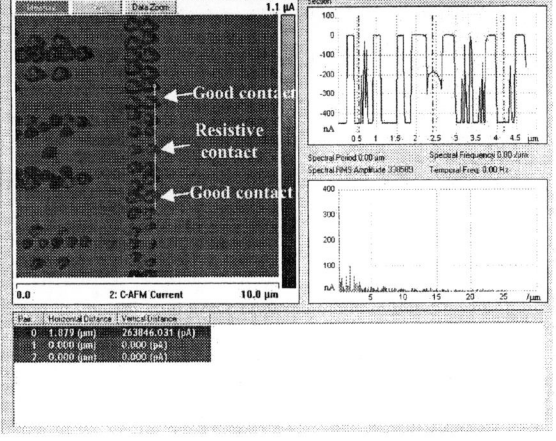

Figure 3c. Current image line scan for resistive substrate contact

Through the current image scan at a bias of -8.5V, the resistive substrate contact was found among a pool of possible contacts as shown in *Figure 3b*. Good contacts for comparison are as displayed due to an area of different structures underneath. In Figure 3c, we see that the resistive substrate contact is conducting approximately 263nA less current than the good contacts.

D. Methodology of Biasing Voltage

In both experiments of different resistive contacts, we find that for the current image scan to be successful, the bias voltage is very important. For the resistive gate contact, we hypothesize that it is the tunneling effect through the transistor gate oxide that amplifies the current difference between the good and the resistive contact, which enables the CAFM to pick it up.

0V is fixed on the CAFM tip which will be in contact with the sample top (including any contacts) while the die substrate has a bias voltage which we were discussing about, that could be varied accordingly.

On the other hand, for the substrate contact, when we make use of -8.5V for biasing, it is believed that the reverse-biasing between the P-Well and the N+ region (*Figure 4a*) again amplifies the current difference for the resistive substrate contact to be picked up visually. In the forward bias region where current quickly saturates when bias is greater than 1.5V, the region of good biasing allowance is too tight. Hence it was unable to be clearly displayed that there is a difference between the good and resistive contacts.

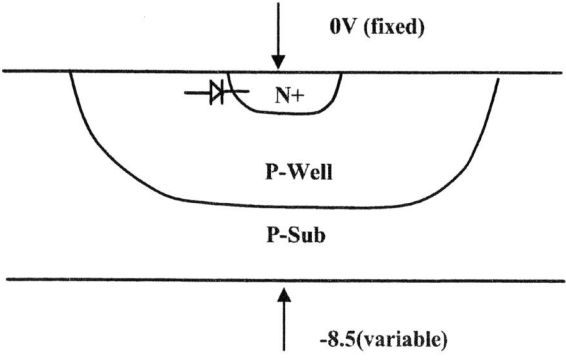

Figure 4a. Structure of substrate contact

III. CONCLUSION

In this paper, we have demonstrated the application of CAFM in the electrical fault isolation of resistive gate contacts and resistive substrate contacts. Through the electrical characterization of CAFM, we can prove that the suspected contact is indeed more resistive compared with the rest, thus

failing the device in package level, which was unable for us to isolate at that level. By understanding the methodology of biasing voltage applied, we deem that CAFM current image can be applied to other structures which will effectively enhance the capability of resistive contacts detection.

REFERENCES

[1] Z.S. Song, J.Y. Dai, S.Redkar, submitted to IEEE transactions of Electronic Devices.

[2] J.Y. Dai, S. Ansari, C.L. Tay, S.F. Tee, Eddie Er, S. Redkar "Failure mechanism study for high resistive contact in CMOS devices", IPFA 2001.

Application of Conductive Atomic Force Microscopy to Study the In-line Electrical Defects

S.L. Toh, Q. Deng, W.T. Tang, V. Lim, F.H. Gn, P.K. Tan, H. Tan, Z.H. Mai and J. Lam

Technology Development Department, Chartered Semiconductor Mfg Ltd
60 Woodlands Street 2, Industrial Park D, Singapore 738406
Phone: +65-6360-4119 Fax: +65-6362-2936 Email: tohsl@charteredsemi.com

Abstract - **Selection of optimized electron beam parameters for in-line monitoring is necessary to eliminate false signals. Application of electron beam to detect electrical defects, particularly leakages, for static random access memory (SRAM) cells poses a great challenge as it requires current measurement tool with nanometer resolution to complement it. By correlating the brightness intensity or the gray-level value to the measured current values, we have shown that conductive atomic force microscopy (C-AFM) can overcome this obstacle and can be used to verify the validity of the voltage contrast (VC) captured by HMI eScan3xx Ebeam inspection tool.**

I. INTRODUCTION

Electron beam inspection (EBI) methodology, implemented through the use of VC to differentiate the failing site, is widely used in industry for in-line electrical defect monitoring [1-3]. It provides timely information on the factors for yield detractors so that proper steps can be taken to rectify the problems promptly. However, optimization of the inspection tool is necessary to reduce surface charging or electron-material interactions so that the number of nuisance defects can be kept to the minimum. Validation of in-line detected signals has to be carried out with failure analysis (FA) or electrical testing. Generally, FA is employed as the wafer is not processed fully.

There are two kinds of FA that can be implemented: physical FA (PFA) or electrical FA (EFA). PFA is related to the detection of physical defects such as bridging materials, under-etching or residues, revealed under high keV electron beam illumination or after deprocessing. EFA focuses more on failure mechanisms relating to leakage or high resistance. Currently, there are few challenges facing the utilization of VC to detect defects associated with current leakages. The severity of the failure has to be measurable. Furthermore, as mentioned earlier, an effective inspection must be capable of capturing the defects of interest predominantly. The prerequisites for this are stable imaging conditions and an appropriate threshold for detection. A good approach to selecting the optimum conditions for in-line inspection therefore involves a detailed electrical analysis of the data collected at the VC-captured sites.

This feedback system between the EBI tool and electrical data collected, if established, benefits the process development greatly. At nanometer technology nodes, with a decrease in the threshold voltage (V_t), the increasing bitline (BL) leakage current is already degrading the performance of circuits. Any drift in process parameter that can result in further increase in BL leakage has to be monitored stringently. If the leakage is too high, it can hinder the read/write operation of an SRAM. The BL may be unable to maintain sufficiently high storage voltage, hence ability to detect BL leakage at the early stage of wafer fabrication is essential. Electrical characterizations have to be carried out on contacts, metals or vias of nanometer size. Characterization tools such as nanoprobing or C-AFM are therefore of paramount importance to the EFA for devices in the smaller technology nodes and these are already well-documented in literature [4-5]. In this paper, C-AFM was used to map the current contrast images and collect the local current-voltage (I-V) curves on in-line identified locations. A correlation of the VC defects, specifically the gray-level magnitude of the bright VC (BVC), with the leakage results measured by C-AFM was also carried out. The conditions applied to capture the C-AFM images were discussed as well.

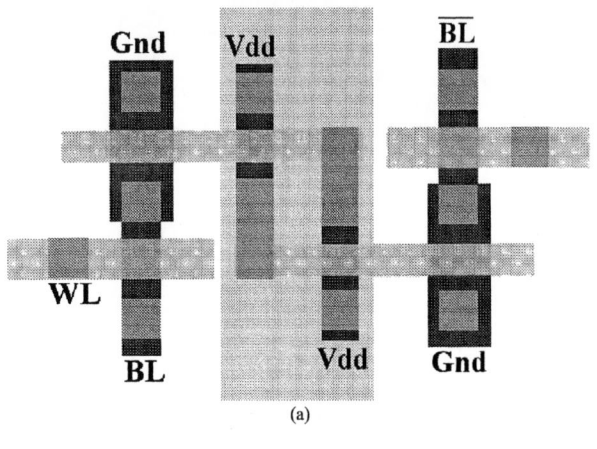

(a)

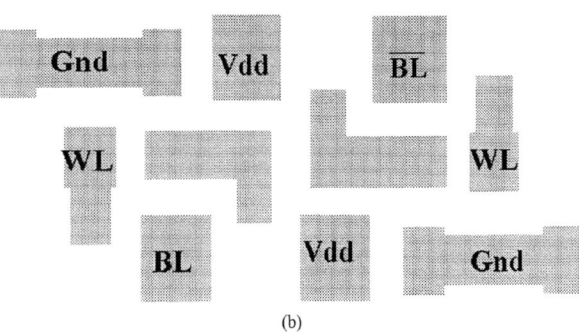

(b)

Fig. 1. Schematic layout of an SRAM cell at (a) contact and active layer, and its corresponding (b) M1 layer.

II. EXPERIMENTAL

C-AFM characterization was carried out using Veeco di Dimension V Scanning Probe Microscope, equipped with Nanoscope V controller and the application module. Using a conductive probe, a doped diamond coated probe tip, to scan across the sample surface, a constant DC bias was applied between the tip and the sample to obtain the current contrast image. To measure the local I-V spectra on a particular position in the scan region, the single probe was fixed at that location and the sample bias was ramped up or down over the range of required voltages. To reduce any discrepancy in the measured current caused by variation in the series resistance during the grounding of the sample, comparative analysis was obtained across defective sites within the same die.

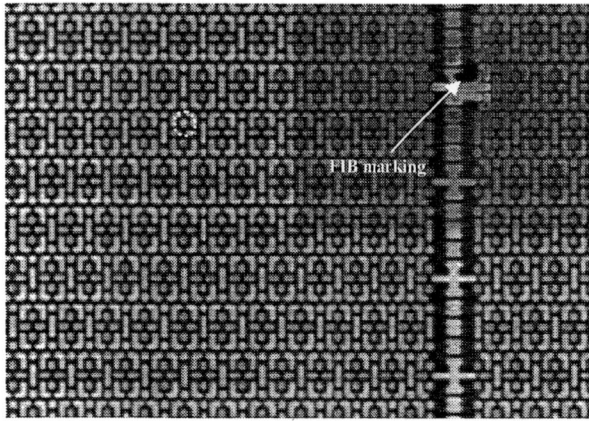

Fig. 2. SEM micrograph at S2 site, in which the BL appears slightly bright and is circled.

The samples analyzed in this paper were nano-SRAM devices at Metal 1 (M1) layer after chemical-mechanical polishing step. Specifically, only samples that showed BVC signals at the BL location (n+) of the SRAM cells were selected for analysis, and a schematic layout is provided in Fig. 1. At M1, the BL is isolated and is directly connected to the p-well via the contact. The inline BVC locations were captured by HMI eScan3xx. FIB markings were created in the vicinity of BVC locations by dual-beam SEM to ease the identification process for further investigation. The bright signals observed were attributed to a breakdown in the n+/p-well junction. Presence of physical defects that might short the p-well to the n-well directly was excluded as these were not observed for few of the dies on BL locations at similar wafer after delayering.

III. RESULTS AND DISCUSSION

An example of the SEM micrograph (S2) from the in-line scan data is shown in Fig. 2 and the abnormal BL site appeared slightly brighter than other BL locations. From the SEM PVC, the contrast difference of p+, n+ or poly-gate was also not obvious. C-AFM was used to compare the differences in current contrasts for the three BVC locations on the same die, in which

the brightness intensity of S1 was the lowest and was labelled L1, and the brightness intensity of S3 was the highest and was labelled L3. Fig. 3 shows (a) the current contrast image and (b) the line-sectioning current profile, obtained for S2 at a sample bias of -0.5 V. Table 1 shows the three BVC sites at BL location and its corresponding junction leakage (n+/p-well) current at -0.5 V.

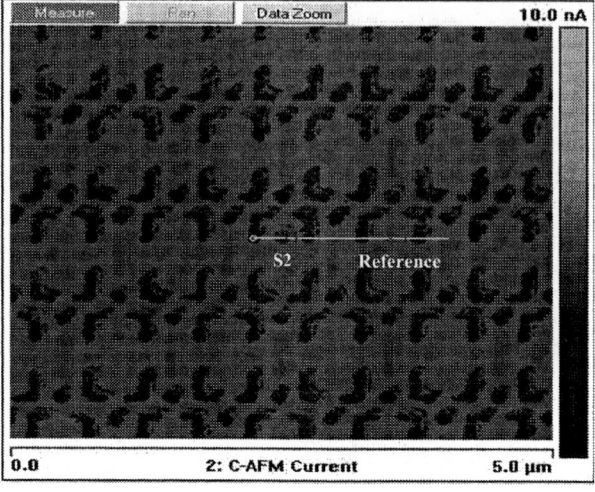

(a) Current-contrast image

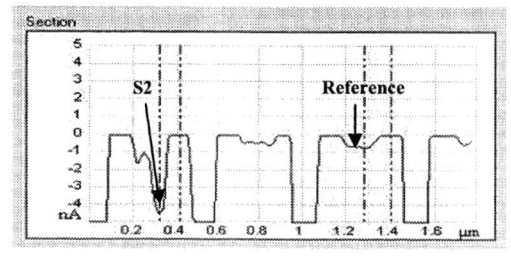

(b) Line-sectioning current profile

Fig. 3. (a) Current-contrast image, and (b) line-sectioning current profile across the abnormal bright BL site, of S2. The defective S2 site depicts a darker contrast and its current value is relatively higher than that of the reference site.

Table I
GRAY-LEVEL COMPARISON OF THE THREE SAMPLES AND ITS CORRESPONDING
LEAKAGE VALUE

Defect	Gray-level Value	I (nA) at sample bias -0.5 V
S1	L1	-1.5
S2	L2	-4.9
S3	L3	-9.1

The contrasts observed depend on the current levels detected. At this applied voltage, darker and lighter contrasts are associated with higher and lower currents respectively. Two different levels of contrasts were mapped: darker spots for p+, and lighter spots for n+ and poly-gate. Contrast on poly-gate was not distinguishable from n+, as it could also be affected by the gate-oxide integrity and the thickness of gate oxide. In Fig. 3(a), the abnormal BL location depicted a darker contrast compared to other reference BL locations. Line sectioning across the defective site in Fig. 3(b) shows that the current was greater than that of the reference locations by about 7 times. With negative sample bias, only the L-shaped M1 and V_{DD} that were located on the n-well were forward-biased and the current detected reached compliance. The BL (n+) that landed on p-well was reverse-biased diode and its colour should be light. The darker contrast observed hence suggested a slight breakdown in the p-n junction.

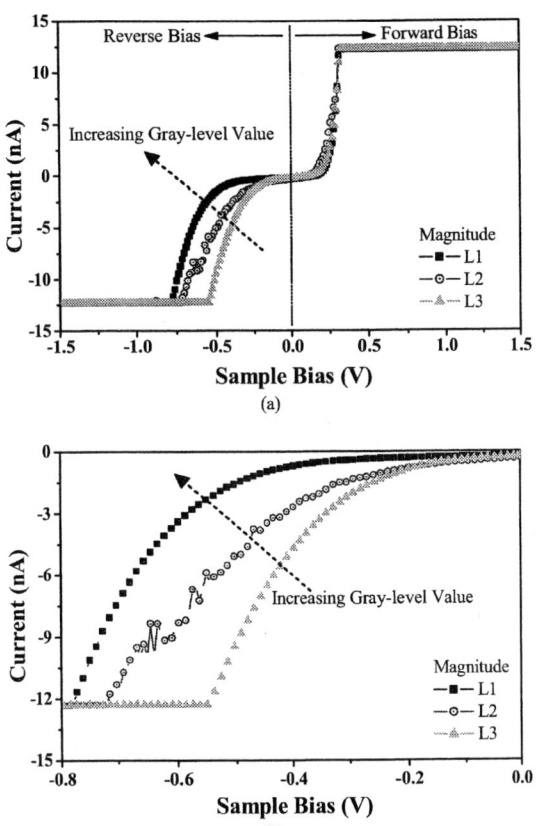

Fig. 4. I-V characteristics of the three BVC locations by C-AFM at sample bias ranging from (a) -1.5 V to 1.5 V, and (b) -0.8 V to 0 V. Higher gray-level magnitude detected by the in-line scan is associated with higher current at reverse biasing condition.

In addition, local I-V measurements were also carried out, in which the probe tip was landed on the location to be measured and the sample bias was swept from -1.5 V to 1.5 V. An overlay of the I-V spectra collected for the 3 samples is shown in Figure 4(a). At the reverse biasing condition (Fig. 4(b)) in which sample bias is less than 0 V, there is a good correlation between the leakage current measured and the brightness intensity. Relationship between the sample bias, measured leakage current

and in-line scan intensity is shown in Fig. 5. Difference in the current measured for different gray levels is only distinguishable when the sample bias is increased greater than -0.2 V. This value is dependent on the built-in potential of the p-n junction. Variation caused by series resistance across the sample stage was considered negligible as the sample was grounded on similar stage. Hence, to evaluate the effectiveness of electron beam parameters used for monitoring, it should be noted that appropriate biasing conditions were applied for C-AFM analysis.

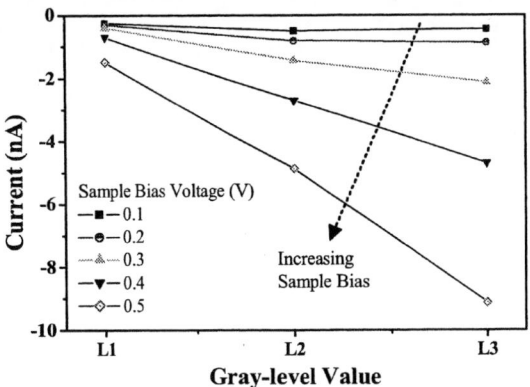

Fig. 5. Relationship between the sample bias, current measured by C-AFM and the inline gray-level value.

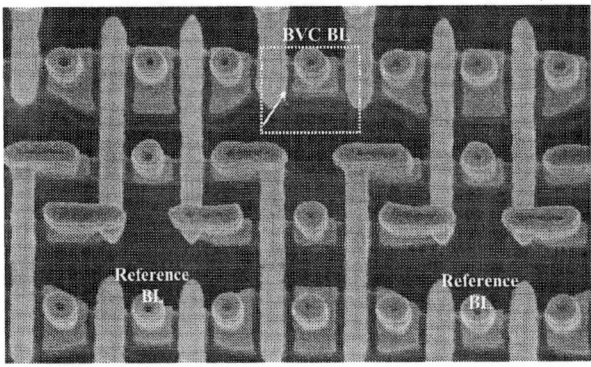

Fig. 6. SEM image of Defect S2 at poly layer. Slight growth of NiSi laterally is observed as indicated by the arrow.

Fig. 6 shows the SEM image of Defect S2 at poly layer after further delayering to contact, removal of dielectric oxide layer by buffered oxide etch and removal of nitride spacer by reactive ion etching sequentially. Slight protrusion of Ni silicide towards poly was observed at the defective BL contact and it was suspected to be the root cause for the leakage observed. A cross-sectional schematic diagram is provided in Fig. 7 to illustrate the failure mechanism, in which high diffusion coefficient of Ni resulted in penetration of Ni towards the channel from the highly doped (n+) to the lightly-doped (n-) region [6-7]. This could lower the barrier potential for carriers to flow, resulting in the degradation of junction leakage when the reverse bias was applied. The BL leakage was therefore determined by the extent of silicide overlap in the n-/p-well region.

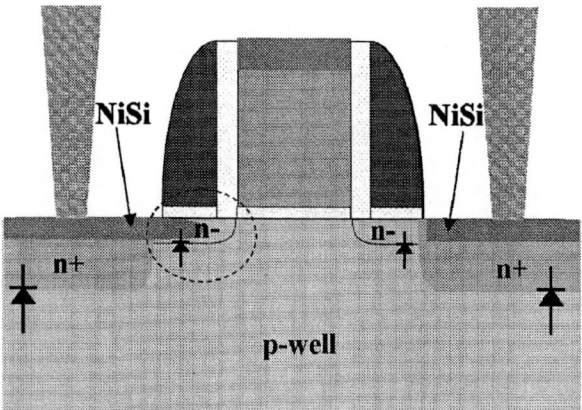

Fig. 7. A schematic diagram illustrating the possible breakdown mechanism of the p-n junction, resulting from slight lateral growth of the NiSi from the heavily doped n+ region into the lightly doped n- region. The abnormal area is highlighted.

IV. CONCLUSION

C-AFM has proven to be effective in characterizing or quantifying current leakages and can serve as viable complementary electrical measurement technique to VC. A detailed analysis of the electrical measurement data from C-AFM and its comparison to the brightness intensity of the in-line VC can be carried out in order to verify the validity of the signals captured under selected electron beam parameters. Any slight variation in process condition can therefore be tackled promptly.

ACKNOWLEDGMENT

The authors would like to thank Dr. Ng Tsu Hau for the preliminary evaluation on the C-AFM approaches, Ms. Yang Tanya for the inline data, and Ms. Xin Hua for her kind assistance in the use of C-AFM.

REFERENCES

[1] A.V.S. Satya, "Microelectronic test structures for rapid automated contactless inline defect inspection", *IEEE Trans. Semiconduct. Manuf.*, Vol. 10, No. 3, pp. 384-389, Aug. 1997.

[2] K. Fujiyoshi, K. Sawai, K. Inoue, K. Saiki and K. Sakurai, "Voltage contrast for gate-leak failures detected by electron beam inspection", *IEEE Trans. Semiconduct. Manuf.*, Vol. 20, No. 3, pp. 208-214, Aug. 2007.

[3] A. Ache and K. Wu, "Production implementation of state-of-the art electron beam inspection", *in Proc. IEEE/SEMI Adv. Semiconduct. Manuf. Conf. ASMC'04*, 2004, pp. 344-347.

[4] J.H. Chuang and J.C. Lee, "Conductive atomic force microscopy application on leaky contact analysis and characterization", *IEEE Trans. Device Materials Reliability*, Vol. 4, No. 1, pp. 50-53, Mar. 2004.

[5] S.L. Toh, P.K. Tan, Y.W. Goh, E. Hendarto, J.L. Cai, H. Tan, Q.F. Wang, Q. Deng, J. Lam, L.C. Hsia, "In-depth electrical analysis to reveal the failure mechanisms with nanoprobing", *IEEE Trans. Device Materials Reliability*, Vol. 8, No. 2, Jun. 2008, *in press*.

[6] T. Yamaguchi, K. Kashihara, T. Okudaira, T. Tsutsumi, K. Maekawa, T. Kosugi, N. Murata, J. Tsuchimoto, K. Shiga, K. Asai and M. Yoneda, "Suppression of anomalous gate edge leakage current by control of Ni silicidation region using Si ion implantation technique", *in Proc. Int. Electron Devices Meet. IEDM*, 2006, pp. 1-4.

[7] C. Lavoie, F.M. d'Heurle, C. Detavernier and C. Cabral Jr., "Towards implementation of a nickel silicide process for CMOS technologies", Microelectronics Eng., Vol. 70, No. 2, pp. 144-157, Nov. 2003.

Investigation of Soft Fail Issue in Sub-Nanometer Devices Using Nanoprobing Technique

E. Hendarto, H.B. Lin, S.L. Toh, P.K. Tan, Y.W. Goh, Z.H. Mai, J. Lam
Chartered Semiconductor Manufacturing Limited, Technology Development Department
60 Woodlands Industrial Park D, Street 2, Singapore 738406
Phone: +65-6360-4923   Fax: +65-6362-2936   Email: erwinhendarto@charteredsemi.com

Abstract – **With the miniaturization of electronic devices, identifying the root cause of soft failures using physical failure analysis (PFA) techniques has become a more challenging task. By characterizing the electrical behavior of malfunctioned devices, nanoprobing precisely locates defects before any PFA is performed and allows for deeper understanding of the root cause of soft failure issues. Two case studies are presented to demonstrate the effectiveness of nanoprobing in investigating the root causes of soft failures.**

I. INTRODUCTION

As electronic devices shrink into the sub-nanometer range and beyond, it brings about corresponding increase in performance and decrease in cost per transistor [1]. However, defect identification becomes more difficult and more advanced failure analysis (FA) techniques have to be employed. Conventional passive voltage contrast (PVC) methodology for fault isolation is often not sensitive or conclusive enough to localize the defective site or identify the defects when devices become smaller and smaller. Some devices may fail marginally, and the PVC technique often fails to determine the physical root cause. In other cases, the root cause of the abnormality in the PVC may not be due to physical defects.

Established techniques, such as liquid crystal analysis, photon emission microscope and scanning laser microscope, are limited by the spatial resolution and are thus capable of giving only a rough estimate of the location of the failure without being able to pinpoint defects to the individual transistor level, which is of increasing paramount interest today. Moreover, the V_{min} issue is getting more severe at the present nanotechnology. Local mismatch in a bit cell caused by slight variations in the transistor parameters such as threshold voltage and saturation current can easily induce soft failure. It is therefore necessary that in-depth analysis of the electrical characteristics of all the transistors within a bit cell be performed.

Nanoprobing has since emerged as a promising tool for defect isolation in the current semiconductor FA methodology [2-4]. This system allows for transistor characterization at the failing site. It comprises at least four probes and can be mounted on a Scanning Electron Microscope (SEM), Atomic Force Microscope (AFM) or Focused Ion Beam (FIB) specimen stage. These electrical analyses are usually carried out at the contact level, where the transistors of a faulty cell are isolated and the defective sites identified from the measured electrical characteristics. With such thorough electrical characterization of failing sites, and usually at different voltages, the failure

mechanisms can be more accurately identified.

In this paper, nanoprobing was used to acquire the necessary I-V characteristics of all the transistors in the failing bit cell prior to the implementation of suitable PFA steps carried out at the failing site. PVC was utilized as a complementary method, while the electrical characterization of failing transistors allowed for deeper understanding of the root cause of soft failures, as well as the effects of such failing transistors on a Static Random Access Memory (SRAM) cell. The aforementioned FA procedures were applied in two cases on SRAM arrays as described in Section III.

II. EXPERIMENTAL SETUP

The devices used in the study were namely Device A and Device B, which have similar GDS layout for memory bit cell. These samples are SRAM chips made up of six transistors, two cross-coupled complementary metal-oxide semiconductor (CMOS) inverters and two pass nMOS transistors connecting the cell to the bitline and the bitline bar as shown in Figure 1. Both samples were deprocessed and inspected layer by layer to the contact level and the PVC technique was utilized to check for any abnormality in the contacts.

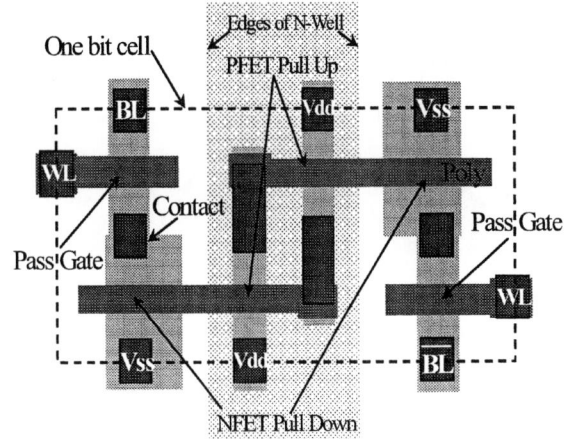

Figure 1. Schematic layout of an SRAM structure.

An SEM-based Zyvex KZ100 Nanomanipulator system was used to perform nanoprobing on the two samples. This tool consists of a manipulator controlling the fine movements of four

978-1-4244-2039-1/08/$25.00 ©2008 IEEE   145   *Proceedings of 15th IPFA - 2008, Singapore*

tungsten nanoprobes, which usually have tip diameter in the 50 nm range or below. Figure 2 shows an SEM micrograph of four nanoprobes at the source (S), drain (D), gate (G) and bulk (B) of a transistor within an SRAM cell.

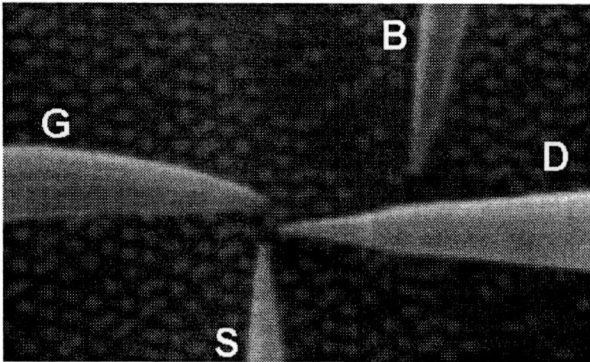

Figure 2. SEM micrograph showing four tungsten nanoprobes landing at the gate (G), source (S), drain (D) and bulk (B) of a transistor in an SRAM cell.

III. RESULTS AND DISCUSSION

Case Study I.

One of the SRAM cells in device A suffered from a soft failure, failing at 1.2 V but passing at 1.32 V. The sample was deprocessed and inspected layer by layer. One bright Word-Line (WL) contact was observed at the failing cell as shown in Figure 3. Some form of leakage was suspected, and in particular, contact-poly short was highly possible.

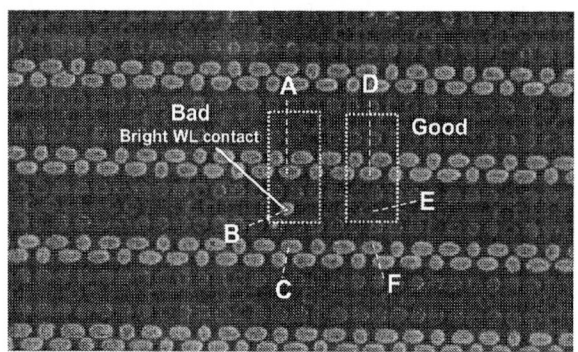

Figure 3. SEM micrograph showing the bad cell and good cell. One of the Word-Line (WL) contacts in the bad cell showed up as bright using PVC technique, and contact-poly short was highly suspected.

Nanoprobing technique was first adopted to further understand the defect before removing the interlayer dielectric (ILD) to expose poly. The I-V characteristics of the pass-gate (PG) transistor with abnormal PVC at contact B revealed that its I_{ON} i.e. the saturation current, was significantly lower than that of a normal PG at the same operating voltage. The values were summarized in table I. At 1.2 V, the I_{ON} of the bad cell PG was 15.1 uA, while that of the good cell PG was 56.0 uA. This translated to a 73% lower I_{ON} in the bad cell PG as compared to

the good cell PG. The findings were also confirmed by the I_d-V_g graph shown in Figure 4. The current values of the bad PG, whether the transistor operated at the linear or saturation mode, were significantly lower than those of the good PG.

Table I.
I_{OFF} and I_{ON} comparisons of bad pass gate (PG) transistor and good PG transistor, and their corresponding I_{ON} mismatches at different drain/gate voltages.

| $V_d=V_g$ | Bad cell | | Good cell | | I_{ON} difference (%) |
	I_{OFF} / pA	I_{ON} / uA	I_{OFF} / pA	I_{ON} / uA	
1.08 V	-1.99	8.24	-2.65	40.3	80
1.2 V	8.23	15.1	-1.19	56.0	73
1.32 V	36.1	20.8	1.64	72.1	71

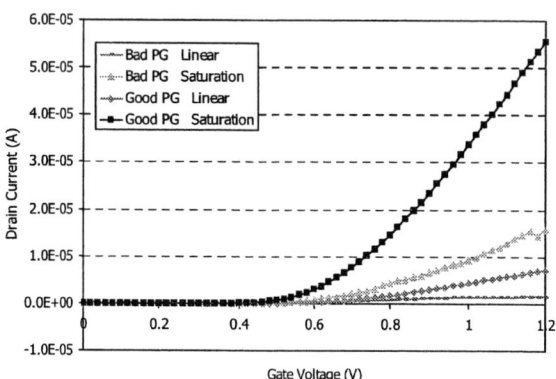

Figure 4. I_d-V_g characteristics of the bad and good PGs operating at the linear and saturation modes. There is significantly lower I_{ON} in bad PG2.

Two-point tests were then conducted between contact B and its adjacent contacts, namely contacts A and C, to investigate why the PG with the bright WL contact had a much lower I_{ON}. Contact E from a neighboring cell was chosen as the corresponding good contact of the PG in the good cell, and two-point tests were also performed with contacts D and F for comparison purposes. The I-V characteristics of contacts A-B, B-C, D-E and E-F are shown in Figure 5. Figure 5(a) shows a short characteristic in the I-V curve of contacts A-B with a current value in the uA range. This value was relatively much higher compared with the current values in the other three cases as shown in Figures 5(b)-(d), which were in the pA range. Therefore, a physical short between contacts A and B was highly suspected.

The sample was then deprocessed by removing the ILD to expose the poly underneath the oxide. Figure 6(a) shows the tilted SEM image and Figure 6(b) shows the cross-sectional TEM micrograph showing the bridging of the two polys. Figure 6(c) shows the TEM cross-sectional micrograph at a reference site. Poly tip-to-tip short was found to be the root cause of this single bit soft failure. Such tip-to-tip short caused the I_{ON} of the PG to be reduced significantly by more than 50%, causing the cell to malfunction.

978-1-4244-2039-1/08/$25.00 ©2008 IEEE

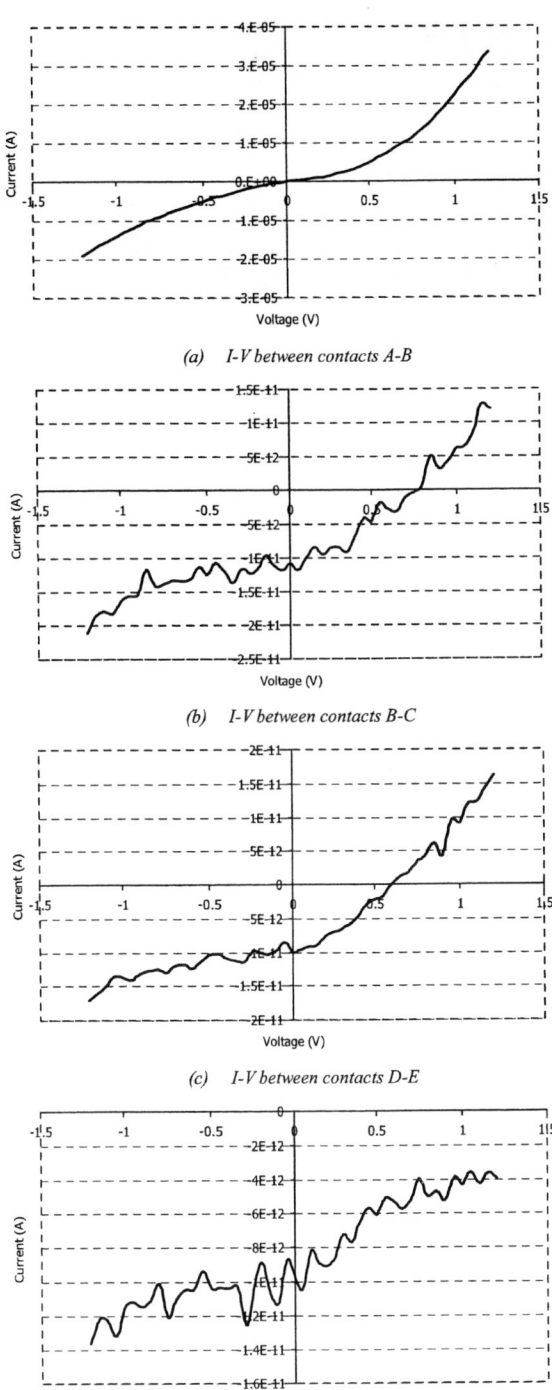

(a) I-V between contacts A-B

(b) I-V between contacts B-C

(c) I-V between contacts D-E

(d) I-V between contacts E-F

Figure 5. I-V characteristics of contacts (a) A-B and (b) B-C (from bad cell), and contacts (c) D-E and (d) E-F (from good cell). The I-V characteristic of contacts A-B shows a short between these two contacts, with a current in the uA range. The current values in the three other cases are in the pA range, showing open circuit behavior.

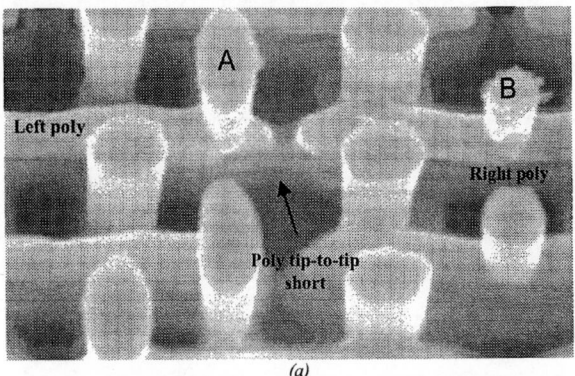

(a)

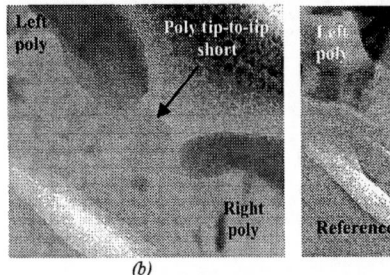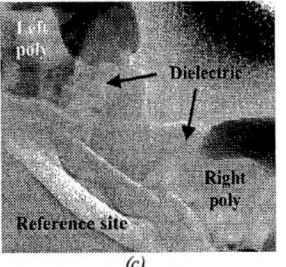

(b)   *(c)*

Figure 6. (a) SEM micrograph (tilted) to show the severity of poly tip-to-tip short. (b) TEM micrograph showing poly tip-to-tip short. (c) TEM micrograph at a reference site.

Case Study II.

The second case was a case study on a single bit soft fail issue from device B. This sample, which passed at voltages above 1.3 V and failed at voltages below 1.3 V, was also deprocessed and inspected layer by layer. A contact with abnormal PVC was observed at the contact layer. Unlike in the previous case where a bright WL contact was found, a bright ground (V_{ss}) contact was found as shown in Figure 7. This V_{ss} contact corresponds to the source (S) of one of the pull-down transistors in the bad cell.

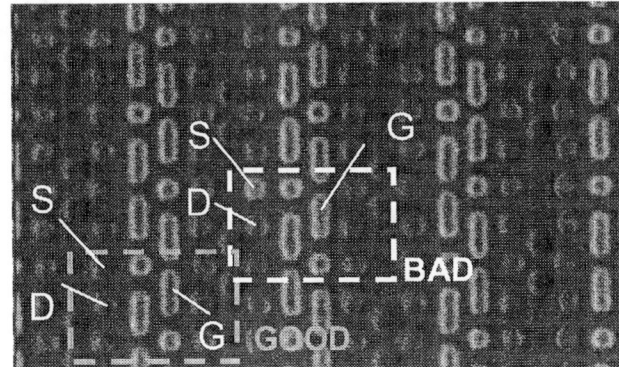

Figure 7. SEM micrograph showing the good cell and bad cell. PVC technique reveals an abnormality in the ground (V_{ss}) contact of the bad cell. This contact is the source (S) of one of the two pull-down transistors in the bad cell.

978-1-4244-2039-1/08/$25.00 ©2008 IEEE   147   *Proceedings of 15th IPFA - 2008, Singapore*

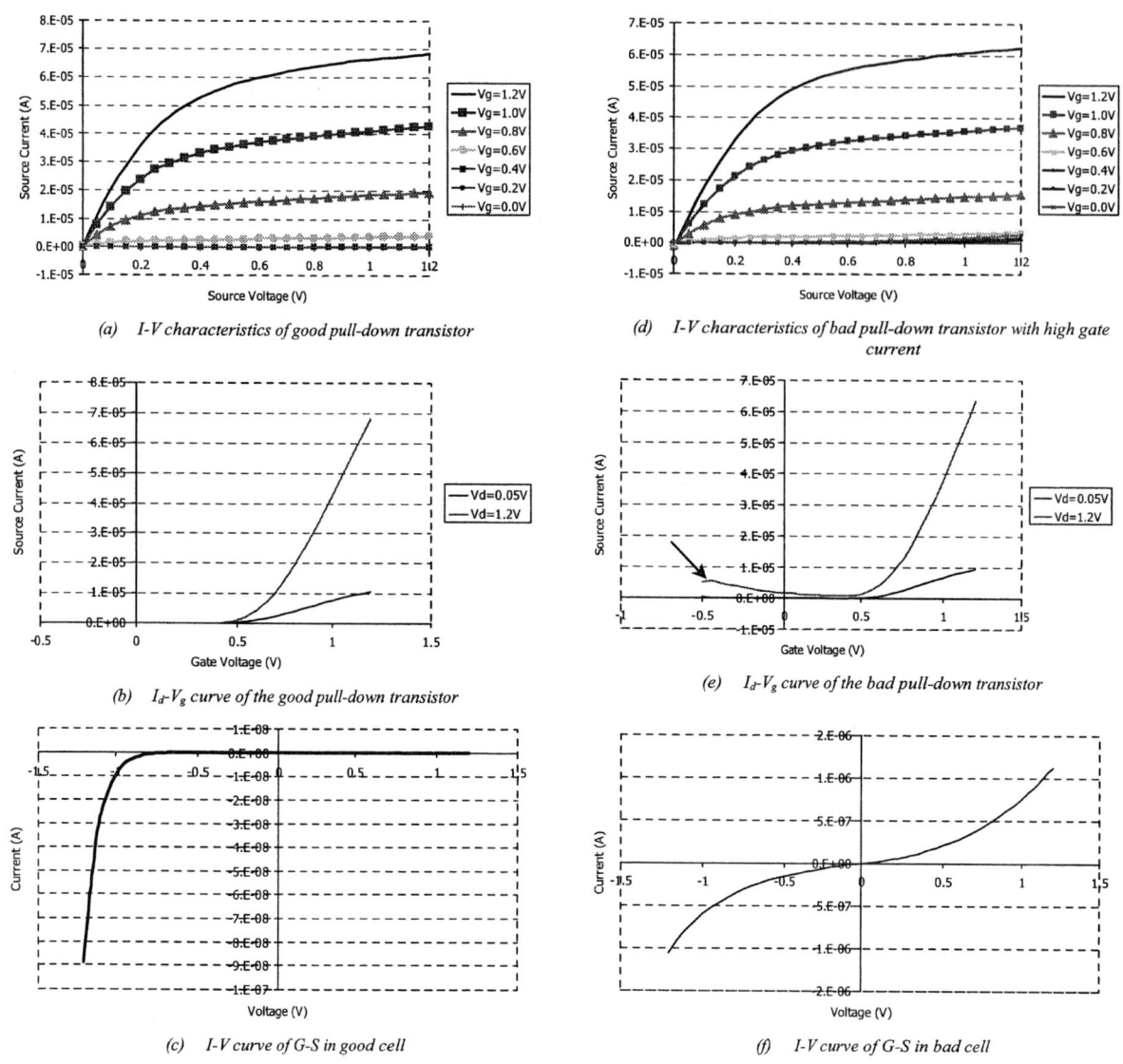

(a) I-V characteristics of good pull-down transistor

(d) I-V characteristics of bad pull-down transistor with high gate current

(b) I_d-V_g curve of the good pull-down transistor

(e) I_d-V_g curve of the bad pull-down transistor

(c) I-V curve of G-S in good cell

(f) I-V curve of G-S in bad cell

Figure 8. Comparisons of I-V characteristics of good and bad SRAM cells

Nanoprobing was thus carried out on the pull-down transistors in the bad and good cells, and all the results were summarized in Figure 8. It was found that the I-V characteristic of the good pull-down transistor was normal with a gate current in the nA range. However, the gate current in the bad pull-down transistor was in the uA range, making it 3 orders of magnitude higher. The corresponding I_d-V_g curve of the bad pull-down transistor also shows some abnormality especially in the negative gate voltage range as shown in Figure 8(e). Two-point tests were performed on the abnormal V_{ss} contact with neighboring contacts to check for any electrical short. It was discovered that the source (S) and gate (G) of the bad pull-down transistor was shorted to each other as shown in the I-V curve of G-S in the bad cell in Figure 8(f), causing a high gate current in the I-V curve of

the bad pull-down transistor. The two-point test in the good cell did not reveal any such short behaviour as shown in Figure 8(c).

The sample was subsequently deprocessed to expose the poly. Figures 9(a) and 9(b) show that contact-poly short was the root cause of the single bit soft failure. The foreign material caused an electrical short between the gate (G) and source (S) of the bad pull-down transistor.

IV. CONCLUSION

Nanoprobing provides more information on the failure mechanism of failing transistors in SRAM cells in cases where bright PVC are found in devices suffering from soft failures. Electrical characterization of failing transistors enables a deeper understanding of the root cause of soft failures, as well as the effects of such failing transistors on an SRAM cell.

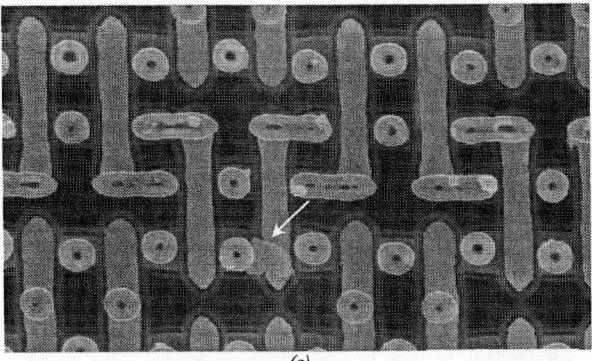

(a)

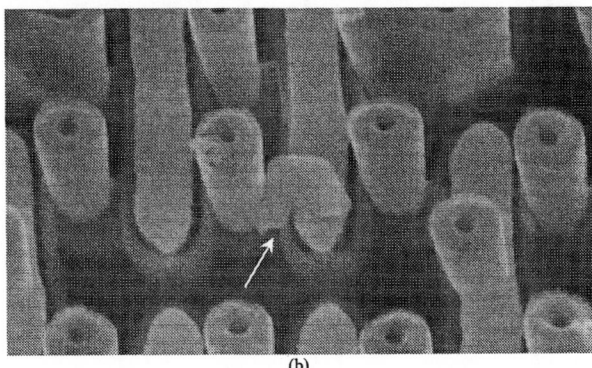

(b)

Figure 9. (a) SEM micrograph (topdown) showing foreign particle causing contact-poly short. (b) SEM micrograph (tilted) showing foreign particle causing contact-poly short.

ACKNOWLEDGMENT

The author and co-authors would like to express their sincere gratitude to the Chartered Corporate Failure Analysis Department for their expertise and assistance in producing the TEM results.

REFERENCES

[1] Semiconductor Industry Association, "International Technology Roadmap for Semiconductors," 2007 Edition, http://public.itrs.net/

[2] E. Hendarto, Z.H. Mai, P.K. Tan, A. Lek, B. Lau, J. Lam, W.K. Chim, "Using Probing Techniques to Identify and Study High Leakage Issues in the Development of 90 nm Process and Below," in 13th Proc. IEEE Int Symp on the Phys and Failure Analysis of Integrated Circuits (IPFA 2006), 3-7 July 2006, Singapore, 2006, pp. 58-62.

[3] S.L. Toh, P.K. Tan, Y.W. Goh, E. Hendarto, J.L. Cai, H. Tan, Q.F. Wang, Q. Deng, J. Lam, L.C. Hsia, "In-Depth Electrical Analysis to Reveal the Failure Mechanisms With Nanoprobing," IEEE Transactions on Device and Materials Reliability, vol 8, no. 2, June 2008, in press.

[4] H. Tan, P.K. Tan, E. Hendarto, S.L. Toh, Q.F. Wang, J.L. Cai, Q. Deng, T.H. Ng, Y.W. Goh, Z.H. Mai, J. Lam, "Salicidation Issue in 65 nm Technology Development," in 14th Proc. IEEE Int Symp on the Phys and Failure Analysis of Integrated Circuits (IPFA 2007), 11-13 July 2007, Bangalore, India, 2007, pp. 44-47.

Failure Analysis of 65nm Technology Node SRAM Soft Failure

Chen changqing, Eddie Er, NEO Soh Ping, Loh Sock Khim, Wang Qingxiao, Teong Jennifer

Chartered Semiconductor Mfg Ltd, Woodlands Industrial Park D, Street 2, Singapore 738406

TEL: 65-63604144, FAX: 65-63622935 Email: chencq@charteredsemi.com

Abstract: **In this paper, a real case of 65nm technology node SRAM failure was studied. The failure of the SRAM is soft failure, so the traditional method was failed to localize the exact position of the failed transistor. To find the root cause, the biased current image-Atom Force Microscopy combined with Atom Force Probing was used to probe the failed cell of the SRAM to find one abnormal pass-gate transistor. Theoretical analysis combined with the probing result was performed to find the failure location. Then current image was used to confirm the failure location. According to the AFP result, TEM and EDX were performed along the active of the pass-gate. Incomplete silicidation was observed under the active contact which correlated well to the electrical analysis result.**

I. INTRODUCTION

As the transistor geometries shrink into nano-scale, more and more challenge was present, not only for manufacturing and design but also for failure analysis. In the FEOL, continued shrink-down of the transistor lead to thinner gate oxide, drain induced barrier lowering (DIBL), short channel effect (SCE) and so on. These ever-changes increase the gate leakage and sub-threshold leakage which make it more possible for the MOS to be soft failing, especially for the SRAM. In the BEOL, the low-k and ultra low-k dielectric material was employed. These porous dielectrics make lower delay in the IC, it also bring along more challenge with itself, not only in the process but also in the failure analysis. Because this porous and soft dielectrics material is easier to be damaged by the E-beam and Ion-beam, while the SEM and FIB are the current major tools for FA engineer, which was designed based on the electron and ion beam. So it is critical for FA engineer to use E-beam or Ion-beam tool to perform failure analysis. Accordingly, the standard SEM Microscopy technology may become incapable of locating most defects at 65nm and beyond sometimes.

Static random access memory (SRAM) is by far the dominant form of embedded memory found today's integrated circuits (ICs), occupying as much as 60-70% of the total chip area and about 75-85% of the transistor count in some IC products [1]. So SRAM is very critical not only in the process monitoring but also in the failure analysis. The commonly used SRAM is six-transistor cell. Because the large embedded SRAM arrays consume a significant portion of the overall power of the ICs. For large arrays, standby power consumption is a major issue.

Therefore, the leakage reduction in large memory arrays has become essential for IC application. So the ultra-shallow and abrupt junction technology, short channel effect, Vth variability caused by random dopant fluctuation are all the challenges we must face. These challenges also bring many soft failures, which is very difficult for us to localize the defect with traditional FA method. Sometimes the traditional method may lead some damage to the device that may mislead us to get a completely different result. For example, according to some study, the Vt of the transistor will drift after the inspection by E-beam or Ion-beam. However, a new method of Atomic Force Microscopy (AFP) combined with Atomic Force Probing was developed to deal with the problem [2]. AFP avoid the risk attendant with highly energetic incident SEM electron beam or FIB gallium ion beams that could rapture the thin gate film or induce charging effects on the silicon layer. The function of the AFP current image is similar to the PVC analysis, but its sensitivity much more than SEM PVC or FIB PVC. It can also bias the scanning tips forwardly or reversely. So both NMOS and PMOS related failure can be detected [3].

II. FA PROCEDURE

In this case, it is a 65nm technology node single bit SRAM failing, and it failed at low voltage. That means the failed bit can pass the memory test at high voltage. It is impossible for the traditional PVC to find the defect. But for comparison, we also performed some conventional failure analysis. The sample was de-processed to the contact layer and low voltage SEM PVC was performed on the failed location. But nothing abnormal was observed as we expected. The Fig. 1 is the low voltage SEM PVC picture.

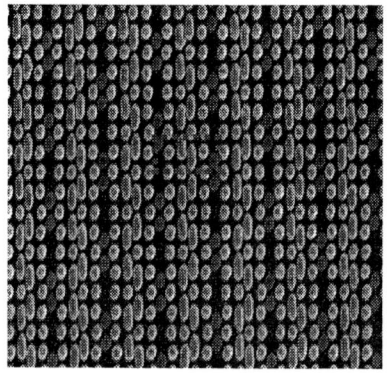

Fig. 1 SEM PVC on the failed location

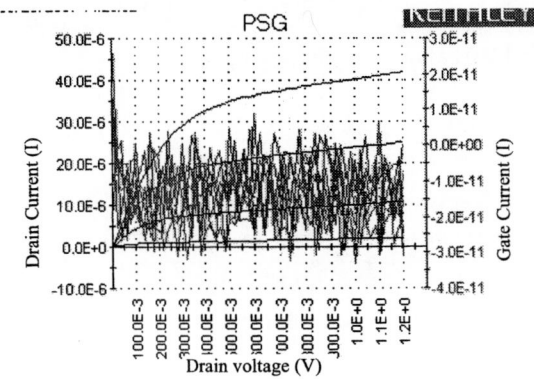

Fig. 2 Family curve of the good transistor

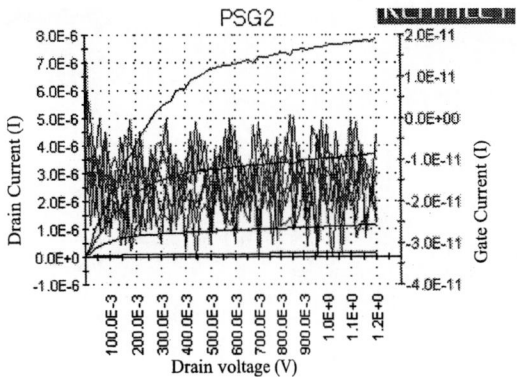

Fig. 4 family curve of the good transistor

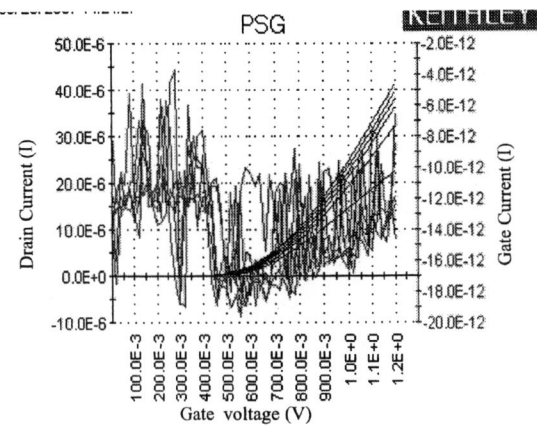

Fig. 3 Vth curve of the good transistor

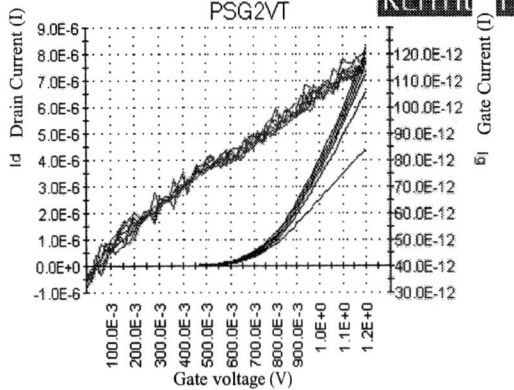

Fig. 5 Vth curve of the failed transistor

Traditionally, the continuous delayering will be performed on the device, or deposit Micro-pad by the circuit edit FIB and probe the I-V curve of the SRAM. But it is very possible to miss the real defect for the first method. For the second method, the success rate continues to decrease as the technology scale down. In addition, main defect for this method is the Vt drift after the Ion-beam inspection. This may mislead us to a different result [4] [5]. Instead, we use the AFP for further analysis, which is developed for 65nm and below. The sample was cleaned by the RIE to remove the carbon on the surface that caused by the SEM inspection, and then expose the contact about 15nm for nano-probing. All the six transistors were probed, nothing abnormal was found in five of the transistors except pass-gate 2. The Fig. 2 and Fig. 3 is the family curve and Vth curve of the good pass-gate.From the curve, it indicates that the Ion of the transistor is about 43u Am, and the Vth of the transistor is about 6.5 volts. These two curves are reference for the failed pass gate family curve and Vth curve.

The Fig. 4 and Fig. 5 are family curve and Vth curve of the failed PSG2. The information we can get from this family curve is that the Ion of the transistor is very much lower than the good one, but no abnormal gate leakage in the failed transistor. From these two curvesit can be deduced that there is a high resistance in the on-state of the transistor. But according to this information, we just know something abnormal happened in the transistor but still can not localize the real defect location in the transistor.

For further analysis, the source and drain was swapped, and the family curve was tested again. The Fig. 6 and Fig. 7 are the family curve and Vth curve of the failed transistor after swap.

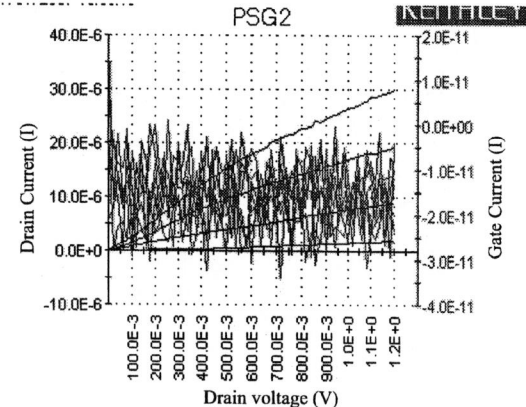

Fig. 6 family curve of the failed transistor after swap

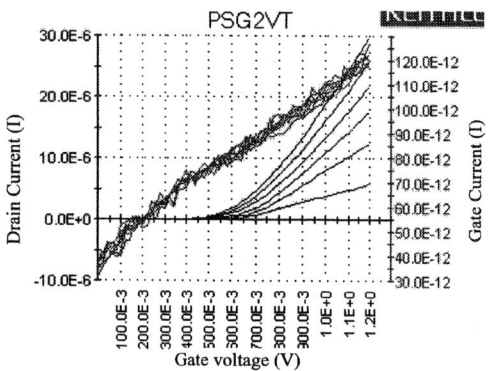

Fig. 7 family curve of the failed transistor after swap

From the family curve, the information we can get is that the gate leakage is normal; the transistor does not pinch off in the test voltage range. That means this transistor works in the linear area in the work voltage. But according to the I-V curve, the transistor will enter the saturation area if the voltage was increased, that may be the reason why the SRAM pass the high voltage test. In addition, we also found that the Ion is still lower than normal one, but bigger than that before swap. That means the Ion is different when we swap source and drain. The reason will be explained in detail in the following electrical analysis [6]..

III. FAILURE MECHANISM

Based on all the nano-probing result and combined with the electrical analysis, we are sure that there is a series resistance in the source of the transistor. There are two possibilities for this failure, the first one is the implantation problem, and second one is contact problem. Fig. 8 shows an equivalent circuit of the transistor. According to our electrical analysis, there is a series resistance in the source of the transistor.

So when the transistor is biased, the transistor is turn on normally. But after the transistor turn-on, the current will be smaller than the normal transistor, not only in the linear area

but also in the saturation area, because there is an additional series resistance in the source. That is the reason why the Ion current is very low in the Fig. 4 and Fig. 5.

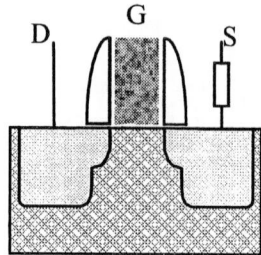

Fig. 8 equivalent circuit of the failed transistor

But when the source and drain are swapped, the voltage stress will be on the source. The transistor turns on normally. But after the transistor is turn on, it will work in the linear area at first, and then enter the saturation area when increasing voltage in source to pinch off the channel. But in this failed transistor, the situation is completely different. Because there additional resistance in the source, the actual potential in the source is lower than the normal one. So it needs much higher voltage in source contact to pull up the potential in the source active to pinch off the transistor. While in the normal work condition, the potential in the source active is not high enough to pinch off the MOS channel, which makes the transistor work in the linear area. That is the reason why the transistor works in the linear area in the Fig. 6.

In addition, why the current is still lower than normal one is because there is an additional series resistance in the source.

For further confirmation, the biased current image was also performed on the failed location. (I) sensitivity of current image function in the AFP is in the pico ampere range. (II) the scanning tip can be biased forward or reversely, thus the leakage and high resistance can be detected not only in n type contact but also in p type contact.

The Fig. 9 is the forward biased current image. It indicates that there is some abnormality happened in the circled contact, which is the source of the failed contact. This result also corresponds well to our electrical analysis [7][8].

From all the analysis above, we are sure that there is an additional series resistance in the source of the pass gate transistor. So we can focus on these two facets which can induce this failing phenomena. The TEM was cut on the failed source contact along the active. The Fig. 10 is the TEM result of the failed contact. According to the TEM result, the silicide below the source contact is abnormal. It seems that the silicide is incomplete. To further confirm the anomaly, the EDX was performed on the abnormal location to check the element. The Fig. 11 is EDX result. The incomplete silicide was confirmed by the EDX result. That means the defect happened during the silicidation process. There may be something that blocked the source of the transistor and lead to the incomplete silicidation.

978-1-4244-2039-1/08/$25.00 ©2008 IEEE

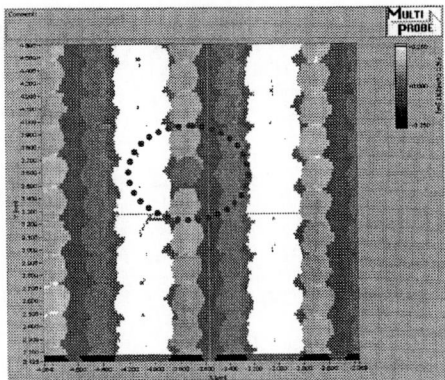

Fig. 9 forward biased current image

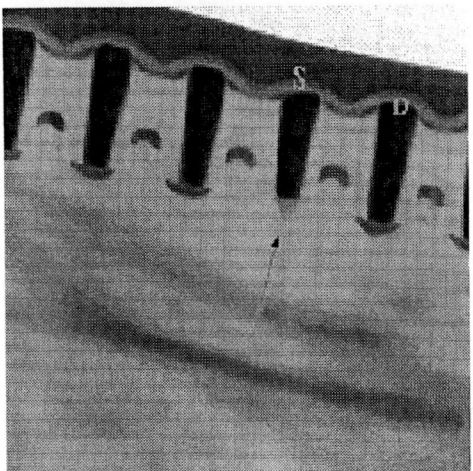

Fig. 10 TEM result of the failed transistor

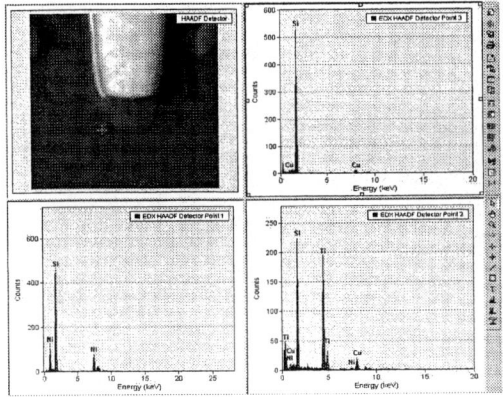

Fig. 11 EDX result of the abnormal location

IV. CONCLUSION

As the technology continues to scale down, the failure analysis is becoming more and more difficult, especially for soft failure. The conventional FA method is becoming incapable in locating the defect accurately. Fortunately, many new methodology and tool were developed, AFM combined with AFP is one of the very useful tool for the 65nm technology node and below, not only for bulk but also for the SOI, for which the traditional method is helpless. In this paper, a real case, which is a soft failure, was studied. It is impossible for the traditional method to localize the defect location. But this new method combined with electrical analysis can accurately and quickly localize the defect location and defect category.

REFERENCES

[1]. L.F.Tz Kwakman, N.B. Lepinay, S.Courtas, "The role of a Physical Analysis Laboratory in a 300mm IC Development & Manufacturing Center", International Conference on Characterization and Metrology for ULSI, 2005

[2]. Yu-Ching Yeh, Chia-Lung Lin, B-Jen Chen, Yuan-Wei Tseng, Jeremy D. Russell, "Application of Atomic Force Probing on 90nm DRAM Cell Failure Analysis" proceeding of 13th IPFA 2006, Singapore, 340-343

[3]. Tom X Tong, A N Erickson," Current Image Atomic Force Microscopy (CI-AFM) combined with Atomic Force probing (AFP) for location and characterization of advanced technology node", proceeding form the 30th international symposium for testing and failure analysis, 42-46,2004

[4]. Terence Kane, Michael P. Tenney, "Electrical Characterization of sub-30nm Gatelength SOI MOSFETs", Proceeding from the 30th International Symposium for Testing and Failure Analysis, 33-37,2004

[5]. Terence Kane, Michael P. Tenney, Andrew Erickson, Sebastian Phan, "Atomic Force Probe Kelvin Measurements of Large MOSFET Device at Contact Level for Accurate Device Threshold Characteristics", Proceeding of the 32nd international symposium for testing and failure analysis, 479-502, 2006

[6]. Chao-Chi Wu, Jon C. Lee, Jung-Hsiang Chuang, Tsung-Te Li, "Single Device Characterization by Nano-probing to Identify Failure Root cause" proceeding of the 31st international symposium for testing and failure analysis, 183-185,2005

[7]. Cha-Ming Shen, Shi-Chen Lin, Chen-May Huang, Huay-Xan Lin, Chi-Hong Wang, "Couple Passive Voltage Contrast with Scanning Probing Micoscope to Identify Invisible Implant Issue", Proceeding of the 31st international symposium for testing and failure analysis, 212-216,2005

[8]. Jon C. Lee, J. H. Chuang, "Fault Localization in Contact Level by Using Conductive Atomic Force Microscopy", proceeding from the 29th international symposium for testing and failure analysis,413-418, 2003

A Simple Method for TEM Sample Preparation Without Carbon Film Background

Meng-Lung Lee ,Ren-De Lin
United Microelectronics Corporation(Singapore Branch), Ltd.
No. 3, Pasir Ris Drive 12,Singapore 519528
Tel: 65-62130018 ext 7655 Fax:65-62130080
Email:m_l_lee@umc.com

*Abstract –***The article describes a simple method to prepare TEM sample put on Cu grid without carbon support film. So that we could remove the background of carbon film while we do componential analysis.**

I. INTRODUCTION

A typical method to prepare TEM sample is FIB lift-out technique and the sample is put on the Cu grid with carbon support film. We could use the method to do profile observation except defect analysis such as Carbon film layer.

It is difficult to tell where carbon film come from?(Carbon support film or defect particle?)

Two typical methods could solve the carbon support film issue :

    1.Pre-thin sample method + FIB milling.(Figure 1)
    2.Use FIB with Omni probe extractor system to prepare the TEM sample.

But we use wax during method 1 sample preparation that will cause the residul of wax on the surface. Or you can use SELA to prepare the pre-thin sample and then use FIB milling. But it costs money to purchase the equipment. And for method 2 ,it also costs much money to purchase a new equipment to solve the issue.

Now we provide a simple method to prepare TEM sample put on the Cu grid without carbon support film.

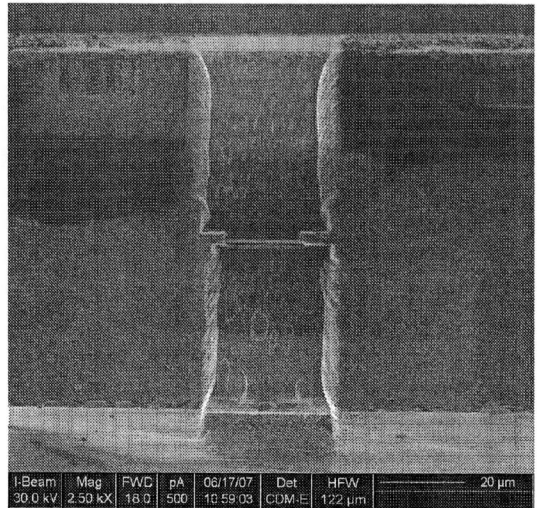

Figure 1:Pre-thin sample + FIB milling

II. SAMPLE PREPARATION PROCEDURE

The first step is to use FIB to prepare the TEM sample .The sample size is about 20um width and 10um height.

The second step is to stick some AB glue on the edge of the Cu grid without carbon support film.(Figure 2)

The third step is to pick up the TEM sample and put it on the edge of Cu grid which is stuck some AB glue .Then wait about 4 hours till the AB glue dry. (Figure 3)

Figure 2:Stick some AB glue on the edge of the Cu grid

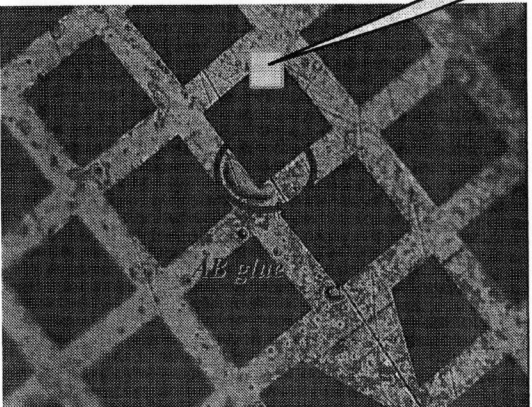

Figure.3A:Pick up the TEM sample and move the needle to the top of the Cu grid.

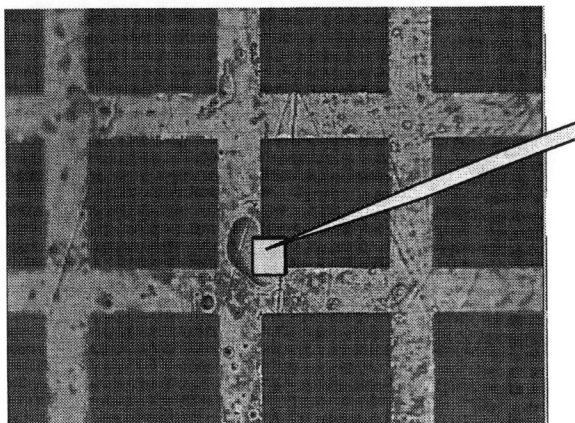

Figure.3B:Rotate the Cu grid and align the TEM sample on the corner of the Cu grid..

III. CASE STUDY AND DISCUSSION

The case is to check an extra abnormal layer contains carbon or not. Some TEM samples were done as traditional method.(Figure4) However the result show that it was difficult to tell where the carbon come from. So we prepared one TEM sample as the new method. (Figure 5).

The Electron Energy Loss Spectrum (EELS) mapping shows that the background is without the carbon signal but the abnormal layer contains carbon. (Figure 6). So we could make sure that carbon exists in the abnormal layer.

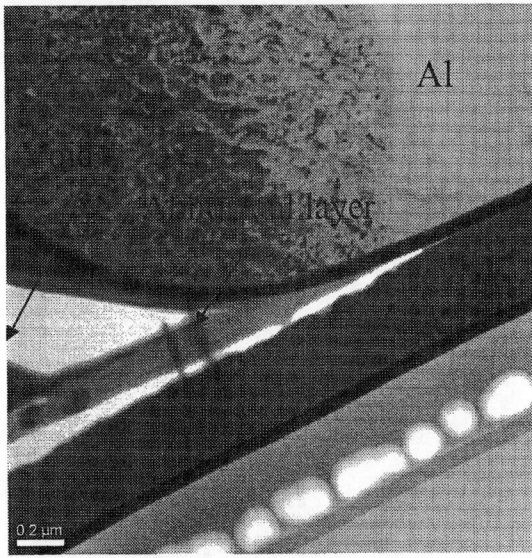

Figure 4A: The sample is put on carbon support film.The high magnification TEM image shows the defect profile.We could see the abnormal layer between Al and Cu

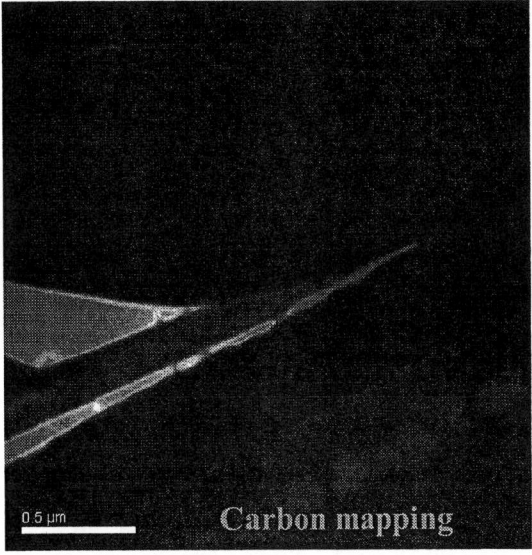

Figure 4B: The EELS data shows that the void (background) is full of carbon signal and the edge of the abnormal layer contains carbon.

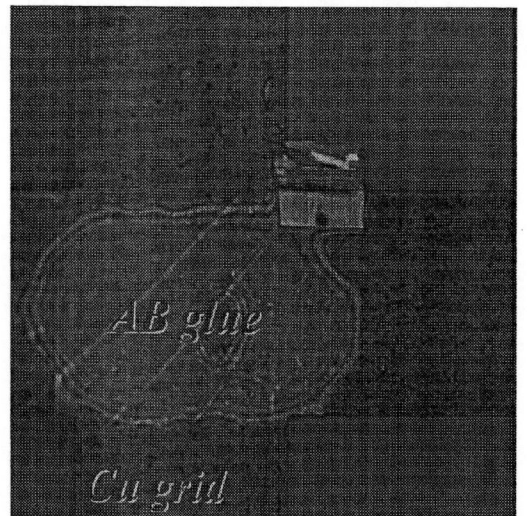

Figure 5A:The sample is put on the edge of the Cu grid without carbon support film

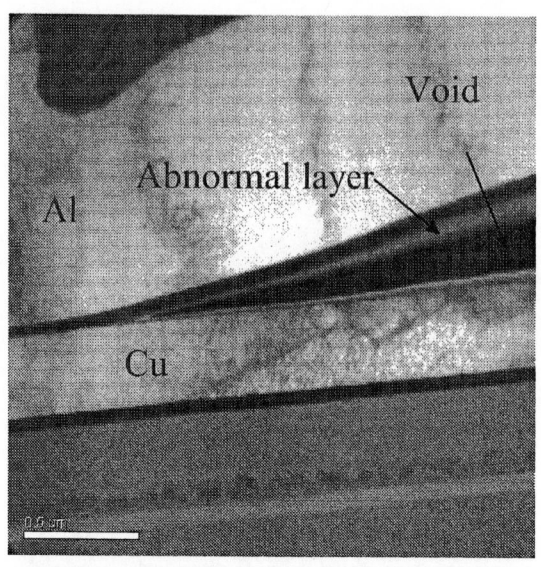

Figure 6A: The high magnification TEM image shows the defect profile.We could see the abnormal layer between Al and Cu

The background is without carbon support film

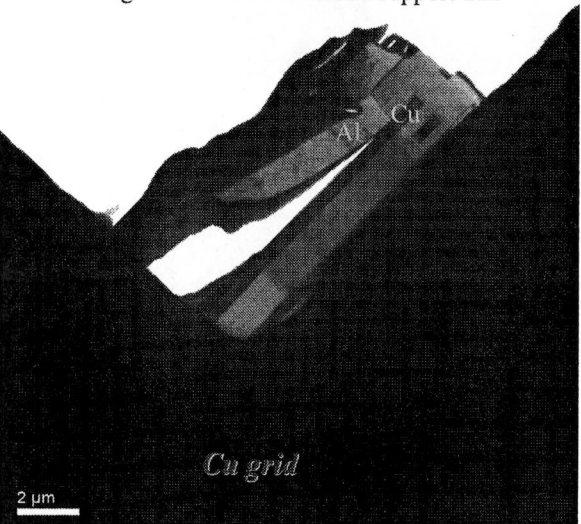

Figure 5B:The low magnification TEM image shows the whole defect profile.

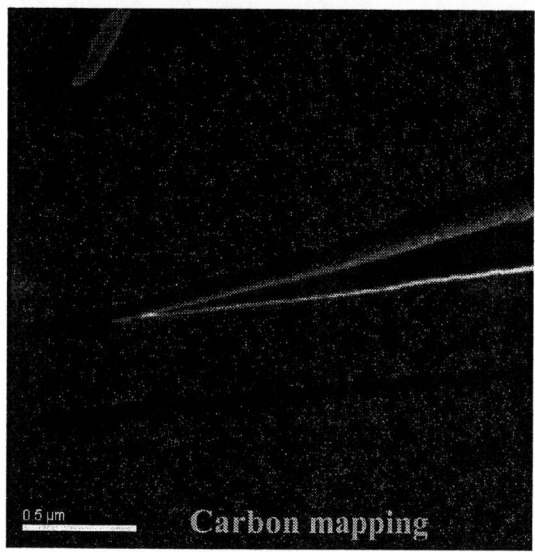

Figure 6B: The EELS data shows that the void (background) is without carbon signal but the abnormal layer contains carbon.

IV. CONCLUSION

A simple and fast technique of the TEM sample preparation without carbon background was introduced. And success rate of sample preparation is very high. For 65nm and 45nm generation, the Low-K materials which are consist of carbon are used as dielectrics .With the technique, we can easily remove the carbon support film background .It is useful for us to analyze some defect cases such as polymer residue.

REFERENCES

[1] Kevin Mcllarath ,et al ,ISTFA 2004 p320~323 "A Novel FIB Method to Prepare TEM Samples for 3D Observation"

Improved Image Processing To Enhance Thermal Laser Stimulation Signal

A. Deyine-Barth[a,b,1], K. Sanchez[a,2], P. Perdu[a,3], F. Beaudoin[b], G. Benetti[a], S. Dudit[c], D.Lewis[d]

[a]CNES, 18 Avenue Edouard Belin, 31401 Toulouse Cedex France
[b]CREDENCE, 1299 Orleans Drive, Sunnyvale, CA, USA
[c]STMicroelectronics, 850, Rue Jean Monnet, 38920 Crolles, France
[d]IMS, 351 cours de la Liberation, 33405 Talence Cedex, France
Phone : [1]33 5.61.28.14.32, [2]33 5.61.27.31.78, [3]33 5.61.28.20.17   Fax: 33 5.61.27.47.32
Email : [1]amjad.deyinebarth@credence.cnes.fr, [2]kevin.sanchez@cnes.fr, [3]philippe.perdu@cnes.fr

Abstract: **Thermal Laser Stimulation images are normally acquired using an averaging scheme to improve signal detection. However it does not work well when strong perturbation signals are present. For these cases correlation methods, which are more effective at extracting weak signals embedded in strong perturbations, become more important for TLS techniques.**

1. INTRODUCTION

Thermal Laser Stimulation techniques have been used in Failure Analysis Laboratories for several years ([1]). Among those techniques, OBIRCH (Optical Beam Induced Resistive Change) is efficient to localize shorts through a change in the IC current consumption upon laser heating ([2], [3], [4]). However for some ICs the current variations may not be due to laser heating, but to the device operating mode itself. One of the key issue is to distinguish the signal of interest from the perturbation signal which maybe stronger by far. The OBIRCH signal associated with a refresh mode, a watch dog or an active PLL typically saturate the display and mask the signal of interests. Averaging is generally used to improve OBIRCH signal detection by accumulating the signal of interest, but also the strong perturbing signals when present. The averaging method reaches its limits when the signal of interest is weaker than the perturbing signal.

The aim of the proposed paper is to demonstrate the efficiency of a Correlation method compared to the averaging method. We demonstrate that with a Correlation method we can keep the whole signal and in the mean time considerably reduce the impact of an out of interest saturating signal.

After a short description of the OBIRCH technique, we will expose a mathematical approach to Averaging and Correlation. We will then illustrate the interest of the Correlation method through cases.

2. THERMAL LASER STIMULATION PRINCIPLES

Thermal Laser Stimulation techniques are based on material's resistive variation upon laser heating. For OBIRCH a voltage source is used and the current variations are monitored while the laser beam scans the Device Under Test (DUT).

$$\Delta I = \left(-\frac{\Delta Ric}{Ric^2} \right) Vs \qquad (1)$$

Where R_{ic} is the IC resistance, ΔR_{ic} is the variation of the IC resistance, Vs the voltage source and ΔI the resulting current variation.

The laser wavelength is chosen by taking in account the material to heat and the substrate doping level. Typically a wavelength of 1340nm is used. A current amplifier is required to detect the faint IC's current consumption variations upon laser heating. Current intensity changes are coded via an acquisition card into an 8-bits image showing the sensitive areas. Finally precise localisation of the laser sensitive areas is obtained by performing an overlay of the OBIRCH image on the laser reflected image.

3. AVERAGING METHOD.

Greyscale level for each pixel (x;y) in a frame i is given by:

$$I_i(x, y) = S(x, y) + B(x, y) + N_i(x, y) \quad (2)$$

Each pixel value I(x,y) can be considered as the sum of a signal S(x,y), a background (noise average) B(x,y) and a random noise (with averaged value 0) N(x,y). S and B are supposed to be frame independent. Averaging method applied to K frames to reduce random noise gives:

$$\overline{I_k} = \frac{1}{K} \sum_{i=1}^{K} \{ S(x, y) + B(x, y) + N_i(x, y) \} \quad (3)$$

$$\overline{I_k} = S(x, y) + B(x, y) + \frac{1}{K} \sum_{i=1}^{K} N_i(x, y) \quad (4)$$

The average value of N(x,y) is 0 and then

$$\frac{1}{K} \sum_{i=1}^{K} N_i(x, y) \xrightarrow[K \to \infty]{} 0 \quad (5)$$

In this case, averaging is well adapted. When we have a strong perturbation P(x,y) related to internal switching of transistors (watch dog, refresh …) there is no synchronisation between this electrical event and laser scanning. The perturbation will affect different areas at random. Let us suppose the perturbation affect only the frame d (on K frames).

$$I_i = S(x,y) + B(x,y) + P_i(x,y) + N_i(x,y) \quad (6)$$

then the signal $\bar{I}$ is:

$$\bar{I}_k = S(x,y) + B(x,y) + \frac{1}{K}P_d(x,y) + \frac{1}{K}\sum_{i=1}^{K} N_i(x,y)$$
(7)

This kind of perturbation P can be 2 order of magnitude higher than the signal S. Perturbation averaged on a reasonable values of K=10 frames could remain one order of magnitude higher than the signal S. For FA purpose, we are mostly interested by the ability to distinguish a signal S from the background B and the noise N. Average value of the noise $\bar{N}$ is 0 and therefore we have kept a picture quality criteria:

$$Q_s = \frac{Max(|S(x,y) - B|)}{B} \quad (8)$$

Now, when we have a perturbation in a frame, the ability to observe the perturbation P from the same background B is Q_p

$$Q_p = \frac{Max(|P(x,y) - B|)}{B} \quad (9)$$

By extension, when we are looking at an averaged signal with K frames.

$$Q_{p,K} = \frac{Max(|\frac{1}{K}P(x,y) - B|)}{B} \quad (10)$$

Finally, we can introduce a quality factor giving us the ability to distinguish a signal from a background and a perturbation $Q_{s/p,k}$ obtained with averaging method on K frames

$$Q_{s/p,k} = \frac{Max(|S(x,y) - B|)}{Max(|\frac{1}{K}P(x,y) - B|)} \quad (11)$$

4. CORRELATION METHOD

For the correlation method the greyscale level for each pixel is given by:

$$\bar{I}_k(x,y) = \frac{\bar{I}_{k-1}(x,y) * I_k(x,y)}{\sqrt{(\bar{I}_{k-1}(x,y))^2 + (I_k(x,y))^2}} \quad (12)$$

Where K is the total number of frames taken for the correlation. Then we can introduce the corresponding quality factor:

$$Q_{s/p,k} = \frac{Max|S_{corr}(x,y) - (B + N(x,y))_{corr}|}{Max|P_{corr}(x,y) - (B + N(x,y))_{corr}|} \quad (13)$$

$$Q_{s/p,k} = \frac{Max|S_{corr}(x,y) - B_N(x,y)_{corr}|}{Max|P_{corr}(x,y) - B_N(x,y)_{corr}|} \quad (14)$$

5. THEORETICAL COMPARISON OF THE 2 METHODS.

The image quality criteria for the Correlation method and the Averaging method are compared when a single perturbation occurs in frame d. The following typical values for the different parameters are used to compute equations 11 and 14.

$S(x,y)=90; B=70; d=5; P_{d\neq5}(x,y)_5=B_N(x,y); P_5=250; K=10$

	Averaging method	Correlation method
Image quality criteria	Q=0,44	Q=5.77

Table1: Comparison of the image quality criteria for Averaging and Correlation methods.

6. DISCUSSION

By comparing the image quality criteria of the Correlation method with the Averaging method (Table 1) we can observe that in cases for which a strong perturbing signal is present, the image quality criteria is clearly improved by Correlation. We could go further and consider cases where perturbation and signal are superimposed for instance.

7. EXPERIMENTAL RESULTS

The OBIRCH examples shown below aims to demonstrate the efficiency of the correlation method.

Case 1:

A simple metal line is biased with a constant voltage. In addition a strong periodic perturbation is introduced in the voltage bias to simulate the strong power supply consumption associated with a refresh mode for instance. Fig.2 shows the OBIRCH signal from the metal line when the perturbation are not present. Fig. 3 shows the OBIRCH signal on the same metal line when periodic perturbations are introduced in the voltage supply.

A 10 frame average of the OBIRCH signal is shown in Fig. 4. It shows that with the averaging method the saturated spots are also superimposed, thus resulting in a poor image quality criteria.

However if we use instead a 10 frame correlation we can observe in Fig. 5 a great improvement in the image quality criteria as opposed to the average method.

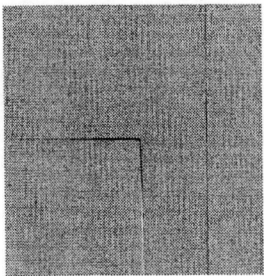

Fig. 2: OBIRCH image without perturbations.

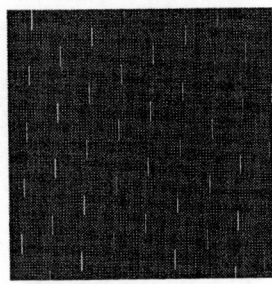

Fig. 3: OBIRCH image with perturbations.

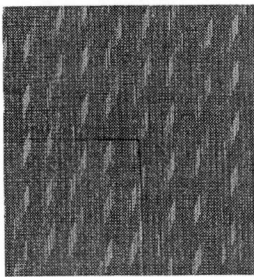

Fig. 4: OBIRCH image after 10 frame averaging.

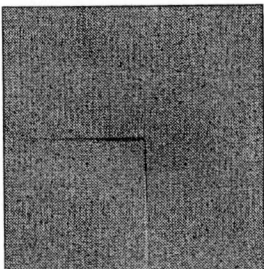

Fig. 5: OBIRCH image after 10 frame correlation

For this first set, we used the *Phemos 1000 (Hamamatsu)* and a Stanford amplifier.

Case 2:

Five resistances are put in serial. We used the ICCDLab 90 nm technology provided by STMicroelectronics (Crolles). The power supply is half the nominal value, so 0.6V, the laser power value is 10 mW and the OBIRCH test had been done in a very noisy environment. The conditions were very similar to a manufacturer failure analysis laboratory. The following set of images shows OBIRH images without perturbations, with perturbations and after Correlation.

Perturbations saturate the OBIRCH image and so some of the OBIRCH signals are invisible [Fig.7]. Once again, 10 frames correlation improved the detection of this OBIRCH signal [Fig.8].

Fig. 6: OBIRCH image without perturbations.

Fig. 7: OBIRCH image with perturbations.

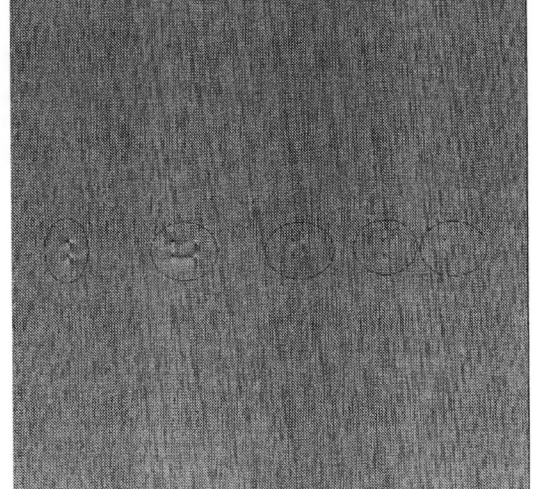

Fig. 8: OBIRCH image after 10 frame correlation.

Case 3:

In this example, we used Correlation and Averaging method on Light Emission Image.

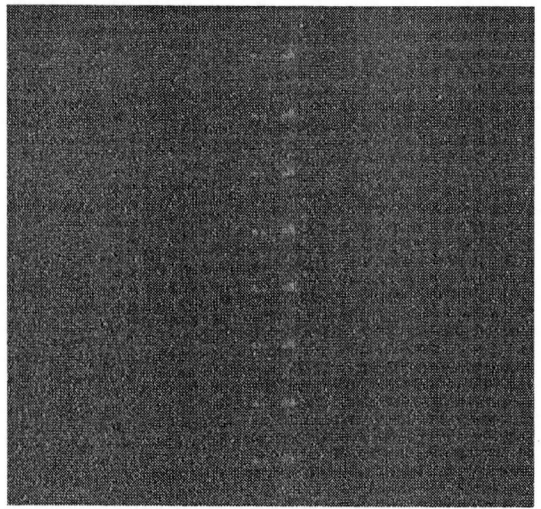

Fig. 9: Emission image without perturbations.

Some images had saturating spots and emission spots of interest are invisible [Fig.10].

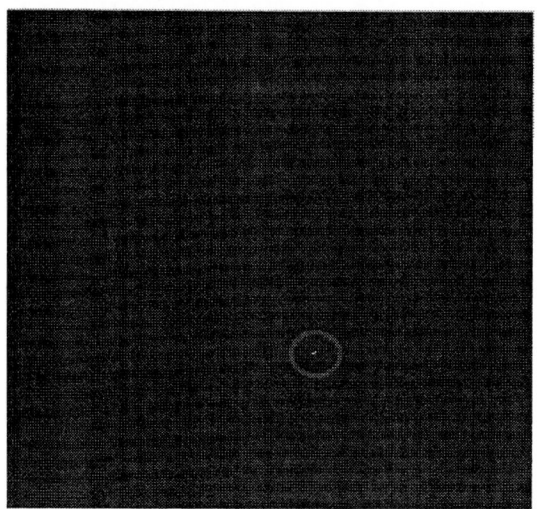

Fig. 10: Emission image with perturbations.

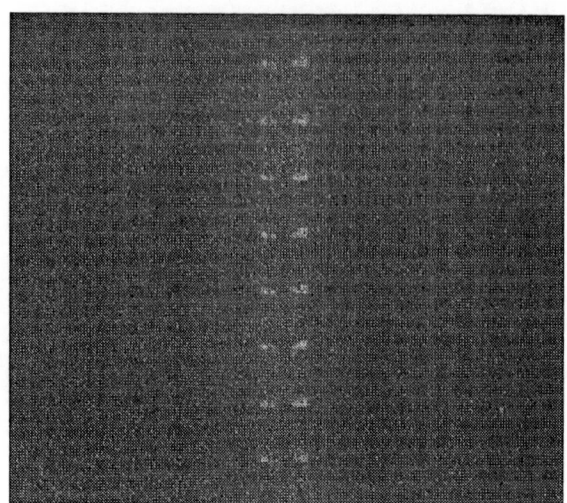

Fig. 11: Emission image after 10 frames averaging.

Fig. 12: Emission image after 10 frame correlation.

In this last case, the most suitable method is the averaging one (Fig.11 vs Fig.12). In Light Emission Imaging, perturbations stay on the same spot, so there are correlated from an image to another, since they are often due to a transistor for instance. For a LSM image, perturbations like watch dog or refresh are not spatially correlated, so Correlation works well.

For the two set of images, OBIRCH and emission, we used the *Meridian (Credence)* and the Amplifier *K-Box* for OBIRCH purposes.

8. CONCLUSION

The laser induced current variation detection is greatly improved by the Correlation method when a strong perturbation is present in the IC power consumption and when the perturbation is not spatially correlated. Moreover, this methodology could be applied to other Laser Stimulation Techniques such as OBIC or Dynamic Laser Stimulation Technique like X variation Mapping, X standing for any analogue parameter (Phase, Delay, etc).

REFERENCES

[1] Beaudoin, F; R Desplats & P Perdu et al. (2004), "Principles of Thermal Laser Stimulation Techniques", Microelectronics Failure Analysis: 417-425.

[2] Cole, E. I; P Tangyunyong & D.L Barton (1998), "Backside Localization of Open and Shorted IC Interconnections", 36th Annual International Reliability Physics Symposium: pp. 129-136.

[3] Falk, R.A (2001), "Advanced LIVA/TIVA Techniques", Proceedings of the 27th International Symposium for Testing and Failure Analysis: pp. 59-65.

[4] Nikawa, K (1999), "Failure Analysis Case Studies Using the IR-OBIRCH Method", IEEE p. 394-399.

Near-Infrared Spectroscopic Photon Emission Microscopy of 0.13 µm Silicon nMOSFETs and pMOSFETs

SL Tan[1], JKJ Teo[1], KH Toh[1], D Isakov[1], DSH Chan[1], LS Koh[2], CM Chua[2], JCH Phang[1,2]
[1]Centre for Integrated Circuit Failure Analysis and Reliability (CICFAR),
National University of Singapore, 10 Kent Ridge Crescent, Singapore 119260
[2]SEMICAPS Pte Ltd, 28 Ayer Rajah Crescent #03-01, Singapore 139959
E-mail: tansoonleng@nus.edu.sg

Abstract-**Near-infrared photon emission spectra were obtained from the frontside of silicon nMOSFETs and pMOSFETs with a gate length of 0.13 µm and biased into saturation. These spectra were obtained using a high sensitivity in-lens spectroscopic photon emission microscope. Frontside NIR photon emission spectroscopy are performed on 0.13 µm saturated nMOSFETs and pMOSFETs at different gate and drain bias. The nMOSFETs photon emission spectra obtained are significantly different from some previously reported photon emission spectra. The NIR photon emission spectra of the nMOSFETs and pMOSFETs have similar peaks and suggest that the electric field condition in the channels of the nMOSFETs and pMOSFETs are similar.**

I. INTRODUCTION

Ever since the first light emission from a Si p-n junction was reported by Newman et al in 1955 [1], there have been many publications on electroluminescence from Si devices [2]. It has been established that MOSFETs emit light when biased at saturation [3-10]. The photon energy distribution can provide information on the energy levels involved in the carrier recombination mechanisms. Most of the work done so far has been in the visible wavelength range where the energy transitions of the carriers are higher than the Si bandgap [3-6]. Recently, publications have reported spectra from MOSFETs in the near-infrared (NIR) wavelength range where the energy transitions of the carriers are lower than the Si bandgap and involve intraband transitions of carriers [7-10]. However, most of the reported photon emission spectra in the NIR range were from nMOSFETs [7,8,10] although photon emission spectra on pMOSFETs has recently been reported in [9]. Moreover, previously reported photon emission spectra of nMOSFETs were significantly different from each other.

II. SPEM SYSTEM

The schematic diagram of the Spectroscopic Photon Emission Microscope (SPEM) System which was used to obtain the photon emission spectra is shown in Fig. 1 [11]. The detector is an Indium Gallium Arsenide (InGaAs) focal plane array (FPA) with a wavelength sensitivity range between 0.9 µm and 1.6 µm. The dispersive element is a three-element prism which is positioned in the infinity corrected path of the optical microscope, between the objective lens and the tube lens. The dispersive element is specially designed to split the light from an emission spot on the device under test into a wavelength spectrum for a wavelength range from 0.9 µm to 1.6 µm with an efficiency of about 70%. The in-lens SPEM system is designed to use one InGaAs FPA for both panchromatic photon emission imaging and spectroscopy and offers ease of switching between the imaging and the spectroscopic modes by moving the dispersive element in and out of the microscope optical axis without the need for moving the detector. Light loss is reduced with the in-lens design as no additional optical elements such as mirrors are needed to redirect the light into the dispersive element or from the dispersive element onto the detector.

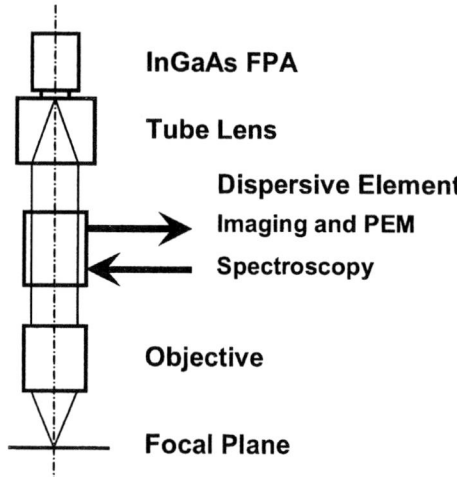

Fig. 1: In-lens near-infrared Spectroscopic Photon Emission Microscope System.

In the spectroscopic mode, the objective lens collects and collimates light from the emission spot. The collimated light passes through the dispersive element and is scattered into its component wavelengths. The spectrum is projected by the tube lens onto the InGaAs FPA detector. Fig. 2 shows the emission spot during panchromatic photon emission imaging and the resulting spectrum after the insertion of the prism for

978-1-4244-2039-1/08/$25.00 ©2008 IEEE

photon emission spectroscopy. The system is calibrated for intensity and wavelength by using a calibrated tungsten halogen lamp source and a set of narrow wavelength bandpass filters [11] to find the wavelength spread of the spectrum and compensate for the difference in sensitivity at different wavelengths. Care is also taken to ensure that the spectrum does not lie on any dead pixels of the detector.

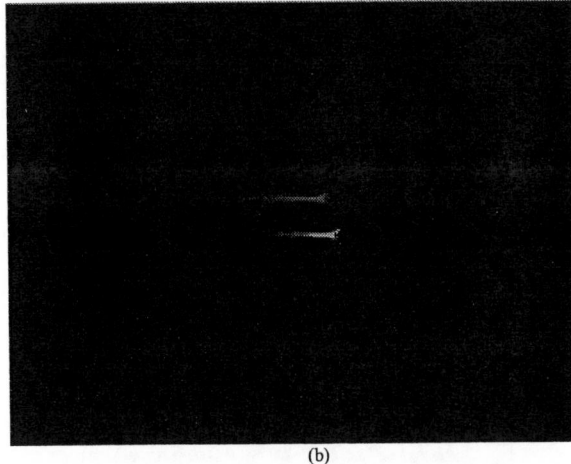

(a)

(b)

Fig. 2: (a) Panchromatic photon emission image and (b) corresponding spectroscopic image after insertion of the dispersive element

III. NIR SPECTROSCOPY RESULTS

NIR photon emission spectra were collected from the frontside of nMOSFETs and pMOSFETs with gate width and gate length of 20 μm and 0.13 μm respectively at V_{gs} = 1.5V, 1.3V, 1.0V and 0.8V. At each V_{gs}, the spectra were collected at V_{ds} = 1.5V to 1.8V in steps of 0.1V.

The I_{ds}–V_{ds} characteristics of the nMOSFETs and pMOSFETs are shown in Fig 3. The nMOSFETs have a higher drive current than the pMOSFETs for V_{gs} = 1.5V and

1.3V. This can be explained by the higher electron mobility in silicon compared to holes. However, at V_{gs}= 1.0V and 0.8V, the pMOSFETs have a higher current drive than the nMOSFETs. This is because the pMOSFETs have a lower threshold voltage than the nMOSFETs.

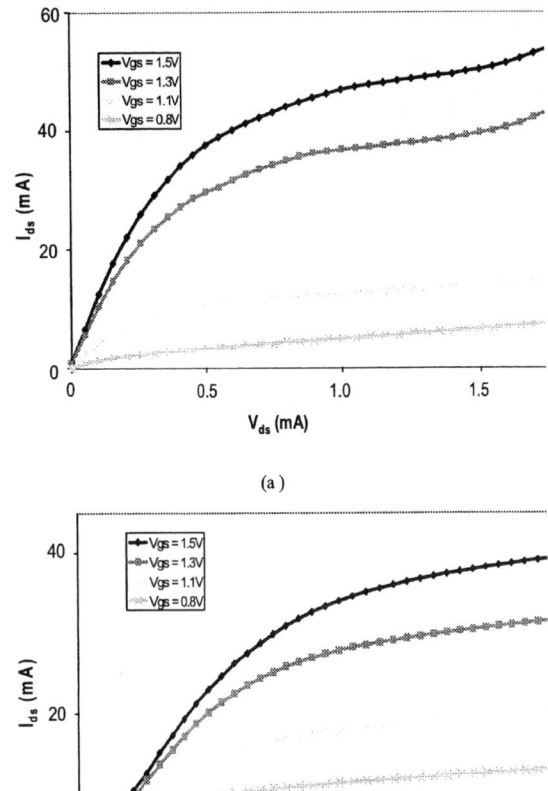

(a)

(b)

Fig. 3: (a) I_{ds}-V_{ds} characteristics of nMOSFETs and (b) pMOSFETs at different V_{gs}

Fig 4 and 5 show the spectra collected from the nMOSFETs and pMOSFETs at V_{gs} = 1.5V and 1.3V respectively, with increasing drain voltage $V_{ds.}$. The spectra of the nMOSFETs collected are similar to those by Herzog et al [7] but significantly different from that by de Luna et al [9]. Both the nMOSFETs' and the pMOSFETs' spectral intensity reduce as V_{ds} decreases even though the current flowing through the MOSFETs did not decrease significantly.

978-1-4244-2039-1/08/$25.00 ©2008 IEEE

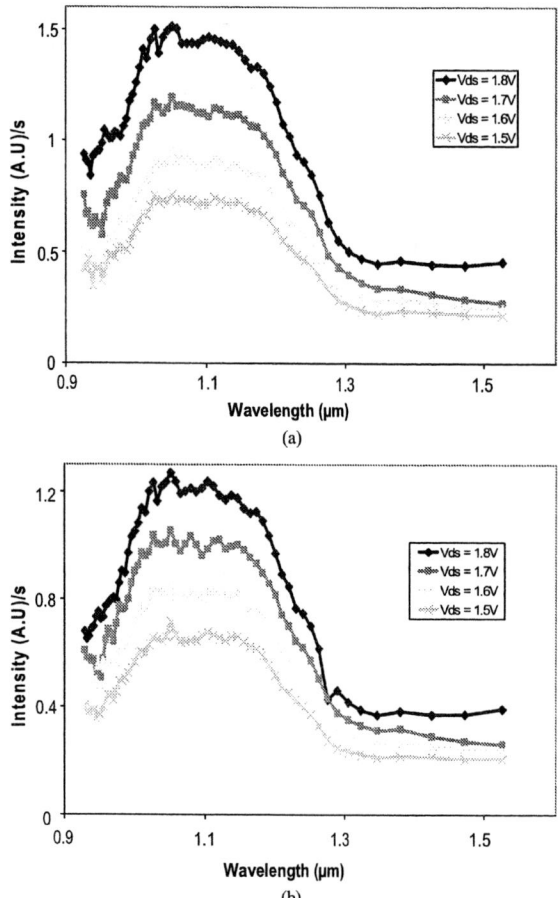

(a)

(b)

Fig 4:NIR photon emission spectra of nMOSFET at (a) V_{gs}= 1.5V, (b) V_{gs}= 1.3V

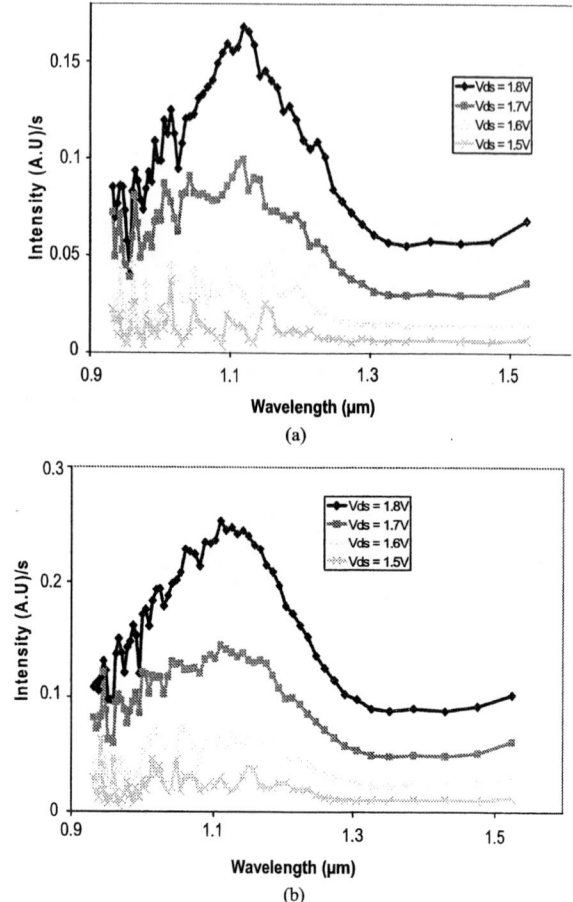

(a)

(b)

Fig 5: NIR photon emission spectra of pMOSFETs at (a) V_{gs}= -1.5V, (b) V_{gs}= -1.3V.

Fig 6 shows the photon emission spectra of the nMOSFETs and pMOSFET V_{gs} =1.3V for different V_{ds}. There are slight differences between the NIR photon emission spectra of the nMOSFETs and pMOSFETs. Both nMOSFET and pMOSFET spectra have one main peak at around 1.1 μm, which corresponds roughly to the energy of the Si bandgap. There are also lower peaks at around 1.0 μm and 1.42 μm. The intensity of the spectra of the nMOSFETs is higher than the pMOSFETs.

Fig 7 shows the full width half maximum of the 1.1 μm peak for the nMOSFETs and pMOSFETs photon emission spectra at different V_{gs} and V_{ds}. An increasing V_{ds} indicates an increasing maximum electric field in the channel. Both 1.1 μm peaks in the nMOSFET and pMOSFET spectra narrow as V_{ds} increases with the pMOSFET peak narrowing more significantly. The nMOSFET 1.1 μm peak is broader than the pMOSFET peak especially at higher V_{ds} biases. The difference gets smaller as V_{ds} decreases.

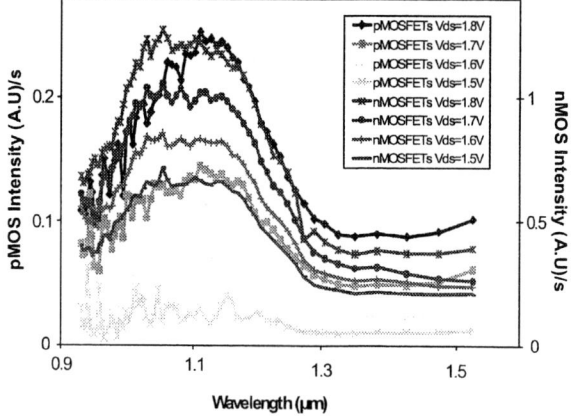

Fig 6: NIR photon emission spectra of nMOSFETs and pMOSFETs at V_{gs} = 1.3V at different V_{ds}

978-1-4244-2039-1/08/$25.00 ©2008 IEEE   164   *Proceedings of 15th IPFA - 2008, Singapore*

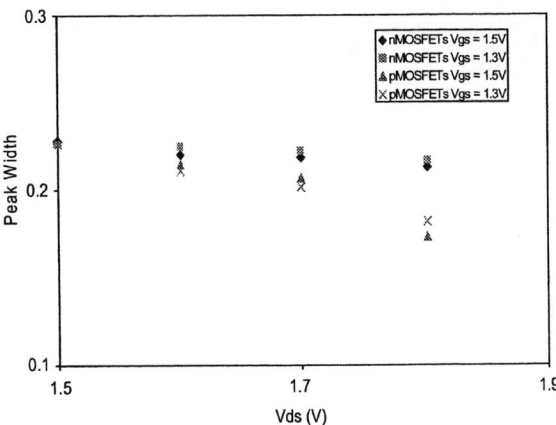

Fig 7: Full width half maximum of the 1.1 μm peak of MOSFET photon emission spectra at different V_{gs} and V_{ds}

The nMOSFETs and the pMOSFETs main wavelength peaks are similar, indicating that photon generation mechanisms in the nMOSFETs and pMOSFETs are similar. However, the nMOSFETs have a broader 1.1 μm peak which extends more into the shorter wavelength region than the pMOSFETs, indicating that at the same bias, there were more high energy carriers in the nMOSFETs compared to that of the pMOSFETs.

The 1.42 μm wavelength peak has been observed in photoluminescence studies on Si and is attributed to recombination of the carriers via energy levels in the Si bandgap [12-13]. These energy levels are formed due to the presence of slip dislocations in the Si, near the drain where the emissions are emitted. The slip dislocations are due to ion implantations induced damage. The 1.42 μm peak which is attributed to recombination via energy levels in Si due to ion implantation induced damage is not so obvious as compared to the spectra obtained from nMOSFETs of longer channel lengths [14]. This can be explained by the reduced energy of the ions used by the source/drain implantations for devices of smaller dimensions and the improvements brought by the annealing conditions leading to fewer defects.

It is interesting to note that the shape of the pMOSFETs photon emission spectra at V_{ds} = 1.7V is similar to the photon emission spectra of the nMOSFETs at V_{ds} = 1.5V. As the emission spectra are dependent on the electric field in the channel, this suggests that the electric field conditions inside the nMOSFETs and pMOSFETs at these two bias conditions are similar.

IV. CONCLUSIONS

Frontside NIR photon emission spectroscopy are performed on 0.13 μm saturated nMOSFETs and pMOSFETs at different gate and drain bias. The nMOSFETs photon emission spectra obtained are similar to some of the photon emission spectra

previously reported but also significantly different from the other photon emission spectra reported.

In this paper, there are only slight differences between the NIR photon emission spectra of the nMOSFETs and pMOSFETs obtained, suggesting that the electric field condition in the channel of the nMOSFETs and pMOSFETs are similar. The position of the peaks of both the nMOSFETs and the pMOSFETs are the same.

ACKNOWLEDGEMENTS

The authors are grateful to Chartered Semiconductor Manufacturing Ltd for providing the samples used in this work. SL Tan, JKJ Teo and D Isakov are supported by a National University of Singapore research scholarship. SL Tan is also supported by a grant from Chartered Semiconductor University Research Program. This work was also partially supported by Singapore Ministry of Education's AcRF Tier 1 Research Project No: R-263-000-252-112.

REFERENCES

[1] Newman R, "Visible Light from a Silicon p-n Junction", Physical Review, Vol 100, No 2, pg 700-703, 1955

[2] Phang JCH, Chan DSH, Tan SL, Len WB, Yim KH, Koh LS, Chua CM, Balk LJ, "A Review of Near Infrared Photon Emission Microscopy and Spectroscopy", Proc Int Symp Physical & Failure Analysis of Integrated Circuits (IPFA 2005), 27 Jun 05 - 1 Jul 05, Singapore, pg 275-281, 2005

[3] Toriumi A, Yoshimi M, Iwase M, Akiyama Y, Taniguchi K, "A Study of Photon Emission from n-Channel MOSFETs", IEEE Transactions on Electron Devices, Vol ED-34, No 7, pg 1501-1508, 1987

[4] Selmi L, Lanzoni M, Bigliardi S, Sangiorgi, "Photon Emission from Sub-Micron p-Channel MOSFETs Biased at High Fields", Microelectronic Engineering, Vol 19, pg 747-750, 1992

[5] Tao JM, Chan DSH, Chim WK , "Spectroscopic Observations of Photon Emissions in n-MOSFETs in the Saturation Region ", J Phys D: Appl Phys, Vol 29, pg 1380-1385, 1996

[6] Rasras MS, De Wolf I, Groeseneken G, Maes HE, "Spectroscopic Identification of Light Emitted from Defects in Silicon Devices", J Appl Phys, Vol 89, No 1, pg 249-258, 2001

[7] Herzog M, Koch M, "Hot Carrier Light Emission from Silicon Metal-Oxide-Semiconductor Devices", Appl Phys Lett, Vol 53, No 26, pg 2620-2622, 1988

[8] Len WB, Liu YY, Phang JCH, Chan DSH, "Near IR Continuous Wavelength Spectroscopy of Photon Emissions from Semiconductor Devices". Proc Int Symp Testing & Failure Analysis (ISTFA 2003), 2-6 Nov 03, Santa Clara, California, pg 311-316, 2003

[9] de Luna NC, Bailon MF, Tarun AB, "Analysis of Near-IR Photon Emissions from 50-nm n- and p-Channel Si MOSFETs", IEEE Trans Electron Devices, Vol 52, No 6, pg 1211-1214, 2005

[10] Tan SL, Ang KW, Toh KH, Isakov D, Chua CM, Koh LS, Yeo YC, Chan DSH, Phang JCH, "Near-IR Photon Emission Spectroscopy on Strained and Unstrained 60 nm Silicon nMOSFETs", Proc Int Symp Testing & Failure Analysis (ISTFA 2007), 4-8 Nov 07, San Jose, California, USA, pg 81-85, 2007

[11] Tan SL, Toh KH, Phang JCH, Chan DSH, Chua CM, Koh LS, "A Near-Infrared, Continuous Wavelength, In-Lens Spectroscopic Photon Emission Microscope System", Proc Int Symp Physical & Failure Analysis of Integrated Circuits (IPFA 2007), 11-13 Jul 07, Bangalore, India, pg 240-244, 2007

[12] Uebbing RH, Wagner P, Baumgart H, Queisser HJ, "Luminescence in Slipped and Dislocation-free Laser-annealed Silicon", Appl Phys Lett, Vol 37, No 12, pg 1078-1079, 1980

[13] Kveder VV, Steinman EA, Shevchenko SA, Grimmeiss HG, "Dislocation-related Electroluminescence at Room Temperature in Plastically Deformed Silicon", Physical Review B, Vol 51, No 16, pg 10520-10526, 1995

[14] Tan SL, Toh KH, Chan DSH, Phang JCH, Chua CM, Koh LS, "Determination of Intrinsic Spectra from Frontside and Backside Photon Emission Spectroscopy", Proc Int Rel Phys Symp (IRPS 2007), 15-19 Apr 07, Phoenix, Arizona, USA, pg 620-621, 2007

IC Package Inspection with Nanofocus X-ray Tubes and NanoCT

Holger Roth [1], Zhenhui He [2], and Thomas Paul [3]

Phoenix|x-ray Systems + Services GmbH

[1] Application Laboratory Stuttgart, Kranstr. 8, 70499 Stuttgart, Germany. Email: HRoth@phoenix-xray.com
[2] Headquarters, Niels-Bohr-Str. 7, 31515 Wunstorf, Germany. Email: ZHe@phoenix-xray.com
[3] Headquarters, Niels-Bohr-Str. 7, 31515 Wunstorf, Germany. Email: TPaul@phoenix-xray.com

Abstract— **Nanofocus tube technology in combination with high resolution CT are the driving forces in X-ray and the leading future technologies for the inspection of IC packages. The paper shows different results of highest resolution failure analysis performed with latest 2D and 3D nanofocus techniques like the first 180 kV nanoCT system.**

I. INTRODUCTION

In qualification and spot checks of IC packaging, X-ray systems are customarily used for the inspection of classical features such bond wires, die bonds, moulding and seals. New challenges to X-ray inspection equipment arise from three trends in package technolgy: miniaturisation, the use of novel materials and increasing device complexity. Package miniaturisation leads to higher density and smaller size of internal structures such as microvias and FlipChip interconnections, demanding for resolution in the sub-micron range at highest magnifications as provided by novel nanofocus® tube technology. The absorption of some materials like non-conductive die adhesives or copper bond wires either is to low to yield sufficient contrast in customary X-ray images or is strong enough to conceal other package features, as observed for copper-tungsten alloy caps. Highly sensitive a-Si 16-bit digital flat panel detectors and image processing techniques now have remarkably improved the detectability of such low contrast features. The complexity of devices with 3D set-ups such as stacked or multiple die including corresponding wire connections leads to confusing overlaps in the two-dimensional X-ray images, in other words, they must be inspected slice by slice or in 3D visualisations as provided by computed tomography (CT). However, up to now, many laboratory CT systems are using X-ray tubes of some microns focal spot size and maximum tube voltages around 100 kV. Since electronic devices contain very fine structures and strongly absorbing materials like gold or copper this results in unsatisfying image resolution and strong image artefacts, respectively. In view of this situation, a nanoCT® system was designed for highest resolution CT of electronic components with an 180 kV nanofocus® tube providing excellent penetration at sub-micron detail detectability. Adjacently some typical failure and quality analysis results are presented.

II. NANOFOCUS X-RAY - METHOD

The following 2D investigations were performed by means of a automatic X-ray shadow microscope with 2 MPixel digital image chain for magnifications and a 180 kV / 15 W high power nanofocus® X-ray tube designed for up to 24.000x magnification and a detail detectability of 200-300 nm (nanome|x 180), see Fig 1. The system can provide highly magnified images also under viewing angles up to 70° (ovhm), by tilting the detector as shown in fig. 2.

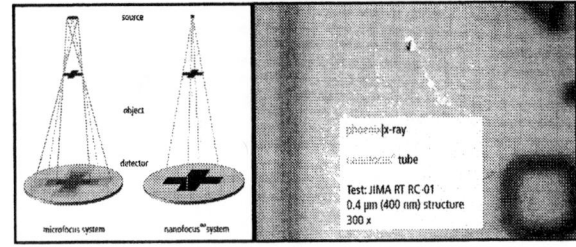

Fig. 1. Influence of focal spot size on image sharpness: The practically punctiform nanofocus® tube increases resolution, so that even *periodic* structures of 0.4 μm are resolved.

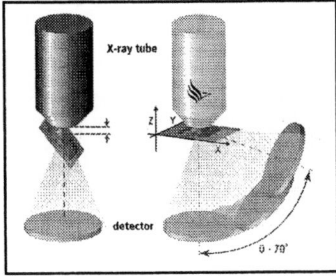

Fig. 2. Oblique views at highest magnification: tilting the detector instead of tilting the sample keeps source-object distance small and avoids the loss of magnification.

III. NANOFOCUS X-RAY - RESULTS

One of the most challenging tasks in 2D X-ray inspection is the proper imaging of the delicate dendrites as created by electromigration: sub-micron spatial resolution at high magnification is required as well as outstanding contrast resolution. Fig. 3 sharply displays dendrites growing out of a copper conductor and causing a short circuit. The measured dendrite width is 10 microns. Another frequent inspection task is the detection of cracks and lifted up wire bonds. If the resulting gap is not only to be found but also measured and clearly displayed for further investigation of its cause, the image resolution must be much smaller than its size. In Fig. 4a a ball bond is shown by nanofocus® X-ray technology to be lifted up about 2 microns. Note that at slightly less resolution the edges of ball bond and bond pad could not be differentiated any more and the gap would be not visible. Due to nanofocus® resolution the ripped of wedge bond displayed in Fig. 4b is not only detected, but also proved to be cracked at the rim of the weld area, since the edge of the bond capillaries footprint is still visible. Note that the images in Fig. 4 and b require precise adjustments of the viewing angle without loss of magnification (ovhm).

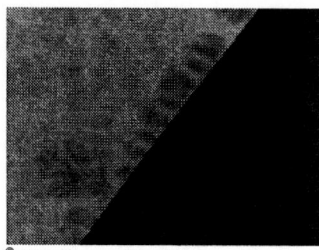

Fig. 3. nanofocus® X-ray image of dendrites growing out of a conductor. The measured dendrite diameter is 10 microns.

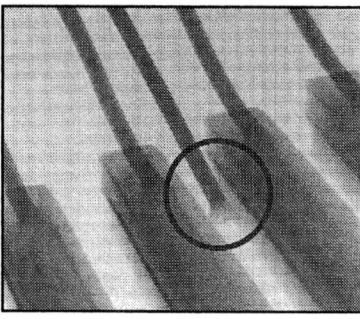

Fig. 4. a) nanofocus® X-ray image of a lifted ball bond at highest magnification. b) Lifted wedge bond at 55° ovhm.

In IC processor packages the internal bonds are flip-chip solder joints which show features and defects similar to area array solder joints as known from electronic assemblies. However, the flip chip interconnection may be more 10 times smaller (25 to 100 µm) and require for an appropriate image resolution to detect defects like voids, opens and shape deviations in the micrometer range, see fig 5.

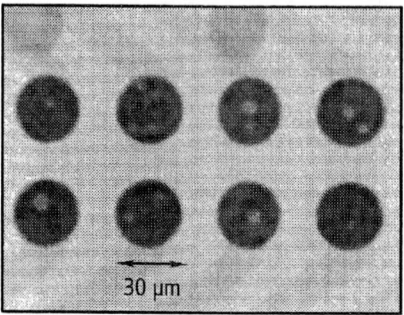

Fig. 5. nanofocus® X-ray image of flip chip solder joints showing micrometer sized voids.

Copper bond wires (when compared to the highly absorbing gold wire as shown above) yield a poor X-ray image contrast which is enhanced not only by using digital imaging but also by high magnification and sharp imaging, cf. fig. 6.

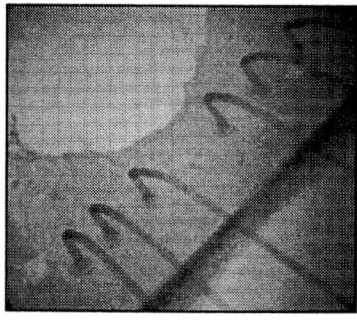

Fig. 6. nanofocus® X-ray image copper bond wires in an electronic device.

IV. NANOCT - METHOD

The nanoCT® scans were carried out with a recently improved very compact laboratory CT system (nanotom 180) dedicated to the analysis of small samples at submicron voxel-resolution, see figures 7 and 8. The system comprises a 180 kV / 15 W high power nanofocus® tube with a focal spot size below 1 micron and a 5-megapixel flat panel detector with an active area of 120 x 120 mm (2300 x 2300 pixels, 50 µm pixel size) and a 3-position virtual detector (up to 360 mm detector width), providing a voxel resolution of 0.5 microns (500 nm) and a detail detectability of 200-300 nm. The influence of vibrations as well as thermal expansion and drift are minimised by a granite based manipulation system, vibration damping, low expansion materials and temperature stabilisation. Minor remaining drifts during long term measurements are monitored and compensated by the CT reconstruction software (datos|x), which utilises efficient algorithms for geometry calibration, detector calibration, noise suppression, beam hardening reduction and region-of-interest-CT.

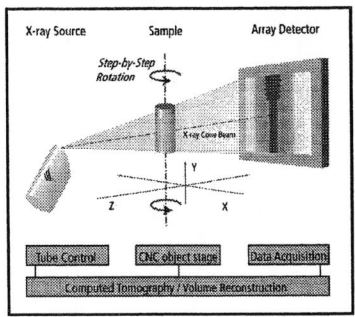

Fig. 7. Function principle of cone beam computed tomography   The system acquires a series of two dimensional X-ray images while progressively rotating the sample step by step through a full 360° rotation at increments of less than 1° per step. These projections contain information on the position and density of absorbing object features within the sample. This accumulation of data is then used for the numerical reconstruction of the volumetric data.

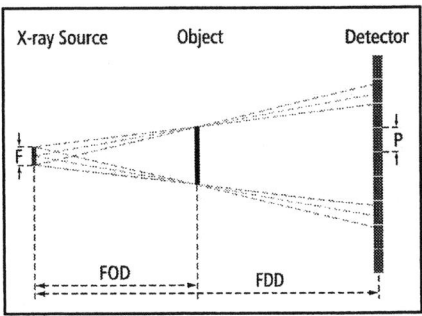

Fig. 8: Resolution of nanoCT®: The geometric voxel resolution V is given by the pixel size P divided by the geometric magnification M=FDD/FDO, ie: V=P/M. With a pixel size P=50 μm and a FDD= 500 mm a voxel size of 0.5 μm can be easily achieved at M=100. The final limitation of resolution is the focal spot size F of the X-ray tube which causes an additional unsharpness on the detector (green lines). Hence, for sub-micron computed tomography an X-ray tube with a focal spot size below 1 μm is required (nanofocus® tube).

V. NANOCT - RESULTS

While 2D images of a memory cube with stacked wires and stacked dies (Fig. 9a) are not suitable for analysis due to overlaying features, 3D nanoCT® virtual slices or sections allow to examine each individual die-attach for voids (Fig. 9b) as well as the flow of the wire bonds. In stacked die and similar devices the wires are arranged in layers so that the wires overlap in the X-ray images and short circuits cannot be clearly told from crossings. In a 3D visualisation as provided

by nanoCT® the spatial arrangement of the wires may be examined slice by slice and their distances may be measured at any position. Further examples for advanced failure analysis with nanoCT® are defects in concealed conductors in package redistribution planes or the spatial localisation of package voids or defects in chip components, see fig. 10.

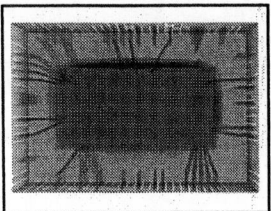

Fig. 9. Frontal 2D X-ray image of a memory cube with stacked dies and b) tomographic section visualising the die attach (by courtesy of 3D-Plus). Size of the sample is about 15 mm x 10 mm x 10 mm.

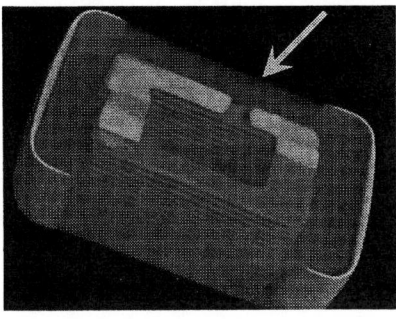

Fig. 10. Tomographic section visualisation of a 0805 chip inductor, clearly revealing a crack in the coil.

VI. CONCLUSION

Digital nanofocus® X-ray inspections systems with the capability of oblique views at highest magnification are a fast and most effective tool for the analysis of electronic packages and detection of defects down to the sub-micron range. nanoCT® widely expands the spectrum of detectable micro-structures in complex electronic devices and packages by 3D visualisation and slice by slice analysis. This opens a new dimension of 3D-microanalysis and will partially replace destructive methods – saving costs and time per sample inspected. nanofocus® tube technology pushes computed tomography systems into application fields that very recently were exclusive to expensive synchrotron techniques.

EMMI Analysis on Silicon Solar Cell

Benjamin Yeh, Russell Huang, Kevin Chung, Alan Chang, and Dr. Chih-Hsun Chu
Materials Analysis Technology Inc./ 1F, No.14, Prosperity Rd. II, Science-Based Industrial Park.
Hsinchu City, Taiwan, R.O.C.
Phone: (886-3) 5635660   Fax: (886-3) 5635661   Email: benjamin@ma-tek.com

Abstract- **This work utilized the EMMI as a tool for investigation of the junction leakage of crystalline silicon solar cell under reverse bias and forward bias. The defected areas were examined by the TEM to reveal the physical structure. In addition, analysis of physical structure and component of the solar cell were demonstrated by using common PFA tools, such as OM, SEM, EDX, FIB, and TEM.**

I. INTRODUCTION

This work tries to develop a systematic methodology in (1) examining low efficiency area of a crystalline silicon solar cell and identifying the physical defects, and (2) analyzing physical structure of a silicon solar cell. Instead of developing a brand new tool, conventional FA tools, such as EMMI, OM, SEM, FIB, and TEM are utilized in this work.

Silicon solar cell is a photovoltaic device. Absorbed radiation with photon energy large than energy gap of Si generates electron-hole pairs inside the silicon solar cell. These electron-hole pairs are separated by the built-in potential across the depletion region of the pn junction and collected to produce photo voltage and photo current. This is called photovoltaic effect.[1]

The silicon solar cell efficiency is defined as ratio of electric power generated to the power of incident radiation. Conventionally, a single number is calculated to represent the efficiency of the whole solar cell system. This work tries to investigate the distribution of conversion efficiency across a silicon solar cell and edge isolation by EMMI. The junction leakage area and poor edge isolation area are located by reverse biasing of the solar cell. The electroluminescence (EL) intensity distribution is revealed by forward biasing of the solar cell. Electroluminescence (EL) is the inverse effect of photovoltaic effect, where light emission caused by electrical current flow through a semiconductor with a pn junction. The junction leakage area and the poor EL area represent the poor conversion efficiency locations of the solar cell.[2] Intensity of emission lights caught by EMMI during EL analysis is assumed to be proportional to the magnitude of silicon solar cell conversion efficiency. The areas with poor conversion efficiency located by either EMMI or EL are investigated by the TEM to reveal the physical defect which causes the low conversion efficiency of the solar cell.

II. EXPERIMENTAL RESULTS, ANALYSIS AND DISCUSSION

A. Analyzing physical structure and component of a silicon solar cell:

Pin holes due to surface texturing can be observed by SEM, FIB and TEM as shown in Fig. 1, Fig. 2, Fig.3a and Fig. 3b. Thicker silicon oxide layer and residual are found at the bottom of a triangle shape pin hole as shown in Fig. 3a.

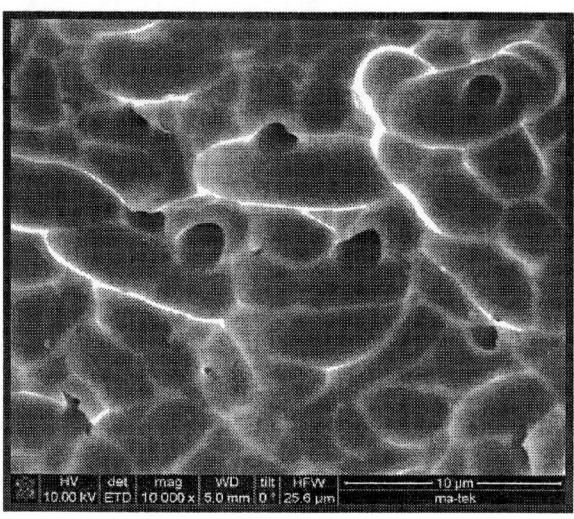

Fig. 1. Pin holes due to surface texturing were observed by SEM.

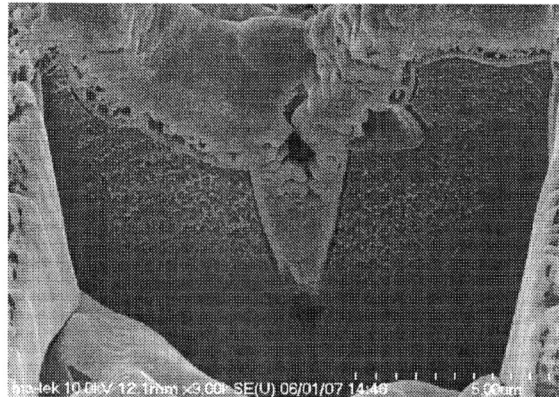

Fig. 2. A triangle shape pin hole was observed by FIB.

978-1-4244-2039-1/08/$25.00 ©2008 IEEE

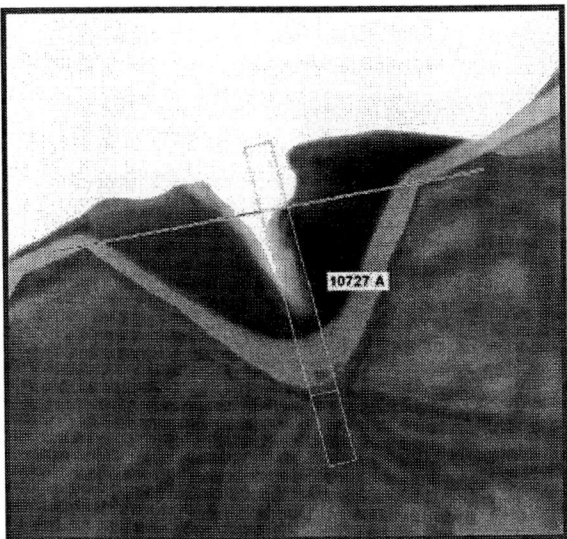

Fig. 3a. Triangle shape pin hole was observed by TEM. Thicker silicon oxide layer and residual were found at the bottom of the pin hole.

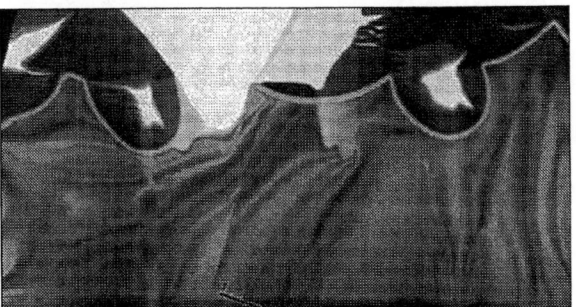

Fig. 3b. Elliptic shape pin hole was observed by TEM. No residual and thicker oxide layer were found at the bottom of the pin hole.

B. Examining distribution of silicon solar cell efficiency

EL analysis by forward-biasing a silicon solar cell results in emitted lights across the solar cell. The emitted light is detected by EMMI. Brighter and warmer color indicates places with higher solar cell conversion efficiency, while darker and colder color indicates areas with lower conversion efficiency as shown in Fig. 4b. Optical image are shown in Fig 4a.

The EL image, as shown in Fig. 5b and Fig. 5c, matches the grain boundary observed by optical microscope as shown in Fig. 5a. Best resolution of the EL image can be achieved by adjusting lens magnification, integration time of EL signal, and forward-biasing current as shown in Table 1.

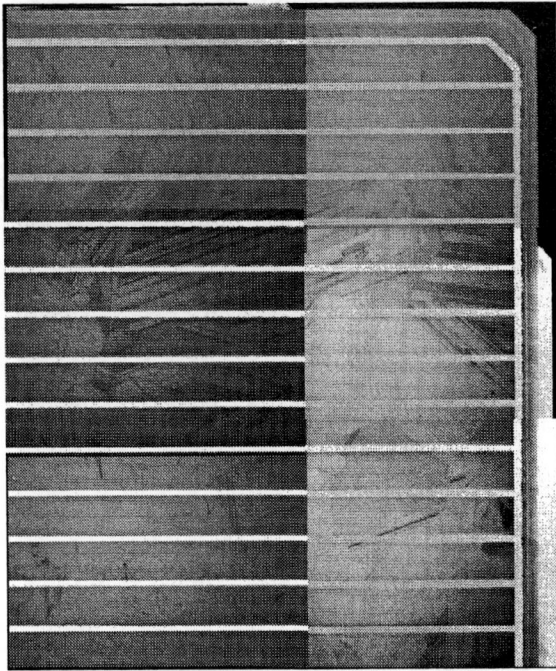

Fig. 4a. Optical image (0.8X) of a silicon solar cell. Grains can be clearly seen.

Fig. 4b. EL image of Fig. 4a. Distribution and intensity of EL signals reflect solar cell conversion efficiency. Brighter and warmer color indicates places with higher solar cell conversion efficiency, while darker and colder color indicate places with lower conversion efficiency.

Fig. 5a. Zoom-in optical image (20X) of Fig. 4a. Darker lines are grain boundaries of the silicon solar cell.

Fig. 5b. EL analysis of Fig. 5a. Change of color reveals variation of solar cell efficiency. Area with blue color suggests that conversion efficiency is lower at the area. The integration time of EMMI is 60 seconds and forward bias current is 600mA. The EL image matches closely with Fig. 5a. Grain boundaries can be clearly observed.

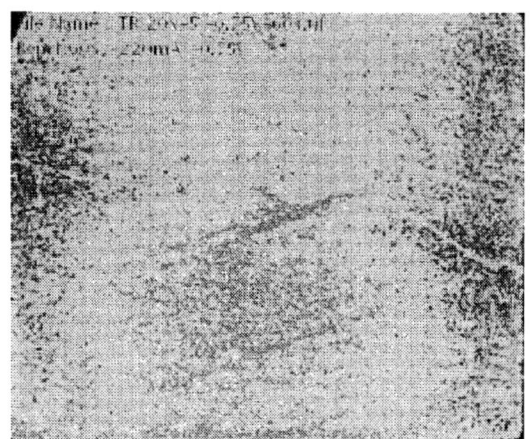

Fig. 5c. EL analysis of Fig. 5a. The integration time of EMMI is 60 seconds and forward bias current is 220mA. With less forward current, the resolution is worse than Fig. 5b.

Table I

Image quality comparison table of EL analysis. Best resolution (3 stars) can be achieved by adjusting lens magnification and integration time of EL signal, and forward-biasing voltage.

Lens Magnification	Integration Time (second)	Current			
		0.22A	0.4A	0.6A	0.85A
5X	10	X X	X X	X	★
	20	★	★★	★★	★★★
	30	★★	★★	★★★	★★★
	60	★★	★★★	★★★	
	80	★★	★★★	★★★	★★★
20X	10	X X	X	★	★★
	20	X	★		
	30	★	★★	★★★	★★★
	60	★★	★★★	★★★	
	80	★★	★★★	★★★	★★★
100X	10	X X	X X	X X	X X
	20				
	30	X X	X	X	★
	60	X		★	★
	80	X	★		★★

C. Identifying physical defects within a silicon solar cell by EMMI analysis

EMMI analysis by reverse-biasing a silicon solar cell results in emission hot spots as shown in Fig. 6a and Fig. 7a. Subsequent X-S TEM analysis reveals a sharp shape structure at the location of emission spot as shown in Fig. 6b and Fig. 7b. The sharp shape structure may result in higher electrical field and thus junction leakage or junction breakdown at that location.

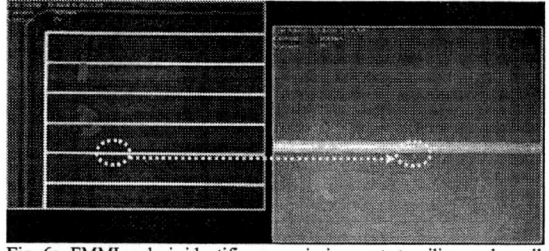

Fig. 6a. EMMI analysis identifies an emission spot at a silicon solar cell on a silicon solar cell.

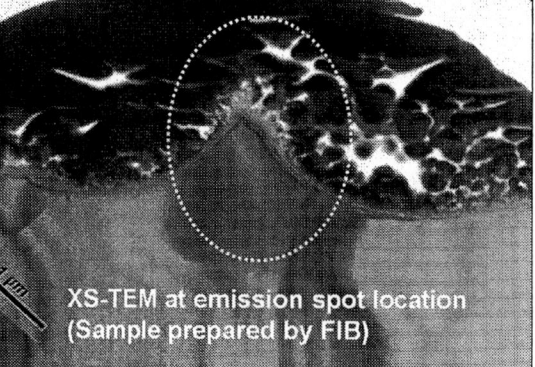

XS-TEM at emission spot location (Sample prepared by FIB)

Fig. 6b. X-S TEM analysis at the emission spot location of Fig. 6a reveals a sharp shape structure.

ACKNOWLEDGMENT

Grateful to TEM team from Ma-tek, who did excellent jobs in preparing TEM samples and providing high quality TEM images.

REFERENCES

[1] Michael Shur, *"Physics of Semiconductor Devices"*, Prentice-Hall, 1990.
[2] T. Fuyuki, *"Luminoscopy" A Versatile Tool for the Diagnosis of Crystalline Silicon Solar Cells Utilizing Electroluminescence*, NAIST, POLYSE 2006.

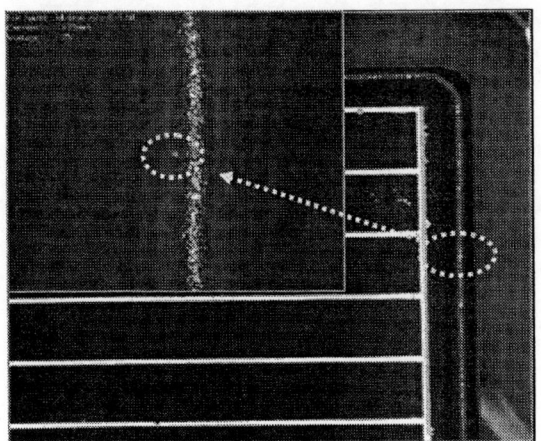

Fig. 7a. EMMI analysis indicates an emission spot near laser edge isolation area of a silicon solar cell.

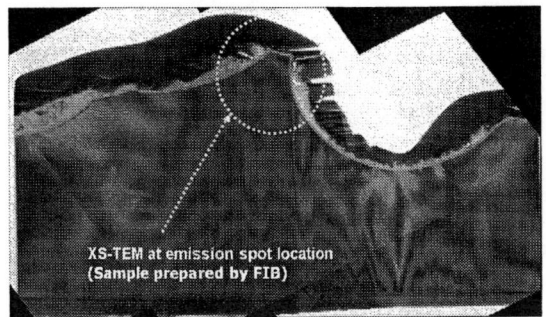

Fig. 7b. X-S TEM analysis at the emission spot location of Fig. 7a reveals a sharp shape structure.

III. SUMMARY

Conventional PFA tools are proven to be useful in analyzing physical structure and component of a silicon solar cell.

EL analysis together with EMMI constructs an effective EFA skill in qualitatively examination of the conversion efficiency of silicon solar cell. It provides a fast way for globally checking on the variation of conversion efficiency of a silicon solar cell. Distribution and intensity of the emission lights during EL analysis reflects variation of efficiency across the whole silicon solar cell. Best image resolution could be achieved by adjusting lens magnification, integration time of EL signal, and forward-biasing voltage.

Conventional EMMI analysis under reverse biased condition is also proven to be useful in identifying junction leakage and edge isolation defects within a silicon solar cell. Sharp shape structure observed at the location of emission spot might result in higher electrical field and thus junction leakage or junction breakdown.

Backside GMR Magnetic Microscopy for Flip Chip and Related Microelectronic Devices

M. Hechtl

Infineon Technologies AG Munich, Germany
Phone +49-89-234-20137, Fax: +49-89-234-713752
email: martin.hechtl@infineon.com

Abstract - **The applicability of GMR Magnetic Microscopy for current paths on flip chip die and interposer is shown. For non thinned samples (200 μm die thickness), interposer currents may be attributed to the interposer signal routing in the 1 mA current regime. With decreasing die thickness the discrimination of on-die currents versus interposer currents becomes possible. Below about 80 μm die thickness, on-die current paths, carrying sub-100μA currents may be identified by comparison with die layout.**

1. INTRODUCTION

Scanning Magnetic Microscopy has become an established failure analysis technique for visualizing current paths in microelectronic devices [1-3]. Due to their high sensitivity, SQUID sensors are particularly suitable to detect weak magnetic fields buried deep inside chip packages, whereas the high resolution of GMR and MTJ sensors is best utilized for tracing current paths by scanning directly on the passivation surface of decapsulated dies [4].

In order to accomplish higher Input / Output density between highly integrated dies and the package and board wiring, flip chip architecture substitutes increasingly the wire bond technique for logic products. As a consequence, the active area of flip chip dies is covered by bump arrays immediately attached to the substrate wiring and is no longer accessible e.g. by GMR or MTJ sensors, if the device has to be fully connected and powered for analysis.

In this work, a systematic backside GMR magnetic microscopy approach is introduced for the first time. For a fully connected device, the resolution of the sensor for the detection of currents in the package and for currents on die level is investigated for different die thickness and current strength.

2. EXPERIMENTAL

GMR sensors utilize the Giant Magneto Resistance effect, discovered in 1988 [5, 6]: A stack of thin ferromagnetic layers, separated by a conducting non-magnetic layer, exhibits a strong dependence of the electrical resistance on the applied magnetic field. One of the ferromagnetic layers has a pinned magnetization, the other is oriented by the external field and acts as sensing layer. The scattering of the conduction electrons passing through the layer stack attains a probability minimum for parallel magnetization of the two ferromagnetic layers and a maximum for antiparallel magnetization.

The measurements shown in this paper were carried out with a GMR sensor integrated in the commercial Magma C30 tool from Neocera Inc. In the failing state of the device with increased current consumption, the GMR sensor is scanning in the x / y plane above the device. Primarily the z (vertical) magnetic field component of the current path is measured. The resulting map of the magnetic field distribution is inverted to a current distribution map by Fourier transformation. By comparison of the current distribution image with the package and die layout, the current path may be identified and the failure may be localized.

The investigated device consisted of a 130 nm technology die, bumped upon a two-layer interposer with 0.5 mm pitch ball grid array. The device was mounted to a socket board. As fixture, a lid was used, which contains a rectangular opening, to enable access to the die backside (fig.1).

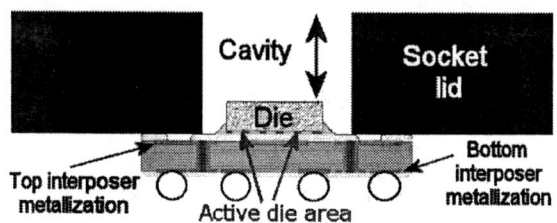

Fig.1: Schematic cross section of a flip chip pressed into a socket by a lid with open window above the die.

3. RESULTS AND DISCUSSION

In a first approach, magnetic microscopy was carried out with a SQUID sensor. Due to the cavity formed by the lid opening and the geometric conditions of the SQUID (cooled to 75K in vacuum containment), an effective working distance of about 1 mm had to be observed. The measurement result shows a broad current loop in the die area of the device (fig.2). This loop however cannot be unambiguously related to the device layout.

978-1-4244-2039-1/08/$25.00 ©2008 IEEE

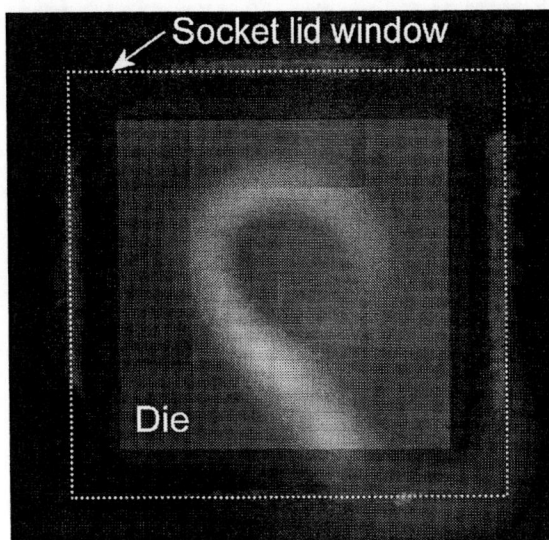

Fig.2: Current distribution detected by the SQUID sensor.

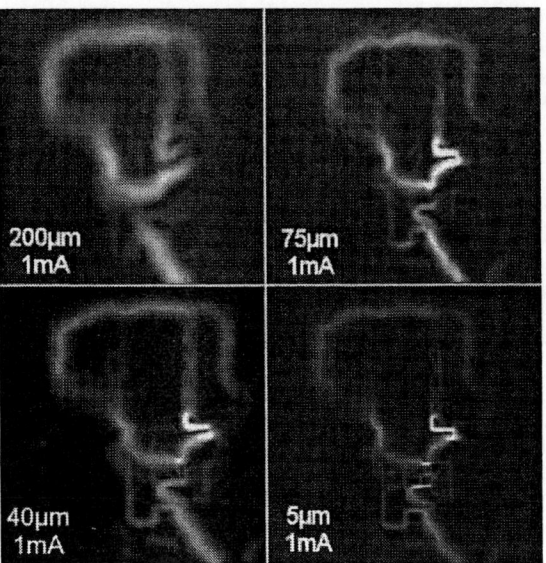

Fig.3: Current path detected by the GMR sensor at a current of 1 mA for different die thickness.

As an attempt to achieve higher resolution, a GMR sensor approach was started. The resolution of a sensor is proportional to

$$\sqrt{4z^2 + w^2} \, ,$$

where z is the sensor-to-current distance and w is the lateral dimension of the sensors active area [7]. The active sensing area of the GMR sensor (about 10nm x 6μm), which is a key factor for the resolution, is much smaller than the SQUID sensor area (circle of about 30 μm diameter). Therefore higher resolution may be achieved with the GMR sensor. Moreover, the sensing GMR thin film layers are deposited on a tip, which is mounted at the end of a cantilever. This tip may be dipped 2.5 mm deep into cavities. Thus the working distance between sensor and sample surface may be reduced to zero and the parameter z corresponds to the die thickness for on-die currents. For the present study, the die thickness was reduced successively by mechanical polishing. Fig.3 shows the current distribution map in the case of a current of 1 mA for different die thickness. The bumped die had an initial thickness of 200 μm. Its current paths appear rather blurred. Nevertheless a large portion of the path structure may already be identified by comparison with the top metallization layout of the interposer (fig.4). Only a short metal line of the bottom interposer layer is involved in the current path. In the current image it appears most blurred, as it is farthest away from the scanning sensor.

With decreasing die thickness (i.e. decreasing scanning distance), all parts of the current path become less blurred. Moreover below about 80μm die thickness, the rectangular x / y directed shape of on-die current paths becomes distinct enough, to be investigated by CAD signal path analysis. For a die thickness of 5 μm, the quality of the on-die current path representation corresponds to GMR sensor images, received by conventional front side scanning upon the passivation of wirebonded dies. The present backside scanning technique, however, not only reveals on-die currents in high resolutions, but also interposer current paths.

Fig.4: Top layer (left image) and bottom layer (right image) of the interposer. Blue and red metal lines are VSS and VDD lines respectively, involved in the current flow. (Small red dots: Bumps; open purple circles: Interposer vias; big black open circles: Balls)

For the device thinned to 5 μm residual die thickness, the dependency of resolution on current strength was investigated (fig.5). For 1 mA, the current path elements resemble distinctly the shape of the corresponding interposer and on-die metal lines. With decreasing current, all parts of the current path become increasingly blurred. Below 100 μA total current, some of the weaker on-die current paths become invisible. At about 30 μA, package current paths begin to vanish in the noise due to their high distance of about 80 μm from the sensor.

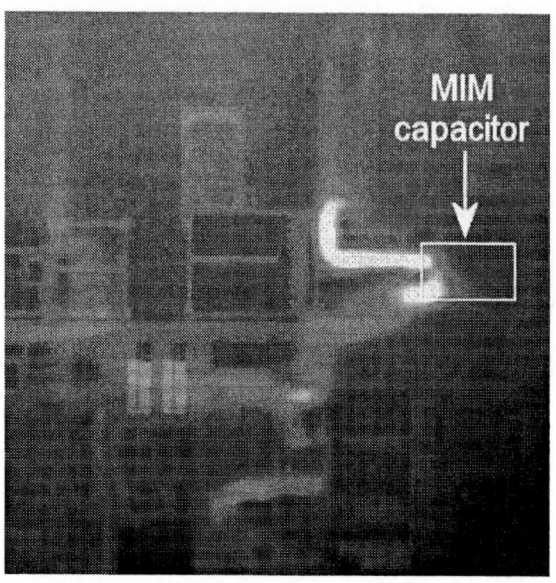

Fig. 6: Detail scan of the current path at the defect MIM capacitor. The current image is overlaid with the IR optical image.

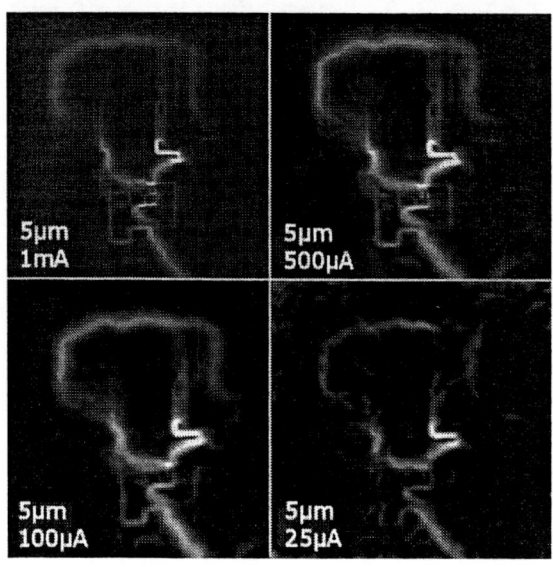

Fig.5: Backside GMR image for decreasing current.

Only two on-die metal lines carrying the highest current remain sharply visible. Actually these two metal lines are connecting the top and bottom plate of a MIM capacitor, located between metal 5 layer and metal 6 layer of the die. The current flow between the plates clearly indicates a dielectric breakdown at the capacitor (fig.6).

SUMMARY

The present case study shows, that for flip chip analysis the backside approach, which is already being used e.g. for emission microscopy, for laser based techniques and for layout modification by FIB, may also be successfully applied to magnetic microscopy with a GMR sensor.

Backside magnetic microscopy visualizes the current distribution in a flip chip device. The resolution is sufficient, to attribute current paths to the package layout. For thinned flip chip dies, on die current paths may be identified and analysed with respect to failure localization.

REFERENCES

[1] L.A. Knauss, S.I.Woods and A. Orozco, "Current Imaging using Magnetic Field Sensors, Microelectronics Failure Analysis Desk Reference", 5th Edition, 2004. Edited by ASM International. ISBN: 0-87170-804-3.

[2] A.Orozco, "Fault Isolation of High Resistance Defects using Comparative Magnetic Field Imaging", Proc. ISTFA, p. 9 (2003).

[3] M. Hechtl, G. Steckert and C. Keller, "Localization of Electrical Shorts in Dies and Packages using Magnetic Microscopy and Lock-In Thermography", Proc. IPFA, p. 252 (2006).

[4] O. Crepel, P. Poirier, P. Descamps, R. Desplats, P. Perdu, G. Haller, A. Firiti, "Magnetic Microscopy for IC Failure Analysis: Comparative Case Study using SQUID, GMR and MTJ Systems", Proc. IPFA, p. 45 (2004).

[5] M.N. Baibich , J.M. Broto, A. Fert, F. Nguyen Van Dau, F. Petroff, P. Eitenne, G. Creuzet, A. Friederich, and J. Chazelas, "Giant Magnetoresistance of (001)Fe/(001)Cr Magnetic Superlattices", Phys. Rev. Lett. **61**, p. 2472 (1988).

[6] G. Binasch, P. Grünberg, F. Saurenbach, and W. Zinn, "Enhanced magnetoresistance in layered magnetic structures with antiferromagnetic interlayer exchange", Phys. Rev. B **39**, 4828 (1989).

[7] B.D. Schrag, M.J. Carter, X. Liu, J.S. Hoftun and G. Xiao, "Magnetic current imaging with tunnel junction sensors", Proc. ISTFA, p. 13 (2006).

Chemical and Physical Characterization Techniques in Highlighting Intermetallic Compound (IMC) Formation

Jean Carla M. Fernandez
Fairchild Semiconductor Philippines, Incorporated
MEZ 1, Lapu-Lapu City, Cebu, Philippines
Phone: (6332) 340-0534 Fax: (6332) 340-0557 Email: jean.carla.fernandez@fairchildsemi.com

Abstract - **Evaluations and qualifications on ball attach process for bump packages, studies on solderability between lead to PCB on leaded packages, wire bond integrity evaluations, these are studies that involves thorough characterization on bonds and intermetallic compound (IMC) layers for different interfaces. For ball attach process in bumped packages, we focus on intermetallic formation between solder ball to the under bump metal (UBM). For solder integrity, studies focus on the lead post to solder to PCB interface.**

Intermetallic Compound (IMC) formations between interfaces, is valuable in characterizations. The formation of intermetallics not just cause good integrity on the interface or bond, but can also cause assembly related problems. Example, intermetallics of Gold and Aluminum are a significant cause of wire bond failures in semiconductor devices and other microelectronic devices.

In analyzing intermetallic compound formations, there is a need for the IMC layer to be highlighted, for interface layer difference, for measurement on the thickness, and to merely highlight. This paper aims to present a study on the manual techniques in highlighting Intermetallics. Chemical etching, delayering, and immersion on chemical solutions, done on different types of samples to highlight intermetallic compound (IMC).

1.0 INTRODUCTION

Requests for analysis were received at FA, for analysis on the Intermetallic compound layers. Evaluations and qualifications on ball attach process for bumped packages, solderability test failure for leaded packages, some concerns were raised on intermetallic formation if it could have attributed to the failure on board-mounting. There were wire bond related issues that needed IMC layers characterization. With the need for thorough analysis on intermetallic formation, techniques were formulated to highlight intermetallic layers for more feasible analysis. Potassium Hydroxide (KOH) is a known chemical compound used to highlight intermetallic formation on Gold (Au) wire to Aluminum bond pad. This technique had also been established for intermetallic formation highlighting on gold wire ball bond to Aluminum bond pad, for non-cross-sectioned units. This technique is not applicable for some other packages with bumps, solder balls, and for different interfaces to highlight. Research was done on specific chemicals to use on highlighting IMCs. And experiments were done on formulating specific etching processes using the specific chemicals. Process rate could be different for every

ratio on the chemical solution. Experiments were done using two different ratio of HCl + Methanol solutions. **1 HCl: 9 Methanol and 1 HCl: 4 Methanol.**

This paper was formulated for highlighting intermetallic compound formation on cross-sectioned units for ball to UBM to solder interface, lead post to solder, wire to solder, die to die attach material, and any other interfaces, using HCl + Methanol.

2.0 EXPERIMENTAL SECTION

2.1 Materials

2.1.1 Chemicals

   a. Hydrochlouric Acid (HCl)
   b. Methanol
   c. De-ionized (DI) Water

2.1.2 Materials and equipments

   d. beakers
   e. plastic droppers
   f. graduated cylinders
   g. vinyl hand gloves
   h. disposable paper / wipes
   i. Techprep Precision Polisher
   j. fume hood

2.1.3 After etching

   a. Optical microscope capable up to 1000x magnification
   b. Polaron Sputter Coater (10seconds coating time)
   c. Scanning Electron Microscope
   d. Elemental Dispersive X-ray (connected to the SEM)

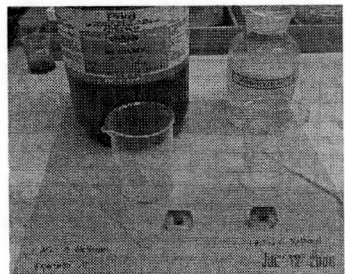

Figure 1: Experimental Set-Up
for the chemical solution

2.2 Procedure

Good and failed units were cross-sectioned using the Techprep Precision polisher, focusing on the interfaces at which intermetallic compound (IMC) formation needs to be highlighted and further characterized. Cross-sectioned units were inspected under high magnification microscope; photos were taken on the interface of interest.

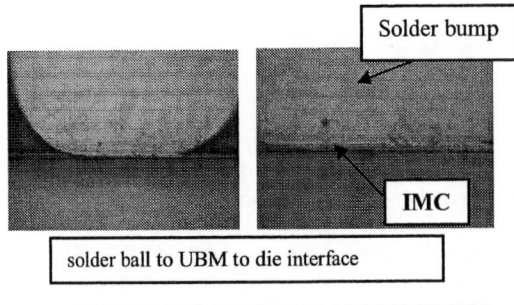

Solder bump

IMC

solder ball to UBM to die interface

lead post to solder interface

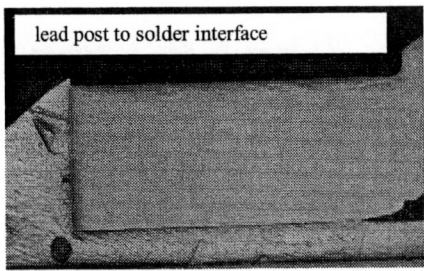

Figure 2. Optical photos of cross-sectioned units before etching.

All chemical processes were made in FA Wet Laboratory using the fume hood. Chemical solutions were then prepared. **1 HCl: 9 Methanol** and **1 HCl: 4 Methanol.** The solutions were placed in beakers that were labeled.

Case Study 1: BGA Ball Attach Process Evaluation

First experiment was done on BGA samples, for ball attach evaluation. Units were then soaked / immersed into the beakers, one for each chemical solution. Times were set for each chemical solution to etch layer under the one with IMC. Different time intervals were used to take the samples from the solution and inspect using optical microscope, to check if the solder had been totally etched out (for the BGA sample). If so,

the units were then cleaned using dampened cotton buds. The units were placed under a lamp to dry up. Units were then subjected to Scanning Electron Microscope (SEM), for inspection and IMC thickness measurement. 1 part HCl + 4 parts Methanol showed faster result in highlighting IMC.

Case Study 2: Solderability test failure on leaded packages

For the second experiment, there was a request on characterization of intermetallic compound formation on leaded packages, solderability test failure was observed. This was used for the second experiment in highlighting intermetallic compound (IMC) formation.

The etchant was prepared. This etchant is to highlight intermetallics, 1 part HCl + 4 parts Methanol solution was placed in a glass beaker. Cotton bud was then used to suave / brush the etchant onto the cross-sectioned units. Only four times one stroke brushing was used. After etching, cross-sectioned units were dried. Etched samples were then inspected under the optical microscope, optical photos were taken. Observed solder was roughened and IMC was highlighted. Unit was further dried up before coating.

Figure 3: Optical photo of lead post to solder interface showing IMC layer highlighted after etch. Arrows point to the IMC layers. (@ 100x magnification)

Sputter coater was set to 10 seconds coating. This is so that the IMC layers will be clear in analysis, not thickly covered by Platinum (sputter coater metal). After total drying, unit was subjected into SEM for interface check and elemental analysis on the IMC layers. SEM setting: magnification = 2,000x to 3,500x, Working Distance = 15, Accelerating voltage = 15kV. Line Scan and elemental mapping EDX analysis were used in analyzing the IMC layers.

3.0 RESULTS AND DISCUSSION

Figure 4: SEM photos of cross-sectioned bumped package units after etching with 1 part HCl + 4 parts Methanol.

Summary of results for Case Study 1 is shown in Table 1 as follows:

1 part HCl : 9 parts Methanol	1 part HCl : 4 parts Methanol
5 minutes	1 minute
10 minutes	1 minute
10 minutes	2 minutes
15 minutes	5 minutes
30 minutes	
Total = 70 minutes	**Total = 9 minutes**

Table 1: List of time intervals for the two different chemical solutions. 1 part HCl + 4 parts Methanol showed faster result.

FA technique on highlighting IMC formation was done. IMC formation was highlighted after soaking cross-sectioned units into HCl + Methanol solution. Soaking time is faster using **1 part HCl : 4 parts Methanol** solution. Intermetallic (IMC) formation between balls to under bump metal (UBM) is one aspect to look at to check on the integrity of the attached ball.

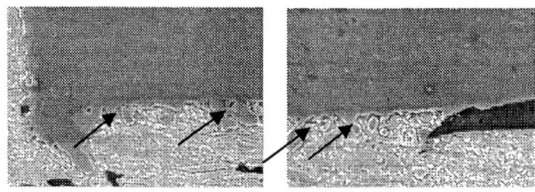

Figure 5: SEM photo of lead post to solder interface showing IMC layer highlighted after etch. Arrows point to the IMC layers. (@ 100x magnification)

Same chemical solution was used for the leaded package. 1 part HCL + 4 parts Methanol. IMC was highlighted by slightly brushing the chemical solution onto the sample using cotton bud. Four times single stroke brushing was used.

4.0 CONCLUSION

Chemical formulation on highlighting IMC formation was established. IMC formation is highlighted using HCl + Methanol solution. Two methods can be done to use the etchant, are soaking method and brushing the solution using a cotton bud, unto the cross-sectioned units. Soaking time is faster using **1 part HCl: 4 parts Methanol** solution. And for brushing, use one stroke and four to ten times brushing. Methods would depend on the sample at which the intermetallic formation be focused.

5.0 RECOMMENDATIONS

It is recommended to use the established chemical solution and method for highlighting IMC on any other packages in FSCP. Intermetallic compound (IMC) formation is one aspect to look at to check on the integrity between specific interfaces. This technique in highlighting IMC should be used. Once established for any other packages, it is recommended to study the use of software such as IMAQ (Image Acquisition) in further characterization and analysis of the highlighted IMC.

ACKNOWLEDGMENT

The author acknowledges QA and R Department Manager, Lester O. Uy, for the encouragement and support, to all the FA Staff for the help and support.

REFERENCES

1) http://en.wikipedia.org/wiki/intermetallics

2) Young and Freedman, "University Physics" ©1998

3) http://semiconfareast.com

4) Thomas Schaffner, "Chemical and Physical Characterizations on Semiconductor Devices" Training Course

5) Morphological Analysis of the Intermetallics in Solder Substrates during Board Level Temp Cycle of MOSFET Units. V. Abellana, J.C. Fernandez, Ma. C Estacio, IERC 2008

ABOUT THE AUTHOR

Jean Carla M. Fernandez is a Failure Analysis Engineer who had been in Fairchild for 4 yrs now. She is a BS in Physics graduate at the University of San Carlos, Technological Center. Currently, she is taking up her Masters of Science degree in Physics.

Influence of Interconnect Dimensions on Electromigration for Cu/Low-k Interconnect Structure: An Analytical Study

Bhavana N. Joshi and A.M. Mahajan*
Department of Electronics, North Maharashtra University, Jalgaon, [M.S.], India
Ph. +91-257-2257476, Fax +91-257-2258403/406
*e-mail: ammnmuj@yahoo.com

Abstract:-**The interconnect capacitance for the deposited low –k material is observed to be lower compared to SiO₂. It reveals from the study that, the MTTF of Cu/ deposited low-k structure is slightly lower than that of Cu/SiO₂ structure due to low thermal conductivity of deposited material.**

I. INTRODUCTION

The scaling of device dimensions increases the functionality and speed of Integrated Circuits (IC's),but with scaling performance of IC get affected by interconnect Resistance – Capacitance delay (RC delay). Thus, the IC technology is engaged in finding new materials for interconnects and dielectrics that will replace the Al/SiO₂, with low resistivity metal/low dielectric (low-k) structures to reduce the problems like RC delay, crosstalk and power consumption prolifically with the miniaturization of the devices [1,2]. But with new materials different issues viz. Electromigration (EM), Time Dependent Dielectric Breakdown (TDDB) etc. escalate which are the concerns for the reliability of devices for longer lifetimes [3, 4].

Electromigration was first identified as failure mechanism in IC in late 1960s [5, 6] and now it has become an important reliability issue related with interconnect/dielectric structure due to the small cross-sectional area of the interconnects used in VLSI/ULSI circuits, the current density at normal operating condition is extremely high and as a consequence the EM induced mass transport may occur through diffusion of metal in dielectric [7]. This mass transport during EM turn out into accumulation of vacancies or atoms, creating voids or hillocks in interconnects giving rise to parametric failure. The hillock formation in the interconnection results in short circuit between adjacent interconnects. The EM induced mass transport is expedited when the temperature of an interconnection is increased. When current passes through the metal lines, thermal energy gets generated because of the collision between electrons and metal atoms. At a lower current density, the generated thermal energy immediately sinks, but at higher current densities joule heating takes place that causes the major failure mechanism of EM. The problem of EM can be minimized by using Cu metal. The Cu metal used for interconnect has lover resistivity and have higher melting point [8-10] than that of Al, due to this it is predicted to have better EM performance. Hence, in the present scenario to satisfy the demand of miniaturization, less resistance–capacitance delay

and cost, the copper metallization has been utilized instead of Al as interconnects along with SiO₂ and most recently with low k materials.

In late 1960's Black carried out set of EM experiments and given widely used Black's empirical equation [11] that relates the median time to failure (MTTF) with respect to the current density and temperature of the metal interconnects. It reveals from literature survey [12] that at elevated temperature device performance degrades due to increased junction leakages and reduced reliability of interconnects because of accelerated EM induced failure mechanisms. According to International Technology Roadmap for Semiconductors 2001 (ITRS), interconnect must bear a constant current density of $1.4-3.7 *10^6$ A/cm² at reference temperature 105°C for 100 nm and 50 nm technology respectively. Due to self heating of interconnect with low-k the temperature of interconnect increases above reference and gives EM because of diffusion of metal ions in dielectric. Thus, the accurate estimation of interconnect temperature is necessary for assessment of interconnect performance and reliability in high speed, multifunctional VLSI/ULSI circuits and the current densities and interconnect temperature should be kept below certain limits to ensure desired lifetime. In the present paper, we are reporting analysis of interconnect capacitance, EM - Median Time To Failure (MTTF) and interconnect temperature for different interconnect dimensions with deposited low-k material. The interconnect material selected here is Cu as it has become important and widely used interconnect in VLSI/ULSI devices.

II. MATHEMATICAL APPROACH

In present work, the dielectric constant of deposited material [13] is used to determine the interconnect capacitance. The dielectric constant of deposited material is 3.1. The interconnect capacitance (Total capacitance) is determined using the equations given by Shyh-Chyi Wong et.al [14]. The parameters of interconnect dimensions used here is being applicable for 45 nm technology node. For estimation of interconnect capacitance the wire thickness is varied from 0.05 to 0.5 micron.

The concept of Schafft [15] is used to study the effect of interconnect dimensions and material s effect on temperature rise of interconnects. The temperature profile of interconnect is

978-1-4244-2039-1/08/$25.00 ©2008 IEEE   181   *Proceedings of 15th IPFA - 2008, Singapore*

studied at different current densities with different wire thickness using the following equation

$$Tc = Ts + \frac{A}{B-C} - \frac{D}{E} \qquad (1)$$

where, Ts is substrate temperature and A,B,C and D are the structural parameters based variable. The reference temperature used here for lower and upper dielectric materials around the metal is 180 and 200°C respectively. The interconnect temperature rise for low-k and SiO_2 material is compared for higher current densities as it is important part of study future technology. The MTTF is estimated for the Cu/SiO_2 and a Cu/low-k structure includes the Joule heating effect. The current density [16] shows dependence on interconnect capacitance, which ultimately depends on interconnect dimensions. Hence, analytical study has been carried out for the interconnect capacitance and MTTF at lower wire thickness for Cu/SiO_2 and Cu/low k based structure.

III. RESULTS AND DISCUSSION

It is observed from fig.1 that, interconnect capacitance (Total Capacitance) of deposited material decreases to 1.1 pF which is less as compared to that of SiO_2 (1.4pF) with shrink wire thickness of 0.1micron. Thus, the deposited material has advantage of reducing the RC delay with miniaturization, which results in enhanced performance and speed of IC.

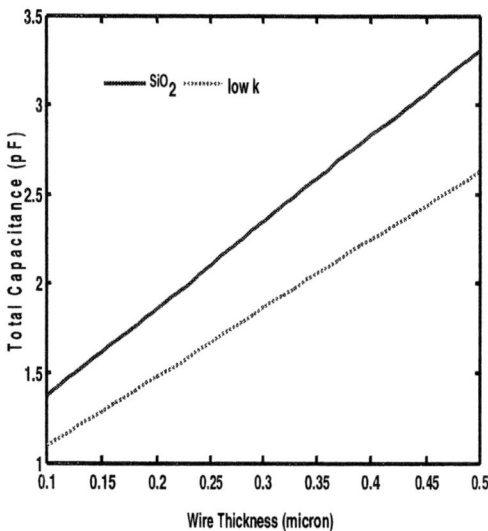

Fig. 1: Interconnect Capacitance Vs Wire Thickness

For reliability assessment the rise in interconnects temperature due to Joule heating is carried out with the help of equation 1. The temperature profile for Cu/low-k with variable wire thickness for different current densities is shown in fig.2. It clearly reveals from figure that up to $11*10^5$ A/cm^2 current density there is not much rise in interconnect temperature but with further increase in current densities above $21*10^5$ A/cm^2

the interconnect temperature increases more than the reference temperature 200°C rapidly, inducing the major concerns for EM effect. It reveals from figure that with lowering in wire thickness the temperature rise in interconnect gets lower. The effective thermal conductivity of deposited low-k material is $0.174*10^{-2}$ W/cm-K.

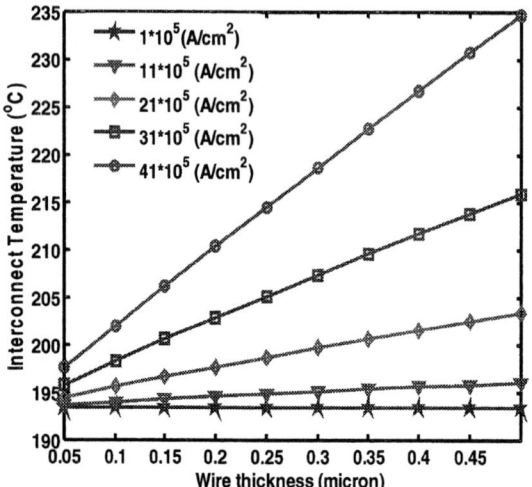

Fig. 2: Interconnect Temperature Vs Wire Thickness for low-k at different current densities

The parameters for SiO_2 and low-k used for this analytical study are presented in table 1. The interconnect dimensions for both dielectric materials are kept same. However, its physical parameters are different such as dielectric constant and thermal conductivity. The interconnect material used for this study is Cu having the resistivity $2.0*10^{-6}$ohm-cm and the temperature coefficient of resistivity of Cu is $6.8*10^{-3}$C^{-1}.

Table 1

Parameters details

Parameters	SiO2	Low-k
Wire width	$0.10*10^{-4}$	$0.10*10^{-4}$
Dielectric thickness	$0.175*10^{-4}$	$0.175*10^{-4}$
Interwire Spacing	$0.15*10^{-4}$	$0.15*10^{-4}$
Current density	$1*10^5$-$41*10^5$ A/cm^2	$1*10^5$-$41*10^5$ A/cm^2
Thermal conductivity	$1.31*10^{-2}$ W/cm-K	$0.174*10^{-2}$ W/cm-K
Dielectric constant	3.9	3.1

The fig.3 shows the comparative rise in interconnect temperature for Cu/low-k structure at higher current density than that of Cu/SiO_2 structure. However, the temperature rise in Cu/low-k and Cu/SiO$_2$ is very negligible. Thus it is clear that the temperature rise for low-k at higher current density is more compare to SiO$_2$ is because of lower thermal conductivity of low-k.

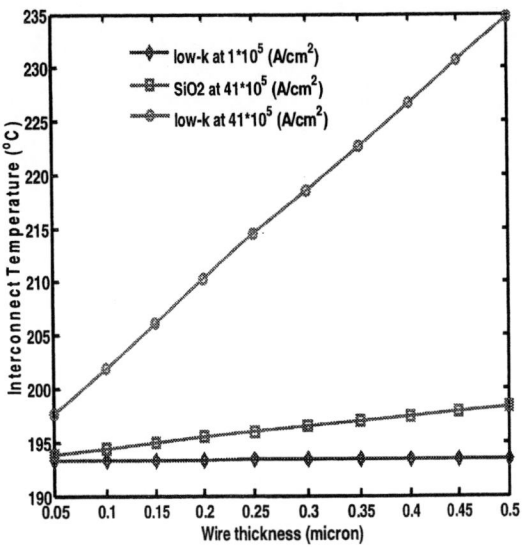

Fig.3: Interconnect Temperature Vs Wire Thickness for low-k at and SiO2 at lower and higher current densities

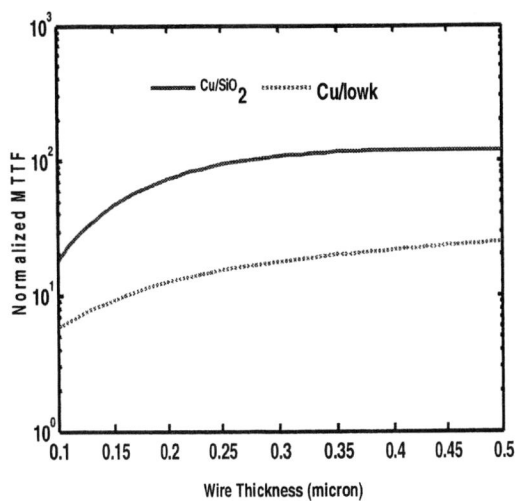

Fig. 4: MTTF Vs wire thickness

The temperature profile study has been further extended for the estimation of MTTF for two interconnect/dielectric structures viz, Cu/SiO$_2$ and Cu/ Low-k. It appears from the figure 4 that, the MTTF for Cu/low-k structure is slightly lower compared to Cu/SiO$_2$ and decreases with shrinking in wire thickness due to lowering in thermal conductivity. Thus, the low k dielectric material deposited by us [13] seems to be preferable for obtaining a compatible reliability in fabricating the ULSI devices.

IV. SUMMARY

The interconnect capacitance for the deposited low –k material is observed to be lower compared to SiO$_2$. The MTTF is determined for Cu/ deposited low-k structure with variation in metal thickness from 0.1 to 0.5µ. It is observed that, the MTTF of Cu/ deposited low-k structure is slightly lower than that of Cu/SiO$_2$ structure due to low thermal conductivity of deposited material. It reveals from this study that, not only materials but the geometry and dimensions of interconnects are also an important issues to determine the EM MTTF. These results can also be used for enhancing backend thermal reliability management performance and reliability for interconnect thermal awareness in interconnect design issues such as power and ground distribution network. On the basis of analytical study of interconnect capacitance and MTTF it can be concluded that, the material and physical parameters utilized in this paper will enhance the efficiency and reliability of the ULSI devices.

REFERENCES

[1] Michael Morgen, E. Todd Ryan, Jie-Hua Zhao, Chuan Hu, Taiheui Cho, and Paul S. Ho, (2000) Annu. Rev. Mater. Sci. 30 pp. 645.

[2] K.Maex, M. R .Baklanov, D. Shamiryan, F. Iacopi, S. H. Brongersma, Z. S. Yanovitskaya, J. Appl. Phys 93,11, (2003), pp. 8793-8841

[3] Guido Groeseneken, Robin Degraeve, Ben Kaczer and Philippe Roussel, Proceedings of 14th IPFA, Bangalore. India (2007), pp. 1-9

[4] Cher Ming Tan, Arijit Roy, Materials Science and Engineering R 58 (2007) pp. 1–75

[5] J. Joseph Clement, IEEE TRANS. on Device and Materials Reliability, Vol. 1, No. 1, (2001), pp. 33-42

[6] H. B. Huntington and A. R. Grone, J. Phys. Chem. Solids, 20, (1961), pp. 76

[7] Oliver Aubel , Wolfgang Hasse , Martina Hommel , Heinrich Koerner, Microelectronic Engineering 82 (2005) pp. 600–606

[8] J.R. Lloyd, J. Clemens, R. Snede, Microelectronics, Reliability, 39 (1999), pp. 1595-1602

[9] Jiang Tao, Nathan W. Cheung, Chenming Hu, IEEE Electron Device Letters, Vol. 14, No. 5, (1993), pp. 249-251

[10] C.-K. Hu, B. Luther, Materials Chemistry and Physics, 41 , (1995), pp.1-7

[11] J. R. Black, IEEE Trans. Electron Devices, vol.-16, (1969), pp. 338-347

[12] R. Streiter, H. Wolf, Z. Zhu, X. Xiao, T. Gessner, Micrelectronic Enginerring, 601 (2002), pp. 39-49

[13] B. N. Joshi, A. M. Mahajan, Optoelectronics and advanced materials – Rapid communications Vol. 1, No. 12, (2007), pp . 659 - 662

[14] Shyh-Chyi Wong, Gwo-Yann Lee, and Dye-Jyun Ma, IEEE TRANS. on Semiconductor Manufacturing, Vol. 13, No. 1, (2000), pp.108-111

[15] Schafft HA, IEEE Trans Electron Dev, 34,(1987), pp.664.

[16] Xiaojun Li, Ph.D. Thesis University of Maryland, (2005)

A Novel Pseudo Tri-Gate Vertical MOSFET with Source/Drain Tie

Jyi-Tsong Lin, Ying-Chieh Tsai, Yi-Chuen Eng, Shiang-Shi Kang,
Yi-Ming Tseng, Hung-Jen Tseng and Po-Hsieh Lin
Dept. of Electrical Engineering, National Sun Yat-Sen University
70 Lien-Hai Rd. Kaohsiung 80424, Taiwan ROC
Phone: (886) 7-5252000-4122 Fax: (886) 7-5254199
Email:jarvis0419@gmail.com

Abstract - **This paper investigates the device behaviours of a pseudo tri-gate ultra-thin-channel vertical MOSFET with source/drain tie. For comparison two transistors are designed. According to the 2D simulation, our proposed structure can effectively enhance the drain current and the thermal stability, mainly due to the ultrathin channel (Tsi = 10 nm). The fabricated device have very low subthreshold swing near 60 mV/dec with channel length 40 nm to 90 nm and excellent G_M of 4 mS/μm with channel length 35 nm owing to its unique features, when compared to its counterpart. Also, the respective discontinuous buried oxide under the channel and the source/drain regions can construct a natural source/drain tie to overcome short-channel effects and self-heating effects as well.**

I. INTRODUCTION

In order to meet the requirement of international technology roadmap for semiconductors (ITRS), process improvement and reliability issue are two key challenges for a scaled-down high-performance and low-power device. As the device's gate-length continues to scale down, the issue of self-heating is incapable of being avoided in silicon-on-insulator (SOI) based devices [1]. Recently, some vertical devices have been proposed to relax these issues and achieve low-cost requirement [2]. Unfortunately, vertical MOSFET also need an ultra-thin body to solve the short-channel effects [3]. Scaling the silicon-film thickness is able to improve the short-channel performance for FD SOI MOSFETs [4]. Alternatively, a quasi-SOI MOSFET with π-shaped semiconductor conductive layer (π-FET) has been proposed to solve these issues [5] but it is not employed for a VMOS. Here, we modify the process of the π-FET to develop a new vertical pseudo tri-gate structure with source/drain tie (S/D tie). Because of the pseudo tri-gate scheme and its S/D tie, the proposed structure can dramatically improve the current drive while still maintaining the desired properties. Also, mainly due to the source/drain-tied scheme, the thermal stability can also be improved, resulting in an improved long-term reliability.

II. PROCESS DESIGN FOR THE TRI-GATE AND S/D TIED FET

ISE-FLOOP is used to verify the processes. The processing schematic diagrams of the π-shaped vertical MOSFET are shown in Fig. 1. First step, the oxide and the nitride layers are deposited on a bulk Si-substrate. Then, the traditional lithography and plasma etch are used to etch the nitride and oxide layers (Figure 1.(a)). Second step, a 10 nm thick nitride is

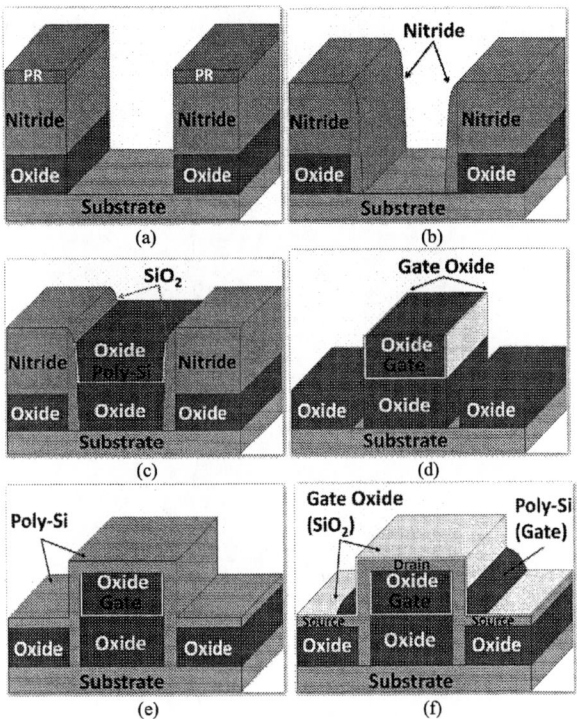

Fig. 1. Key fabrication steps of the π-shaped vertical MOSFET. (a) Nitride deposition on oxidized Si wafer, (b) spacer nitride formation, (c) oxide and poly-Si are deposited and planarized by CMP, (d)-(e) stripping away the nitride, than the π-shaped active region with S/D-tied are formed by polysilicon deposition, (f) the sidewall-gate spacers are formed after the formation of the thin-gate oxide by thermal oxidation.

deposited and etched using a dry-etch process (Figure 1.(b)) in order to contact substrate. After the formation of the nitride spacer, the buried oxide is deposited, planarized by chemical mechanical polishing (CMP), and wet-etched in the trench. Then, a thin 2-nm gate oxide is deposited. After that, a poly-Si mid-gate layer is formed by deposition, planarization with CMP, and then wet-etching sequentially (Figure 1.(c)). The poly-Si mid-gate is then implanted by arsenic ion (Arsenic, 3e15cm^{-2} dose, 8 KeV energy) before a 20-nm-thick oxide layer is deposited and planarized to perform the trench filling in order to reduce the drain – mid-gate overlap. After we stripping away the nitride from the wafer (Figure 1.(d)), the pseudo tri-gate and S/D tie MOSFET can be produced as follows: the active region is formed by depositing a poly-Si layer (Figure 1.(e)), then the polysilicon sidewall-gate spacers are formed after the formation of the thin gate oxide by a thermal oxidation on the active polysilicon layer (Figure 1.(f)).

III. CHARACTERIZATION AND DISCUSSION

We will compare the new device with the other two transistors as listed in the following. Fig 2. to Fig. 4. show the cross-section views of the pseudo tri-gate with S/D-tied VMOS, the double-gate VMOS with S/D-tied, and the double-gate VMOS, respectively.

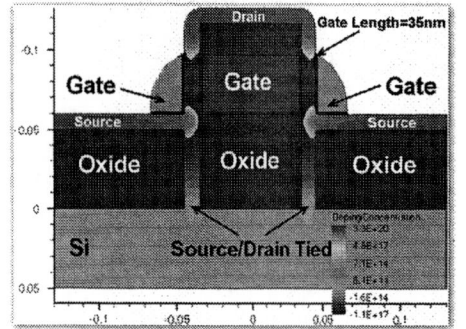

Fig. 2. Schematic cross-section of the pseudo Tri-Gate VMOS with S/D tie.

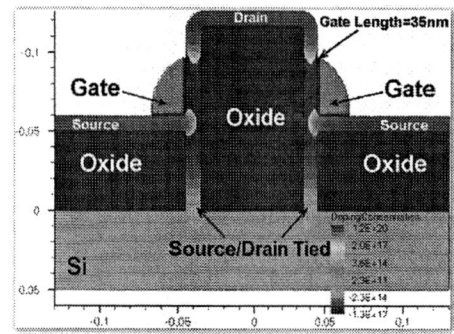

Fig. 3. Schematic cross-section of the D-G VMOS with S/D tie.

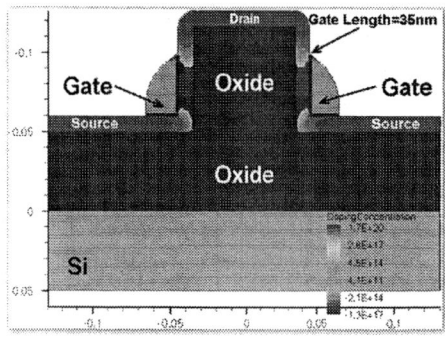

Fig. 4. Schematic cross-section of the D-G vertical MOSFET.

However, the influence of the channel surface orientation on the MOSFET performance should be carefully examined especially in the case of an ultrathin channel. As Si film becomes very thin, lucky hot-carriers may be injected into buried-oxide (BOX) and back interface may sustain damage during hot-carrier (HC) stress at front-channel [6-7]. Here, we mainly discuss the S/D tied for thermal stability and the influence of pseudo tri-gate on the drain current in the same condition of the S/D doping concentration of 3e20 cm^{-3} (3e15cm^{-2} dose, 8 keV energy). Beside, we target a body doping of 1e17 cm^{-3} (1e13 cm^{-13} dose, 1.2 keV energy) for a fully depleted operation with channel length of 35nm.

Fig. 5 shows the drain-induced barrier lowering (DIBL) as a function of gate length. The DIBL is calculated from the difference between the linear threshold voltage (at $V_{DS}=0.05V$) and the saturation threshold voltage (at $V_{DS}=1.0V$) in the subthreshold curves. The new device has low DIBL because of shielding of the drain-source strong electric field between the four vertical channels. It shows the best result of DIBL compared with the other VMOS transistors. The reason is that the drain electric field encroachment to the channel region can be well controlled by using the pseudo tri-gate scheme.

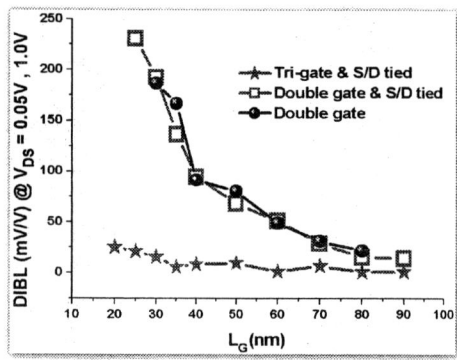

Fig. 5. DIBL as a function of gate length (L_G).

Fig. 6 shows that our proposed structure can achieve the highest ON/OFF current ratio among them. It retains high current drive while reducing short channel effects significantly due to the large gate-to-gate strong coupling in the ultra-thin channel. The subthreshold leakage current of our pseudo tri-gate device is significantly reduced because the punchthrough and DIBL are effectively improved compared to those of the other VMOS device.

Since the short-channel effects are significantly suppressed, the Subthreshold swing (SS) of our pseudo tri-gate S/D-tied VMOS are improved as shown in Fig. 7. Only a small gate-voltage variation is needed to achieve a ten-time drain current. That is why our proposed structure can exhibit the desirable ON/OFF current ratio. In addition, the 3rd gate also helps to turn the device more on and off easily than the other devices do.

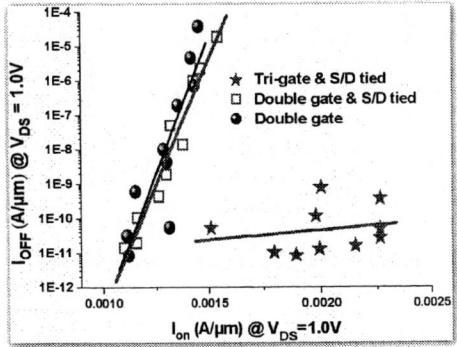

Fig. 6. I_{OFF} - I_{ON} characteristics of the three devices.

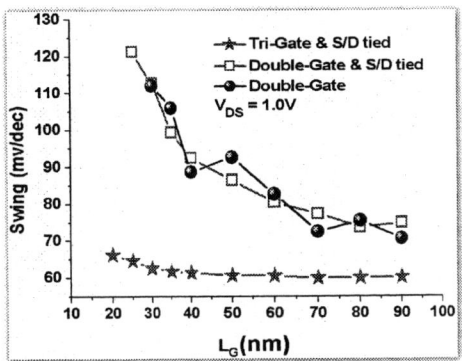

Fig. 7. Subthreshold swing as a function of gate length (L_G).

Fig. 8 shows the G_M characteristics. The same phenomenon and result of the tri-gate S/D-tied VMOS can also be achieved due to the same reason as discussed above. At the low drain bias, three types of transistors show similar results and the pseudo tri-gate one is still the best one. Also, at the high drain bias, the G_M of the pseudo tri-gate one shows very high $G_{M,MAX}$ because the tri-gate four-channel structure is manifested to gain a high desired G_M.

Fig. 9 shows the lattice temperature profiles along the channel to the substrate. For respective 20, 40, and 80 nm gate lengths, three types of vertical transistors clearly exhibit that the S/D tie can reduce the lattice temperature significantly. Moreover, the short-channel device having higher drain current implies that it has higher lattice temperature than the long channel devices. We also found out that the S/D-tied devices shows lower lattice temperature profiles than that of non-S/D-tied device. Hence, we can conclude that the S/D tie indeed can improve the thermal stability of the device.

Due to the presence of the S/D tie and the pseudo-trigate four channels, our proposed structure can get highest drain saturation current among the three structures as shown in Fig. 10. Furthermore, the negative output conductance is also alleviated by the S/D tie. However, because of very high drain current, the small negative output conductance is still observed in the I_D-V_D curve of the pseudo-trigate structure. While on the contrary, the

non-S/D-tied devices are incapable of reducing the self-heating effectively at all, which leads to an evident negative output conductance.

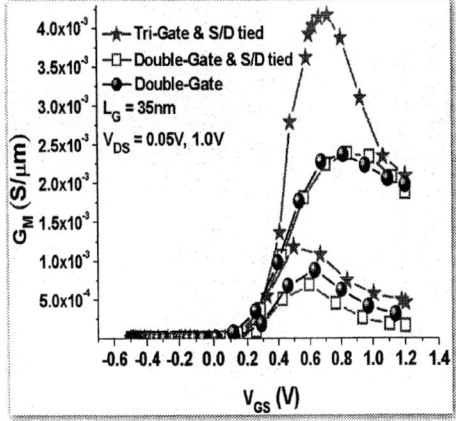

Fig. 8. Comparison of G_M characteristics of the three devices.

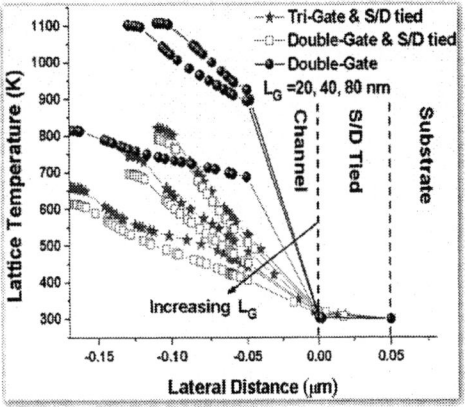

Fig. 9. Lattice temperature (K) profiles of the three devices (V_{GT} = 1.2 V).

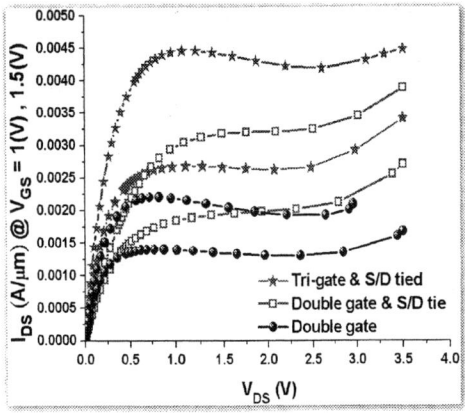

Fig. 10. I_D-V_{DS} characteristics of the three devices.

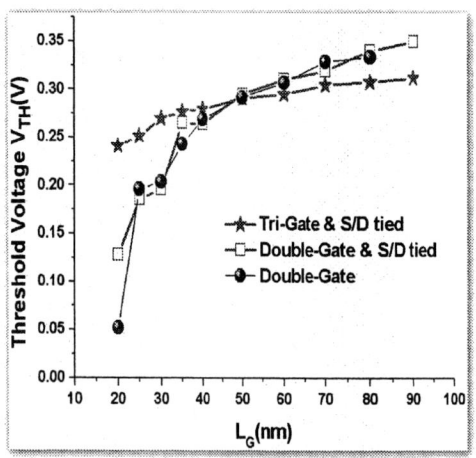

Fig. 11. Threshold voltage characteristics of the three devices.

Fig. 11 shows the SCE of the three devices with a gate oxide of 1.4 nm and a thin film of 10 nm, and the threshold voltage is lifted upward as compared to the tri-gate and the double-gate between gate length 20nm-40nm. When the channel length shrinks, the controllability of the tri-gate over the channel depletion region increases because the charge sharing from the source/drain is reduced. The predominating reliability problems associated with SCE are a lack of pinch-off and a shift in the threshold voltage with decreasing channel length as well as the drain-induced barrier lowering (DIBL) and the hot-carrier effect at increasing drain voltage [8]. Finally, shown in Fig. 12, the $G_{M,MAX}$ of the pseudo tri-gate vertical MOSFET is just higher than that of the double gate vertical MOSFET.

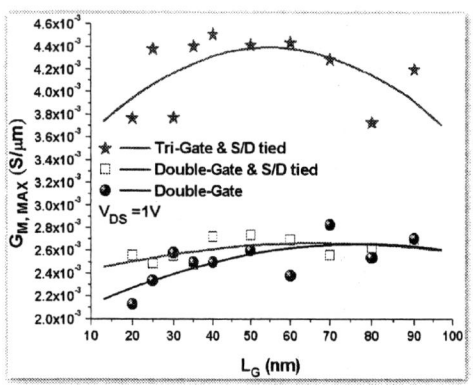

Fig. 12. $G_{M, MAX}$ as a function of gate length (L_G).

IV. CONCLUSION

We have designed and demonstrated a novel pseudo tri-gate vertical MOSFET which improves both the short-channel effects and the current drive significantly. It is good for the future device scaling applied in a high-performance and low-power device and system. Although the high drain current in the device results in a slight self-heating effect observed in the output characteristics, the thermal stability is still better than that of the non-S/D-tied devices. Additionally, the subthreshold leakage current of the new device is significantly reduced because the DIBL and punchthrough can be well controlled by the pseudo tri-gate scheme due to the strong electrical coupling and reduced charge sharing controlled by the drain.

REFERENCES

[1] Jyi-Tsong Lin, Kao-Cheng Lin, Tai-Yi Lee, and Yi-Chuen Eng, "Investigation of the Novel Attributes of a Vertical MOSFET with Internal Block Layer (bVMOS): 2-D Simulation Study," in *MIEL, Nis*, p488, May 2006.

[2] Thomas Schulz, Wolfgang Rösner, Lothar Risch, Adam Korbel, *Student Member, IEEE*, and Ulrich Langmann, *Senior Member, IEEE*, "Short-channel Vertical Sidewall MOSFETs," *IEEE TED*, vol. 48, NO.8, p1783, AUGUST 2001.

[3] Masahara, M. Yongxun Liu Hosokawa, S. Matsukawa, T. Ishii, K. Tanoue, H. Sakamoto, K. Sekigawa, T. Yamauchi, H. Kanemaru, S. Suzuki, E, "Ultrathin channel vertical DG MOSFET fabricated by using ion-bombardment-retarded etching," *IEEE TED*, vol. 51, NO. 12, p2078, 2004.

[4] Bin Yu, Zhi-Jim Ma, George Zhang, and Chenming Hu, "Hot-Carrier Effect in Ultra-Thin-Film (UTF) Fully-Depleted SOI MOSFET's," DRC, p22, 1996.

[5] Yi-Chuen Eng, Jyi-Tsong Lin, "Advanced π-FET Technology for 45 nm Technology Node," *IPFA*, p185, JULY, 2007.

[6] Akira Yoshino, T.-P. Ma, Senior Member, Koichiro Okumura, *IEEE,* and Koichiro Okumura, "Hot-Carrier Effects in Fully Depleted Submicrometer NMOS/SIMOX as Influenced by Back Interface Degradation," IEEE Electron Device Lett., 13, p.522, 1992.

[7] S. Cristoloveanu, S. M. Gulwadi, D. E. Ioannou, G. J. Campisi, H. L. Hughes, "Hot-Electron-Induced Degradation of Front and Back Channels in Partially and Fully Depleted SIMOX MOSFET's," IEEE Electron Device Lett., 13, p.603, 1992.

[8] Anurag Chaudhry and M. Jagadesh Kumar, "Controlling Short-Channel Effects in Deep-Submicron SOI MOSFETs for Improved Reliability: A Review," IEEE TRANSACTIONS ON DEVICE AND MATERIALS RELIABILITY, VOL. 4, NO. 1, p. 99, MARCH 2004.

Simulation of the Multi-Source/Drain SOI MOSFET

Po-Hsieh Lin*, Shiang-Shi Kang, Jyi-Tsong Lin, and Yi-Chuen Eng
Dept. of Electrical Engineering, National Sun Yat-Sen University
70 Lien-Hai Rd. Kaohsiung 80424, Taiwan ROC
Phone: (886) 7-5252000-4122 Fax: (886) 7-5254199
Email: *phlin_39@yahoo.com.tw

Abstract-In this paper, for the first time, a novel devise-architecture namely multi-source/drain SOI MOSFET is proposed and compared with a conventional SOI MOSFET. According to the simulation result, our proposed transistor not only maintains the desirable short channel behaviour, but also enhances the on/off current ratio due to the multi-source/drain scheme.

I. INTRODUCTION

Several kinds of silicon-on-insulator (SOI) [1-4] metal-oxide semiconductor field-effect transistors (MOSFETs) had been proposed in the recent years [5]. This is because the conventional bulk MOSFET structure are incapable of reducing the short channel effects (SCEs) significantly and there are too much PN junction area in the device, which would produce a large parasitic capacitance and leakage current. Those undesirability effects would affect the reliability of the devices, resulting in a degraded performance for the bulk MOSFET. That is why SOI is one of the most important devises for further scaling of a complementary MOS (CMOS). However, multi-source/drain (S/D) based transistor has not yet been studied so far. Therefore, in this paper, a new novel device structure called the multi-S/D SOI MOSFET (as shown in Fig. 1) is presented to investigate the device behaviors and reliability phenomena. Due mainly to its multi-S/D scheme, a special current flow of electron is also presented in later figures. Compare with a conventional MOSFET produced using SOI technology, the new structure can achieve the aim of overcoming the above issues and reveals advantages as listed in the following paragraphs. The device performance such as off-state vs. on-state currents plot, drain-induced barrier lowering (DIBL), subthreshold swing (SS), and threshold voltage (V_{TH}) will be discussed at the next section.

II. DEVICE STRUCTURE

Three-dimensional (3-D) process simulation tool DEVISE was used to build the proposed structure. Fig. 2shows the main process steps of the multi-S/D SOI MOSFET. The process starts an SOI wafer (Fig.2.(a)) which is used as substrate (boron, $5x10^{15}/cm^3$). Then, a photolithography process is carried out in order to form the 55 nm-thick recessed oxide (Fig.2.(b)), after that, we firstly define the active region. Second, a 1.2nm-thick gate oxide is thermally growth on the surface of the SOI substrate. Then, a 50 nm-thick layer of polysilicon (poly-Si) and

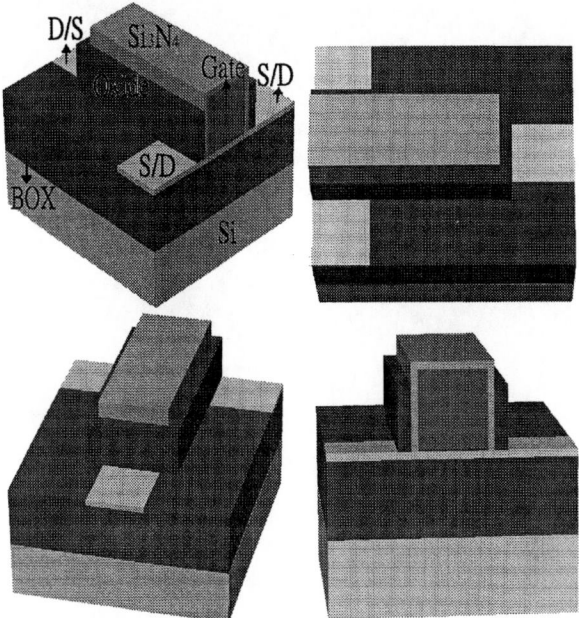

Fig. 1. Perspective views of the proposed SOI MOSFET.

10nm-thick nitride are deposited for the gate structure. After the gate patterning (Fig. 2(d)), we define the active region again for the multi S/D structure. Next, a nitride layer is deposited and etched back to form the sidewall spacers on the sides of the poly-Si gate as shown in Fig. 2(e). Before the S/D doping (arsenic, $2x10^{19}/cm^3$), the screen oxide is deposited and etched back to form the oxide spacer as shown in Fig. 2(f). After that, the proposed structure multi-S/D SOI MOSFET can be done. Table 1 lists the simulation parameters of the multi-S/D and the conventional SOI MOSFETs. In this paper, because of the device`s scaling, SCEs has turned out to be a critical issue. Hence, we use ultra thin body (UTB) [6] [7] structure to promote the device's performance and the control capability of the gate on the depletion region. Meanwhile, it can also reduce the leakage current as well. Additionally, we have two layers of spacer for the novel device, the nitride spacer and the oxide spacer. The gate is covered by the nitride spacer, and the oxide spacer is deposited around the nitride spacer, the different of heights between two spacers will affect the device's ability. The oxide spacer acts as an implantation screen-oxide, which is used for lowering the channel effect cause by the following

Table 1: Simulation parameters of the conventional SOI MOSFET and the novel multi-S/D SOI MOSFET.

nm Device	T_{GOX}	T_{Si}	L_G	T_{BOX}	W_{SiN}	W_{OX}	T_{Gate}
Conv. & Multi-S/D	1.2	5	10 ~ 100	50	5	10	50

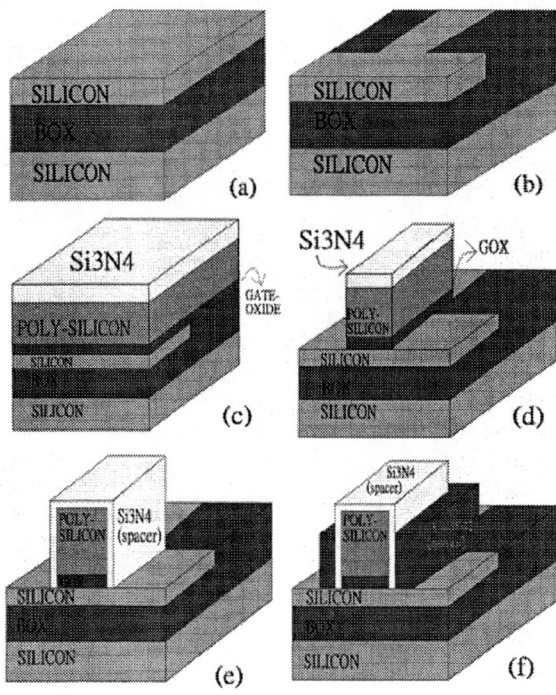

Fig.2: The main processes of the proposed SOI MOSFET.

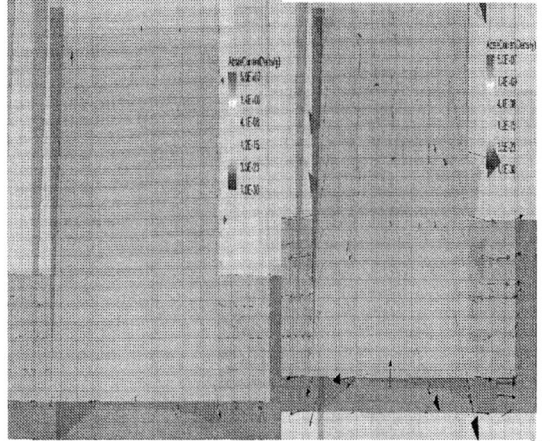

Fig. 3: Top view of the vector plot of electron current density (A/cm^2) for multi-source SOI MOSFET.

Fig. 4: Top view of the vector plot of electron current density (A/cm^2) for multi-drain SOI MOSFET.

implantation. As for increase the thickness of spacer [8], it can lower the Miler capacitance, but it will increase some resistance at the same time. Figs. 3 & 4 are the top views of the vector plots of the electron current density (A/cm^2) and the current flow direction for the proposed multi-S/D SOI. In Figs 3 & 4, the vector plots on the left hand side are the original one, and we magnify the original figures 4X that are shown on the right hand side. We observe that the drain current (I_D) flows to the silicon body region between the two source regions; the electron currents are separated from the middle of the silicon body, and then respectively flow into the both sides of source regions.

III. CHARACTERIZATION AND DISCUSSION

For performance comparison, two SOI MOSFETs, the conventional SOI MOSFET, and the new novel multi-S/D SOI MOSFET are designed as shown in Fig. 5. Figs. 6-9 are the characterization of the novel multi-S/D and conventional SOI MOSFETs.

Fig. 6 shows the on-state/off-state currents (I_{OFF} /I_{ON}) characteristics of the two devices from the gate length (L_G) 100 nm down to 10 nm. Since the multi-S/D scheme is applied, our proposed structures show better performance than the conventional one.

Fig. 5: Pictures of the two SOI MOSFETs: (a) the conventional SOI MOSFET, (b) the novel multi-S/D SOI MOSFET.

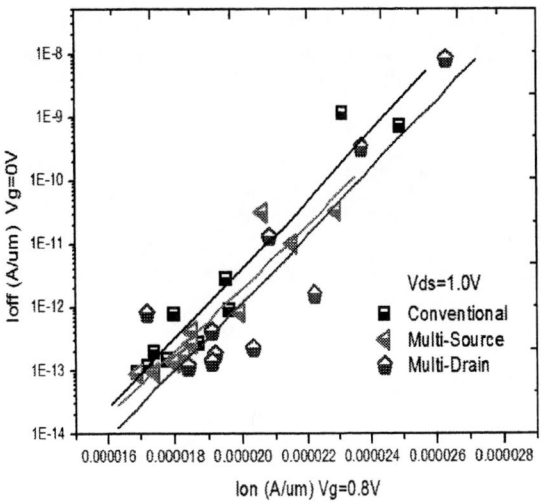

Fig. 6: I_{ON}/I_{OFF} characteristics of the three SOI devices.

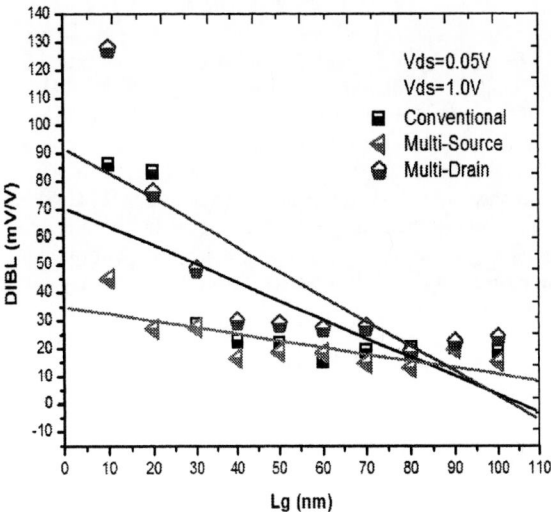

Fig. 8: DIBL as a function of gate length (L_G).

In Fig. 7, the V_{TH} roll-off is shown. It can be seen that both the multi-S/D schemes show better V_{TH} roll-off than the conventional one. Compare with the two multi S/D SOI MOSFETs, the multi-drain scheme had a worse performance in its on-off current ratio. Because its multi–drain schemes structure, the electric field of the drain is much bigger than the electric field of the source.

Fortunately, because of the UTB and a part of lightly doped drain (LDD) n- extension area, the results are still acceptable. Therefore, that is also why the results of SS are quite similar to the behaviours of DIBL as shown in Fig. 9. This study demonstrates that if the thin silicon active regions as well as multi-S/D are applied, the reliability of the device and the SCEs as well will be also affected, which may be employed for the IC applications.

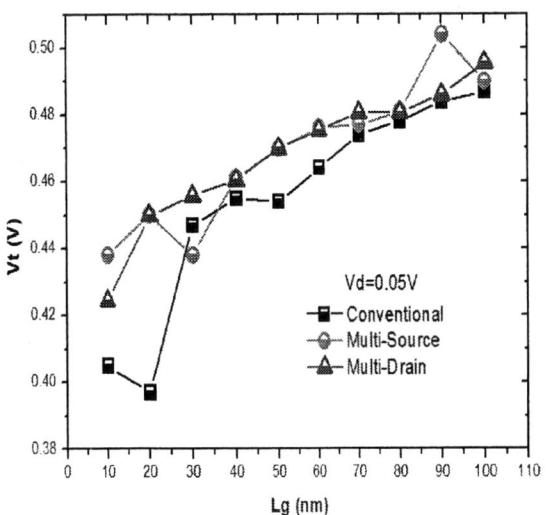

Fig. 7: V_{TH} as a function of gate length (L_G).

Fig. 8 shows the DIBL as a function of L_G. Our proposed device (multi-source) shows the best result of DIBL among them. Additionally, chiefly owing to the multi-drain scheme, the proposed multi-drain structure shows the worst DIBL. In other words, the additional drain terminal will enlarge the drain electric-field encroaching on the channel depletion region, resulting in large DIBL.

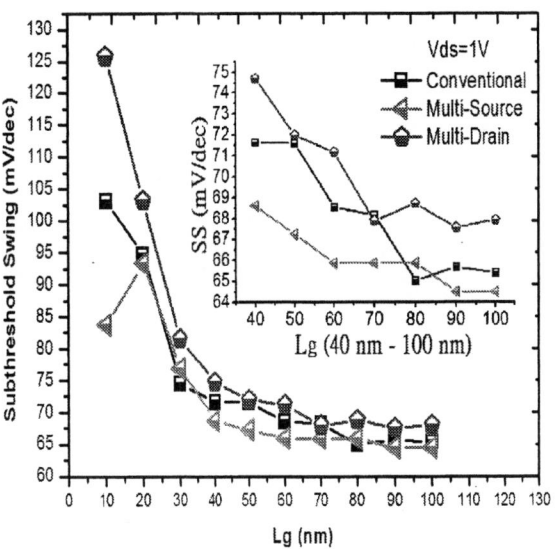

Fig. 9: SS as a function of gate length (L_G).

978-1-4244-2039-1/08/$25.00 ©2008 IEEE

IV. CONCLUSION

We have designed and demonstrated a novel multi-S/D SOI MOSFET for reliability improvement. The multi-S/D SOI MOSFET can be fabricated by a simple process sequences being similar to that of the conventional SOI MOSFET. According to the simulation results, our proposed structure (e.g. multi-source) can not only show the better result than that of a SOI MOSFET, but also reveal the full support of the advantages of SOI technology for future device scaling, which can be applied to a high-performance and low-power device and system with a low-cost mass production.

ACKNOWLEDGMENT

The Authors would like to sincerely thank Ying-Chieh Tsai, Yi-Ming Tseng from the SOI laboratory in National Sun Yat Sen University for the contribution of this paper.

REFERENCES

[1] Bolam, R.; Shahidi, G.; Assaderaghi, F.; Khare, M.; Mocuta, A.; Hook, T.; Wu, E.; Leobandung, E.; Voldman, S.; Badami, D.;Electron Devices Meeting, 2000. IEDM Technical Digest. International, 10-13 Dec. 2000 Page(s):131 - 134

[2] Chen, Jiun-Yu; Wang, Chen-An; Yeh, Wen-Kuan; Electron Devices and Solid-State Circuits, 2007. EDSSC 2007. IEEE Conference on 20-22 Dec. 2007 Page(s):593 – 596.

[3] Chiang, T. K.; Chen, M. L.; Wang, H. K.; Electron Devices and Solid-State Circuits, 2007. EDSSC 2007. IEEE Conference on 20-22 Dec. 2007 Page(s):597 – 600.

[4] Alam, M. K.; Khosru, Quazi D. M.; Electron Devices and Solid-State Circuits, 2007. EDSSC 2007. IEEE Conference on 20-22 Dec. 2007 Page(s):601 - 604.

[5] Bin Yu, IEEE ICSICT, 2006, p19.

[6] Yu TIAN et al, IEEE ICSICT, 2004 vol. 1, p283.

[7] Deleonibus, S.; de Salvo, B.; Clavelier, L.; Ernst, T.; Faynot, O.; Poiroux, T.; Vinet, M.; Solid-State and Integrated Circuit Technology, 2006. ICSICT '06. 8th International Conference on Oct. 2006 Page(s):51 - 54.

[8] Zhikuan Zhang; Shengdong Zhang; Mansun Chan; Electron Device Letters, IEEE, Volume 25, Issue 11, Nov. 2004 Page(s):740-742.

Stress-Induced Degradation in Strain-Engineered nMOSFETs

T. K. Maiti, S. S. Mahato, M. K. Bera, M. Sengupta, P. Chakraborty, C. Mahata, A. Chakraborty, and C. K. Maiti

Dept. of Electronics & ECE, Indian Institute of Technology Kharagpur, 721302, India

Phone: +91 3222 281475   Fax: +91 3222 255303   Email: tkm.iitkgp@yahoo.com

Abstract-**Effects of electrical stress on DC performance of strain-engineered nMOSFETs are investigated using simulation. The applicability of technology CAD (TCAD) for the prediction of MOSFET reliability is demonstrated.**

I. INTRODUCTION

Tremendous progress in the wireless and handheld personal communications in recent years has necessitated for more effective, compact and economical RF integrated circuits. For these applications, CMOS is essential, as it then becomes possible to integrate the RF front-end and base band building blocks of communication circuits onto a single chip. In the 90-nm technology node, strain engineering are being used as a vector for improving CMOS performances in parallel to technology scaling. By using strain engineering, improvements on the drive capability of up to 60% have been demonstrated [1]. Basically, induced strain alters the band structure of silicon by lowering the effective mass of channel carriers, which results in a higher mobility and thus the drive capability. Very few reports are available in the literature on the hot-carrier degradation of strain-engineered MOSFETs in the literature. It is thus important to study the effects of electrical stress on the strain-engineered MOSFETs and to evaluate their analog/RF potential in CMOS technology. Also, an increase in the oxide or interface trap density is expected due to the extra processing steps necessary for the implementation of the mechanical stress.

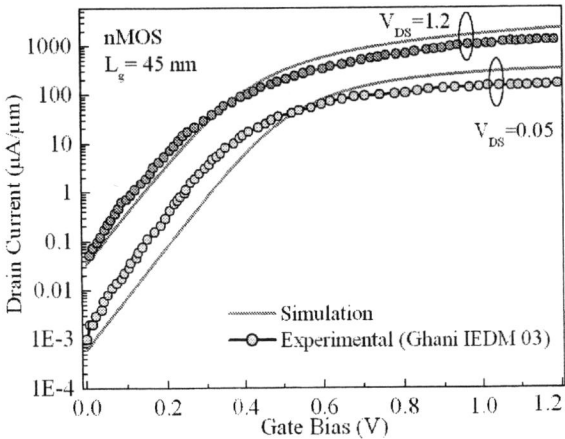

Fig. 1. Calibration of TCAD simulation. Comparison of the output characteristics with reported experimental device.

In this paper, DC hot-carrier induced degradation has been studied. Trap formation kinetics in the channel of nMOSFETs has been studied using simulation and it is shown that the traps are generated due to the breaking of Si-H bonds.

II. DEGRADATION MECHANISM

Device reliability is a serious concern and includes various degradation mechanisms: trapping of injected charge, trap generation, oxide breakdown, hot carrier effects, ion drift, inter-diffusion of metals, stress migration, and mechanical effects. It is well-known that hydrogen (H) plays a critical role in the fabrication of high quality Si/SiO_2 interfaces where these dangling bonds are compensated by hydrogen atoms. The experimental data for the kinetics of interface trap formation shows that the time dependence of trap generation can be described by a simple power law:

$$N_{it} - N_{it}^0 = \frac{N_{hb}^0}{1+\left(\gamma t\right)^\alpha} \qquad (1)$$

where N_{it} is the concentration of the interface, and N_{hb}^0 and N_{it}^0 are the initial concentration of Si-H bonds and interface traps, respectively. Considering $N = N_{hb}^0 + N_{it}^0$ total Si bonds at the interface, the remaining number of Si-H bonds at the interface after stress is $N_{hb} = N - N_{it}$ and follows the power law:

$$N_{hb} = \frac{N_{hb}^0}{1+\left(\gamma t\right)^\alpha} \qquad (2)$$

and the Si-H concentration during stress is given by

$$\frac{dN_{hb}}{dt} = -\gamma . N_{hb} \qquad (3)$$

where γ is a reaction constant and is given by $\gamma = \gamma_0 \exp\left(-\varepsilon_A / k_B T\right)$ in the Arrhenius approximation. ε_A is the Si-H activation energy and T is the temperature. The activation energy needed to release hydrogen from the interface can be expressed as

$$\varepsilon_A = \varepsilon_A^0 + \left(1+\beta\right)kT \ln\left(\frac{N-N_{hb}}{N-N_{hb}^0}\right) \qquad (4)$$

where ε_A is the energy needed to break a Si-H bond and the last term represents the potential energy needed to go over the potential barrier of the 2-D potential system with prefactor β.

III. RESULTS AND DISCUSSIONS

A reported MOSFET structure and experimental data for interface trap generated were taken from [2] and used in simulation for calibration purposes.

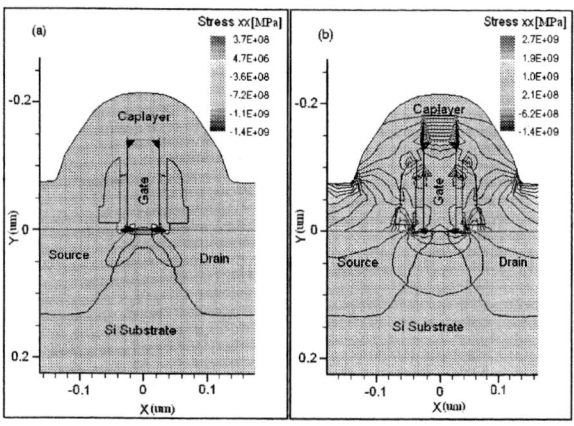

Fig. 2. Comparison of xx component of stress tensor between NMOS transistor with (a) no stress and (b) highly tensile process-induced

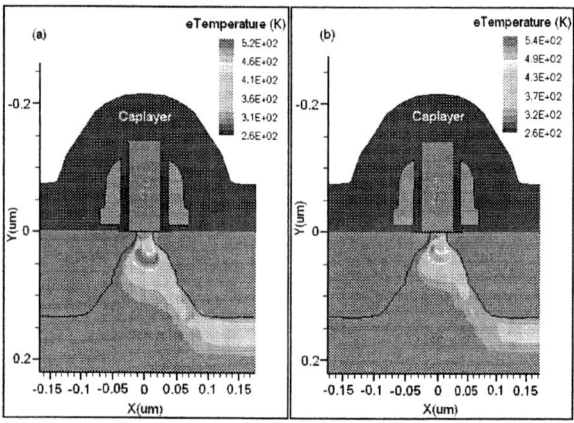

Fig. 3. Electron temperature distribution in strain-engineered n-MOS after hot carrier stressing.

The MOSFET has a 45 nm gate length and 1.2 nm thick oxides. Benchmarking is done for two different drain biases and the measured drain current vs. gate bias is plotted in Fig. 1. A highly tensile cap layer with initial stresses of 1.8×10^{10} dyn/cm^2 was deposited at the end of the of the process flow. Fig. 2 shows the mechanical stress distribution in the simulated devices with a highly tensile silicon nitride cap layer and relaxed (no stress) cap layer. A comparison shows that the stress in the cap layer leads to a tensile stress (mechanical) in the channel area.

The constant voltage stress (electrical stress) conditions for which the simulations are performed are gate voltage (V_g) in 1.5 to 2.8 V range, and the device is kept under the stress for 10^5 s. When the degradation simulation is finished, the device is set to

normal operating conditions and the I_{ds}-V_{gs} sweep is performed again. The electron temperature distribution due to the electrical stressing is presented in Fig. 3 for both due to the electric field stressing and hot-carrier stressing for devices with highly tensile cap layer and relax cap layer. A time dependent interface trap generation profile at strained-Si/SiO2 interface is shown in Fig. 4. As expected, electrically stressed devices show higher interface traps.

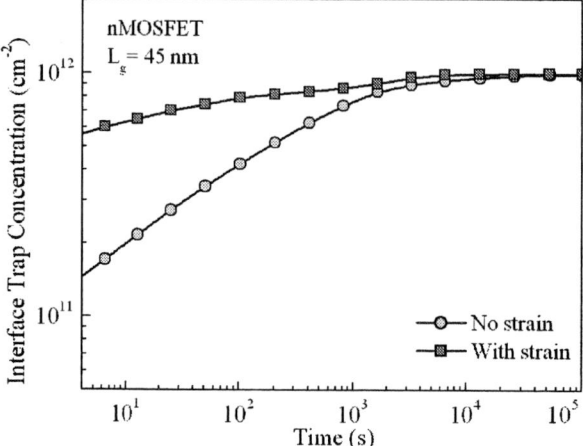

Fig. 4. Trap distribution with time at the interface

Reliability comparison for strained devices vs. unstrained ones was systematically addressed in simulation by applying different gate and drain voltages, and a combination of them. Important simulation results are presented below.

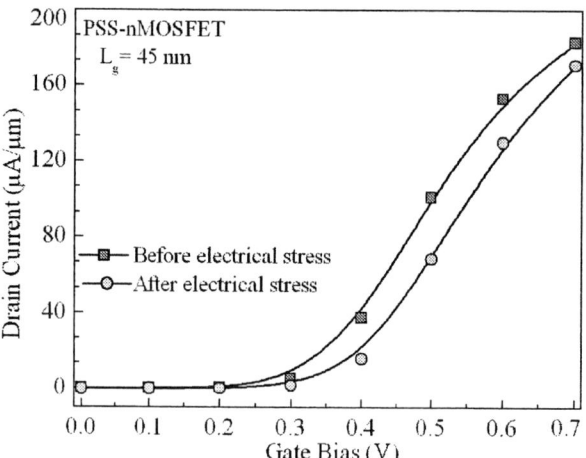

Fig. 5. Simulated output characteristics showing device degradation. Decrease in drain current is shown

The simulated drain current as a function of the gate voltage, transconductance vs. gate voltage, and drain current as a function of drain voltage before and after stressing are shown in Figs. 5, 6, and 7, respectively. Trapped charges at interface

978-1-4244-2039-1/08/$25.00 ©2008 IEEE   194   *Proceedings of 15th IPFA - 2008, Singapore*

states can reduce Coulomb mobility due to Coulombic scattering and electrostatically repel channel carriers (i.e., shift V_{th}), overall reducing the drain current (I_{ds}).

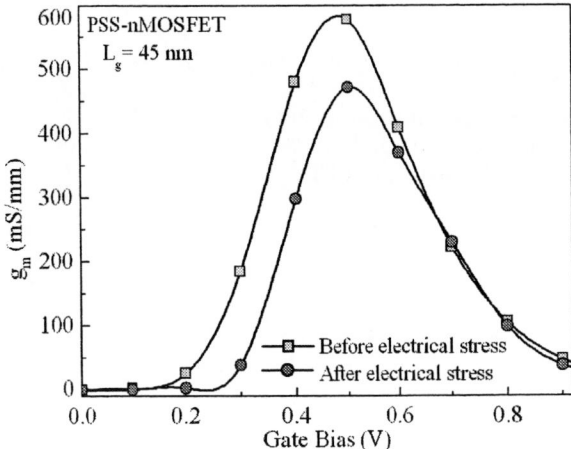

Fig. 6. Simulated transconductance showing device degradation. Simulated transconductance degradation is higher at lower gate voltage since inversion layer screening is less.

Fig. 6 shows the degradation in transconductance $\left(\partial I_{ds} / \partial V_{gs}\right)$ extracted from simulation data. In accordance with the theory, the degradation is higher at lower V_{gs} because of smaller screening.

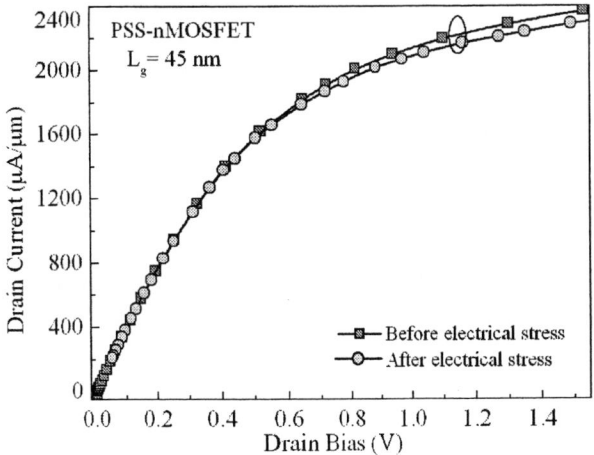

Fig. 7. Simulated I_{ds}-V_{ds} characteristics.

A decrease in drain current and conductance is observed due to the decrease in channel mobility. The threshold voltage extracted using maximum transconductance method, before and after degradation, is shown in Table 1.

Table I
COMPARISON OF THRESHOLD VOLTAGE

Type	Threshold voltage before degradation (V)	Threshold voltage after degradation (V)
Without strain	0.331	0.447
With strain	0.305	0.328

IV. CONCLUSION

The hot-carrier degradation for 45 nm gate length strain-engineered nMOSFETs with ultra-thin (1.2 nm) gate oxide under the gate voltage stress has been investigated using simulation. The impact of the uniaxial tensile strain induced by the SiN cap layer on the performance and reliability of nMOSFET is shown. It is found that the drain current decreases and the threshold voltage increases after the stress.

ACKNOWLEDGMENT

The work has been supported by DST, N. Delhi.

REFERENCES

[1] C. K. Maiti, S. Chattopadhyay, and L. K. Bera, Strained-Si Heterostructure Field Effect Devices, CRC Press, Boca Raton, 2007.

[2] T. Ghani et al., "A 90 nm High Volume Manufacturing Logic Technology Featuring Novel 45 nm Gate Length Strained Silicon CMOS Transistors," in IEDM Tech .Dig., pp. 978-980, 2003.

DIBL in Short-Channel Strained-Si n-MOSFET

S. S. Mahato[1*], P. Chakraborty[1], T. K. Maiti[1], M. K. Bera[1], C. Mahata[1], M. Sengupta[1], A. Chakraborty[1],
S. K. Sarkar[2] and C. K. Maiti[1]

[1]Dept. of Electronics & ECE, Indian Institute of Technology Kharagpur, 721302, India
[2]Electronic and Communication Engg. Dept., Jadavpur University, Kolkata-700032, India
Phone: (+91) No. 3222281475 Fax: (+91) No. 3222255303 *Email: satyamahato@yahoo.com

Abstract-**Drain-induced barrier lowering in substrate-induced strained-Si n-MOSFETs has been investigated. The variation of subthreshold swing as a function of both the gate length and gate to source voltage has also been examined.**

I. INTRODUCTION

As the classical silicon CMOS will become more and more difficult for the ultra-large-scale integrated circuits below 30 nm technology, several new high-mobility channel materials such as strained-SiGe and strained-Si and new device structures are being studied for improving the performance and also reliability [1]. As the device is miniaturized, the influence of the subthreshold swing increases.

The electric flux from the drain affects the channel potential and degrades device characteristics, such as the threshold voltage and the subthreshold swing (S factor). For some analog circuits and for low-voltage applications, model parameters, describing the behavior of MOSFET's in weak inversion, have to be known accurately. The most important parameter in this regime is the subthreshold swing S=$\dfrac{1}{\left(\partial \log I_d / \partial V_{gs}\right)}$. For very high-gain analog amplifiers [2] and for differential transconductors [3], amongst others, MOSFET's are operated in the weak inversion regime, because in this regime, the transconductance to current ratio is highest. Gate-source bias dependence of the subthreshold swing causes nonlinear gain and distortion, (see e.g. [3, Fig. 3] and [4, Fig. 8]). Other applications are imaging devices for which the output voltage varies logarithmically with the incoming light intensity [5] or current references used in ultra-low-power circuits, like watches [6].

The change in the threshold voltage due to the variation of drain voltage (ΔV_t), is calculated as an index of Drain induced barrier lowering (DIBL). For a low-power and high-speed operation, S factor reduction (a steep transition from OFF to ON) and DIBL reduction (reducing the V_t change caused by VD variation) is important to obtain a high ON/OFF rate.

Drain induced barrier lowering (DIBL) is one of the fundamental limitations in VLSI MOSFETs. Its effect on the drain current of a short-channel MOSFET is well known. Innovative device structures and new materials must be considered to continue the historic progress in information processing and transmission. As a promising MOSFET channel material candidate, strained-Si offers numerous advantages over Si. Thus, it is very important to study the DIBL effects in short-channel substrate-induced strain-engineered MOSFETs. In this work, we have addressed the classic problems of DIBL in advanced strained-Si MOSFETs appropriate for nanoscale technologies.

II. EXPERIMENTAL

The n-MOSFETs used in this study were fabricated on a p-type Si (100) substrate. However, the basic criteria for the growth of a strained-Si layer are to obtain a good quality SiGe buffer layer on p-Si substrates. For that reason a relaxed $Si_{0.80}Ge_{0.20}$/graded-SiGe films were grown at 900°C using ultra low pressure chemical vapor deposition technique. Next the strained-Si channel was pseudomorphically grown by lowering the reactor temperature to 650°C. Boron (in-situ) doping was designed in such a manner that a low p-type doping concentration profiles of ~10^{15} cm^{-3} for all layers may be obtained. Strained-Si n-MOSFET devices were fabricated using a 1.0 μm standard poly-gate CMOS process. A 10 nm thick thermal oxide was grown at 800°C. In-situ doped n+-poly-Si, defined the gate stack. S/D implantation was followed by metallization. Rapid thermal annealing was limited to 15 sec at 950°C. In addition, overall thermal budget and etch steps were carefully controlled in order to minimize strain relaxation, Ge out-diffusion, and strained-Si consumption. A schematic presentation of the drain induced barrier lowering in MOSFETs is shown in Fig. 1.

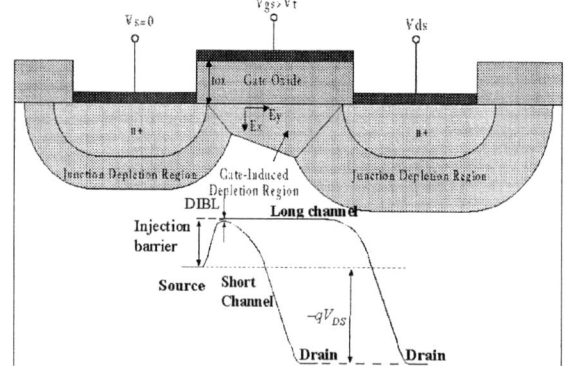

Fig. 1. Schematic showing the physical mechanisms of DIBL in MOSFETs.

The drain current vs. gate voltage measurements with varying drain voltages at room temperature were made using an automated Agilent 4156C Precision Semiconductor Parameter Analyzer.

III. RESULTS AND DISCUSSION

The subthreshold swing (S) for the strained-Si (SS) device is given as:

$$S \cong (60mV)\left(1 + \frac{t_{ox}\varepsilon_{Si_{1-x}Ge_x}}{t_d \varepsilon_{ox}}\right) \quad (1)$$

where ε_{ox} and $\varepsilon_{Si1-xGex}$ are the permittivities of the gate oxide and relaxed $Si_{1-x}Ge_x$ layer of strained-Si device, respectively, and t_d is the depletion width in strained-Si device. Also, t_d may be derived by solving one dimensional Poisson's equation with the boundary conditions [7], and it can be intuitively described as:

$$t_d \cong \sqrt{\frac{2\varepsilon_{Si_{1-x}Ge_x}\Psi_s}{qN_{Si_{1-x}Ge_x}}} \quad (2)$$

from the depletion approximation, where $N_{Si1-xGex}$ is the body doping density in the relaxed $Si_{1-x}Ge_x$ layer. It may be noted that the dielectric constant of the strained-Si layer is not too much different from that of the SiGe layer in the full relaxation. For 20% Ge content, strained-Si device may have ~7.5% higher dielectric constant based on the linear interpolation:

$$\varepsilon_{Si_{1-x}Ge_x} \cong \varepsilon_{Si} + (\varepsilon_{Ge} - \varepsilon_{Si})x \quad (3)$$

where ε_{Si} and ε_{Ge} are the permittivities of Si and Ge, respectively. However, td of strained-Si device is not too much different from that of bulk-Si device due to the lower ψ_s in (2), which tends to compensate for the higher dielectric constant. For scaled t_{ox} ($\ll t_d$), S could be almost the same for Si and strained-Si MOSFETs. The DIBL is derived by solving the two-dimensional Laplace's equation with Gauss's law and physical approximations [8]. For bulk-Si devices, the DIBL is expressed as:

$$DIBL \cong 180mV\left(\frac{t_d t_{ox} V_{DS}}{S \cdot L_{eff}^2}\right) \quad (4)$$

where t_d is the depletion width described in (2) and L_{eff} is the effective channel length. From (4), DIBL is found to be different in bulk-Si and strained-Si devices. However, DIBL-induced V_t shift, ΔV_t(DIBL) is almost similar based on Eqns. (1) and (4), since ΔV_t(DIBL) = DIBL*(S/60).

The dependence of the subthreshold swing on channel length at room temperature is shown in Fig. 2.

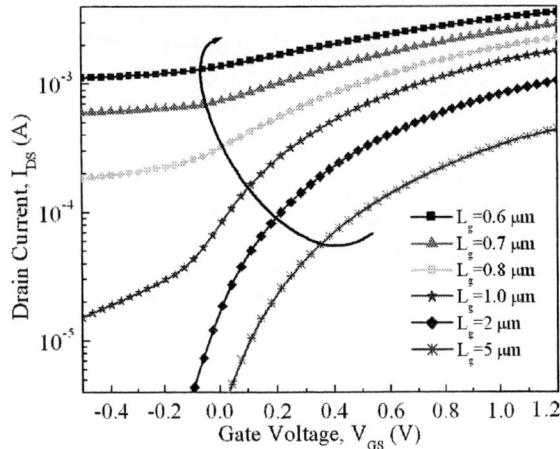

Fig. 2. Subthreshold characteristics for strained-Si n-MOSFETs (W=5μm).

The results indicate that the subthreshold swing increases sharply as the channel length is reduced to submicron range. For channel lengths greater than 1 μm, the subthreshold swing decreases at a slower rate and it becomes almost zero for channel lengths greater than 2 μm. For such channel lengths, the device behaves as a long channel device and the effect of drain voltage on the threshold voltage becomes negligible.

Saturation drain current, I_{Dsat} as a function of device gate length and width for strained-Si n-MOSFETs is shown in Fig. 3. It can be noted that I_{Dsat} increases with increasing device width whereas it decreases with increasing gate length.

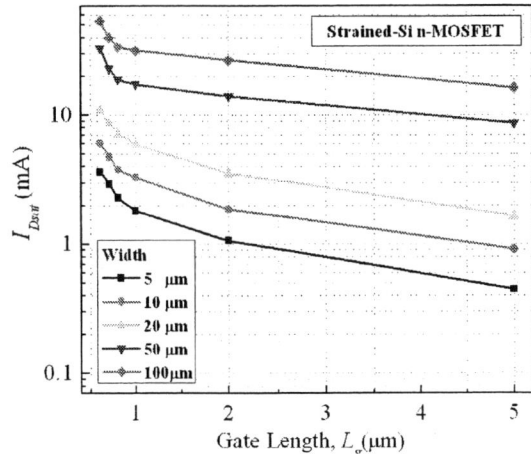

Fig. 3. Saturation drain current as a function of Gate length and width for strained-Si n-MOSFETs.

Fig. 4 depicts the enhancement in the drain current with respect to gate length of the devices for a fixed width (5μm).

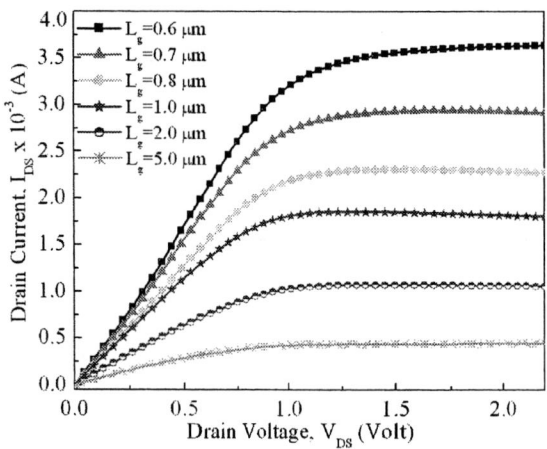

Fig. 4. Drain characteristics as a function of Gate lengths for strained-Si n-MOSFETs (W=5μm, VGS=1.2V).

Figs. 5 and 6 show the variation of subthreshold swing as a function of gate length and gate to source voltage, respectively. Influence of Gate length is investigated. The subthreshold swing is an important parameter for high-gain analog applications, imaging circuits, and low-voltage applications.

It is shown by experiments that the subthreshold swing varies with gate bias and exhibits a global minimum. The gate-source voltage for which minimum subthreshold swing is reached, is linearly related to the voltage at which moderate inversion starts.

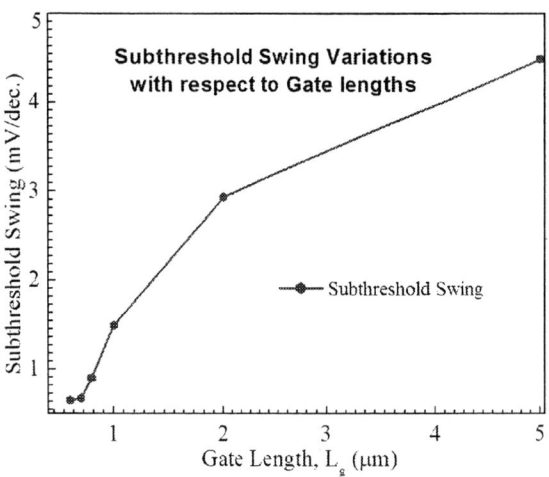

Fig. 5. Subthreshold swing variations with respect to gate length for strained-Si n-MOSFETs (W=5μm).

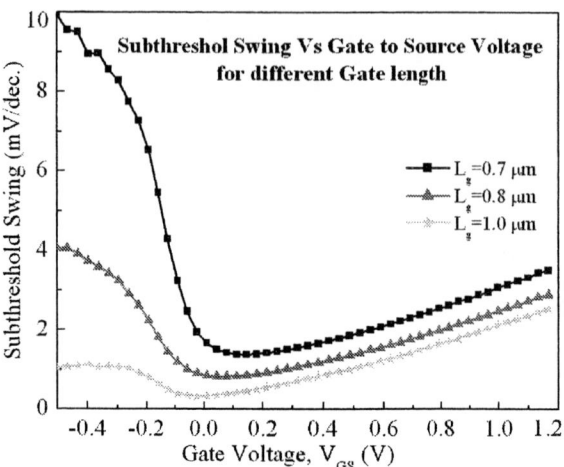

Fig. 6. Subthreshold swing as a function of Gate to Source Voltage at room temperature (300K).

SUMMARY

In this paper, the effect of drain bias on the threshold voltage is analyzed in terms of lowering the barrier between the source and the channel in subthreshold regime. Innovative device structures and new materials must be considered to continue the historic progress in Si CMOS technology. DIBL effects in short-channel substrate-induced strain-engineered MOSFETs have been reported.

ACKNOWLEDGMENT

The work has been financed by Defense Research and Development Organization (DRDO), New Delhi, India

REFERENCES

[1] C.K. Maiti, N.B. Chakrabarti and S.K. Ray, *Strained Silicon Heterostructures: Materials and Devices.* London, U.K.: Institute of Electrical Engineers, 2001.
[2] K. Laker and W. Sansen, *Design of Analog Integrated Circuits and Systems.* New York: McGraw-Hill, 1994.
[3] P. Furth and A. Andreou, "Linearized differential transconductors in subthreshold CMOS," *Electron. Lett.*, vol. 31, no. 7, pp. 545–547, 1995.
[4] S. Tedja, J. Van der Spiegel, and H. Williams, "A CMOS low-noise and low-power charge sampling integrated circuit for capacitive detector/sensor interfaces," *IEEE J. Solid-State Circuits*, vol. 30, pp. 110–119, Feb. 1995.
[5] N. Ricquier, "*Study and development of intelligent and flexible image sensors.*" Ph.D thesis, Katholieke Universiteit, Leuven, 1995.
[6] E. Vitoz and J. Fellrath, "CMOS analog integrated circuits based on weak inversion operation," *IEEE J. Solid-State Circuits*, vol. SSC-12, pp. 224–231, Mar. 1977.

978-1-4244-2039-1/08/$25.00 ©2008 IEEE

[7] J. B. Roldán, et al., "Strained-Si on Si1-xGex MOSFET inversion layer centroid modeling," *IEEE Trans. Electron Devices*, vol. 48, pp. 2447-2449, Oct. 2001.

[8] S. Veeraraghavan and J. G. Fossum, "A Physical shortchannel model for the thin-film SOI MOSFET applicable to device and circuit CAD," *IEEE Trans. Electron Devices*, vol. 35, pp. 1866-1875, Nov. 1988.

Influence of Hydrogen Annealing on NBTI Performance

L. J. Jin, H. P. Kuan, D. Sim and M. Mukhopadhyay

Systems on Silicon Manufacturing Co. Pte. Ltd., 70 Pasir Ris Industrial Drive 1

Phone: 65-6248 7296 Fax: 65-6248 7606 Email: li.juan.jin@nxp.com

Abstract- **The effect of hydrogen annealing after Local Interconnect Layer (LIL) formation and before Contact/Metal was evaluated with negative bias temperature instability (NBTI), electrical testing and yield in 0.14μm embedded flash. No significant difference was observed in electrical testing and yield amongst all splits. However, difference was observed from NBTI test. This paper mainly summarizes the influence of hydrogen annealing on NBTI performance.**

I. INTRODUCTION

As MOSFET size is continuously scaled down without reducing supply voltage, reliability becomes more and more important with tests such as hot carrier injection (HCL) and negative bias temperature instability (NBTI) etc [1]. NBTI in PMOS devices represent a major concern in modern CMOS technologies. The defects responsible for negative bias instability are either interface states [2-4] or oxide bulk traps [2, 5, 6]. The NBTI performance can be affected by a lot of process steps. These process steps include gate oxide thickness change, incorporating fluorine in the gate oxide [7], and oxide nitridation etc [8]. In addition, hydrogen-annealing effect on NBTI in thin gate oxide for Multi-Metal-layer CMOS process is also discussed by Lee et al.[9]. Hydrogen or forming gas annealing is often used at the end of the CMOS process in order to passivate the oxide dangling bonds and neutralize the oxide defects [9]. In this paper, besides annealing at the end of the process, hydrogen annealing after Local Interconnect Layer (LIL) formation and before Contact/Metal is also performed for good device performance, yield and NBTI. The role of hydrogen annealing with different annealing temperatures, annealing time after LIL formation is studied in order to find the optimized conditions for good NBTI performance, yield and electrical parameters, as well as effective productivity from manufacturing point of view.

II. EXPERIMENTAL DETAILS

The devices studied in this work were fabricated using 0.14μm embedded Flash Memory technology with shallow trench isolation (STI) gap-filled by high-density plasma (HDP) oxide. Firstly, all transistors were formed, which includes high performance, low leakage (29Å gate oxide thickness), and thick gate transistors (~75Å gate oxide thickness) for I/O, flash access transistors (FAT) and high voltage (HV) transistors. Secondly, cobalt salicide process was carried out, followed by nitride, ILD (SACVD & PETEOS oxide), LIL, Contact, inter-metal

dielectric (IMD), and Metal layers. Finally forming gas annealing (10% hydrogen+90% Nitrogen) is done at the end of the process in order to improve the metal and interconnect quality. In our process, besides final forming gas annealing, additional annealing after LIL formation before Contact is also carried out. In this experiment, the annealing effect was studied. The split items are summarized in Table I.

Table I Split condition of LIL annealing.

Split	Anneal Temp	Anneal gas	Anneal time
Split 1	410°C	H_2	1.5hr
Split 2	410°C	H_2	1.0hr
Split 3	410°C	H_2	0.5hr
Split 4	450°C	Forming gas(H_2+N_2)	0.5hr

Electrical testing was done for transistors under typical biasing conditions in order to obtain subthreshold, I_{sat}, V_t and I_{off} etc. It was reported that thick gate transitors are the most critical devices with respect to NBTI. Thick gate transistors degraded faster than the thin gate transistors during the circuit operation because the thick gate transistor fabrication is with double oxidization and the dual gate oxidation process may not be optimized for stress hardness [10]. As a result, in our paper, the NBTI (bias-temperature) test was mainly studied on 3.3V PMOS transistor with dimensions 9/0.322 and 9/0.376 (um/um, W/L) and the stressed gate voltage is equal to −3.6V and $V_s/V_d/V_{sub}$ is grounded. The stress temperature is 125°C and is under 5.2MV/cm stress.

III. RESULTS AND DISCUSSION

Yield shows no significant difference among all the splits, which verified that, all the wafers are functioning well. This is shown in cumulative plot of Fig. 1. Electrical testing for the transistors with different annealing conditions shows comparable results. Fig. 2 shows that the typical threshold voltage (Vt) of the 3.3V PMOSFET with dimensions 9/0.322 (um/um, W/L) is comparable among the splits.

It is widely known that hydrogen or forming gas annealing will passivate the oxide dangling bonds, so no apparent difference will be observed from yield and electrical testing with pure H_2 or less H_2, or annealing time (0.5 to 1.5hrs), as long as some annealing exists to neutralize the interface dangling bonds.

978-1-4244-2039-1/08/$25.00 ©2008 IEEE

Proceedings of 15th IPFA - 2008, Singapore

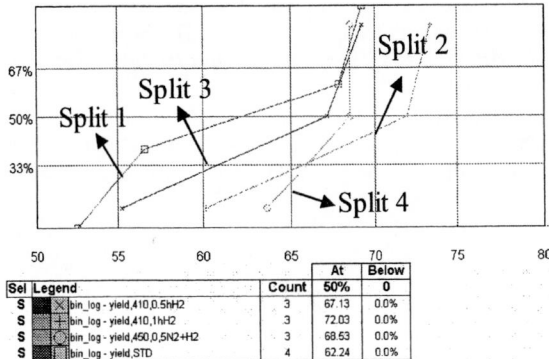

Fig. 1. Cumulative distribution of yield comparison among all the splits

Sel	Legend	Count	At 50%	Below 0
S	bin_log - yield,410,0.5hH2	3	67.13	0.0%
S	bin_log - yield,410,1hH2	3	72.03	0.0%
S	bin_log - yield,450,0.5N2+H2	3	68.53	0.0%
S	bin_log - yield,STD	4	62.24	0.0%

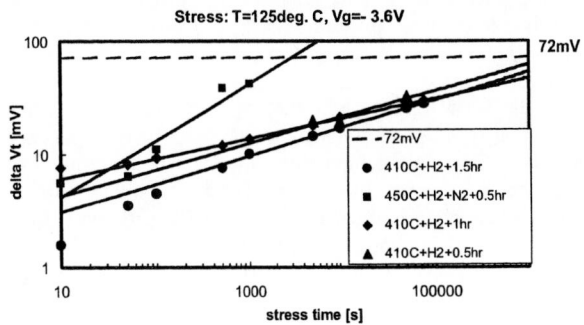

Fig. 3. Threshold voltage shift (delat Vt) versus lifetime for different LIL annealing splits (Measured on PMOS transistors (9/0.322)).

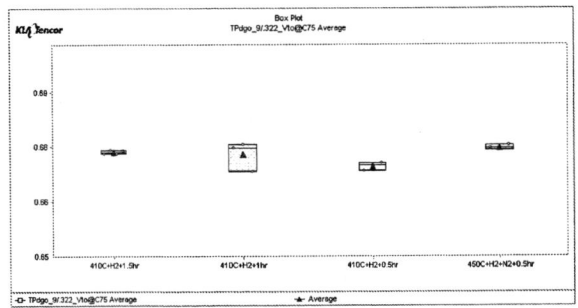

Fig. 2. Box plot of Vt (unit: V) distribution for PMOS transistor (9/0.322) with different LIL annealing splits.

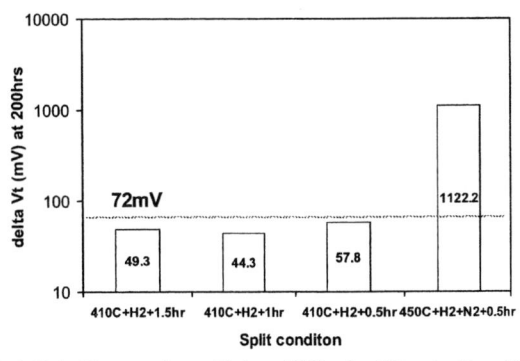

Fig. 4. Delta Vt comparison at lifetime of 200hrs for different splits on PMOS transistors (9/0.322).

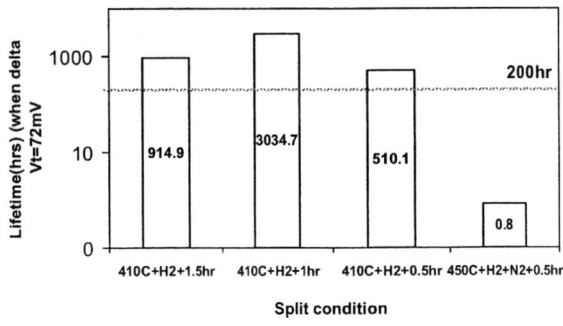

Fig. 5. Lifetime (when delta Vt=72mV) comparison for different splits on PMOS transistor (9/0.322).

Fig. 3 shows that V_t shift versus stress time among all the splits for PMOS transistor with W/L 9/0.322(um/um). Significant difference of pMOSFET NBTI performance is observed between forming gas and H_2 gas annealing. Similar findings were also observed when measuring PMOS transistor with W/L 9/0.376(um/um) (graph not shown). Extracted V_t shift (at lifetime 200 hrs) amongst all the splits is shown in Fig. 4. 72mV is set as upper limit of Vt shift. It is clear that all the hydrogen gas annealing splits show V_t shift less than 72mV while forming gas annealing split shows >72mV V_t shift (~1120mV) after 200 hours stress. Comparing V_t shift for splits with different pure hydrogen annealing time, 1hr annealing split shows least V_t shift, followed by 1.5hrs annealing, then 0.5hr annealing. Fig. 5 compares the lifetime difference when Vt shift reaches a value of 72mV. 200hrs is set as lowest limit of lifetime. Split with forming gas gives the worst performance (~0.8hr), much less than 200hrs. Splits using pure hydrogen with different annealing time meet the required lifetime 200hrs and split with 1-hour pure hydrogen annealing shows longest NBTI lifetime (~3034hrs).

978-1-4244-2039-1/08/$25.00 ©2008 IEEE

Two main findings are summarized from Fig. 3 to Fig. 5. One is that pure H_2 annealing gives much less NBTI degradation than forming gas annealing. The other is that appropriate hydrogen annealing time gives optimum NBTI performance, which also means that longer annealing time does not improve NBTI further.

For the first finding, two factors were considered when comparing splits using forming gas and pure hydrogen annealing. One is the temperature; the other is the amount of hydrogen or nitrogen. It is worth mentioning that another experiment using forming gas annealing at 410°C shows similar poor NBTI performance as annealed at 450°C (results not shown in this paper). That means higher annealing temperature is not main factor to affect NBTI performance. The only possible reason for different NBTI performance is the amount of hydrogen or nitrogen. There are a lot of papers about the role of hydrogen in NBTI and lots of debates exist due to the complexity of hydrogen annealing mechanism. Some believed that hydrogen mobile species diffuse and reach the Si/SiO_2 interface, where they create Pb centres, which are precursors for NBTI [11]. In our paper, we believe that the diffusion of hydrogen to Si/SiO_2 interface depends on the origin of hydrogen, the distance and the path they take to the oxide interface [12]. In our fabricated devices, borderless nitride film was grown before ILD layer as stop layers during oxide contact etch. Hydrogen mobility in nitride is very small (~3eV bond strength of N-H bond), so nitride is considered as a good barrier to H migration [13]. For forming gas, there is less hydrogen (~10%) compared with pure hydrogen gas, so it is possible that the less hydrogen passing through nitride film to neutralize the interface dangling bonds. As a result, worst NBTI performance is observed when using forming gas annealing. But it is believed that improved NBTI performance can be possibly obtained by increasing forming gas annealing time (e.g. 1h, 2h, or even longer) so that more hydrogen can passivate the dangling bonds. In addition, nitrogen in forming gas may also possibly induce the strain on the Si/SiO_2 interface and creates more P_b centers that induce bias NBTI degradation.

For the second finding, adequate hydrogen annealing time can improve the NBTI performance significantly but excessive hydrogen may not improve NBTI performance further because the passivation of the oxide dangling bonds is possibly already saturated when annealing reaches certain optimum time such as 1 hour in our experiment. It was believed that there is a close link between excessive hydrogen and interface state formation as reported by Lee et al.[9]. Si-O bulk defect in the gate oxide is associated with positive charge. The annihilation of positive charge at defect sites involves electron capturing process and the drift of hole to Si/SiO_2 interface. The drifting hole is speculated as atomic hydrogen. When too much drifting hydrogen moves to interface, it bonds to hydrogen that passivates an existing dangling bond. So H_2 forms at the interface, the dangling bond are exposed and the interface state is formed. As a result, the pMOSFET bias-temp is degraded.

IV. CONCLUSION

The hydrogen annealing effect after LIL formation before Contact/Metal on NBTI performance was studied. Hydrogen annealed wafers show better NBTI performance than forming gas-annealed wafers. For hydrogen annealing, appropriate H_2 annealing time is required for optimum NBTI performance. But for manufacturing, 0.5hour pure hydrogen annealing will be the better choice.

ACKNOWLEDGMENT

The authors are grateful to Puat Hwee Tua and Teck Cheong Yeo for their help with the annealing experiments. The authors would also like to express their gratitude to Systems on Silicon Manufacturing Co Pte. Ltd (SSMC) for the support given in running this experiment.

REFERENCES

[1] D. K. Schroder and J. A. Babcock, "Negative bias temperature instability: road to cross in deep submicron silicon semiconductor manufacturing," *J. App. Phys.*, vol. 94 (1), pp. 1-18, 2003.

[2] V. Huard, F. Monsieur, G. Ribes, and S. Bruyere, "Evidence of Hydrogen-Related Defects during NBTI Stress in p_MOSFETs," *in Proc. Int. Reliab. Phys. Symp.*, Dallas (USA), pp. 178-182, April 2003.

[3] J. M. Soon, K. P. Loh, S. S. tan, T. P. Chen, W. Y. Teo, and L.Chan, "Study of Negative-Bias Temperature instability induced defects Using First-principle Approach," *Appl. Phys. Lett.*, vol. 83 (15), pp. 3063-3065, October 2003.

[4] S. Fujieda, Y. Mura, M. Saitoh, E. Hasegawa, S. Koyama, K. Ando, E. Hasegawa, S. Koyama, and K. Ando, "Interface Defects Responsible for Negative-bias Temperature Instability in Plasma-nitride SiON/Si (1000) Systems," *Appl. Phys. Lett.*, vol. 82 (21), pp. 3677-3679, May 2003.

[5] V. Huard, and M. Denais, "Hole Trapping Effect on Methodology for DC and AC negative Bias Temperature Instability Measurements in pMOS Transistors," *in Proc. Int. Reliab. Phys. Symp.*, pp. 40-45, April 2004.

[6] S. Tsujikawa, T. Mine, K. Watanabe, Y. Shimamoto, R. Tsuchiya, K. Ohnishi, T. Onai, J. Ygami, and S. Kimura, "Negative Bias Temperature Instability of pMOSFETs with Ultra-thin SiON Gate Dielectrics," *in Proc. Int. Reliab. Phys. Symp*, pp. 183-188., April 2003.

[7] T. B. Hooek, E. Adler, F. Guarin, J. Lukaitis, N. Rovedo, and K. Schruefer, "The Effect of Fluorine on Parametrics and Reliability in a 0.18μm 3.5/6.8 nm Dual Gate Oxide CMOS Technology," *IEEE Trans. Elecctron Devices*, vol. 48 (7), pp. 1346-1353, July 2001.

[8] K. Kushida-Abdelghafar, K. Watanabe, J. Ushio and E. Murakami, "Effect of Nitrogen at SiO_2/Si Interface on Reliability Issues – Negative bias temperature Instability and Fowler-Nordheim-stress Degradation," *Appl. Phys. Lett.,* vol. 81 (23), pp. 4362-4364, December 2002.

[9] Y.H. Lee, R. Nachman, K. Seshan, D. –C. Kau and N. Mielke, "Role of hydrogen anneal in thin gate oxide for multi-metal-layer CMOS process," *in Proc. Int. Reliab. Phys. Symp,* pp. 186-190, 2000.

[10] A. Scarpa, D. Ward, J. Dubois, L. V. Marwijk, S. Gausepohl, R. Campos, K. Y. Sim, A. Cacciato, R. Kho and M. Bolt, "Negative Bias Temperature Instability Cure by Process Optimization," *IEEE TRANSACTIONS ON ELECTRON DEVICES,* Vol. 53 (6), pp. 1331-1339, June 2006.

[11] M. M. de Nijs., K. G. Druijf, V. V. Afanas'ev, and P. Balk, " Hydrogen Induced Donor-type Si/SiO_2 Interface States," *Appl. Phys. Lett.,* vol. 65 (19), pp. 2428-2430, August 1994.

[12] M. Ichimura and Y. Sasajima, "Diffusivity and solubility of hydrogen in grain-refined aluminum," *Material Transactions, JIM,* vol. 34 (5), pp. 404-409, 1993.

[13] W. M. Arnoldbik, C. H. M. Marée, A. J. H. Maas, M. J. van den Boogaard, F. H. P. M. Habraken and A. E. T. Kuiper, "Dynamic behavior of hydrogen in silicon nitride and oxynitride films made by low-pressure chemical vapor deposition," *Physical Review B.,* vol. 48 (8), pp. 5444-5456, 1993.

Reliability of ZrO_2/GeO_xN_y Stacked High-k Dielectrics on Ge under Dynamic and Pulsed Voltage Stress

C. Mahata[1*], M. K. Bera[1], P.K. Bose[2], A. Chakraborty[1], M. SenGupta[1] and C. K. Maiti[1]

[1]Dept. of Electronics and ECE, Indian Institute of Technology Kharagpur, India
[2]Mechanical Engineering Dept., Jadavpur University, Jadavpur, Kolkata 700032, India
Phone: (+91) No. 3222 281475 Fax: (+91) No. 3222 255303 Email: chandreswar@gmail.com

Abstract- **Charge trapping/detrapping in Al/ZrO_2/GeO_xN_y/p-Ge MIS capacitors have been studied under dynamic voltage stresses of different amplitude and frequency in order to analyze the transient response and the degradation of the oxide as a function of the stress parameters. The current transients observed in dynamic voltage stresses have been interpreted in terms of the charging/discharging of interface and bulk traps. The evolution of the current during unipolar and bipolar voltage stresses shows the degradation being much faster at low frequencies than at high frequencies.**

I. INTRODUCTION

Stress induced leakage current (SILC) produced under electrical stress before the onset of soft or catastrophic breakdown is one of the major reliability concerns in deep submicron technology. Accelerated life-test of MOS capacitors are conventionally performed by applying a constant gate voltage (CVS) or injecting a constant gate current (CCS) over a period of time, with a periodic interruption to allow electrical measurements to monitor the oxide degradation [1-2]. However, MIS devices experience time-varying bias during normal operation inside IC's and, as a consequence; it is worth studying their degradation and wearout using dynamic stress condition [3-4]. In this paper, we report on the reliability issues of ZrO_2/GeO_xN_y stacked gate dielectric on p-Ge under dynamic and pulsed voltage stress (PVS) at different frequencies, and the results have been compared to those obtained with conventional stress procedures (*i.e.*, CVS and CCS). Attempt has been made to find out correlation between the transient current characteristics at pre-tunneling voltages, low tunneling voltages and high voltage stressing under different pulse sequences. The trap charge density, Q_t and charge trapping centroids have been measured using the bidirectional I-V technique. The stress (1000s long) was periodically interrupted at 100 sec intervals to measure the I-V characteristics for both polarities so as to obtain the trapped charge distribution in the oxide. We also examine the transient voltage response under sequential dynamic current stressing and discuss whether the obtained results can enlighten some aspects of the degradation and breakdown processes in stacked layers.

II. EXPERIMENT

The Ge substrate (100) used was B-doped p-type wafers with a resistivity of 25-30 Ω-cm. The Ge-substrates were held in highly concentrated HF vapor for 5-10 s prior to loading in the deposition chamber. ZrO_2 thin films (~14 nm) were then deposited on p-Ge using a metallorganic compound, ZTB $[Zr(OC(CH_3)_3)_4])$ in a microwave (700 W, 2.45GHz) system at 500 mTorr for 1 min. For oxynitride formation, deposited ZrO_2 films were exposed to NO (99.999%) plasma for 1 min. MIS capacitors were fabricated for electrical characterization with Al metal contacts having an area of 1.96×10^{-3} cm^2. Charge trapping characteristics and trap centroid of gate stack on Ge was carried out by applying pulsed unipolar and bipolar stress voltage to the gate. In order to investigate the current characteristics of the devices CVS with different pulse sequences were carried out. The dynamic current stress, dynamic voltage stress, unipolar and bipolar pulsed voltage stresses was performed by Agilent 4156C precision semiconductor parameter analyzer. Square wave PVS, produced by Agilent 41501B PGU expander, was applied to the MIS capacitors with the signal frequency varying from 1 Hz to 10 kHz in conjunction with a fully automated computer aided measurement.

III. RESULT AND DISCUSSION

As a measure of oxide reliability the 63% (cumulative time-to-breakdown) values of t_{bd} was determined with the aid of constant voltage stressing. Figure 1 shows the measured current during a sequence of high voltage pulses at -12V (E_{ox}=8MeV/cm), demonstrating either a soft and/or catastrophic breakdown of the ZrO_2/GeO_xN_y high-k gate dielectric stack. The dependence of SILC on the stress voltage frequency is shown in Fig. 2, where the leakage current is shown for different injected fluences in the range from 5×10^{19} to 5×10^{20} e/cm^2. It has been observed that positive SILC is maximum at 1kHz stress frequency for all injected fluences. Charge trapping characteristics of oxynitride films have also been investigated by pulsed unipolar and bipolar stressing where a 50% duty-cycle pulse voltage was applied to the gate when the substrate was grounded [5]. The experimental setup for pulsed stresses was carefully designed and assembled to measure the gate current under PVS and to avoid any voltage overshoot/undershoot

978-1-4244-2039-1/08/$25.00 ©2008 IEEE

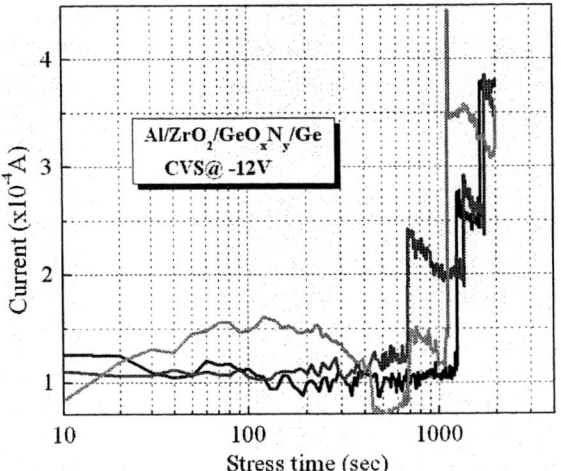

Fig. 1: Soft breakdown characteristics of GeOₓNᵧ/ZrO₂ stacked dielectric during CVS at -12V.

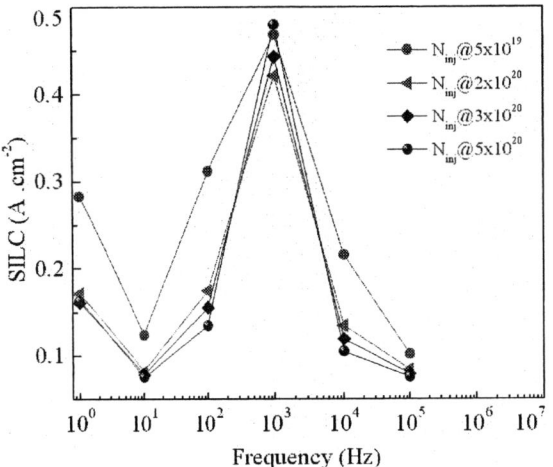

Fig. 2: Positive SILC measured at different frequencies for the PVS injected charges (N_{inj}) from 5×10^{19} to 5×10^{20}.

>100mV when the 50% duty cycle pulsed wave produced by Agilent 41501B PGU expander was applied to the MIS gate with the signal frequency varying from 100 Hz to 100 kHz. The trap charge density, Q_t and charge trapping centroid, X_t has been measured using the bidirectional I-V technique [6-7], which is given by:

$$X_t = t_{ox}\left[1 - \frac{\Delta V_{FN}^-}{\Delta V_{FN}^+}\right]^{-1} \quad (1)$$

$$Q_t = \frac{\varepsilon_{ox}}{t_{ox}}\left[\Delta V_{FN}^- - \Delta V_{FN}^+\right] \quad (2)$$

where a voltage shift is observed under unipolar or bipolar square voltage pulse stressing.

Figure 3 shows the trap charge distribution for unipolar as well as bipolar PVS with voltage amplitudes of +9V and -9V to +9V, with the frequency as a parameter, for a total stress time of 1000 sec. The fast increase in the beginning is due to the filling of native traps and a high rate of generation of new traps is followed by a slower evolution, mainly caused by the reduction of generation rate due to trapping. These distributions have been assumed to be indicative of the degradation level of the oxide i.e. charge trapped in the defects that have been generated during the stress [8]. At high frequency, the trap charge generation probability or degradation level is lower because low stress time leads to lower trap charge occupation (the lower the number of generated defects). It has also been noticed that in case of bipolar stress, trap charge distribution as shown in inset of Fig. 3, is lower than that of unipolar stress. However, it may be observed that the dependence of trapping level or position of trap centroid is lower at high frequency for both unipolar and bipolar stress (see Fig. 4) due to the lower generation rate of traps at higher frequency.

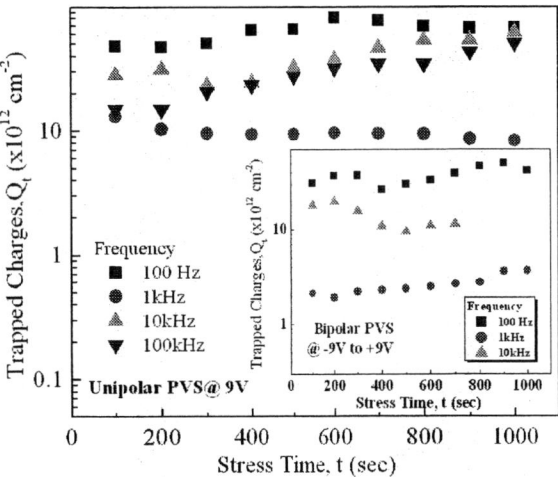

Fig. 3: Magnitude of number of trapped charges as a function of stress time at different frequencies.

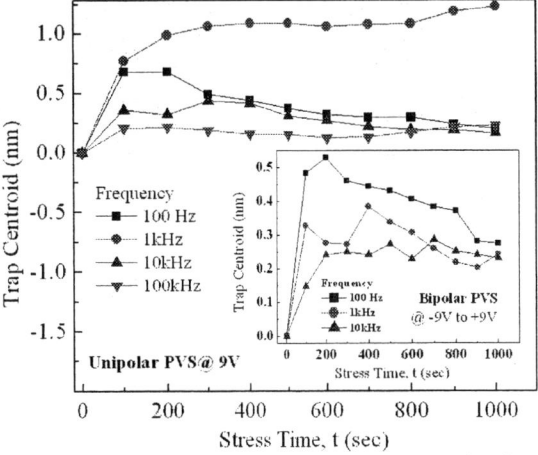

Fig. 4: Distribution of trap charge centroid in ZrO₂/GeOₓNᵧ gate stack on Ge as a function of stress time for different frequencies.

In an attempt to find out if a correlation exists between the current characteristics at low-tunneling and high-tunneling voltage stressing, we applied an alternate sequence of pulses at high-tunneling voltage 9V and low-tunneling voltage (4V) to the MIS capacitor as shown in Fig. 5. Figure 6 shows the current characteristics at alternating voltage of -9V and -4V.

In Fig. 5 both in low positive tunnelling and high positive tunnelling voltage from I-t response it is seen that negative traps are generated. In case of negative tunnelling voltage positive traps are seen to be created. In both set of I-t characteristics initial part of every pulse started with lower value of current level which is due to the detrapping of charges after each pulses.

initial current component is smaller for every pulse voltage which may come from the detrapping of the trapped charge of previous pulse.

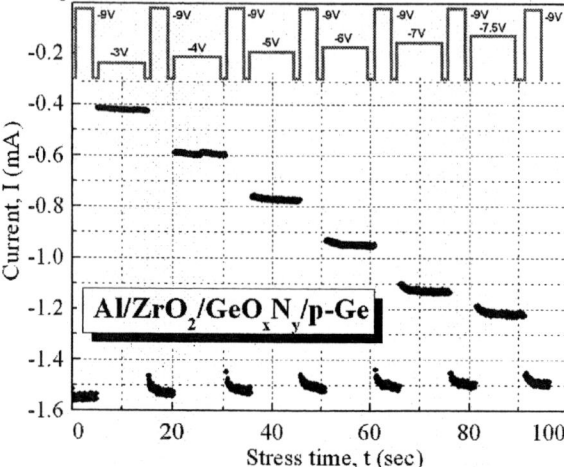

Fig. 7: Current response to alternating voltage pulses at different voltage heights for -9V and -3 to -7.5V.

It is observed that a characteristics variation of the current with time changes from a negative to a positive value in the initial phase. The influence of the pauses between two voltage pulses on the current response during a voltage pulse has also been investigated. A pulse sequence consisting of an initial pulse at -9V followed by pulsed at fixed but lower voltage (-4V) along with increasing pause time in between have been applied as has been shown in Fig. 8.

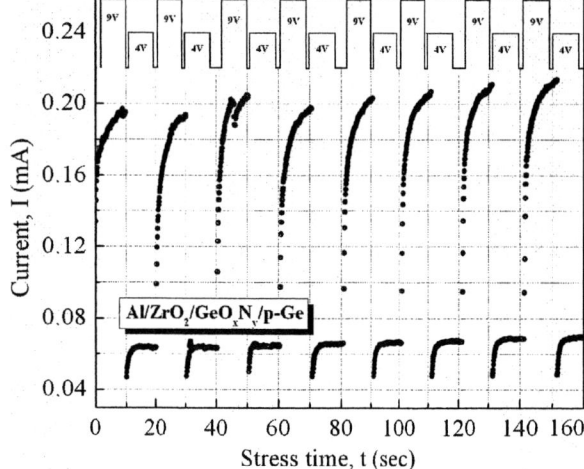

Fig. 5: Current response during CVS at low stress levels of 9V and 4V in different scale.

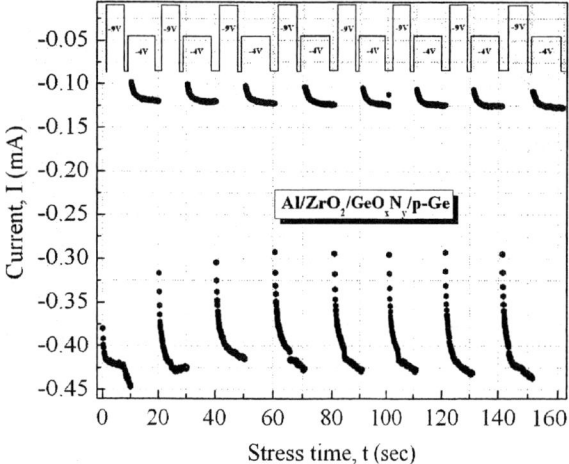

Fig. 6: Current response during CVS at low stress levels of -9V and -4V in different scale.

In Fig. 7 a high voltage pulse at -9V is followed by pulses with varying amplitude corresponding to pre-tunnelling, low tunnelling and high-tunneling. From I-t characteristics it is noted that the transient current increases with stress time for all tunneling limits. Varying the pulse height from -3V to -7.5V the

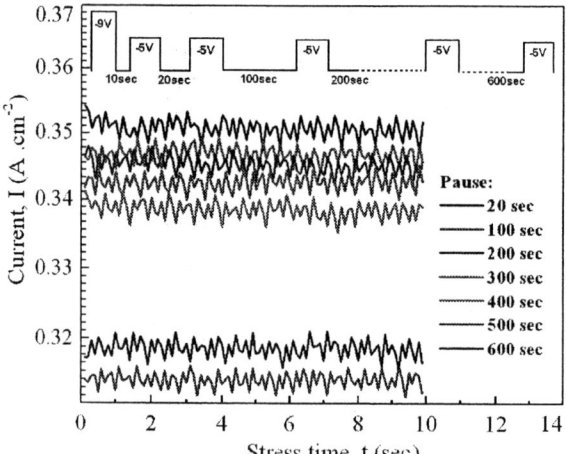

Fig. 8: Current response of 8 pulses with -5V pulse height after one -9V pre-stress pulse. Between the -5V pulses the interval was varied from 10 to 600 sec.

To investigate the nature of charge trapping and de-trapping, we used dynamic current injection measurements to find out the generation of trapped charges and its relaxation in the oxide and at its interfaces. Trapping/de-trapping of stacked layers under sequential dynamic current stresses was studied under different current density from $J_{inj} = -2.5$ mA/cm^2 to $J_{inj} = -12.5$ mA/cm^2 shown in Fig. 9. In lower current density ($J_{inj} = -2.5$ mA/cm^2) during the first pulse, positive charge trapping was observed during the stress in gate stack. After all next pulses electron trapping is observed. Thereafter, an decreasing negative gate voltage shift has been noticed during each pulse. Although, a rapid decrease in gate voltage shift is being observed for higher current density shows dominant trapping of positive charges in stack layers. However, between two successive pulses, detrapping of the charge occurs, leading to the shift in gate voltage at the beginning of the next pulse than at the end of the preceding. This repetitive gate voltage behavior shows the relaxation characteristics that are related with the detrapping of the previously trapped charges [9]. From these observations, it may be concluded that trapping/detrapping behavior is strongly dependent on the gate stack thickness and chemical structure of the high-k and interfacial layer.

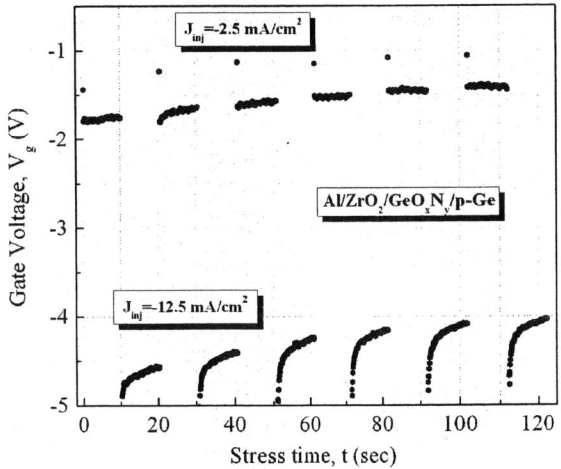

Fig. 9: Change in gate voltage during pulsed CCS at low stress levels of -2.5mA.cm^{-2} and -12.5mA.cm^{-2} in different scale.

IV. CONCLUSION

In summary, the SILC and transient current characteristics through thin ZrO_2/GeO_xN_y high-k gate dielectric stack on p-Ge have been investigated by applying dynamic and pulsed voltage and current stress at different frequencies (100Hz-100kHz) as well as at various injected fluences (10^{19}-10^{20} e/cm^2). Under varying pulse voltage from pre-tunneling to high-tunneling level (in small increments), it has been observed that steady transient current depends not only on the generation rate of traps but also on the pulse height. An increase in life-time has been observed for all MIS capacitors when subjected to pulsed voltage conditions.

ACKNOWLEDGMENT

The work has been supported by DST, Govt. of West Bengal, India.

REFERENCES

[1] E. Rosenbaum, Z. Liu, and C. Hu,"Silicon dioxide breakdown lifetime enhancement under bipolar bias conditions," IEEE Trans Electron Dev 1993;40(12):2287–95.

[2] A. Cester, A. Paccagnella, and G. Ghidini, Stress induced leakage current under pulsed voltage stress," Solid-State Electronics 46 (2002) 399–405.

[3] S. Haddad and M.S. Liang, "The nature of charge trapping responsible for thin-oxide breakdown under a dynamic field stress," IEEE Electron Device Lett., vol.8, p. 524, 1987.

[4] M.S. Liang, S. Haddad, W. Cox, and S. Cagnina, "Degradation of very thin gate oxide MOS devices under dynamic high field/current stress," IEDM Tech. Dig., p. 394, 1986.

[5] M. Nafria, D. YClamos, J. Sufiic, and X.Aymerich, "Relation between degradation and breakdown of thin SiO, films under AC stress conditions," Microelectronic Engineering 28 (1995) 321-324.

[6] D. J. DiMaria, E. Cartier, and D. Arnold, "Impact ionization, trap creation, degradation, and breakdown in silicon dioxide films on silicon," J. Appl. Phys., 1993;38:344-54.

[7]Z.H. Liu, P. T. Lai, Y.C. Cheng, IEEE Trans Electron Dev Vol. 38,pp. 344-354, 1993.

[8] R. Rodriguez, M. Nafria, J. Sufi6 and X. Aymerich, IEEE Trans. Electron Dev., Vol. 45, No. 4, pp. 881-888 (1998).

[9]R. Choi, B. H. Lee, K. Matthews, J. H. Sim, G. Bersuker, L. Larson, and J. C. Lee, Proc. Device Res. Conf. (2004)17–18.

978-1-4244-2039-1/08/$25.00 ©2008 IEEE

SESSION 8:

NOVEL DEVICES I -
SOLAR CELLS & MEMORY

INVITED PAPER

Trends in Solar Cell Research

R. Mertens
IMEC and K.U.Leuven
Kapeldreef 75, B-3001 Leuven, Belgium
Phone No. +32 16 281 280. Fax No. +32 16 281 501. Email: Robert.Mertens@imec.be

Abstract- **Photovoltaics (PV) is becoming an important industrial product. Today the PV market is dominated by wafer based crystalline Si cells. The cost of these cells will decrease in the future by using thinner wafers and devices with higher conversion efficiency. It is expected that nanotechnology will play an important role in the longer run in order to further lower the PV cost. PV can also profit from cross fertilization with micro- and nano-electronics. PV modules have shown to be very reliable in the field but more reliability research is needed in fields such as ultra-thin Si wafer based cells, concentrator PV and organic solar cells.**

I. INTRODUCTION

Since many years institutes around the world are active in the field of renewable energy more specifically in the field of photovoltaics (PV): the direct conversion of solar energy in electrical energy using solar cells. The main motivation for this comes from the fact that energy is most probably humanity's top problem. A 2003 study from the famous MIT forum has indeed shown that energy comes first on the list of the world problems for the next 50 years and that it even precedes water and food as the top priority problems.

Already 40 years ago it became obvious that conventional energy sources such as oil, coal and gas will be unable to meet the future energy demand of the world as in the long run their production will decrease rather than increase. Therefore new sources such as nuclear and especially renewable energy sources such as solar, wind, biomass and geothermal will have to used on a massive scale. It has been stated several times that the transition from a conventional to a sustainable global energy system is one of the biggest challenges that mankind has ever faced.

Historically since their invention 50 years ago solar cells have first been used in space but also in small scale stand alone systems in remote areas in developing countries e.g. for lighting and vaccine cooling applications. In these applications solar cells are used in combination with rechargeable batteries. In more recent years we also see these small scale solar cell applications in the industrialized world e.g. for telecommunication and signalization systems. But as the cost of solar cells has come down by an order of magnitude during the last twenty years, they are now more and more used in large power applications in grid connected applications. In these grid connected applications no batteries are used but the energy produced by the cells is sent onto the grid.

We can distinguish central applications where additional land is needed to install the solar cells and building or structure integrated systems where existing buildings or structures such as sound barriers along the highways are used. We believe that especially these building or structure integrated systems are important for Europe.

The solar market has grown very strongly during the last 10 years, as shown by Fig. 1, with average growth rates of more than 40%. In 2007 the world annual production of solar cells was more than 4GW peak as indicated by the right bar in Fig. 1; peak power refers to the power produced under standard illumination of 1kW per square meter. The result is that the solar cell industry has become an important industry worldwide with a turnover of more than 20 billion euro. The European vision is that the cost of photovoltaic systems will continue to go down.

By 2030 it is estimated that the PV industry will produce 1000 GWp globally from which 200 will be manufactured in Europe, thereby creating 200000 jobs in Europe.

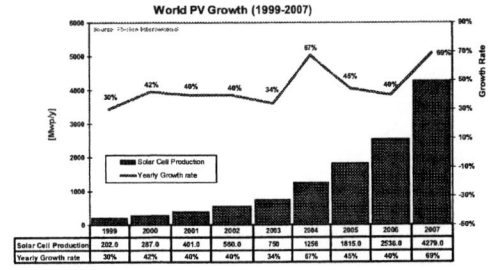

Fig. 1. World solar cell market 1999-2007 in MWp/year.

II. OVERVIEW OF PV TECHNOLOGIES

PV can operate under light intensities that vary from a small percentage of 1 sun up to 1000 suns (Fig. 2). The very low intensities are found in indoor applications such as e.g. sensor networks for ambient intelligence.

Today more than 90 % of PV is used as flat plate modules designed for 1 sun operation. In some parts of the world with more than 2500 hours of sunshine per year there is an increasing interest in the use of 1-axis or 2-axis PV concentrator systems where the sunlight is concentrated using mirrors or lenses onto the cell. In such systems the required cell area is reduced by the concentration factor.

INVITED PAPER

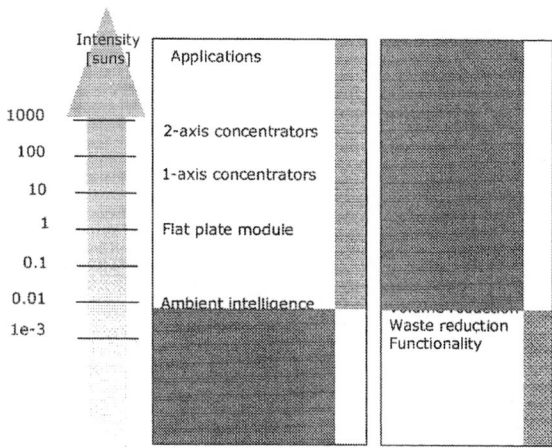

Fig. 2. PV: from nanoW to GigaW scale.

Today more than 90 % of PV is used as flat plate modules designed for 1 sun operation. In some parts of the world with more than 2500 hours of sunshine per year there is an increasing interest in the use of 1-axis or 2-axis PV concentrator systems where the sunlight is concentrated using mirrors or lenses onto the cell. In such systems the required cell area is reduced by the concentration factor.

The characteristics of the most important present PV technologies for 1 sun operation are summarized in Fig. 3, including the type of junction, the best laboratory and industrial efficiency and the market share.

It is clear that the actual market is dominated by bulk crystalline or wafer based Si solar cells. The other three technologies are thin film based. In recent years also organic solar cells using organic semiconductor material received much attention. Organic solar cells can be made very thin (100 nm) and are therefore potentially cheap.

Cell Technology	Type of junction	Lab efficiency [%]	Industrial efficiency [%]	Market share [%]
Bulk crystalline Si solar cells	p-n homojunction	24.7	13 – 17	92
a-Si:H (a-Si:H; a-SiGe:H, µc-Si)	p-i-n homojunction multijunction	13	6-7 single junction 9-10 multijunction	5
CuIn(Ga)Se$_2$(S$_2$) =CIS	p-n heterojunction with CdS	18.8	9 – 13	3
CdTe	p-n heterojunction with CdS	17	9 – 12	

Fig. 3. Present PV-technologies for one sun application.

III. IMEC'S ROADMAP IN PHOTOVOLTAICS

Fig. 4 shows IMEC's roadmap in photovoltaics.

The two main activities at IMEC are dealing with silicon based and organic solar cells respectively. Silicon solar cells are most widespread today and represent more than 90% of the global solar cell market. Today IMEC's largest effort is therefore in silicon solar cells. This includes the development of industrially viable processes for thin wafer based cells. Here

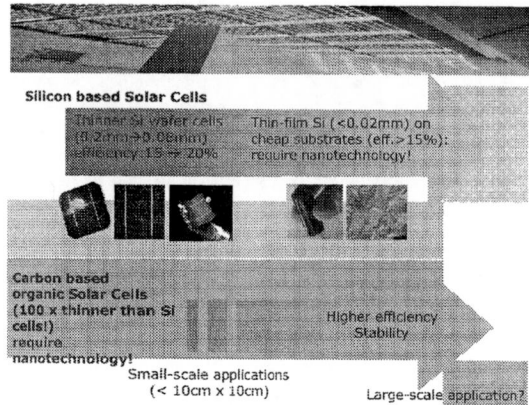

Fig. 4. IMEC's roadmap in photovoltaics.

cells are made in Si wafers that are not as pure as those used by the microelectronic industry. Since the cost of these cells strongly increases with the amount of silicon used in the wafer a strong cost reduction is possible by using thinner wafers. Today the wafer thickness is typically 0.2mm and the plan is, in the longer run, to use wafers that are only 0.08 mm thick. At the same time the conversion efficiency should increase from 15 to 20%. The long term goal is to go to thin film crystalline cells that are less than 0.02 mm thick, that are not mechanically self supporting and that must be deposited on cheap substrates. This approach requires nanotechnology as will be explained in the next section.

In parallel we are working on organic solar cells using organic layers of only a few 100 nanometers (100 to 1000 times thinner than Si cells). Therefore these cells are potentially much cheaper than Si cells. As the efficiency and stability of these cells are, for the time being, too low we see these cells in the short term only being used in small scale applications where the total cell area is smaller than 100 cm^2. Only when higher efficiencies and better stability will have been obtained these cells will be used in large scale grid connected power applications. But the development of stable and highly efficient organic solar cells also requires nanotechnology.

IV. NANOPHOTOVOLTAICS

Thin film crystalline solar cells epitaxially grown on cheap upgraded metallurgical grade Si as schematically shown in Fig. 5 can compete with conventional wafer based Si solar cells if the light absorption in the thin epi layer of a few microns thick

INVITED PAPER

can be enhanced. Recently, an important breakthrough was achieved on this type of solar cells at IMEC. We introduced a porous Si stack that acts as intermediate Bragg reflector, based on nanotechnology, enhancing light absorption and therefore yielding higher efficiencies.

Our analysis shows that by inserting a number of nanoporous silicon layers with alternating refractive index before epi growth the absorption of longer wavelengths can strongly be enhanced.

Carrier = Low-cost upgraded metallurgical silicon. Active layer deposited epitaxially by chemical vapor deposition. Intermediate reflector, based on nanotechnology, is needed to enhance optical path length.

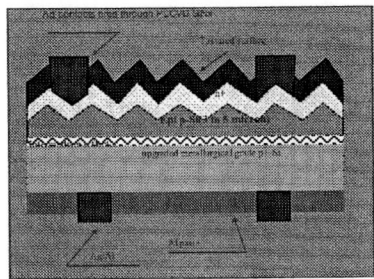

G. Beaucarne et al., 21st European Photovoltaic Conf., Dresden, 2006

Fig. 5. Thin-film Si cells on low-cost Si substrates.

The analysis of reflection measurements of epitaxial layers on such porous Si reflectors indicates that a high internal reflectance can be reached, exceeding 80% for a 15 layers stack.

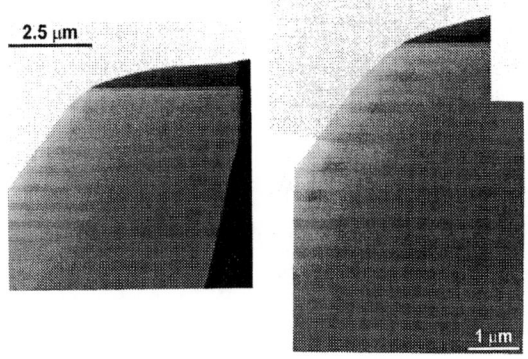

Fig. 6. Intermediate porous Si reflector: 15 porous Si layers-as anodized.

The stack (Fig. 6) is formed by an electrochemical etching process (anodization) in an HF solution, during which the current density is alternated between a high and a low value, resulting in a stack of multiple porous silicon layers with low and high porosities. If the appropriate conditions are used, epitaxial deposition with high temperature CVD is possible on such layers.

Annealing this stack at high temperature (1130 C) leads to reorganization of the porous silicon transforming the nanometer sized pores into large voids, some exceeding 0.1 µm width. However, the alternating porosity structure is maintained, so that the stack still functions as a Bragg mirror. It was shown that high quality epitaxial growth can be achieved when the porous silicon stack is completely reorganized into large voids with a closed surface.

Integrating the porous silicon buried reflector into epitaxial solar cell has been successfully carried out. On a highly doped single crystalline substrate, the buried reflector led to a short-circuit density increase of 3 mA/cm^2 and a conversion efficiency increase of 1.5 % absolute. Efficiencies close to 14 % have been achieved at IMEC with an industrial type process (POCl3 diffusion with high P surface concentration and screen printed contacts).

The fact that, in order to obtain efficient organic solar cells, nanotechnology is required comes from the low value of the exciton diffusion length. In a planar organic solar cell (Fig. 7)

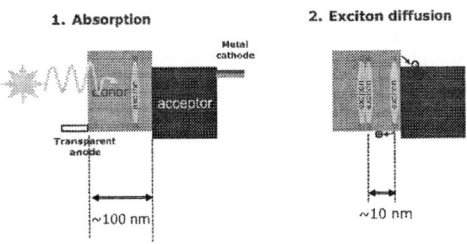

Problem: very low exciton diffusion length -> poor collection efficiency
Solution: **use of nanotechnology**
  -3D bulk heterojunctions
  -Absorption enhancement at the junction by metal nanoparticles

Fig. 7. Organic solar cells: efficient operation requires. nanotechnology.

the light enters through a transparent electrode and creates hole electron pairs, called excitons, in the diode consisting of a donor and acceptor material forming a junction. In an inorganic solar cell the exciton binding energy is very low and a created exciton immediately splits in an electron and hole that can be individually collected by the junction. In an organic solar cell, on the other hand, the exciton binding energy is large and the electron-hole pair, created by phonon absorption, must diffuse to the junction where the exciton can be splitted by the field and the electron can be collected by the acceptor material. Unfortunately the exciton diffusion length is extremely short, typically 10 nm; only the electron-hole pairs created within 10 nm of the junction are collected. Therefore the collection efficiency and also the global conversion efficiency of a conventional planar junction is small.

Fortunately nanotechnology can be used to solve this problem. Two possibilities exist: 3D (3 dimensional) junctions rather than planar ones and absorption enhancement at the junction by plasmonic effects in metal nanoparticles.

INVITED PAPER

The first approach is based on the use of 3D junction structures between the donor and acceptor type materials as schematically shown in Fig. 8. In such a structure the excitons

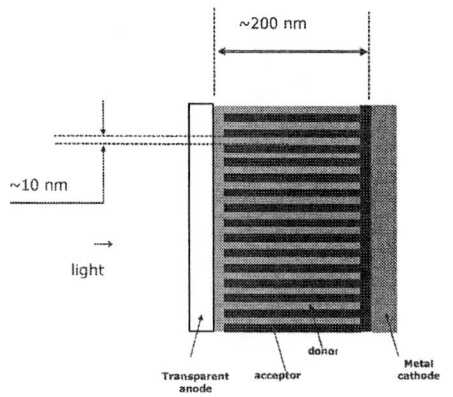

Fig. 8. First approach to high efficiency organic solar cells: ideal 3D bulk donor-acceptor heterojunction.

are everywhere created within a distance of the junction smaller than the diffusion length (typically 10 nm) and therefore the collection efficiency will strongly increase. In practice this approach can be implemented by the structure of Fig. 9.

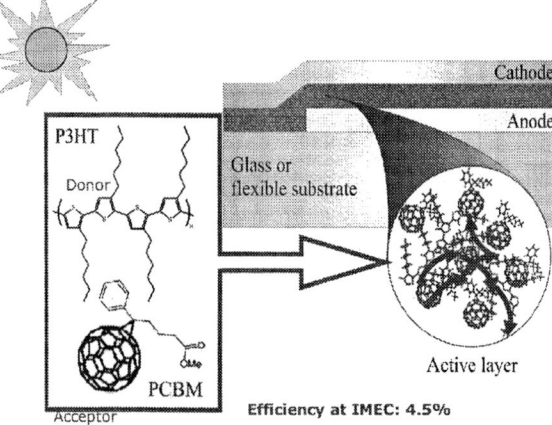

Fig. 9. Organic donor-acceptor bulk solar cells developed by IMEC.

A nanocomposite layer (Fig. 9) is sandwiched between two contact layers with a different work function one of which is transparent. The nano composite layer consists of two intermixed nano sized donor and acceptor type phases to enhance the dissociation of excitons into free charge carriers which then move by drift and diffusion to the respective contacts. In this example the nano composite active layer is a blend of a P3HT (PolyTriHexylThiophene) donor material and PCBM (Phenyl C61 Butyric acid Methyl ester) acceptor material which is derived from the C60 buckminster fullerene but is more soluble than the original C60. The C60 molecules

are in close contact in the PPV matrix such the electrons can hop from one molecule to the other to reach the metal cathode.

The second approach to high efficiency organic solar cells is combining organic PV with plasmonics. Organic solar cells suffer from low efficiency, mainly because of the low collection efficiency of the generated carriers, and the embedding of metal nano particles, around the junction, enhancing absorption close to collecting junction results in a strong increase in collection and conversion efficiency [1]. It is important that the metal nano particles are oxidized since without oxide layer the excitons will immediately recombine at the metal surface.

V. CROSS-FERTILIZATION BETWEEN MICROELECTRONICS AND PV

There are many areas where PV can learn from microelectronics. The first important area is that of environmentally benign processing. It is clear that, as PV will be the most material intensive semiconductor technology and certainly needs a green image.

The fact that crystalline Si based PV is a material intensive technology can be illustrated by looking at the Si consumption of PV and by comparing it with the Si consumption of electronics (Fig. 10).

	2007	2030
• Photovoltaics		
• Gram of Si/peakWatt(g/Wp)	10	2
• Yearly production (peak Giga Watt)		
• Tons of Si used	3	10^3
• Electronics (*)		
• Tons of Si used	3×10^4	2×10^6
	2×10^4	10^5

(*) based on 7% growth per year

Fig. 10. Si consumption for crystalline Si based PV and for electronics.

The Si consumption for crystalline Si based photovoltaics is indeed very large. In 2007, for the first time in history, the Si consumption of the PV industry exceeded that of the electronics industry. The 2007 number for the PV Si consumption comes from a Si use of 10 gram per Watt peak and a production level of 3 GWp. If we now extrapolate these numbers to the year 2030 we see that the PV industry will consume 20 times more Si than the electronics industry. This is based on the assumption of a 7% annual growth for the electronics industry and of 30% for the PV industry but at the same time assuming that that industry will be able to reduce the Si consumption from the actual 10 gram per Wp to 2 gram per Wp.

The second area is surface passivation because cell thickness decreases continuously and therefore surface passivation becomes increasingly important.

INVITED PAPER

VI. DEGRADATION AND FAILURE MODES IN PV

Today Si based PV has excellent reliability and lifetime characteristics. Most manufacturers specify a lifetime of more than 20 years. Possible degradation mechanisms can be classified as follows.

1. Semiconductor material related.
- Light induced lifetime instability in p-type Cz material.

Oxygen is unavoidable in Cz grown Si. Therefore B-O complexes can be formed. For a long time it has been observed that the performance of cells made from B-doped Cz-Si wafers degrades under illumination to stabilize at an efficiency one or two absolute percent below the initial efficiency [2]. The reason is a substantial decrease of the minority carrier lifetime due to light induced formation of defects associated with B-O complexes.

- Staebler-Wronski effect in amorphous Si.

It was already discovered in 1977 [3] that the illumination of amorphous Si films with sunlight lowered the conductivity. Today this effect is not yet completely understood but there is general agreement that hydrogen, critical to the passivation of dangling bonds, plays an important role.

2. Hot spot heating.

Current mismatching, e.g. resulting from shadowing a cell, in a string of cells will result in a hot spot since part of the power produced by the non shadowed cells will be dissipated in the shadowed cell. That cell can be overheated which may lead to destructive effects such as cell cracking or local melting of the solder. This problem can be avoided by adding bypass diodes over strings of maximum 12 cells.

3. Front surface soiling.

The accumulation of dirt can reduce module performance. In practice the resulting performance degradation is limited to 10% due to the cleaning by wind and rain.

4. Module optical degradation

Optical degradation of encapsulating materials can result in an important drop in efficiency.

Future interesting research topics in the field of PV reliability are the reliability effects of the use of extremely thin Si wafers, of concentrator systems and of organic PV. It can be expected that the use of ultrathin Si wafers (< 100 micron) will introduce additional reliability challenges on the module level. Concentrator PV systems track the sun with a high precision and therefore have moving parts. The long term reliability of these mechanical components introduces new challenges. Also the long term stability of the concentrating optics must be guaranteed. Organic semiconductors are known to be very humidity sensitive. Therefore the encapsulation of these cells will need special precautions.

VII. CONCLUSIONS

Photovoltaics is becoming an important industrial product. Crystalline Si PV dominates the actual PV market. The cost of crystalline Si solar cells will further decrease by the use of thinner wafers and of more efficient devices. Nanotechnology will play an important role in the longer term to obtain both cheaper Si and more efficient and stable organic solar cells. Several examples of cross fertilization between nanoelectronics and PV exist. Although the reliability of PV modules has been demonstrated to ensure lifetimes of 20 years in the field research is needed in areas such as the reliability of ultra-thin Si wafer based cells, concentrator PV and organic solar cells.

REFERENCES

[1] B. Rand, P. Peumans and R. Forrest, Journal of Appl. Physics 96, p.7519, 2004.
[2] H. Fischer and W.Pschunder, Proc. 10th Photovoltaic specialists Conf., p.13, 1973.
[3] D. Staebler and C. Wronski, Appl. Phys. Lett., Vol. 31, p. 292, 1978.

Au Nanocrystal Flash Memory Reliability and Failure Analysis

Pawan K Singh[1,2], Kaushal K Singh[2], Ralf Hofmann[2], Karl Armstrong[2], Nety Krishna[2], Souvik Mahapatra[1]

[1]Dept. of Electrical Engineering, IIT Bombay, Mumbai – 400076. INDIA
[2]Applied Materials, 3225 Oakmead Vill. Dr., Santa Clara, CA, 95054. USA
Ph: +1-408-986-3353, FAX: +1-408-563-6311, Email: pawan@ee.iitb.ac.in

Abstract- **In this work we investigate the memory performance and reliability of Au nanocrystal memory devices. We analyze the Au NC formation process and fabricate actual test wafers for electrical characterization. With reference to good Pt NC devices, poor performance of Au NC devices is investigated in detail by analytical and electrical methods.**

I. INTRODUCTION

Conventional floating gate NAND Flash is unlikely to scale below the 3X node [1], and TANOS/CTF [2]-[3] and metal NC [4]-[6] devices are considered as possible alternatives. So far, reported CTF devices have shown good memory window but poor retention, while NC devices show poor memory window. While poor retention of CTF is linked to shallow trap depth of the nitride storage layer and is difficult to control, the choice of tunable workfunction of the metal-NC storage layer is advantageous for achieving good retention, provided optimal NC area coverage is obtained.

A significant amount of work on nanocrystal based flash memory structures has been done using semiconductor nanocrystals like Si [7] and Ge [8]. A few issues with semiconductor based NCs that have been noted very widely are poor workfunction, quantum confinement leading to Fermi level splitting and poor control over the NC size which results in poor overall performance of the devices. Use of metal as NC offers significant advantage over the semiconductor as some metals have very large workfunction, very little to no quantum confinement and accurate control of NC size. A large workfunction can improve the retention of the device without compromising on P/E speeds. The metals that have been studied till now include Pt, Au, Ni, Co, W etc. Au nanocrystal based NC devices have been reported as a suitable candidate for NC flash application owing to the large workfunction (5.1eV) and good chemical stability of Au metal [4], [9]-[11].

In this work we report memory operation and reliability of Au nanocrystal embedded in optimized Al_2O_3 blocking dielectric using industry standard tools and processes for more accurate evaluation. The fabrication process is chosen to closely resemble the standard CMOS process flow to accurately evaluate the performance of these devices. We have used industry standard tools (PVD, RTP etc) from Applied Materials to complete the fabrication.

II. FABRICATION

Fabrication of the devices is performed on p-type wafers with doping level of $\sim 1 \times 10^{15}/cm^3$. After Surface clean by HF, 40Å ISSG tunnel oxide is grown with H2 and O2 gas flow in the RTP chamber. Nanocrystal deposition is performed via PVD sputter in a 200mm Endura tool. Au target is chosen with 99.999% purity with <5ppm Cu. Au metal deposition followed by anneal forms the metal nanocrystals. 120Å optimized Al_2O_3 is then deposited as the blocking dielectric. Gate metal is chosen to be Pt (1000Å) because of the large workfunction reducing the top electrode injection during erase. For Pt NC devices, Au deposition was replaced by Pt deposition and other steps were kept identical.

III. RESULTS

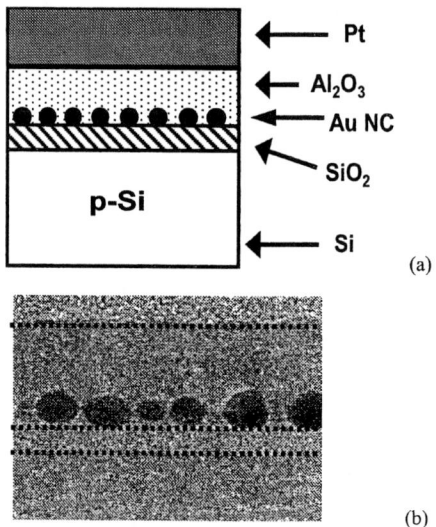

Fig. 1 a) Schematic of NC device showing the position of nanocrystals in the stack b) cross-section of actual fabricated device stack showing the corresponding materials.

A. Physical Analysis

Au nanocrystals are embedded in the deposited Al_2O_3 film as shown in the schematic (Fig. 1) which is confirmed by cross-section TEM images. Plane view TEM of samples with 5Å and 10Å gold deposition, with and without NC anneal (Fig. 2), are used to extract the number density, areal coverage and size distribution of nanocrystals in the gate stack. Table-1 summarizes device details and extracted parameters.

We find that maximum NC number density is observed for samples with 5Å gold deposition but with small area coverage of ~20%. Samples with 10Å gold show an optimal area coverage of ~30% [12] but the number density is small and the size distribution of NCs is very wide. Also NC anneal is found to decrease the number density of the Au NCs. The trend for area coverage is not very clear. We extract the size distribution of the NCs from the TEM images and these are plotted in Fig. 3. It is observed that, un-annealed 5Å samples have NC size in the range of 3 nm - 5 nm (average of ~3.2 – 3.7nm), which is very desirable. After anneal the distribution get tightened even further. Although the distribution moves slightly towards larger sizes, the width of the distribution is reduced. On the other hand, 10Å gold samples show a very wide distribution of NC sizes, with average size of ~5.5nm. Annealing tightens the distribution by causing coalescence of some smaller dots into larger ones.

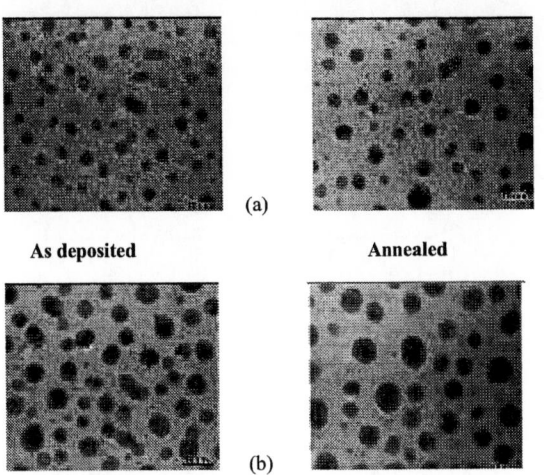

As deposited **Annealed**

Fig. 2 Plane view TEM Images of a) 5Å as deposited and b) 10Å as deposited and Annealed. Annealed samples have larger size of nanocrystals.

Table I
DEVICE SPLITS USED IN THIS WORK ILLUSTRATING THE EXTRACTED PHYSICAL PROPERTIES OF THE NANOCRYSTAL DEVICESW

ID	Initial Dep	Anneal	Area Coverage	Max size (nm)	Avg size (nm)	Min Size (nm)	Density (#/cm²)
S11	5Å	N	20.7%	5.4	3.2	2.5	2.50E+12
S12	10Å	N	25.3%	9.5	5	1.9	1.20E+12
S2	5Å	Y	17.2%	9.7	3.7	2.7	1.60E+12
S5	10Å	Y	28%	9.2	5.4	3.1	1.20E+12
Pt	5Å	Y	26%	5.5	2.8	2.2	3.40E+12

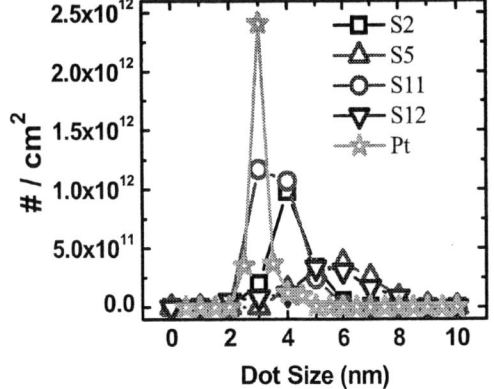

Fig. 3 Size distribution of Au nanocrystals extracted from TEM images. Cumulative of all gives the total number density.

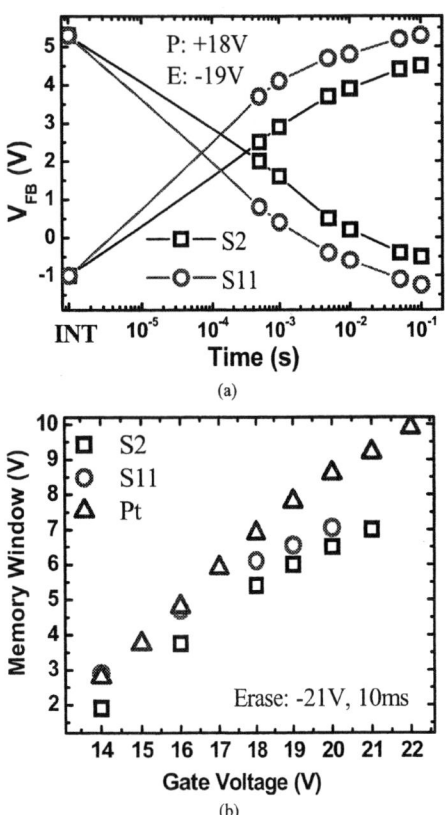

Fig. 4. a) Program/Erase transients and b) Memory window at different gate voltages for S2 and S11 samples. Faster P/E and larger window is observed for S11 compared to S2.

978-1-4244-2039-1/08/$25.00 ©2008 IEEE

B. Electrical Analysis

Electrical analysis of the devices is performed on capacitor with diameter = 100μm. We use a constant capacitance method for the extraction of flatband voltage of the devices. QSCV measurements performed using a 4156C semiconductor parameter analyzer gives the V_{FB} values for the devices. We perform Program/Erase transient, gate leakage, retention and endurance studies on these structures.

Program/Erase transient for 5Å samples (S2 and S11) are shown in Fig. 4 (a). Unannealed sample (S11) shows a significant improvement in the P/E speeds when compared to the annealed sample (S2). When we compare the extracted analytical data for these two devices we find that sample S2 has lower area coverage and number density compared to S11. This could result in the smaller P/E speed and memory window observed in the transients. When we compare the memory windows at different programming bias for S2 and S11 we find that the trend is continued (Fig. 4(b)). A comparison is made to optimized Pt sample with area coverage of 26% and number density of 3.4×10^{12} cm^{-2}, which shows a very significantly improved memory window with increasing gate bias. We can therefore conclude that number density and area coverage are essential to obtain a good performance for the NC devices. Though lower when compared to Pt, we were able to achieve memory window of ~7V in Au NC devices. We observe a saturation setting in the memory window at higher biases. We also observe a significant overerase in the devices at negative gate biases. We were able to achieve negative V_{FB} for the devices under erase. Use of optimized low leakage Al_2O_3 with a high workfunction metal gate (Pt) suppresses the injection of electrons from the top gate during erase and thus negative V_{FB} can be achieved.

While performing similar experiments with 10Å samples we found ourselves unable to measure the QSCV characteristics for any of the 10Å devices (S5, S12). The cause of this was found to be the high gate leakage observed from the 10Å Au gate stacks. As observed in Fig. 5 (a), the gate leakage of S12 is orders of magnitude higher than sample S11. Also the breakdown voltages were significantly lower. The high gate leakage was making CV measurements impossible using our system. No further measurements were performed using the 10Å samples in this work.

In a separate set of measurements we also found that the gate leakage of the NC devices was also dependent on the temperature of the NC anneal as observed in Fig 5 (b). Slightly larger leakage of the S2 sample compared to the S11 sample can also contribute to the lower observed memory window for S2 (Fig 1.).

Retention (Fig. 6) experiments were performed on 5Å samples to asses the reliability. These experiments were

performed at room temperature with 0V applied at the gate

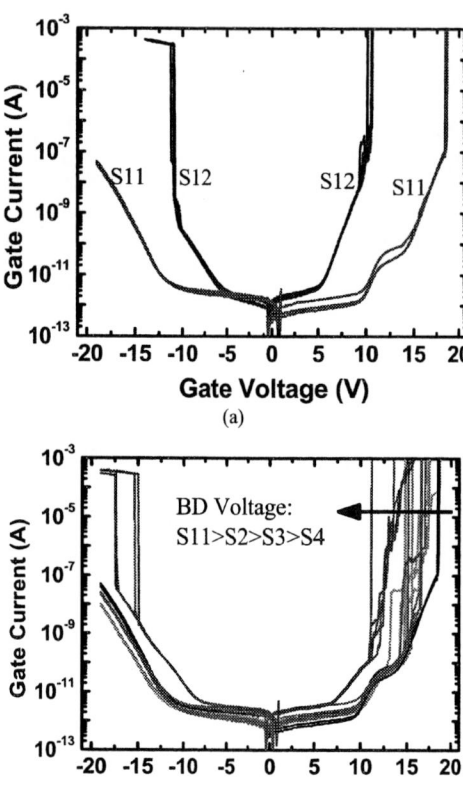

(a)

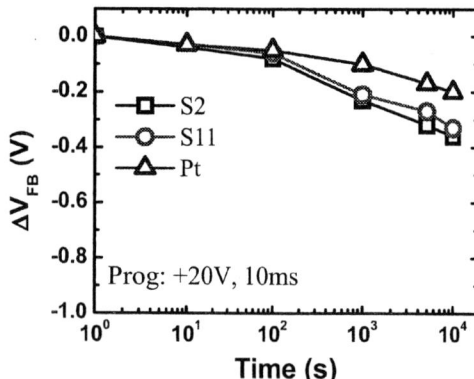

(b)

Fig. 5 Gate Leakage comparison between a) S11 and S12. Very large leakage is observed for S12 stack. b) 5Å metal samples with increasing anneal temperature. (S3 and S4 have higher NC anneal temperature than S2).

Fig. 6 Pre-Cycling retention for S2 and S11 devices. S11 device shows marginally better retention than S2.

during the retention stress. The retention stress was stopped and the V_{FB} measured at pre-determined time intervals. Pre-cycling retention loss of these devices after 10K seconds was found to be close to 0.35V and 0.4V for S11 and S2 respectively.

Retention loss was found to be larger for S2 sample than S11. The retention loss of both Au samples was found to be larger than Pt sample which showed a loss of only about 200mV after 10K seconds. These observations are consistent with our observation on gate leakage.

A bigger issue is observed with device cycling endurance (Fig. 7). For a memory window of 4.2-4.4V the devices could only be cycled for ~2K cycles before break down. Breakdown was observed at ~1K cycles for initial memory window of 5.1V (not shown). Similar observation is made for both S2 and S11 samples. The poor cycling characteristics are due to breakdown of dielectric film in the gate stacks.

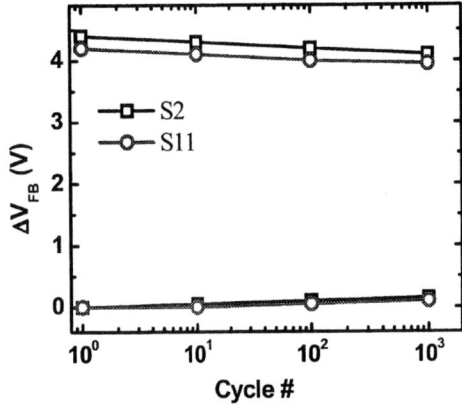

Fig. 7 Cycling endurance characteristics for S2 and S11 devices. A window closure of ~350mV and 400mV is observed for S11 and S2 respectively.

C. Failure Analysis

Gate leakage and breakdown voltage dependence on initial metal thickness and NC anneal temperature suggest a temperature activated mechanism. We speculate the diffusion of Au atoms during the PDA to be the dominant mechanism causing the device failure. Cross-section TEM along with EDX (Fig. 8) shows the presence of Au and Cu in the dielectric close to NC (sites 2 and 4) whereas the signal is almost unobservable at sufficient distance from the NC (site 1).

Low melting point of Au and Cu causes diffusion of metal atoms into the gate dielectric during NC anneal and PDA enhancing the leakage. Trace metal amounts were also found in Al_2O_3 BD film. This could not have been possible during the NC anneal as the blocking dielectric is deposited after NC anneal. Post- Al_2O_3-deposition-anneal is also suspected to increase the diffusion of the metal atoms in the film. Further analysis is required to fully understand the reason for failure of Au NC devices with anneal temperature increase.

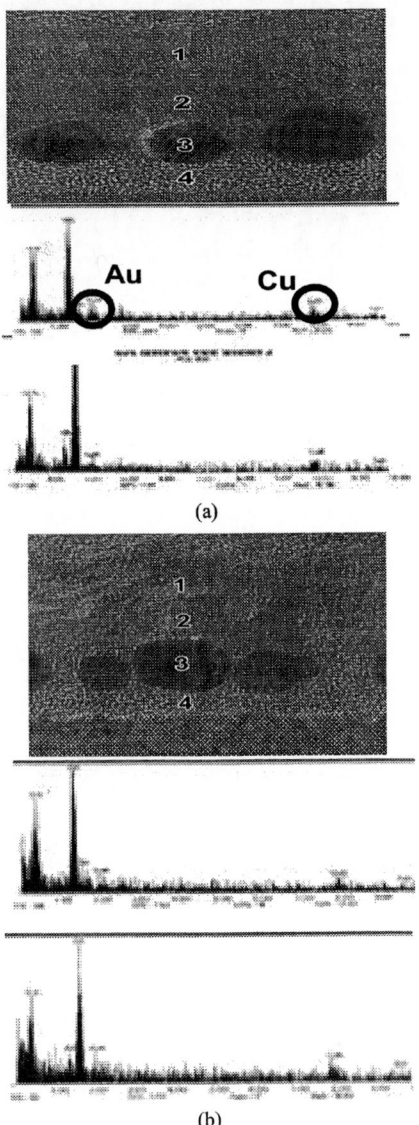

Fig. 8. TEM and EDX of a) S5 and b) S12 samples at position 2 and 4. EDX analysis shows presence of trace amounts of Au and Cu in the dielectric. No significant metal is seen in position 1 in both samples.

IV. CONCLUSION

In this work we fabricate and analyze Au metal nanocrystals for use in metal NC flash EEPROM applications. It is observed that Au (and Cu impurity in the target) metal atoms diffuse into the tunnel and blocking dielectric causing serious reliability issues like lowering of breakdown voltage, poor retention and cycling endurance in the memory devices. In conclusion, Au is shown to be unsuitable for NC flash application irrespective of a good workfunction and chemical stability. Other metals such as Pt need to be investigated for the application.

REFERENCES

[1] K. Kim, "Technology for sub-50 nm DRAM and NAND flash manufacturing," *IEDM Tech. Dig.*, 2005, pp. 333–336.

[2] Y. Park, "Highly Manufacturable 32Gb Multi – Level NAND Flash Memory with 0.0098 μm^2 Cell Size using TANOS (Si -Oxide - Al$_2$O$_3$ - TaN) Cell Technology, *IEDM*, 2006, pp. 54-55

[3] Z. Huo *et al*, "Band Engineered Charge Trap Layer for highly Reliable MLC Flash Memory", *VLSI Tech. Symposium 2007*, pp 138-139

[4] Z. Liu, C. Lee, V. Narayanan, G. Pei, and E. C. Kan, "Metal Nanocrystal Memories—Part I: Device Design and Fabrication" *IEEE Trans. Electron Devices*, Vol. 49, p. 1606-1613.

[5] S. K. Samanta *et al*, "Enhancement of memory window in short channel non-volatile memory devices using double layer tungsten nanocrystals", *IEDM Tech Dig 2005*, pp. 170-173

[6] P. K. Singh, A Nainani, "Extensive Reliability Analysis of Tungsten Dot NC Devices Embedded in HfAlO High-k Dielectric under NAND (FN/FN) Operation", IPFA 2007, pp. 197-201

[7] B. De Salvo *et al*, "Performance and Reliability Features of Advanced Nonvolatile Memories Based on Discrete Traps (Silicon Nanocrystals, SONOS)", IEEE Trans. Device, Materials and Reliability, Vol. 4, No. 3, 2004. pp. 377 - 389

[8] Jing Hao Chen *et al*, "Nonvolatile Flash Memory Device Using Ge Nanocrystals Embedded in HfAlO High-k Tunneling and Control Oxides: Device Fabrication and Electrical Performance", IEEE Trans. Electron Devices, Vol. 51, No. 11, 2004. pp. 1840 - 1848

[9] C. C. Wang *et al*, "Memory characteristics of Au nanocrystals embedded in metal–oxide–semiconductor structure by using atomic-layer-deposited Al$_2$O$_3$ as control oxide", *J. Phys. D: Appl. Phys.* **40** (2007) 1673–1677,

[10] W Guan *et al*, "Fabrication and charging characteristics of MOS capacitor structure with metal nanocrystals embedded in gate oxide", *J. Phys. D: Appl. Phys.* **40** (2007) 2754–2758

[11] C-H Lee *et al*, "Metal nanocrystal/nitride heterogeneous-stack floating gate memory", *Proc. DRC* 2005, pp 97-98

[12] A. Nainani *et. al.*, "Development of A 3D Simulator for Metal Nanocrystal (NC) Flash Memories under NAND Operation", *IEDM Tech. Dig.* 2007

Characterization and Modeling of Program/Erase Induced Device Degradation in 2T-FNFN-NOR Flash Memories

Guoqiao Tao, Helene Chauveau, Dick Boter, Erik van der Vegt, Do Dormans and Rob Verhaar
NXP Semiconductors, Gerstweg 2, 6534 AE Nijmegen, The Netherlands
Corresponding author: Guoqiao Tao Phone: +31 24 353 4549 Guoqiao.Tao@nxp.com

Abstract – In this paper, we report the program/erase degradation mechanisms in two transistor (2T) Fowler-Nordheim (FN) tunneling operated flash memories, based on extensive experimental study of the degradation characteristics of such 2T-FNFN test memory arrays and reference transistor arrays from several generation process technologies. A quantitative model has been established describing the degradation characteristics under various stress conditions (*i.e.* degradation due to program/erase cycling at various voltages and temperatures). A software tool has been developed to estimate the reliability performance of various products under different use conditions. This model (and tool) can also be used to estimate the product reliability performance in future process generations.

I. INTRODUCTION

As the complexity of electronic devices continues to increase, more and more system on chip (SOC) devices have found their way in a wide range of application areas: from consumer to automotive, from digital right management to secured e-banking/e-passport, from office environment to applications like Mars exploration missions. At the same time, the overall quality and reliability requirements are increasing. For products with embedded non-volatile memories (NVM, flash or EEPROM memory), the most important parameters are the data retention and program/erase endurance cycles. To meet the ever-increasing application and quality requirements, it is essential to characterize, understand the reliability performances of these devices, establish a sound model for that, predict the device performance under particular use conditions, and providing customer-tailored solutions for special applications.

In this paper, we report the characterization of degradation mechanisms in two-transistor (2T) Fowler-Nordheim (FN) tunneling operated flash memories. Based on extensive experimental characterization of the 2T-FNFN test memory arrays and flash transistor arrays from several process technology generations, it has been found that the degradation mechanism is due to electron trapping in the tunnel oxide and the interface states generation at the Si/SiO2 interface. A quantitative model has been established describing the degradation characteristics under various stress conditions (*i.e.* degradation due to program/erase cycling at various voltages and temperatures). This model also takes into account of the cell-to-cell variations in large memory arrays. A software tool has been developed based on this model. This tool enables us to predict the reliability performance of various products under

different use conditions. It can also be used to estimate the product reliability performance in future process generations.

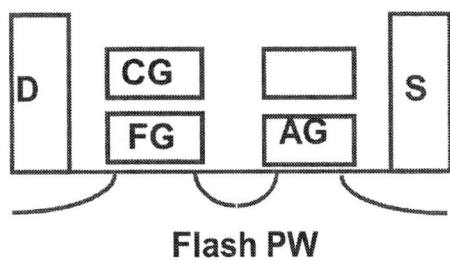

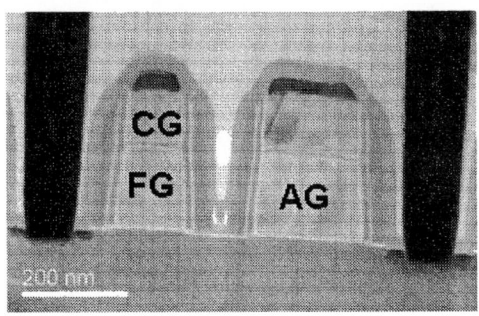

Figure. 1: *Upper graph: A schematic cross section of the two-transistor cell. Control gate (CG) and floating gate (FG) are on the side near the drain (D) while the access gate (AG) is on the side near the source (S). Lower graph: TEM cross-section of the two-transistor cell in 90nm technology node.*

II. EXPERIMENTAL SET-UP

The flash memory devices in a multi-purpose 2T-FNFN NOR embedded Flash technology in 90nm node

Figure 1 shows a schematic cross-section of a (N-channel based floating gate) 2T-FNFN flash memory device, and a corresponding TEM cross-section of such a device made in 90nm generation node [1]. The cells used in this study typically have a 8.5nm ISSG (In-Situ Steam Generation) RTO tunnel oxide with post oxidation nitridation. The inter-poly-dielectric (between CG and FG) is a 6/5/5 nm ONO stack (13.5nm oxide equivalent). The AG in the cell ensures that uniform FN tunneling can be applied for both programming and erasure when the cells are configured in a NOR array (2T-FNFN-NOR). The threshold voltage (Vt) of such a device is modulated by the net charge on the floating gate, as expressed by equation (1):

$$V_t = V_{t0} - \frac{Q_{fg}}{C_{ipd}} \qquad (1)$$

Where the V_{t0} is the "neutral threshold voltage (V_t)" of the cell when the floating gate is neutral, Q_{fg} is the charge on floating gate and C_{ipd} is the capacitance between control gate and floating gate, all assuming that the dielectrics (tunnel oxide and inter-poly-dielectric) are charge free. Typically, a 1 ms pulse of +15V is applied to the CG to program the cell (V_t high state), and 100ms pulse of –15V is applied to erase the cell (V_t low state).

The flash memory test array

In order to study the memory cell degradation due to program/erase cycling, and to study the cell-to-cell variations, a 2.7Mbit random accessible cell array has been used. The cells are arranged in an NOR array with multiplexer for the bit-lines (Drain) and word-lines (AG), while the control gates are all connected together. In this way, all the cells can be programmed/erased simultaneously, while they can be measured individually.

Array of flash reference transistors

This reference transistor array is similar to the flash memory test array described above, but with the floating gate and control gate shorted. The transistors in this array can be electrically stressed in parallel, and every transistor being measured individually. This allows us to separate the transistor degradation mechanisms under positive gate stress (emulating programming stress) and negative gate stress (emulating erasing stress) conditions. The transistor-to-transistor spread under the same stress condition can be studied as well.

III. EXPERIMENTAL RESULTS

Vt evolution under FN programming/erase cycling.

Programming/erasure cycling tests have been performed on cell arrays with single-shot pulses (without program-verify or erase-verify). Figure 2 shows typical write/erase characteristics of cell threshold voltages. Under uniform FN tunneling programming/erasure cycling, both programmed and erased V_t increase as a function of number of cycles [2], due to the generation of interface states at the tunnel oxide to Si interface and due to electron trapping in the tunnel oxide (indicated as "a" on the right graph in figure 3). Additionally, the trapped electrons in the tunnel oxide will degrade the tunneling efficiency for both programming and erasure (indicated as "b" on the right graph in figure 3), causing a reduction of the Vt window ($V_{tp} - V_{te}$), or an effective increase of tunneling barrier as shown in the left part of figure 3. From figure 2, it is clear that the erased state is the most

critical state in terms of reliability, because the difference between the read condition and erased V_t (V_{te}) is getting smaller after P/E cycling.

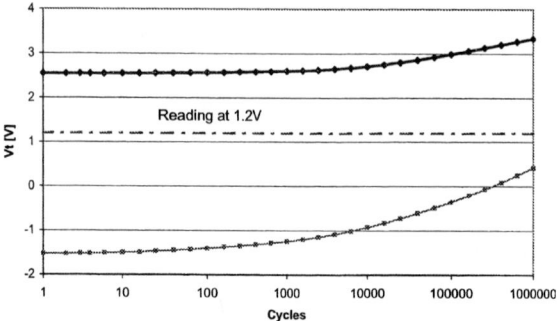

Figure 2. *Typical endurance characteristics of a flash cell (median cell in a cell array) with single-shot P/E pulses.*

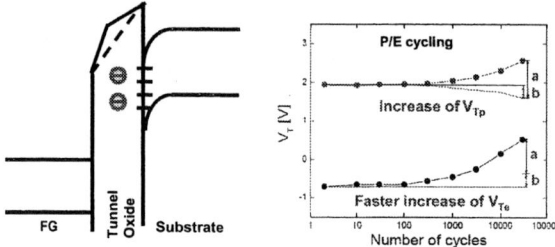

Figure 3. *The effect of P/E cycling stress causes generation of interface states and negative oxide charge. Left: Simplified energy-band diagram; Right: effect on endurance characteristics (after [2]).*

Detailed studies reveal that the degradation of erased V_t follows a square-root law for conventional tunnel oxides [3] and a cubic-root law for appropriately nitrided tunnel oxides [4] as depicted in figure 4. In general, it can be described by the following power-law equation:

$$V_{te} = V_{te0} + D_e * (cycles)^n \qquad (2)$$

Where V_{te0} is the initial value of the erased V_t; and D_e is the degradation coefficient; and n is the exponent in the power-law. We recently reported that both V_{te0} and D_e are functions of the V_t window, thus the program/erase conditions (pulse duration, CG voltage during programming/erasure and temperature) [4]. As a first order approximation, a "linear" relationship can be used (around the default P/E conditions):

$$D_e = D_{e,0} + D_{e,pulse} * \ln(PulseE) + D_{e,V} * V_{e,CG} + D_{e,t} * T \qquad (3)$$

Where the *PulseE* is the erase pulse duration, $V_{e,CG}$ is the erase voltage, T is the temperature, and [$D_{e,0}$, $D_{e, pulse}$, $D_{e,V}$, $D_{e,T}$] are all experimentally determined coefficients. Similar equations can be established for programmed state V_{tp}, with the corresponding parameters for programming:

$$V_{tp} = V_{tp0} + D_p * (cycles)^n \qquad (4)$$

$$D_p = D_{p,0} + D_{p,pulse} * \ln(PulseP) + D_{p,V} * V_{p,CG} + D_{p,T} * T \qquad (5)$$

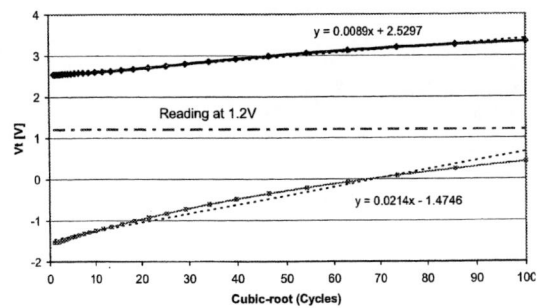

Figure 4. *Data taken from figure 2, but plotted against cubic-root of cycles on X-axis.*

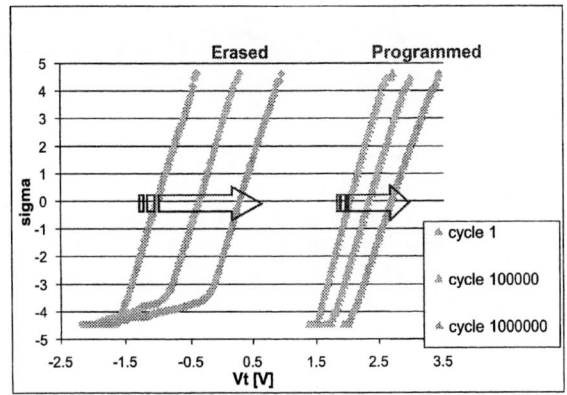

Figure 5. *Cumulative distributions of Vte and Vtp at various P/E cycles on a normal plot (with standard deviation sigma as the unit of Y-axis).*

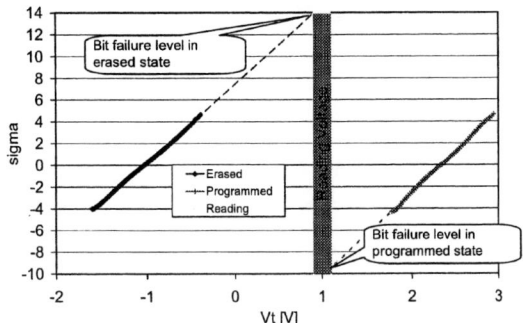

Figure 6. *A simplified graphical presentation of the method used in determining bit-failure levels from the Vte and Vtp distributions.*

Cell-to-Cell Variance, and scaling from cell array to products

Real products can contain huge amount of cells. By nature, there are cell-to-cell variations (caused by variations in cell geometry, variations of local dopant atoms in the channel, and some statistical variations in the cell degradation). Figure 5 shows distributions of programmed and erased threshold voltages, before and after programming/erasure cycling. The V_t values increase (the lines are moving to the right) as a result of cycling – the same as presented in figure 2 and figure 3. Also the distribution widths slightly increase as a result of cycling.

Keeping in mind that the weakest cell determines the product reliability; we can estimate the bit level reliability as graphically shown in figure 6. It should be noted that when a bi-nominal distribution is present [5], the most critical part (close to the reading condition) should be used for determining the bit-failure level. For products with error-correction code (ECC) on board [6,7], a "random single bit" failure can be corrected and will not cause a product failure. In that case, the product failure rate depends on the product size, the ECC scheme and the bit failure level [6,7].

Reliability simulation software tool

Taking all the above effects into consideration, after having experimentally determined the in the equations (2) to (5), a NVM reliability simulation tool has been developed. Figure 7 shows an example of the output of the simulated results when the cycling temperature is varied. Similar outputs can be easily presented when varying the product memory size, the cycling voltage and/or pulse duration.

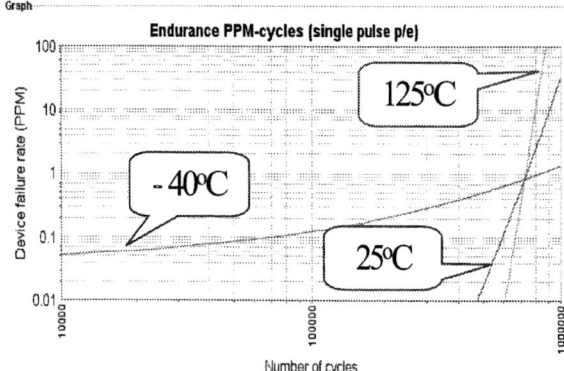

Figure 7. *Simulated program/erase endurance reliability performances of a 16Mbit chip when P/E cycled at various temperatures. The increase of the device failure rate is partially due to the increase of the erased Vt and partially due to the slight increase of the distribution width.*

Array of flash reference transistors

Both interface states and trapped charge are responsible for the endurance V_t window degradation. Two questions arise: (A), how to separate the interface states from the trapped oxide charge? (B), how do programming and erasing stresses degrade the transistor separately? By stressing the flash reference array, as shown in figure 8, it can be observed that the programming stress (positive voltage on the gate) directly causes electron trapping in the tunnel oxide (V_t continuously increases without significant trans-conductance degradation under +8V gate stress), while the erasing stress (negative voltage on the gate) initially involves hole trapping in the oxide before electron trapping dominates (V_t slightly decreases first and then increases definitively). More detailed study of the transistor trans-conductance shows that the easing stress causes significantly (deep) interface states (shown as significant trans-conductance degradation, to be published elsewhere).

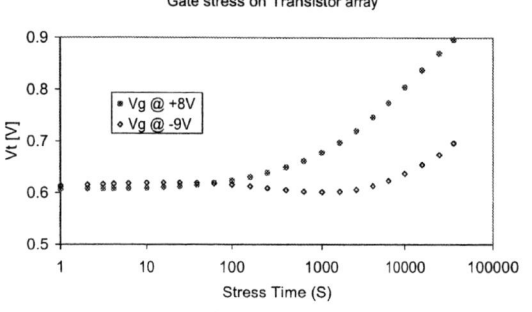

Figure 8. *Vt evolution of the reference transistor array under emulated programming stress (+8V) and erasing stress (-9V).*

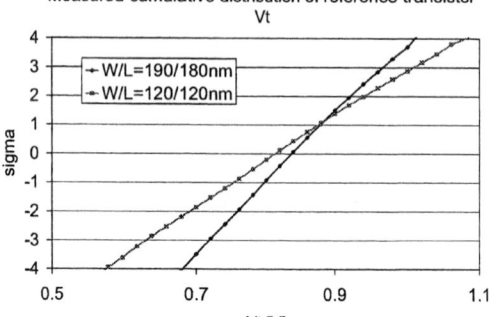

Figure 9. *Vt distribution of the flash reference array widens as the technology scales (marked by the W/L values).*

Figure 9 shows the measured Vt distribution from the flash reference array from two process generations (marked by the W/L values). It can be seen that the distribution widens as the technology scales, as expected from the transistor matching theories [8] due to scaling of the transistor geometry.

IV. DISCUSSIONS ON SCALING

When scaling the technology, the transistor dimensions (W & L) will become smaller. As a result, the natural cell-to-cell variation will increase. Figure 10 shows the V_t variation dependence on the transistor active area, given the gate oxide being fixed at 8.5nm. The solid line presents the values extracted from figure 9 and the dashed line shows projected values based on theory [8]. In figure 11, it is schematically shown for the flash cell V_t distributions for reference process technology and for that technology plus two generations (thus with W & L roughly half of the reference values). When the median V_t window is kept the same (small window, thin purple and thin red lines, that are crossing the thick-lines at sigma=0, but with a half of the reference slope= 2x wider distribution), the wider distribution (twice as wide as the reference) will lead to a higher bit failure level (the level where these distributions are crossing the read-conditions). Since the wider distribution is inherent to technology scaling, the solution should come from device/circuit/product/system design. In order to lower the bit failure level (for single pulse P/E, as mentioned above), a higher median window will be needed, as indicated by the thin green lines in figure 11 (by shifting the thin purple and red lines further away from the read-condition).

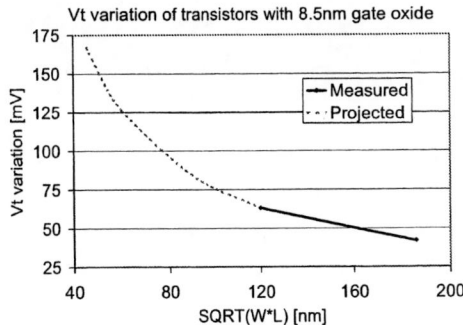

Figure 10. *The transistor (the gate oxide kept at 8.5nm in this graph) Vt variation will increase when technology scales (decreasing W & L). .*

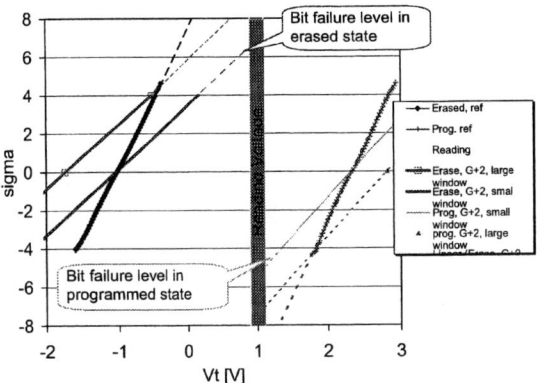

Figure 11. *Vt distribution will become wider in scaled technologies.*

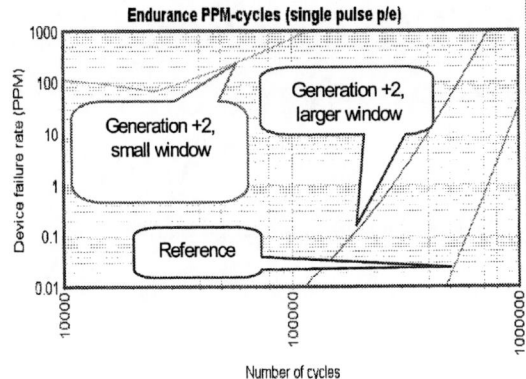

Figure 12. T*he simulated impact on product reliability with scaled technologies (wider Vt distribution).*

V. SUMMARY

The uniform FN programming/erase cycling induced Flash memory device degradation has been studied. The generation of interface states and negatively charged traps in the tunnel oxide causes the increase of erased Vt (of N-channel device), forming the critical state for product reliability. With such quantitative knowledge, a model has been developed, helping to estimate the NVM reliability performance under various use conditions and helping to estimate the scaling effect in future processes.

Assuming that the degradation characteristics (cubic-root law, dependence on initial window and temperature, and increase of distribution width) in scaled technologies are the same as in reference technology, the program/erase endurance reliability in scaled technology can be simulated. Figure 12 shows the impact on 16Mbit chip when the technology is scaled for two generations further, with two scenarios: "small window" and "larger window", together with the reliability performance of reference technology. "Small window" means the same median window as the reference technology. "Larger window" corresponds to the same initial lines in figure 9, the scenario with a larger window.

Alternatively, intelligent P/E algorithms should be applied to artificially narrow the cell-to-cell variation (to achieve a lower bit level failure rate.

Yet another possibility would be to employ a much heavier error correction code (ECC) scheme (to tolerate a higher bit level failure rate [7]).

In scaled technologies, a combination of more than one measure in device/circuit/product/system level would be a realistic scenario.

REFERENCES

[1]. G. Tao , *et al.*, "A low voltage, low power, highly reliable, multi-purpose, cost-competitive embedded non-volatile memory in 90nm node", in ICMTD 2007, pp 113-115.

[2]. A. Modelli, "Flash memory reliability", IRW 2005 Tutorial.

[3]. H. Belgal, *et.al.*, "A New Reliability Model for Post-Cycling Charge Retention of Flash Memories", in IRPS 2002, pp 7 – 20.

[4]. G. Tao , *et al.*, "A quantitative study of endurance characteristics and its temperature dependence of embedded Flash memories with 2T-FNFN NOR device architecture", In IEEE TDMR Vol 7, No. 2, (June 2007), pp 304 –309.

[5]. A. Scarpa, *et al.*, "Tail bit implications in advanced 2T flash memory device reliability", in Microelectron. Eng., vol. 59, pp. 183–188, 2001.

[6]. G. Tao, *et.al.* " On intrinsic failure rate of products with error correction", IEEE IRW 2005 final report, pp 71-73

[7]. N. Mielke, *et.al.* "Bit Error Rate in NAND Flash Memories", IEEE IRPS 2008, pp 9-19.

[8]. M. Vertregt, "The analog challenge of nanometer CMOS", IEDM 2006, pp 1-8

The Effect of Band Gap Engineering of the Nitride Storage Node on Performance and Reliability of Charge Trap Flash

Sandhya C.[a], U. Ganguly[b], K.K. Singh[b], C. Olsen[b], S. M. Seutter[b], G. Conti[b], K. Ahmed[b], N. Krishna[b], J. Vasi[a]
and S. Mahapatra[a]

[a] EE Dept, IIT Bombay, Mumbai 400076, India;
[b] Applied Materials, Santa Clara, CA, USA
Email: sandhyac@ee.iitb.ac.in

Abstract – **The effect of nitride composition, i.e. Si-rich (Si$^+$) and N-rich (N$^+$) nitride bi-layers separated by an oxynitride (SiON) layer on memory performance and reliability is studied. Bottom Si$^+$ layer and top N$^+$ forms the Si$^+$/N$^+$ bi-layer that is compared to the opposite configuration of N$^+$/Si$^+$ bi-layer to reveal large impact on memory performance and reliability. Si$^+$/N$^+$ bi-layers exhibit superior P/E windows and endurance characteristics but worse retention charge loss compared to N$^+$/Si$^+$ stacks. The oxynitride layer composition and position play a dominant role in trap generation as evident from endurance performance. A low energy-threshold degradation mechanism with higher degradation of the SiON layer with greater H-content is observed. A Si-H bond breaking mechanism is proposed as trap generation mechanism during endurance cycling. Retention is primarily bottom nitride composition dependent as tunnel oxide is shown to be the dominant charge loss path.**

I. INTRODUCTION

Charge Trap Flash (CTF) demonstrates negligible inter- cell interference, superior scalability, planar structure and E:\Classified\Web Jobs\CMOS compatible process flow and presents a promising replacement to floating gate for scaled NAND Flash beyond the 3X node [1]. Multi-level cell (MLC) operation in CTF requires maintenance of large memory window during cycling endurance and data retention. Band-gap engineered silicon nitride (Si_XN_Y) trap layer has recently attracted significant attention to optimize CTF performance and reliability [2-4]. Single layer nitride composition engineering of silicon nitride i.e. Si-rich (Si$^+$) vs. N-rich (N$^+$) shows performance trade-offs of program speed versus retention while offering poor endurance window insufficient for MLC operation [2]. Recently, negligible cycling degradation using a composite charge trap layer consisting of NAN (N: nitride, A: Al_2O_3) storage material in an advanced TANOS stack was demonstrated [4]. Previously, we have shown that a Si$^+$/N$^+$ nitride bi-layers separated by a controlled oxy-nitride (SiON) interface in a SONOS stack provides comparable performance enhancement with simpler process integration [2]. This work compares the performance and reliability of Si$^+$/N$^+$ and N$^+$/Si$^+$ bi-layer storage material. The importance of the position and composition of the interface SiON layer is demonstrated and plausible underlying physical mechanisms are discussed.

II. DEVICE STRUCTURE

The fabrication process flow for the SONOS gate stacks is illustrated in Table-1. The devices characterized were

- **N substrate wafer clean**
- **Tunnel oxide (3.5nm), RTO**
- **LPCVD nitride storage layer**
- **HTO deposited top oxide (8nm)**
- **Post HTO anneal**
- **Poly deposition (10nm)**
- **Boron Implant**
- **Metal contact formation**
- **Backside metallization (Al)**

Table. 1. Fabrication process sequence for SONOS Flash capacitor structure

SONOS capacitor squares of length 100μm. The gate stack is composed of rapid thermal oxidation (RTO) based tunnel oxide of 3.5nm, LPCVD silicon nitride of varying thickness and composition (see Table-2 for details) and high temperature oxide (HTO) top oxide of 8nm thickness. The Si$^+$ and N$^+$ single layers were obtained by varying the Si_2H_6/NH_3 flow rates [2]. Two types of charge trap layer were fabricated in nitride-1/interface-oxynitride/nitride-2 scheme. Firstly, a nitride stack where Si$^+$ nitride is deposited on the tunnel oxide, the surface is oxidized into an oxynitride and finally N$^+$ nitride is deposited on the top is denoted as the Si$^+$/N$^+$ stack. Secondly, the reverse stack with N$^+$ nitride in the bottom and Si$^+$ nitride on the top separated by an oxynitride layer is denoted as N$^+$/Si$^+$.

978-1-4244-2039-1/08/$25.00 ©2008 IEEE

Split	Nitride layer type	Thickness (nm)	EOT (nm)
D1	Si+	6	14.20
D2	N+	6	14.67
D3	Si+ bottom/ N+ top	4/4	15.89
D4	Si+ bottom/ N+ top	6/2	14.81
D5	N+ bottom/ Si+ top	4/4	17.55
D6	N+ bottom/ Si+ top	6/2	18.098

Table 2. Split conditions of the SONOS devices with different nitride charge storage layer compositions, 3.5nm tunnel oxide and 8nm blocking oxide.

In these stacks, the lower nitride was partially oxidized under identical conditions to obtain ~2nm of SiON interface layer [5-7]. The interfaces layers are denoted as Si$^+$ON or SiON$^+$ depending on the nitride layer composition that was oxidized. The distinction is necessary as surface-oxidation process of Si$_X$N$_Y$ is nitride composition dependent [8]. For example, Si$^+$ nitride oxidizes more readily i.e. thicker and higher O-content, compared to N$^+$ nitride under identical conditions. This is consistent with XPS analysis. In addition, the Si$^+$ON layer may contain less hydrogen than SiON$^+$ layer with similar trends as Si$^+$ vs. N$^+$ nitrides [8]. The nitride layer hydrogen content trend is verified by Hydrogen Forward Scattering (HFS). After the trap layer deposition, HTO blocking oxide was deposited followed by anneal in N$_2$ ambient. The process was completed with poly deposition, p$^+$ implant and activation, gate metallization and gate patterning. The HTO anneal was responsible for partial oxidation of the top nitride layer resulting in formation of ~2nm thick SiON layer. It consists of ~2nm of Si$^+$ON or SiON$^+$ depending on the top nitride composition consistent with previous nomenclature.

The schematic of the gate stack cross section, trap distribution and band diagram of single layer is shown in Fig.1 and both the bi-layer stacks are illustrated in Figs. 2 and 3. The Si$^+$ layer has high trap density and low band-gap [3] while N$^+$ layer has lower trap density with relatively higher band-gap. Higher band-gap oxynitride layers are depicted with a low trap density [7, 9]. To study the impact of the position of the oxynitride interface between the two nitrides on performance and reliability, the top and bottom nitride thicknesses were systematically varied as shown in Table-2.

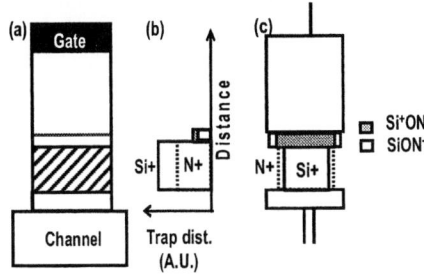

Fig. 1. Single layer nitride SONOS stack (a) gate stack (b) trap profile (c) band structure

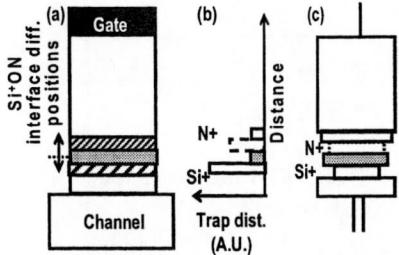

Fig. 2. Si+/N+ bi-layer nitride stack (a) gate stack (b) trap profile (c) band structure

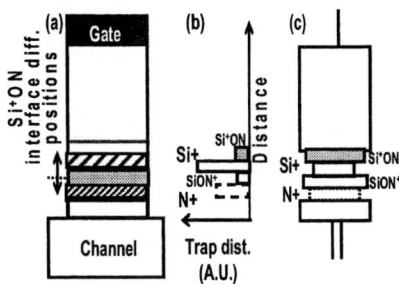

Fig. 3. N+/Si+ bi-layer nitride stack (a) gate stack (b) trap profile (c) band structure

III. EXPERIMENTAL DETAILS

To assess the impact of nitride engineering on memory performance, program/erase (P/E) transients, retention and P/E cycling measurements were conducted on SONOS MOS capacitors squares of length 100μm. The single layer Si$^+$ and N$^+$ nitrides results are presented in detail elsewhere [2] while the results from the nitride stacks comprising of Si$^+$/N$^+$ and N$^+$/Si$^+$ nitride layers are discussed here. However, a short summary of the performance of single layer nitrides is essential to set the frame of reference.

A. Program/Erase transients

Fig. 4 contrasts the properties of Si$^+$ and N$^+$ nitride through the Program/Erase (P/E) transients. Devices with Si$^+$ nitride show fast program and erase along with low program saturation of 3.5V and low fixed charge (0.2V) owing to high charge mobility in the Si$^+$ film. The capability of over-erase (-0.5V) gives a larger effective memory window of 4V. On the other hand, the N$^+$ nitride exhibits slow charging with high saturation V$_{FB}$ of 7.5V but cannot be erased completely beyond V$_{FB}$ of 3.5V, thus again resulting in an effective memory window of 4V. N$^+$ nitride also shows high initial fixed charge in a fresh device (2.1V). A major conclusion from our previous work [2] is that the permanent trapping in N$^+$ single nitride layer splits does not scale with nitride layer thickness, which indicates that the top SiON$^+$ layer is responsible for the permanent trapping.

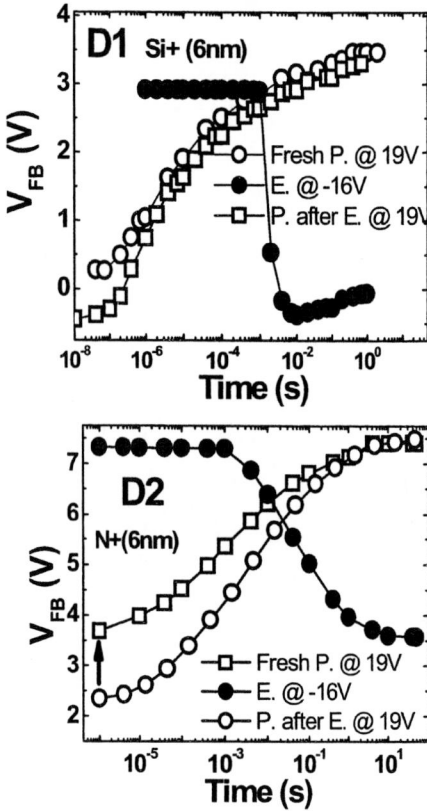

Fig. 4. P/E transients of D1 showing low initial fixed charge and capability of over erase while the P/E transients of D2 show incomplete erase and high saturation V_{FB}

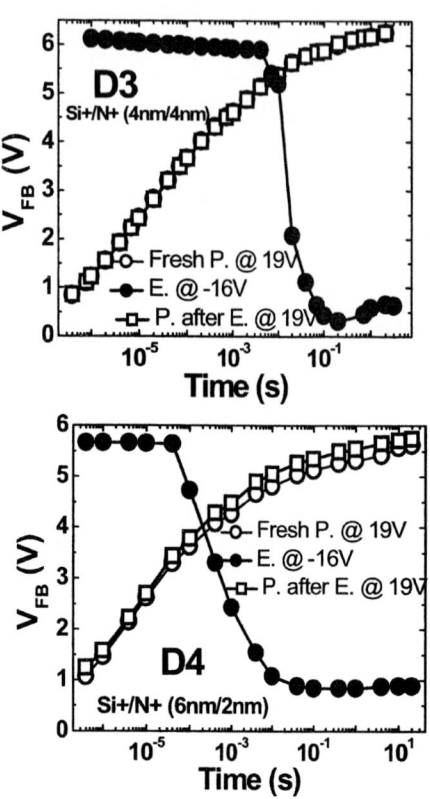

Fig. 5. P/E transients of the Si+/N+ stacks, D3 and D4, showing complete erase with no permanent trapping.

The P/E transients of bi-layer stacks are shown in Figs. 5 and 6. The Si^+/N^+ bi-layer stacks exhibit complete erase with no permanent trapping, as shown in Fig. 5. This is consistent with the property of the lower Si^+ layer. This indicates that charges are confined strictly to the bottom Si^+ layer and are unable to interact with the N^+ nitride layer which exhibits permanent trapping. In Fig. 6, both the bi-layer stacks with N^+/Si^+ show incomplete erase and permanent trapping after the first P/E operation, similar to single layer N^+ film. This observation further establishes that the layer formed on conversion of N^+ to $SiON^+$ is responsible for permanent trapping.

A further contrast in performance between Si^+ON versus $SiON^+$ needs to be understood from materials differentiation. A plausible explanation is that $SiON^+$ films show higher H-content than Si^+ON which maybe responsible for permanent (deeper) traps in the oxynitride. A detailed comparison of the program transients measured at program gate bias of 19V in the bi-layer stacks as the position of the oxynitride interface layer between the nitride layers is varied is shown in Fig. 7. The transients indicate a low fixed charge for Si^+/N^+ while the N^+/Si^+ stacks exhibit high initial fixed charge. An interesting contrast is that the saturation V_{FB} for Si^+/N^+ increases with the decrease in the bottom nitride (Si^+ nitride) layer thickness while the opposite is true for N^+/Si^+ splits. The improvement in saturated V_{FB} in the Si^+/N^+ splits with the thinning of the high trap-density Si^+ nitride provides two insights. Firstly, the trapped charge is confined nearer to the channel by the Si^+ON layer which results in a larger V_{FB} shift. Secondly, the charge trapped in the Si^+ nitride layer is not proportional to the Si^+ nitride thickness i.e. charge trapping is not limited by trap density.

978-1-4244-2039-1/08/$25.00 ©2008 IEEE

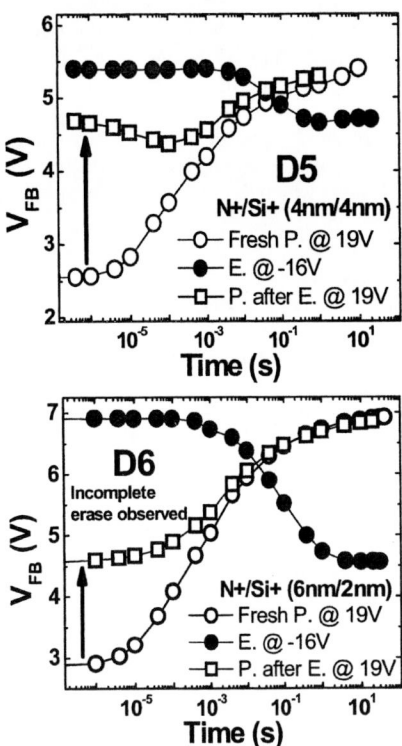

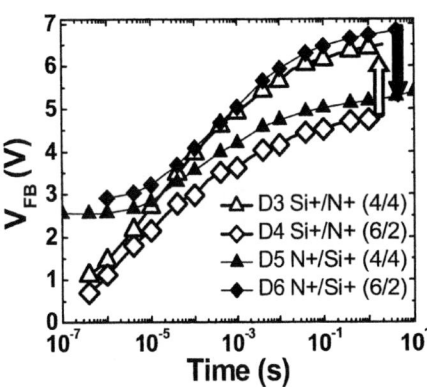

Fig. 6. Program transients of the N+/Si+ stacks, D5 and D6, showing permanent trapping, suggesting SiON$^+$ layer to be the possible cause of the trapping.

Fig. 7. Comparison of the program transients of D3-D6. Si$^+$/N$^+$ stacks show better P/E window and speed over N$^+$/Si$^+$ stacks.

nitride split to achieve the enhanced V_{FB} shift (38% increased). The magnitude of V_{FB} shift cannot be solely explained by invoking the "proximity of the charge to the channel" argument even if the amount of trapped charge is equal in Si$^+$ layers in both splits (irrespective of Si$^+$ layer thickness). Larger saturated V_{FB} indicating greater charge storage with Si$^+$ layer thickness reduction can be understood as follows. Saturation V_{FB} depends on a dynamic equilibrium between current in (J_{IN}) and out (J_{OUT}) in the charge storage node i.e. Si$^+$ nitride layer [11]. J_{OUT} is a function of electric field in the Si$^+$ON layer. From simple 1-dimensional electrostatics as depicted in Fig 8, the electric field reduces in the Si$^+$ON layer with the reduction in the Si$^+$ layer thickness for the same V_{FB} shift (for equivalent charge stored at the Si$^+$/Si$^+$ON layer interface). This implies that the charge storage can be achieved in thinner Si$^+$ layer before J_{OUT} becomes significant- consistent with experimental observation. In contrast, as N$^+$ nitride thickness is reduced, an approximately proportional reduction in saturated V_{FB} is observed. This indicates that the total trapped charges in N$^+$ nitride scales with nitride thickness. Charge trapping is limited in N$^+$ nitride trap density which is consistent with the low trap density but deep traps in N$^+$ nitride. A performance evaluation of the program transients of the Si$^+$/N$^+$ and N$^+$/Si$^+$ (4nm/4nm) bi-layer stacks in Fig. 7, indicates a low fixed charge of ~0.68V, a high saturation V_{FB} of 6.5V with a program speed of 3.4V V_{FB} shift achieved in 100μs in the Si$^+$/N$^+$ stack. The N$^+$/Si$^+$ (4nm/4nm) bi-layer stack shows higher fixed charge of ~2.6V, a relatively lower saturation V_{FB} of 5.3V with a slower program speed of 1.0V V_{FB} shift achieved in 100μs. These characteristics affirm a strong influence of the lower nitride layer on the device performance [2].

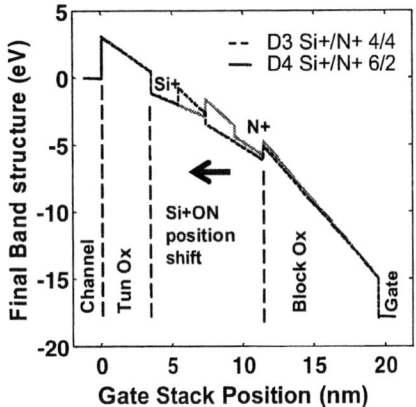

Fig 8. For same charge stored (10^{13}/cm^2) at the Si$^+$/Si$^+$ON interface, electric field in Si$^+$ON layer in greater in D3 than in D4 indicating that D3 can store more charge than D4 before saturation V_{FB} occurs.

The scenario of the high charge mobility in Si+ nitride which is characteristic of shallow traps and high trap density is consistent with observation of high interface charge density and low bulk charge density in charge trap nitride by Ishida et al [10]. In fact, trapped charges must be greater in the thinner Si$^+$

B. Retention

The room temperature retention characteristics in the bi-layer stacks are illustrated in Fig. 9. The Si$^+$/N$^+$ bi-layer stacks show high retention losses ranging between 0.8V-1.6V while the

N^+/Si^+ stacks demonstrate excellent charge retention with a loss range of only 5.5mV-32mV. This implies that the material properties of the lower nitride layer in the bi-layer stack, essentially governs the charge loss owing to its proximity to the tunnel dielectric. This is consist with our earlier work [2] which associates N^+ films with better retention capability over the Si^+ stack largely due to the relatively shallow traps in the latter [8, 12].

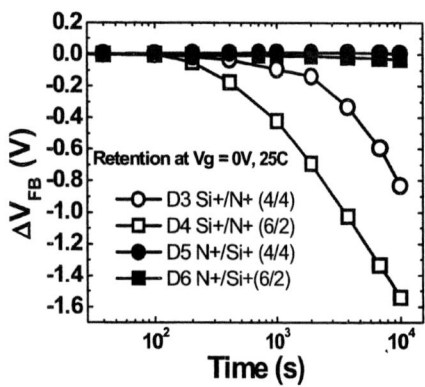

Fig. 9. N+/Si+ shows an overall improved retention when compared to Si+/N+ stacks, owing largely to the good retention capability of the N+ film [7]. All stacks were initially programmed to $V_{FB} = 6V$.

To estimate the primary path of retention loss i.e. to gate or the channel, retention measurements were performed by varying the gate bias during retention, as shown in Fig. 10. Under positive gate bias stress, the trapped electrons will be attracted towards the gate for charge loss through the blocking oxide. A negative gate bias stress would serve to repel the electrons towards the tunnel dielectric. The strong retention loss under low negative bias suggests the loss through the tunnel dielectric as the major charge loss mechanism in Si^+/N^+ and N^+/Si^+ stacks.

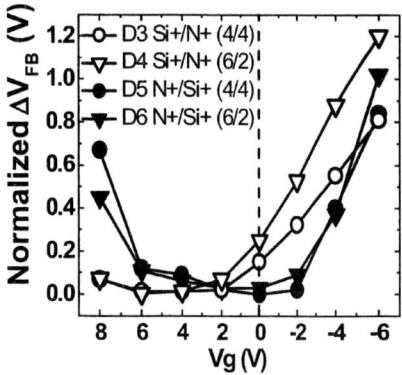

Fig. 10. Retention at varying bias suggest the charge loss through tunnel dielectric in Si+/N+ stacks and through the blocking oxide and the tunnel oxide in N+/Si+ stacks.

This confirms the assumption that tunnel oxide is the dominant path for retention loss and justifies the impact of the

lower nitride layer on retention performance. In comparison, charge loss was to the gate was observed through the HTO at higher biases (>6V) N^+/Si^+ stack. This is explained by the eventual charge transfer from N^+ to the top Si^+ layer (due to relatively lower offset between the conduction bands of N^+ and $SiON^+$ layers) resulting in the eventual charge loss through the blocking dielectric under high positive gate bias stress. For Si^+/N^+ case, retention in the gate direction is dominated by the top nitride i.e. N^+ layer which does not leak even at 8V bias [2].

C. Endurance

The endurance characteristics in the two bi-layer stacks show contrasting trends. As shown in Fig. 11, the Si^+/N^+ stacks show excellent endurance characteristics with negligible erase state degradation while maintaining a constant memory window. The enhanced performance is attributed to the interfacial oxynitride (Si^+ON) barrier layer between the Si^+/N^+ layers which serves to scatter the injected high energy carriers, thus preventing trap generation at the N^+/HTO interface [2].

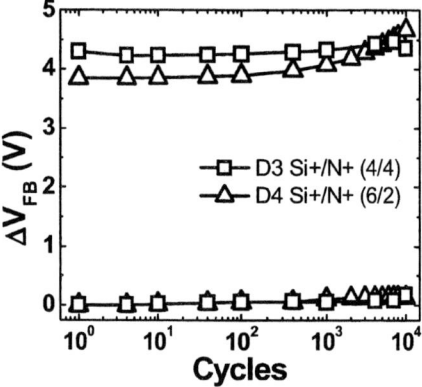

Fig.11. Superior endurance characteristics of Si+/N+ stacks with negligible erase state degradation with P/E cycling.

In contrast, the N^+/Si^+ stacks can be cycled at a significantly smaller memory window of ~1.8V as shown in Fig. 12. Typically the N^+/Si^+ (4nm/4nm) stack exhibits a dramatic increase in the erase state degradation of 1.2V and window closure by nearly 50% at the end of 6000 P/E cycles. In N^+/Si^+ splits, positioning the $SiON^+$ interface closer to the channel enhances the degradation of the erase V_{FB} by 2x which cannot be explained solely by assuming equal damage (trap generation) irrespective of $SiON^+$ position.

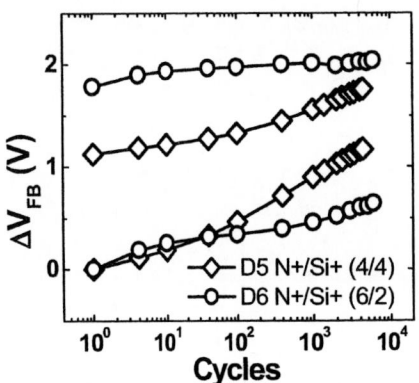

Fig.12. Relatively poor endurance characteristics of N+/Si+ stack which exhibits significant trap generation with window closure with P/E cycling.

The degradation of $SiON^+$ layer compared to the resilience of the Si^+ON layer needs to be explored. To understand this observation, two effects need to be considered here. Firstly, kinetic energy of high energy 'unscattered' electron (derived from the high electric field during program) impinging on the $SiON^+$ layer increases with distance from the tunnel oxide. Secondly, the flux of "unscattered" electron reduces strongly with distance from the channel (e.g. possibly depending on scattering length in part). For damage to occur close to channel, the threshold energy for defect generation must be very low. In this regard, more Si-H bonds in $SiON^+$, can be broken by relatively at low energy of 2eV [13, 14], which would be key to explaining this behavior. At this low energy-threshold of damage, increased degradation as $SiON^+$ gets closer to the channel is possibly due to impingement of a larger number of high energy "unscattered" carriers closer to the channel. This result is consistent with N^+ single layer results [2]. In comparison, the Si^+ON layer with lower H content is more resistant to degradation and demonstrates superior endurance performance. Thus, the Si^+ON layer with reduced H and better oxidation demonstrates reliable barrier action.

IV. CONCLUSION

To summarize, the impact of band gap engineered Si_XN_Y trap layer on the performance and reliability of CTF is studied. The P/E window, endurance and retention of Si^+/N^+ bi-layer stacks are compared to that of N^+/Si^+ bi-layers. The role of the composition of the SiON layer (i.e. Si^+ON vs. $SiON^+$) and position of the interface between the two nitride layers, show a strong impact of device performance and reliability. In terms of performance trade-off, Si^+/N^+ bi-layers show enhanced P/E window (negligible fixed charge), better cycling endurance, but poor retention compared to N^+/Si^+ bi-layer stack. The poor P/E window and cycling endurance of N^+/Si^+ bi-layers can be attributed to presence of fixed charge in, and the degradation of the $SiON^+$ interfacial layer. The low threshold-energy requirement for defect generation in $SiON^+$ indicates breaking Si-H bonds as a possible mechanism. In comparison, the presence of robust, low H-content Si^+ON interface for Si^+/N^+ bi-layers acts as a reliable interface barrier. It demonstrates low defect generation and therefore provides better P/E window and cycling endurance. The dominant path for charge loss during retention is shown to be through the tunnel oxide, and hence retention performance can be improved by using a thicker tunnel dielectric.

ACKNOWLEDGMENT

IIT Bombay would like to acknowledge SRC (GRC) for financial support.

REFERENCES

[1] K. Kim, "Technology for sub-50 nm DRAM and NAND Flash manufacturing," *International Electron Device Meeting Technical. Digest*, pp. 333, 2005.

[2] Sandhya C., U. Ganguly, K.K. Singh, P.K. Singh, C. Olsen, S. Seutter, G. Conti, K. Ahmed, N. Krishna, J. Vasi, and S. Mahapatra; "Nitride engineering and the effect of interfaces on Charge Trap Flash performance", *46th Annual International Reliability Physics Symposium*, pp. 406, 2008.

[3] H.-C. Chien, C.-H. Kao, J.-W. Chang, and T.-K. Tsai, "Two-bit SONOS type Flash using a band engineering in the nitride layer," *Microelectronic. Engineering*, v.80, pp. 256, 2005.

[4] Z. L. Huo, J. K. Yang, S. H. Lim, S. J. Baik, J. Lee, J. H. Han, I.-S. Yeo, U-I. Chung, J. T. Moon, and B.-I. Ryu, "Band Engineered Charge Trap Layer for highly Reliable MLC Flash Memory," *VLSI Technical Symposium*, pp. 138, 2007.

[5] V. J. Kapoor, D. Xu, and R. S. Bailey, "The Combined Effect of Hydrogen and Oxygen Impurities in the Silicon Nitride Film of MNOS Devices," *Journal of Electrochemical Society*, v. 137, pp. 3589, 1990.

[6] K.-H. Wu, H.-C. Chien, C.-C. Chan, and T.-S. Chen, and C.-H. Kao, "SONOS device with tapered bandgap nitride layer," *IEEE Transaction on Electron Devices*, v. 52, pp. 987, 2005.

[7] S. J. Wrazien, Y. Zhao, J. D. Krayer, and M. H. White, "Characterization of SONOS oxynitride nonvolatile semiconductor memory devices," *Solid State Electronics*, pp. 885, 2003.

[8] H. Wong, M. C. Poon, Y. Gao, and T. C. W. Kok, "Preparation of Thin Dielectric Film for Nonvolatile Memory by Thermal Oxidation of Si-Rich LPCVD Nitride," *Journal of Electrochemical Society*, v. 148, pp.G275, 2001

[9] S.C. Bayliss, and S. J. Guman, "Silicon oxynitride as a tunable optical material," *Journal of Physics Condensed Matter*, v. 6, pp. 4961-70 ,1994.

[10] T. Ishida, Y. Okuyama, and R. Yamada, "Characterization of Charge Traps in Metal-Oxide-Nitride-Oxide-Semiconductor (MONOS) Structures for Embedded Flash Memories," *44th Annual International Reliability Physics Symposium*, pp. 516 2006.

[11] A. Furnemont, M. Rosmeulen, A. Cacciato, L. Breuil, K. De Meyer, H. Maes, and J. Van Houdt, "A Consistent Model for the SANOS Programming Operation," *22nd IEEE NVSMW*, pp.

96, 2007.

[12] G. Rosenman, M. Naich, Ya. Roizin, and R. van Schaijk, "Deep traps in oxide-nitride-oxide stacks fabricated from hydrogen and deuterium containing precursors," *Journal of Applied Physics,* v. 99, 023702 2006.

[13] P. E. Nicollian, W. R. Hunter, and J. C. Hu, "Experimental Evidence for Voltage Driven Breakdown Models in Ultrathin Gate Oxides,"*38th Annual International Reliability Physics Symposium,* pp. 7, 2000.

[14] D. A. Buchanan, A. D. Manvick, D. J. DiMaria, L. Dori, "Hot-Electron Induced Hydrogen Redistribution and Defect Generation in Metal-Oxide-Semiconductor Capacitors", *Journal of Applied Physics,J. Appl. Phys.* v.76, 3595 1994.

A Novel Dual-BBHH Erasing Scheme to Improve Endurance and Retention Performances for Localized Charge Trapping Memories

Guangjian SHI [1], Liyang PAN* [1,2], Romain RITZENTHALER[1], Lei SUN [1], Zhigang Zhang [1,2] and Jun XU [1,2]

Institute of Microelectronics of the Tsinghua University, Beijing 100084, China [1];
Tsinghua National Laboratory for Information Science and Technology [2];
Phone: 0086(10)62789192 Fax: 0086(10)62771130 Email*: panly@tsinghua.edu.cn

Abstract-**The mismatch of trapped electrons and holes is the main mechanism causing reliability degradation for localized charge trapping memory devices. This paper proposes a novel dual-BBHH erasing scheme to alleviate the mismatch effect, therefore improve the endurance and retention performances simultaneously.**

I. INTRODUCTION

Charge trapping memories (CTM) [1] exhibit great advantages in scaling capability, better immunity to defects in the tunnel oxide and higher storage density compared with conventional floating gate flash memories. Localized CHE and BBHH methods are usually utilized in CTM for programming and erasing operations [2], resulting in special reliability issues, such as the localized charge trapping during program operation and the mismatch of electron and hole profiles under P/E cycling stress [2-6]. It's also found that the accumulated holes have shallow trap energy level, therefore are easier to redistribute and migrate to recombine with electrons, leading to the retention degradation [2-5]. The localized charge distribution makes the long-term charge retention the bottleneck of the CTM devices for real application. In order to reduce the mismatch effect of the trapped electrons and holes, researchers tried to optimize the operating conditions [3] or developed some new operating methods [7] to improve the endurance and retention performances.

This paper firstly studies the reliability characteristics of tested CTM devices with different operating conditions, demonstrating the difficulty of making trade-off between endurance and retention performances. A novel dual-BBHH erasing scheme is then proposed to greatly improve the V_T window after high temperature baking. The mismatch is alleviated with the novel operating scheme and both of the endurance and retention behaviors are remarkably optimized.

II. EXPERIMENTAL RESULTS AND ANALYSIS

The tested SONOS-type CTM devices are fabricated with a 1-poly/4-metal CMOS logic compatible technology, whose TEM cross-section is shown in fig. 1. The designed Width/Length ratio of the cell is 0.28μm /0.18μm, with the printed gate length and the physical channel length being about 0.15 μm and 0.1 μm, respectively. The total equivalent oxide thickness (EOT) of the ONO (Oxide/Nitride/Oxide) stack is

13.7 nm and the tunnel oxide thickness is around 4.5nm.

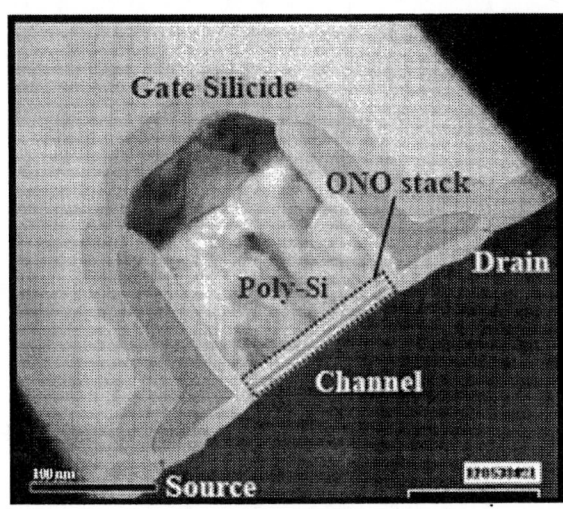

Fig. 1. TEM cross-section of the SONOS memory device with 0.1μm physical channel length and 4.5nm tunnel oxide thickness.

Considering the localized operation induced mismatch problems, P/E cycling conditions should be optimized for better reliability characteristics. Fig. 2 (a) and (b) shows the endurance and retention characteristics with different erasing voltages, wherein CHE programming conditions are optimized to V_d=3.6V, V_g=8V and t_e=20μs and BBHH erasing condition are V_d=4V, V_g=-8V and t_e=10ms with V_d=4.3V, V_g=-6V and t_e=10ms for comparison. From fig. 2a, the positive shift and shrinkage of the V_T window can be greatly improved with higher gate erasure voltage, owing to more holes captured during BBHH erasing to neutralize CHE injected electrons. However, more hole accumulation will result in larger V_T loss (as shown in fig. 2b), caused by the hole redistribution and recombination with electrons [2], [4]. It is also found that the electron-field induced by the accumulated electrons and holes will speed up the migration of the holes and make the retention degradation more serious. By using smaller gate and higher drain erasure voltages, better retention characteristic could be achieved with 510 mV decrease of V_T loss (fig. 2b) because the injected holes move towards the channel center and match the electrons profile better, but the endurance performance becomes worse due to more electrons accumulated after long term P/E cycles, resulting in the closure of V_T window as shown in fig. 2a. Therefore, the optimization of the traditional programming and erasing conditions is rather difficult to make tradeoff between better endurance and higher retention performances.

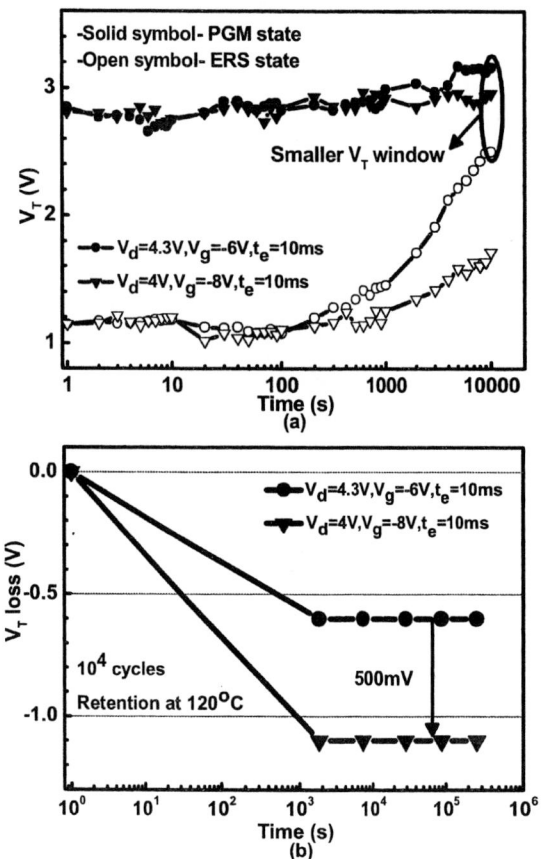

Fig. 2. Endurance (a) and 120°C retention (b) characteristics under different erasing conditions which are V_d=4V, V_g=-8V, t_e=10ms and V_d=4.3V, V_g=-6V, t_e=10ms. The optimized program condition is V_d=3.6V, V_g=8V and t_e=20µs. It is difficult to achieve better endurance and retention performances simultaneously by simply optimizing the traditional programming and erasing conditions.

Based on the study of reliability performances with different operating conditions, this paper firstly proposes a novel dual-BBHH erasing scheme, which adds another erasing operation within each P/E cycle, to alleviate the mismatch effect. As illustrated in fig. 3, the scheme combines two steps of BBHH erasing operation. The 1st one has smaller drain voltage and lower gate voltage, forming holes mainly trapped above the drain junction. While the 2nd erase operation has higher drain voltage and wider holes distribution to match the injected electrons after CHE programming. Fig. 3c shows the potential along channel after long term P/E cycles in the program state. The holes' potential decreases with the new erasing operation, indicating that fewer holes accumulate compared with conventional operation scheme. By using this dual-BBHH

erasing scheme, the accumulated holes in the program state can be greatly reduced without shrinking the V_T window, and the charge redistribution induced retention degradation can also be alleviated.

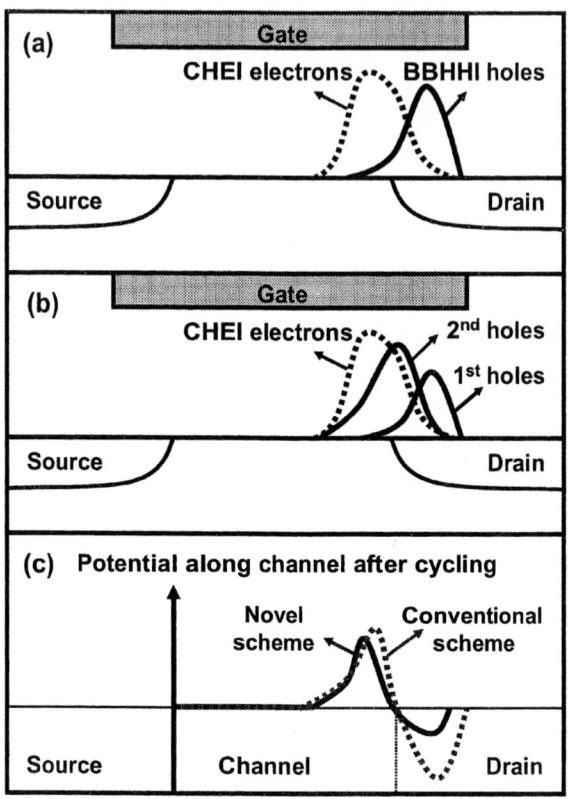

Fig. 3. Illustration of trapped charges with conventional BBHH erasing (a) and dual-BBHH erasing (b), and potential along channel in the program state after P/E cycling (c). The novel erasing scheme could match the injected electron and hole profiles better.

By comparing the different combinations of dual erasing conditions, an optimized dual-BBHH erasing scheme has been achieved with V_d=3.6V, V_g=-7V and t_e=5ms followed by V_d=4.3V, V_g=-7V and t_e=2ms within each P/E cycle. Fig. 4 (a) and (b) compare the endurance and retention performances of the conventional method and the optimized dual-BBHH erasing schemes. From fig. 4, the V_T windows after 10^4 P/E cycling are nearly the same with the two methods, while the V_T loss after 72 hours baking (at 120°C) is -0.69 V and -1.10V respectively, with 410 mV improved for the proposed scheme. The novel operation scheme not only maintains the V_T window after long term P/E cycles, but also decreases the charge loss when baking at high temperature, indicating that a better match of the trapped electron and hole profiles has been achieved as shown in fig. 3 (b).

978-1-4244-2039-1/08/$25.00 ©2008 IEEE

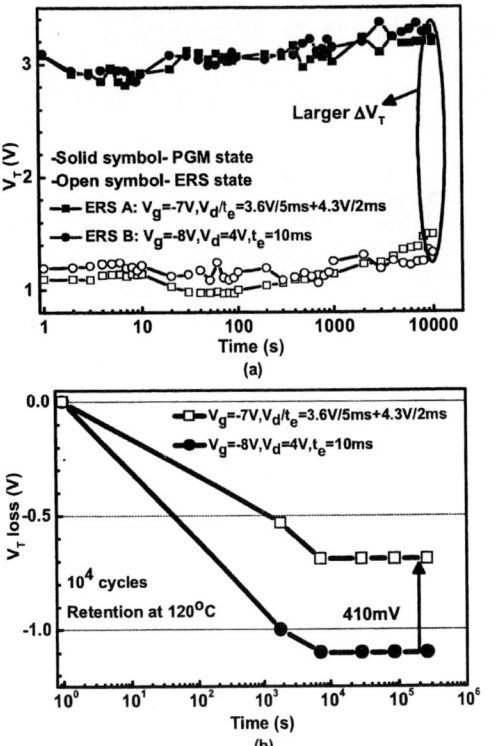

Fig. 4. Endurance (a) and retention (b) characteristics with proposed dual-BBHH erasing and conventional erasing schemes. The novel operating scheme can realize a better tradeoff between endurance and retention performances.

Fig. 5 shows the final V_T window curves of the samples stressed with the two erasing methods. A 0.69V V_T window (after 10^4 cycles followed by 72 hours baking at 120°C) is finally obtained with the new scheme, 490 mV higher than the traditional method and acceptable for the real application. From the results, it can be seen that the proposed erasing scheme can realize a better tradeoff between endurance and retention characteristics, and features with higher erase speed, lower voltage stress and simple control circuit design.

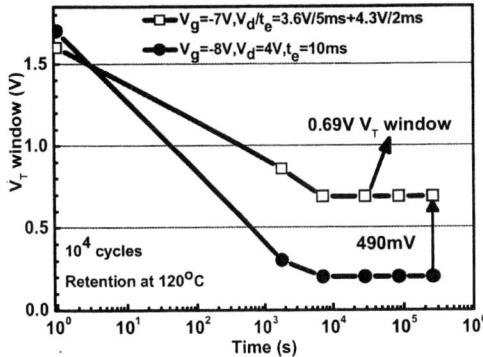

Fig. 5. V_T window versus time curves with proposed dual-BBHH erasing and conventional erasing schemes. Obvious improvement of V_T window after high temperature baking has been observed

III. CONCLUSION

A novel dual-BBHH erasing scheme is presented to alleviate the mismatch of trapped electrons and holes, which is the main mechanism causing reliability degradation for localized CTM devices. From the tested results, the final V_T window is greatly increased to 0.69V, demonstrating that the proposed scheme can improve the endurance and retention performances simultaneously.

ACKNOWLEDGMENT

This work is supported by the National Basic Research Program of China (No. 2006CB302700) and the Basic Research Foundation of TNList.

REFERENCES

[1] B. Eitan, P. Pavan, I. Bloom, E. Aloni, A. Frommer, and D. Finzi, "NROM: A novel localized trapping, 2-bit nonvolatile memory cell," *IEEE Electron Device Lett.*, vol. 21, no. 11, pp. 543-545, Nov. 2000.

[2] M. Janai, B. Eitan, A. Shappir, E. Lusky, and I. Bloom , "Data retention reliability model of NROM nonvolatile memory products," *IEEE Trans. Device Mater. Rel.*, vol. 4, 2004, pp. 404-414.

[3] A. Furnemont, M. Rosmeulen, K. van der Zanden, J. Van Houdt, K. De Meyer and H. Maes, "New Operating Mode Based on Electron/Hole Profile Matching in Nitride-Based Nonvolatile Memories," *IEEE Electron Device Lett.*, vol 28, no. 4, pp. 276-278, 2007.

[4] E. Lusky, Y. Shacham-Diamand, A. Shappir, I. Bloom, G. Cohen and B. Eitan, "Retention loss characteristics of localized charge-trapping devices," *Proc. of the 42nd Annual International Reliability Physics Symposium*, pp. 527-530, 2004.

[5] N. K. Zous, M. Y. Lee, W. J. Tsai, Albert Kuo, L. T. Huang, T. C. Lu, C. J. Liu, Tahui Wang, W. P. Lu, Wenchi Ting Joseph Ku and Chi-Yuan Lu, "Lateral migration of trapped holes in a nitride storage flash memory cell and its qualification methodology," *IEEE Electron Device Lett.*, vol. 25, no. 9, pp. 649-651, Sep. 2004.

[6] Huiqing Pang, Liyang Pan, Lei Sun, Dong Wu, Jun Zhu, "Trapped charge distribution during the P/E cycling of SONOS memory," *Proceeding of the International Symposium on the Physical and Failure Analysis of Integrated Circuits (IPFA)*, pp. 84-87, 2006.

[7] Hang-Ting Lue, Yi-Hsuan Hsiao, Yen-Hao Shih, Erh-Kun Lai, Kuang-Yeu Hsieh, Rich Liu and Chih-Yuan Lu, "Study of charge loss mechanism of SONOS-type devices using hot-hole erase and methods to improve the charge retention," *Proceeding of the International Reliability Physics Symposium (IRPS)*, pp. 523-528, 2006.

Study on SRAM Soft Failure Using Planar-View Transmission Electron Microscopy Techniques

P. Liu, *IEEE Member*, K. Li, Y. Li, C.Q. Chen, E.Er, J.Teong
Failure analysis department, Chartered Semiconductor Mfg Ltd
Woodlands Industrial Park D, Street 2, Singapore 738406
Phone: (+65) 63603430. Fax: (+65) 63604592. Email: liupan@charteredsemi.com

Abstract- **Soft failure in static random access memory (SRAM), where there are several mechanisms related to it, is a kind of major obstruction to improve the yield. Transmission electron microscopy (TEM) is a powerful failure analysis tool, which has a high spatial resolution and is widely used in IC failure analysis with the shrinkage of integrated circuit to a nano-level transistor. Planar-view TEM techniques have great advantages in finding failure location and mechanism. In this paper, two kinds of soft failure root causes are identified by the planar-view TEM techniques.**

I. INTRODUCTION

With the scaling-down of the critical dimension of IC, the feature size of SRAM have become smaller and smaller. Failure analysis on the SRAM is a powerful method in that it can isolate the problem, find the process margin and improve the yield. The reasons why SRAM is popularly selected to analyse are: firstly, SRAM area usually obeys the tight design rule; secondly, SRAM test program is standard and easy to be revised; lastly, it is easy to isolate the failure location to do the physical failure analysis according to the test map [1]. Therefore, it is very important to find the root cause of SRAM failure modes and mechanism.

From the viewpoint of failure mode in bit map, the SRAM failure is usually divided into several modes: Single bit (SB), pair bit (PB), quaternary bit (QB), single full column /row, multi-bits and so on. Among all the failure modes, it is most difficult to find the root cause in SB because one bit in SRAM includes 6 transistors as shown in fig. 1. Any abnormality in these 6 transistors or connects would lead to the single bit failure. Soft failure is defined as follows: the transistor does not meet certain specification, but it would work on condition that there is a higher operation voltage.

Planar view TEM technique [2] is different from the cross section TEM sample preparation technique. Take the SRAM single bit failure for example, planar view TEM is easy to cover all six transistors compared with 2~3 transistors covered by the cross section TEM technique. There are four kinds of planar view TEM sample preparation methods.

Atom force probe (AFP) [3-8] is a latest technique, which develops from AFM and nano-probe and combines these techniques together in a machine. The functions of AFP include the current images and family I-V curves of probed transistors, the former of which are more powerful than passive voltage contrast (PVC) [9]. Furthermore, it is easy to get the failure bit electric parameters using probing function without depositing any pads [10].

In this paper, the authors introduce two SRAM soft failure cases using planar view TEM technique to find the root causes.

II. CASE STUDY I:
Single bit soft failure in SRAM

a) FA procedure and results

Generally, after SRAM finial test, one bit map can be obtained, which gives the detailed information of failure modes in the SRAM. Fig. 2 is failure bit map of SB and failure address is shown as the red point.

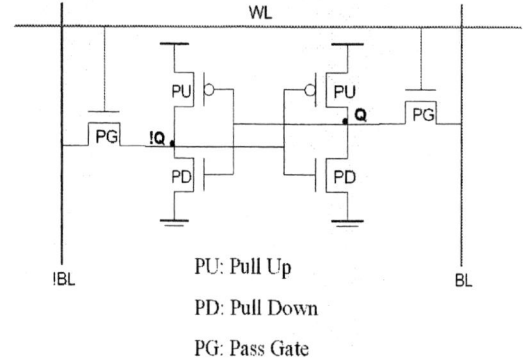

PU: Pull Up

PD: Pull Down

PG: Pass Gate

Fig. 1. 6T SRAM cell circuit.

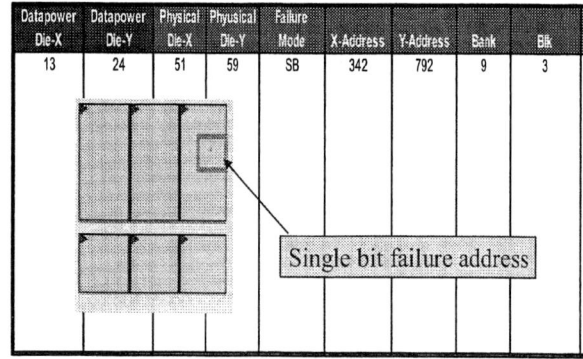

Datapower Die-X	Datapower Die-Y	Physical Die-X	Physical Die-Y	Failure Mode	X-Address	Y-Address	Bank	Blk
13	24	51	59	SB	342	792	9	3

Fig. 2. Single bit failure bit map and address

978-1-4244-2039-1/08/$25.00 ©2008 IEEE   234   *Proceedings of 15th IPFA - 2008, Singapore*

Based on the failure address, the physical location of the failure bit is easy to count. After confirmation of the physical failure address, the standard procedure is to de-process the sample layer by layer so as to review the failure location using optical microscope or scanning electric microscope (SEM), and try to find any abnormality in the failure location.

However, no abnormality is found till the sample was de-processed to contact level layer by layer. Combined with the SRAM layout, it is less likely that the single bit failure is above the contact level. Therefore, we focus on the contact level and below.

Fig. 3 is the SEM image at failure bit. From this picture, all six transistors can be observed. At the same time, the contact and poly gate are described on the image as reference as the 6T SRAM cell circuit layout.

The AFP has been applied to obtaining the results of the current image and the I-V curves at the failure bit location. As shown in fig. 4, fig. 4 (b) and (c) show that the current images of the PMOS contacts and NMOS contacts on conditions of positive bias and negative bias respectively. Any abnormality induced leakage on the positive or negative condition will be observed on the current images. At the same time topography image as shown in fig4. (a) gives a good reference to contact locations.

Compared with the current images of marked failure bit and the neighbourhood good bits, however, no abnormal contrast is found. That is, all contacts on poly silicon and on active area are normal. Therefore, it is better to use the probing function of the AFP to measure the turn-on current (I_{on}) of the failure bit, and the good bit I_{on} is taken as reference. Table 1 gives the result of I_{on} on the condition of Vs=0 (ground) and Vd= Vg=1.2V. The I_{on} of PD1 in failure bit (11.2uA) is much lower than that of the good bit (96.8uA), while other transistors of Ion are comparable.

To confirm the low I_{on} PD1 in failure bit, the family (Id-Vd) curves are measured as fig. 5 shows. And Fig. 6 shows the family curves of good bit.

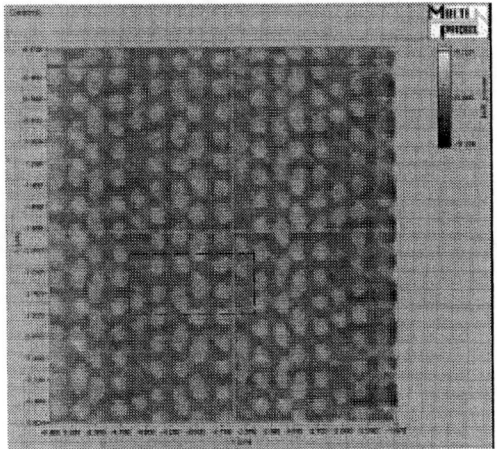

(a)Topography image, No bias

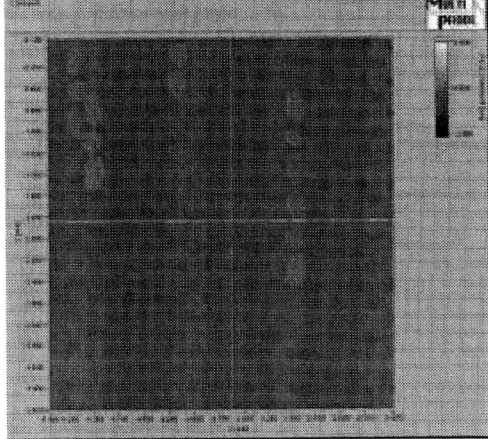

(b) Positive bias (+0.5V)

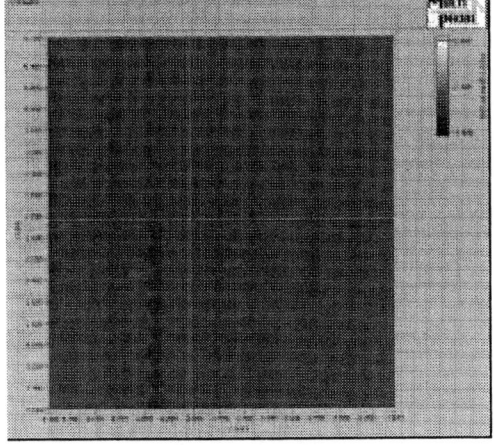

(c) Negative bias(-0.5V)

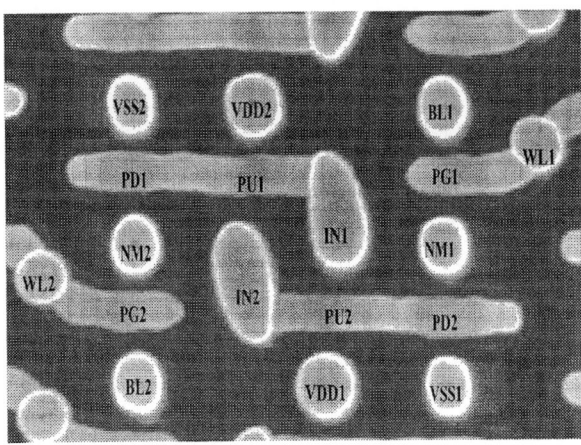

Fig. 3. The SEM image of failure bit at the contact level.

Fig. 4. Current images at different bias condition, (red marked location is the failure bit)

Table I
Turn-on currents (I_{on}) of six transistors in one bit

	I_{on} (uA) @ Vd=+/-1.2V Vg=+/-1.2V					
	PG1	PG2	PD1	PD2	PU1	PU2
Good bit	59.2	54.0	96.8	91.3	-29.6	-24.8
Bad bit	67.2	70.1	**11.2**	87.4	-26.0	-22.1

The family curves of failure bit PD1 are abnormal. The Id values are low in different Vg and Vd (different colour curves show different Vg). After switching the source and drain as shown in Fig.5 (a) and (b), we find that the PD1 transistor is not symmetric. Therefore, high resistance of transistor channel and unbalanced S/D are the reasons of SB failure based on the results of low I_{on} and asymmetric Vd-Id curves.

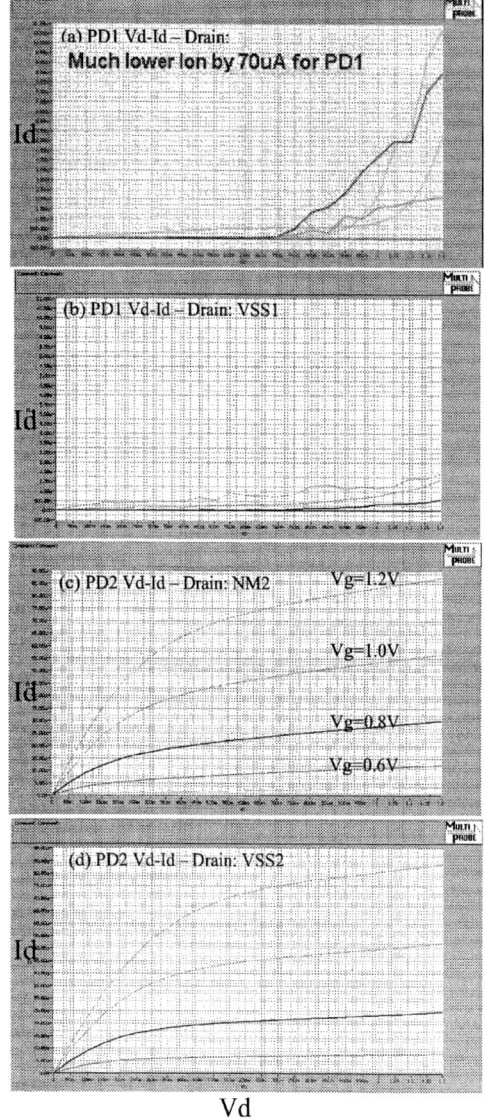

Fig. 5. Failure bit PD1/PD2 family (Id-Vd) curve and switch the source/drain.

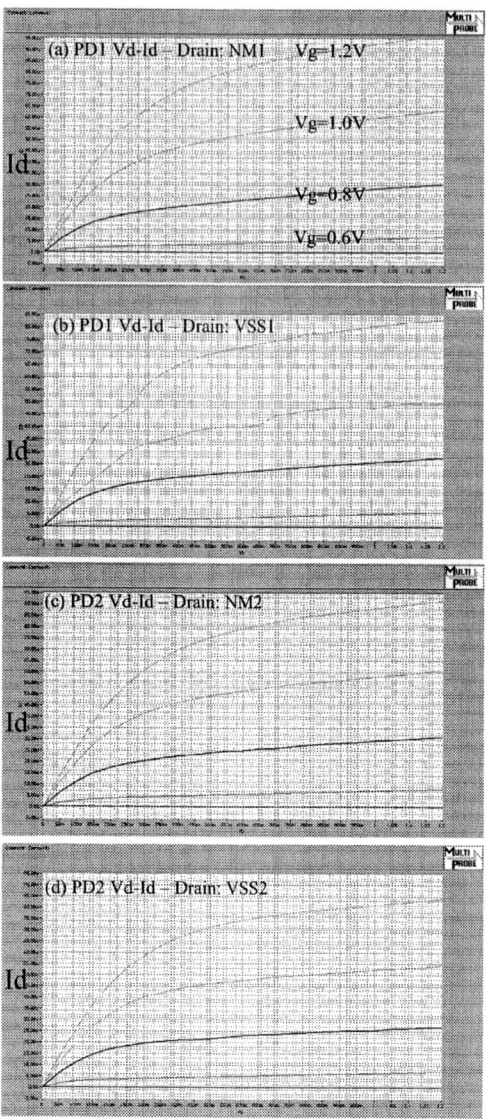

Fig. 6. Good bit PD1/PD2 family (Id-Vd) curve and switch the source/drain.

b) Planar-view TEM and root cause discuss

It is difficult to find defect with the cross-section TEM (X-TEM) if the defect cannot be observed in focus ion beam (FIB). In some worse conditions, even if the one transistor failure location is exactly known, the X-TEM still cannot find any defect due to the constraint of TEM sample thickness and small defect size. In fact, before the planar-view TEM cut, several X-TEM cuts were done, but no abnormality was found.

Planar-view TEM technique can observe big area and obtain the PD1 whole channel profile. But more attention should be paid to sample preparation. We need to control the cut location and sample thickness carefully to make sure the existence of the poly layer and active layer in the sample.

Brief introduction of the sample preparation procedure is as follows: coat Pt and mark the failure bit in FIB; polish the cross-section near the failure bit less than1 um; flip up the sample 90 degree in FIB chamber and mill the sample following standard TEM procedure; last pick up the sample on the carbon film coated copper grid.

Fig. 7 shows the PD1 planar-view TEM image. The abnormal spacer is found in the image. Some defect makes the nitride spacer far away from the poly gate as the arrow instruction shows.

As viewed from the process, the defect at the poly gate sidewall blocks the LDD implant because we cannot see the liner spacer at the defect location.

The mechanism of defect blocked the LDD implant is shown in Fig. 8. The blocked LDD implant makes the channel resistance increase, which leads to lower I_{on} current. At the same time, random liner spacer defect of poly sidewall is asymmetric. This matches the fact that the family I-V curves are different if the source and drain are switched.

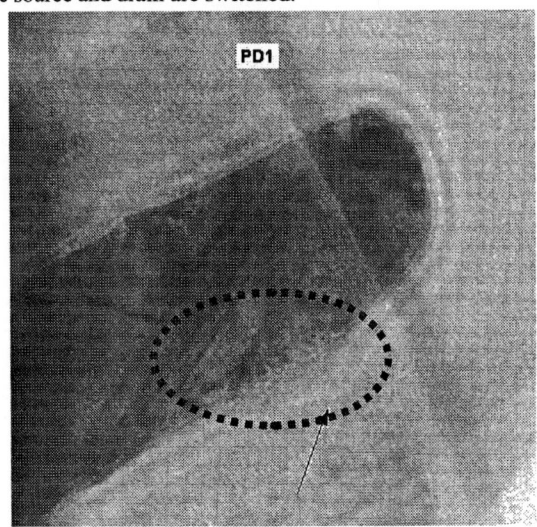

Fig. 7. Planar-view TEM of PD1 in failure bit.

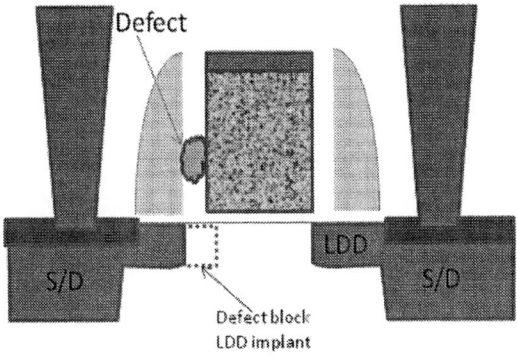

Fig. 8. Schematic of the defect block the LDD implant.

III. CASE STUDY II:
Built in self test (BIST) soft failure and IDDQ high issue in SRAM chip

It is popular to apply BIST structure to the modern logic or SRAM test. In this case, we find that the major failures are single bit and the chip IDDQ high leakage at the final test of the whole wafer. The test wafer failure map is shown in Fig. 9.

One die from this wafer was selected to do the failure analysis. The standard FA procedure is the same as that of the previous case (electric data not shown here). After that, planar-view technique is used to analyse the root cause. Fig. 10 gives the planar-view TEM result. Generally, it is popular to use the bright-field TEM (BF-TEM) because of its high resolution.

The transistors of failure bit and reference bit were observed carefully. No obvious abnormality was found because the contrast between poly and salicide was very weak in BF-TEM [11]. In order to distinguish between poly and salicide, the dark-field TEM (DF-TEM) technique was applied to the case. Fig. 11 gives the DF-TEM result.

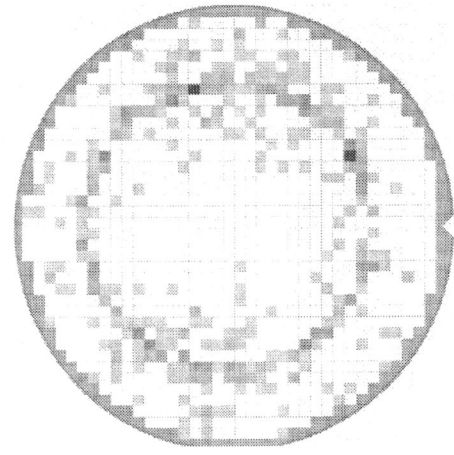

Fig. 9. The wafer yield sort map. The red points are failure dies.

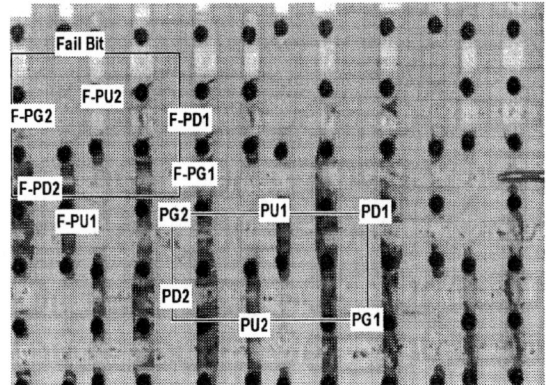

Fig. 10. Bright filed TEM image of the failure bit and reference bit.

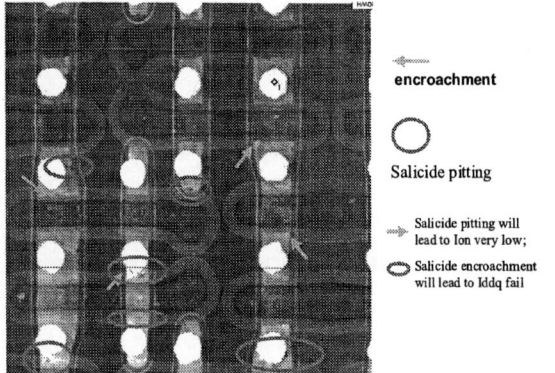

Fig. 11. Dark-field TEM image on the failure bit location. And different defects are marked.

Two kinds of defects are found in the sample. One is salicide pitting and another is salicide encroachment. Salicide pitting leads to low I_{on} current while salicide encroachment leads to high leakage because the salicide penetrates into the gate channel. Therefore, the salicide encroachments are the killer defect in the case.

DF-TEM has good contrast because the dark field detector is only sensitive to the Z-contrast. That is, bigger number atom or denser material has a higher scatter angle and is more easily to be accepted by the dark field detector.

The root cause was found after our discussion of the above results with the process owners. The salicide anneal usually adopts the rapid thermal anneal (RTA) process. However, the temperature of RTA was not uniform within this wafer. Actually, the temperature in middle of the wafer was higher than that of center and edge in this wafer. Therefore, the map of sorted yield wafer shows a circle-like pattern, which matches the non-uniformity of the RTA temperature.

IV. SUMMARY

In this paper, two kinds of SRAM soft failure cases are investigated. Planar-view TEM technique including BF/DF-TEM methodology gives the detailed explanations of electric phenomenon. Furthermore, in combination with the process, the failure mechanism and root cause are obtained.

ACKNOWLEDGMENT

The authors would like to thank EFA team for their help & deep discussion on the electric results and thanks FAB7 PI team strong sprit of teamwork and continue to give us effective feedback. The authors would also like to thank our FA colleagues, Irene Tee, Jili Wang, Qingxiao Wang, Chiwen Soo, for warm support.

REFERENCES

[1]. P.Egger, C.Burmer, "SRAM Failure Analysis Strategy", Proc. 29th ISTFA (2003), pp177-183 , 2003

[2]. P.Liu, K. Li, E.Er, S.P. Zhao, " Pane-view Transmission Electric Microscopy for advanced Integrated circuit" Proc. ICSE 2006, pp630-633, 2006

[3]. L.F.Tz Kwakman, N.B. Lepinay, S.Courtas, "The role of a Physical Analysis Laboratory in a 300mm IC Development & Manufacturing Center", International Conference on Characterization and Metrology for ULSI, 2005

[4]. Yu-Ching Yeh, Chia-Lung Lin, B-Jen Chen, Yuan-Wei Tseng, Jeremy D. Russell, "Application of Atomic Force Probing on 90nm DRAM Cell Failure Analysis" proceeding of 13th IPFA 2006, Singapore, pp340-343, 2006

[5]. Tom X Tong, A N Erickson," Current Image Atomic Force Microscopy (CI-AFM) combined with Atomic Force probing (AFP) for location and characterization of advanced technology node", proceeding form the 30th international symposium for testing and failure analysis, pp42-46,2004

[6]. Terence Kane, Michael P. Tenney, Andrew Erickson, Sebastian Phan, " Atomic Force Probe Kelvin Measurements of Large MOSFET Device at Contact Level for Accurate Device Threshold Characteristics", Proceeding of the 32nd international symposium for testing and failure analysis, pp479-502, 2006

[7]. Chao-Chi Wu, Jon C. Lee, Jung-Hsiang Chuang, Tsung-Te Li, "Single Device Characterization by Nano-probing to Identify Failure Root cause" proceeding of the 31st international symposium for testing and failure analysis, pp183-185,2005

[8]. Jon C. Lee, J. H. Chuang, "Fault Localization in Contact Level by Using Conductive Atomic Force Microscopy", proceeding from the 29th international symposium for testing and failure analysis, pp413-418, 2003

[9]. Cha-Ming Shen, Shi-Chen Lin, Chen-May Huang, Huay-Xan Lin, Chi-Hong Wang, "Couple Passive Voltage Contrast with Scanning Probing Micoscope to Identify Invisible Implant Issue", Proceeding of the 31st international symposium for testing and failure analysis, pp212-216,2005

[10]. Z. G. Song*, S. K. Loh, S. P. Neo, X. H. Zheng and H. T. Teo, "Application of FIB Circuit Edit and Electrical characterization in Failure analysis for Invisible Defect Issues", Proceedings of 13th IPFA 2006, Singapore, pp187-191, 2006

[11]. David B. Williams and C. Barry Varter, Transmission electrion microscopy. Plenum Press, 1996, pp351-364

SESSION 9:

FEOL

INVITED PAPER

Review of Reliability Issues in High-k/Metal Gate Stacks

R. Degraeve[1], M. Aoulaiche, B. Kaczer, Ph. Roussel, T. Kauerauf, S. Sahhaf, G. Groeseneken[2]

[1]IMEC, Kapeldreef 75, B-3001 Leuven, Belgium, Phone +32-16-281269, email: Robin.Degraeve@imec.be

[2]KU Leuven, Belgium

Abstract-This paper reviews some of the recent learning at IMEC in reliability research on high-k gate stacks. We show how measurement, characterization techniques and physical degradation models can be transferred from SiO_2 (or SiON) single layers to SiO_2(SiON)/high-k stacks. In a first part, Negative Bias Temperature Instability (NBTI) is discussed. We show how interface states created at the SiO_2 (or SiON)/substrate interface determine to a large extend the NBTI. Nitridation has a strong negative impact on NBTI, while thickness or composition of the high-k layer have nearly no influence. In a second part, we discuss the effect of bulk traps in the high-k layer. These traps cause fast V_t-instability and hysteresis, as well as significant Positive Bias Temperature Instability (PBTI). Additional bulk traps are created under electrical stress and form percolating paths of two or more traps causing Soft Breakdown (SBD). At low voltage and with metal gates, the SBD-leakage path develops into a Hard Breakdown (HBD) after some further wear-out time. We summarize the methodology to come to a complete reliability prediction that includes multiple SBDs and HBD. In high-k stacks, the leakage current increase due to multiple SBDs can be a reliability threat for some applications.

I. INTRODUCTION

Due to the exponential increase in leakage current when scaling down the gate oxide thickness of MOSFETs, an urgent need emerged to replace SiON by a different material with a higher dielectric constant. This allows a further increase of the gate oxide capacitance, while keeping the physical layer thickness sufficiently large to avoid excessive leakage current.

In the past years, many companies, universities and research organizations spent major efforts screening a wide range of high-k material candidates for their suitability to be integrated in CMOS technology. The main criteria in this screening process were manufacturability, integration capabilities, and, obviously, transistor performance. Looking back at this extensive screening effort, it is concluded that, quoting M. Green et al [1], "most dielectric materials have only one good property, namely their k-value, while SiO_2 had only one bad property, namely its k-value".

At present, most research focuses on Hf-based dielectrics, as these show the most promising properties. Pure HfO_2 as well as HfSiO(N) are, however, incompatible with Si, and further modifications to the gate stack are therefore unavoidable. At the Si-substrate/dielectric interface, the incompatibility issue is solved by maintaining a very thin (0.5-1 nm) SiO_2 layer. This layer is mandatory to attain sufficient channel mobility. At the dielectric/poly-Si gate interface, the incompatibility issue is solved by replacing the poly-Si gate by a metal gate, with as additional beneficial effect the elimination of the poly-Si depletion.

Once sufficient transistor performance can be guaranteed, the reliability questions are raised: will the performance improvement obtained with high-k stacks remain within specification for the entire lifetime of the device?

In this paper, we review the main findings of the reliability research on SiO_2/high-k/metal gate stacks done at IMEC in the past years. The overview is restricted to (Negative and Positive) Bias Temperature Instability (sections 2 and 3), leakage current increase (section 4) and Time-Dependent Dielectric Breakdown (section 5). It is not our intention to discuss details of specific processes, but rather concentrate on the generic findings common for all stacks.

The reliability research in high-k stacks did not start from an empty plate. In the past decades, the degradation mechanisms that cause BTI and TDDB in SiO_2 have been studied extensively, and physical pictures have been proposed to explain these phenomena. One of the first questions that arose when high-k stacks were introduced was: can these physical pictures be transferred to high-k? Luckily, the answer to this question is largely affirmative and therefore the learning curve on high-k has been very steep.

II. NEGATIVE BIAS TEMPERATURE INSTABILITY (NBTI)

Phenomenological, Negative Bias Temperature Instability is the shift of the threshold voltage in a pmos transistor when a negative voltage is applied at the gate at elevated temperature (typically 125C). NBTI has received continuous interest from the research community both in SiON as well as in high-k stacks. It was identified as one of the most limiting reliability issues in scaled CMOS technologies [2]..

Measurement methodology

One of the main issues in addressing NBTI concerns the measurement procedure. Tracking the threshold voltage shift as a function of stress time seems simple enough, but an important complication in the measurement, analysis and prediction of NBTI comes from the so-called *relaxation* or *recovery* of the degradation when the stress is removed [3]. As a consequence, the monitored threshold voltage shift depends strongly on the measurement timing, making a comparison of published results from different research groups very unreliable, if not impossible.

978-1-4244-2039-1/08/$25.00 ©2008 IEEE   239   *Proceedings of 15th IPFA - 2008, Singapore*

INVITED PAPER

In the past years, several publications [4,5] discuss new methodologies that aim at reducing the time between the removal of the stress and the V_t-determination in order to minimize the relaxation effect. It has, however, become clear that even in a few milliseconds the relaxation process already sets in, and therefore it is virtually impossible to directly measure the 'relaxation-free' degradation directly after stress. Only pulsed IV-techniques offer a solution in this matter.

In recent work [6], we show that more insight into the NBTI mechanism can be gained by abandoning the as-fast-as-possible approach and instead recording a short portion of recovery during every measurement phase as shown in Fig. 1. In many samples, the time dependence of the recovery can be described by a universal relationship that allows to extract a permanent damage component and a recoverable component. Both components can be independently evaluated and extrapolated to operating conditions.

The introduction of high-k stacks and metal gates did not bring about any modifications in the measurement and analysis methodology. The degradation phase follows a power law as a function of time, similar as for SiO_2 and SiON. Furthermore, as illustrated in Fig. 1, the time dependence of the recovery phase can be described by the same universal relationship as for SiON. These observations already indicate that the degradation and recovery mechanism do not substantially change. A detailed study of the permanent and recoverable component in various SiO_2/high-k stacks is still ongoing and will be reported in later publications.

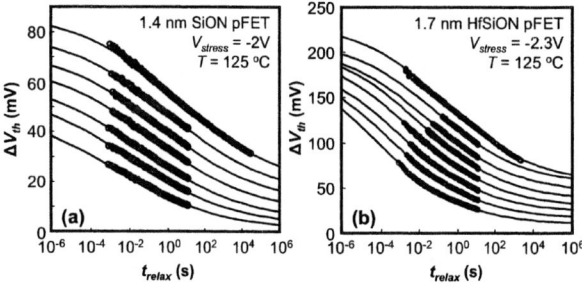

Fig. 1: The recovery of the degradation after removing NBTI stress is fitted as $(1+B.(t_{relax}/t_{stress})^\beta)^{-1}$ with β a dispersion parameter and B a scale factor. The same expression succeeds in describing both SiON and high-k stack recovery. Each line corresponds to different t_{stress}.

Most of the data on high-k stacks in this paper are either collected with a 'classical' stress-measure-stress sequence, or with a fast 'on-the-fly' V_t-evaluation, designed to minimize the recovery effect. One should never compare data taken with different methodologies.

In the 'classical' stress-measure-stress method, the stress is interrupted for measuring a complete I_D-V_G characteristic, from which V_t is determined. In some cases, we also take a charge pumping base level curve in order to monitor the increase of the interface trap density.

In the fast as-fast-as-possible method, the V_t-shift is obtained by switching the stress voltage configuration momentarily to a sense configuration with $V_D = 50mV$ and $V_G =$

$\sim V_t$. From the measured source-to-drain current and an initial I_D-V_G characteristic the V_t is calculated. In our measurement set-up, this method allows to determine V_t in $\sim 100ms$. A second I_D-V_G measurement and the end of the complete stress cycle is used as a control to make corrections for changes of sub-threshold voltage slope, if needed.

NBTI in SiO_2/high-k stacks

We monitor the NBTI by plotting the effective trap density as a function of the electric stress field. Both quantities are defined as:

$$N_{eff} = dV_t(C_{ox}/q) \quad \text{and} \quad E_{ox} = (V_G - V_t)/EOT$$

with C_{ox} the stack capacitance, q the elementary charge, V_G=stress gate voltage and EOT=effective oxide thickness.

In Fig. 2, data are shown for $\sim 1nm/\sim 2nm$ SiON/HfSiO(N) stacks with different Hf-content ranging between 23% to 100% (=pure HfO_2). All curves coincide indicating that the high-k composition has no impact at all on the NBTI reliability.

Furthermore, charge pumping measurements on the same samples, shown in Fig. 3, reveal that the generation of interface states is also not affected by the high-k layer composition. These observations suggest that NBTI is caused by degradation of the interface layer only while the high-k dielectric has negligible impact.

Further confirmation of this picture comes from the data in Fig. 4 that show the generated traps as a function of the oxide field after NBTI-stress for a range of processes that aim at incorporating nitrogen into the layer. The same 2 nm HfSiO(N) dielectric is maintained. Two groups of data are distinguished: the samples with nitridation show systematically a worse NBTI-reliability compared to the non-nitrided samples. Furthermore, the nitrided SiON/high-k stacks have identical degradation as single layer SiON samples as is also shown in Fig. 4.

Note that nitrogen mainly accumulates in the SiO_2 interface layer and, therefore, the change of NBTI-degradation when a nitridation process step is introduced again indicates that NBTI is mainly an interface reliability issue.

The observations in Figs. 2-4 are consistent with the generally accepted physical model of NBTI-degradation. Summarized, this model describes NBTI as a hydrogen depassivation reaction at the Si/SiO_2 interface that leads to creation of interface states causing degradation (= 'instability') of the transistor performance. It has already been shown that also trapping centers in the bulk of the oxide have an impact on the V_t-shift during an NBTI-test, but this effect reduces when extrapolating to operating voltage.

INVITED PAPER

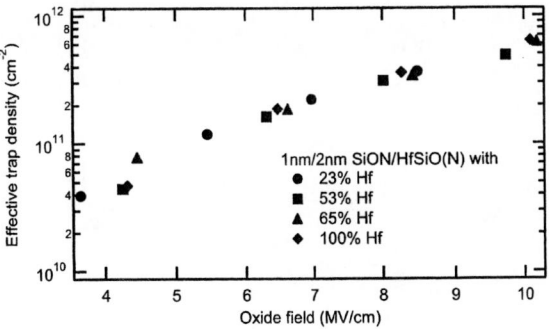

Fig. 2: The effective hole trap density as a function of stress field during NBTI testing at 125C. The degradation is independent of the Hf-content. These data were taken by interrupting the stress periodically to measure a complete I_D-V_G characteristic.

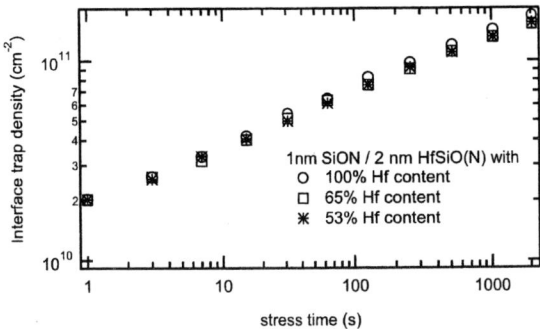

Fig. 3: The generated interface trap density monitored by Charge Pumping for the three SiON/HfSiO stacks with different Hf-content (same samples as in Fig. 2). The generation of interface states caused by NBTI is independent of the Hf-content.

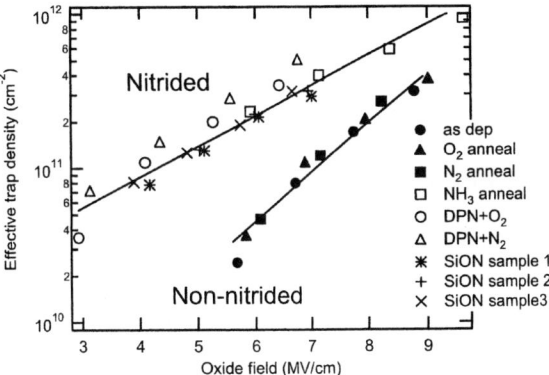

Fig. 4: The effective trap density generated after NBTI stress as a function of the stress field. The six first samples have a 2 nm HfSiO(N) dielectric with 53% Hf-content. The last three samples have SiON layers without any high-k layer. Two clusters of data appear: all treatments that aim at incorporating nitrogen in the dielectric have worse NBTI reliability compared to the non-nitrided samples. These data were taken with the as-fast-as-possible method (details in text).

We have published a number of dedicated experiments to further demonstrate the importance of the substrate/SiO_2 interface for NBTI reliability. First, in [7] we show that devices with SiO_2/high-k stacks processed on (110) surface have increased trap generation after NBTI-stress compared to those on (100). This is consistent with the physical picture of H-depassivation since (110) surfaces have a higher concentration of Si-H bonds to start with. Second, in [8] we show that D_2-annealed surfaces in SiO_2/high-k stacks are more NBTI-robust than H_2-annealed surfaces, consistent with the picture that the Si-D bond requires more energy to break than the Si-H bond. The improvement with D_2-anneal can, however, not compensate the detrimental effect of nitridation.

Conclusions

On SiO_2/high-k or SiON/high-k stacks, the degradation during NBTI-stress as well as the recovery of the V_t after stress removal are described by the same functions as for single layer SiO_2 and SiON. This implies that all developments on measurement methodologies can be directly transferred to SiO_2/high-k stacks.

Since high-k gate materials are only used with a thin interfacial SiO_2 layer we can state in general that the depassivation reaction at the Si/SiO_2 interface still dominates NBTI in high-k stacks. The effect of the high-k layer is introduced only through the EOT-calculation. In other words, the high-k affects the NBTI only through its impact on the field in the interface layer.

As a consequence, the processing of the interface layer is the key factor in controlling NBTI in high-k stacks. Nitridation of the dielectric strongly reduces the NBTI-lifetime, similar as in gate dielectric without high-k component.

III. POSITIVE BIAS TEMPERATURE INSTABILITY (PBTI)

The threshold voltage shift induced by Positive Bias Temperature stress in nmos transistors with SiO_2 or SiON is negligible. In SiO_2/high-k stacks, however, it can be an important reliability issue as illustrated in Fig.5 for an SiO_2/HfO_2 stack.

The mechanism for PBTI has been identified as filling of initial traps in the high-k layer [9]. All Hf-based dielectrics suffer from a defect band close to the conduction band, especially when poly-Si gates are used [10]. Besides PBTI, these defects also give rise to V_t-instabillity and C-V hysteresis. Using metal gates, scaling down the high-k physical thickness and lowering the Hf-content in the dielectric reduces both V_t-instability and PBTI to acceptable levels. The correlation between bulk trap density and PBTI lifetime is illustrated in Fig. 6.

INVITED PAPER

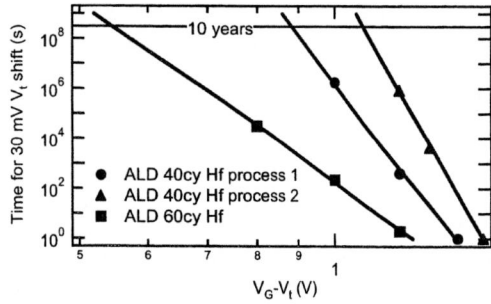

Fig. 5: The time till 30 mV V_t-shift after PBTI stress as function of V_G-V_t for three SiO_2/HfO_2 splits processed with Atomic Layer Deposition (ALD). PBTI strongly depends on the HfO_2 thickness and quality.

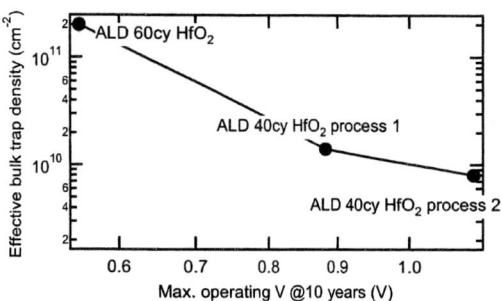

Fig. 6: The bulk HfO_2 trap density measured by variable $t_{charge}/t_{discharge}$ charge pumping (details in [11]) reveals the correlation between bulk traps in the HfO_2 and the PBTI reliability.

IV. LEAKAGE CURRENT GENERATION

Besides the initially present defects, additional bulk traps in Hf-based dielectrics are created during positive Constant Voltage Stress (CVS), leading to dielectric breakdown when a critical trap density is reached. Although the details of the trap generation mechanism might be different, the concept of bulk trap creation up to a critical density at breakdown is similar as in SiO_2 [12].

In this paper, we will illustrate the generation of traps in a 0.9 nm EOT SiO_2/ALD-HfO_2/TaN layer [13]. The trap generation can be observed in several ways:

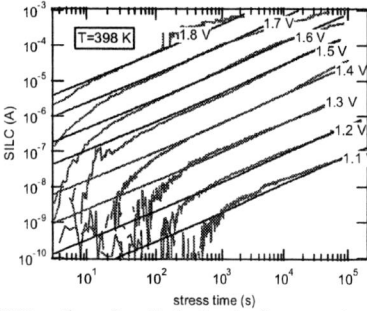

Fig. 7: SILC vs. time after eliminating the charge trapping effect (only shown at 398 K). After the transient part, a power law can fit the curves with a constant exponent of 0.7 (298 K) or 0.8 (398 K).

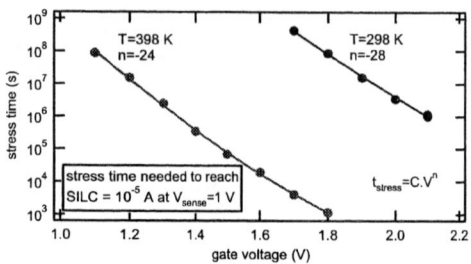

Fig. 8: The SILC voltage acceleration can be fitted well with a power law. Especially at 398 K, the data are in a sufficiently wide V_G range to allow a clear discrimination between an exponential t~exp(-γV_G) and power law.

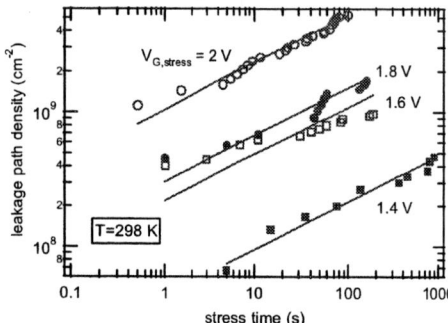

Fig. 9: The generation of leakage paths vs. stress time can be fitted with a power law with exponent 0.35.

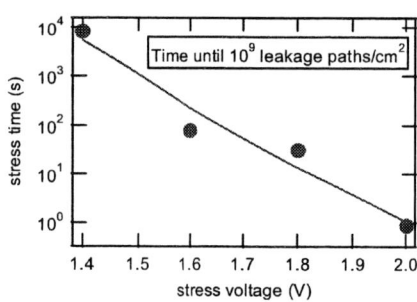

Fig. 10: A power law with exponent n=-24 fits the voltage acceleration of the generation of leakage paths.

1) First, the generated traps give rise to Stress-Induced Leakage Current (SILC) (Fig. 7) that shows a power law increase with time with exponent ~0.7 (298 K) or ~0.8 (398 K). At 398K, the SILC voltage acceleration at a sense voltage of 1 V (Fig. 8) can be fitted well with a power law [14] ($t=t_0.V_G^n$) with n=-24 (or n=-28 at 298 K).

2) On very small samples (1.25×10^{-9} to 5×10^{-9} cm^2) we can actually observe the creation of an individual leakage path corresponding to the creation of an individual trap [15]. If we 'count' the number of generated traps as a function of stress time a power law is found with exponent ~0.35 (Fig. 9), half of the value of the SILC power law exponent in Fig. 7, while the voltage acceleration exponent n=-24 is within error bar identical to the SILC voltage acceleration exponent (Fig. 10).

INVITED PAPER

3) The bulk HfO_2 trap generation can also be measured by variable frequency Charge Pumping [12]. Again a power law increase of the trap density vs. time is seen (Fig. 11) with exponent ~0.4. The V_G-acceleration has in this case a power exponent n=−31 (Fig. 12).

Linking the results of the three characterization methods discussed above, a consistent pictures emerges of the degradation of HfO_2. Indeed, from the similarity in both time and voltage dependence of the trap generation (fig. 11-12) and the generation of leakage paths (Fig. 9-10), it can be concluded that the creation of each single HfO_2 bulk trap is causing a trap-assisted leakage path in the dielectric, which gives rise to the small single trap current increases (1-100pA). After more time, more conducting (~10nA-1mA) two-trap percolating clusters form, causing the SILC in Figs 7-8. This is confirmed by the fact that SILC has the same voltage acceleration as the trap generation (Fig. 8 and 12), but twice the time power law exponent (0.7 for SILC vs. 0.35 for trap generation).

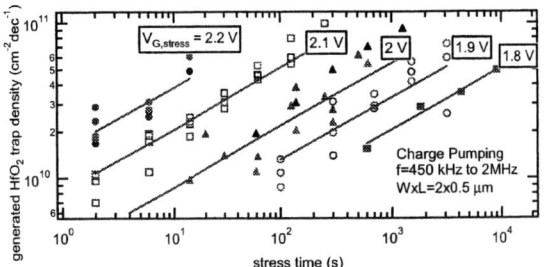

Fig. 11: The increase of the HfO_2 bulk trap density extracted from variable frequency charge pumping measurements. A power law with an exponent of approximately 0.4 fits the data satisfactorily.

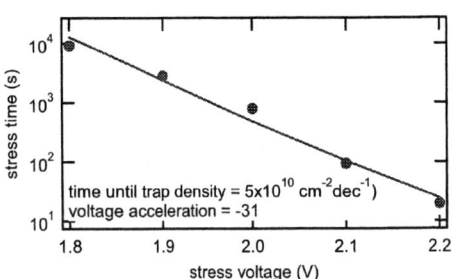

Fig. 12: The voltage acceleration of the trap generation process measured by charge pumping. A power law fit results in exponent n =-31.

V. TIME-DEPENDENT DIELECTRIC BREAKDOWN (TDDB)

In very small devices the creation of a two-trap percolation cluster is interpreted as a soft BD (SBD). This means that the SILC in Fig. 7 can also be interpreted as being caused by *multiple* SBDs. Consistent with theory [16], the SBD Weibull distribution (Fig. 13) has a slope of ~0.7, identical to the SILC power law exponent. All observations in Figs 7-13 are also seen in single layer SiO_2 or SiON and demonstrate that the physical and statistical models developed in the past can be applied to high-k stacks.

In stacks with *metal* gates the SBD develops after some wear-out time into an abrupt hard BD (HBD) [17]. This is different to the case of *poly-Si* gates where a slow wear-out ends with a gradual current runaway [18]. Both cases are illustrated in Fig. 14. The different evolution of the BD-path conductivity in metal and poly-Si gated devices is probably due to the difference in the ballasting resistance that limits the positive feed-back mechanism controlling the SBD to HBD transition.

We found that the wear-out time *between* SBD and HBD can also be described by a Weibull distribution. Consequently the HBD distribution is *not* Weibull distributed. In a first order attempt, the HBD-distribution is described as a convolution of the Weibull SBD-distribution with the Weibull wear-out distribution [19]. This approach is, however, only correct if a single SBD develops into its corresponding HBD.

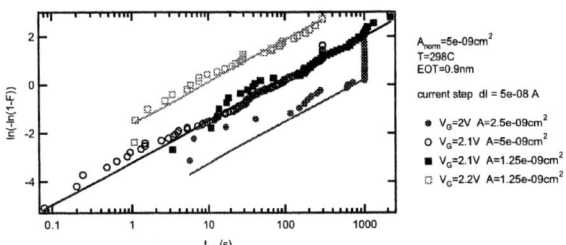

Fig. 13: The distribution of the first created conduction path with current >5x10⁻⁸A at stress condition (i.e. soft BD). These distributions have a slope of 0.7-0.8, indicating the formation of two-trap clusters. Vertical columns of data are time-outs (and not BDs), which are consistently included in the Weibull fitting algorithm.

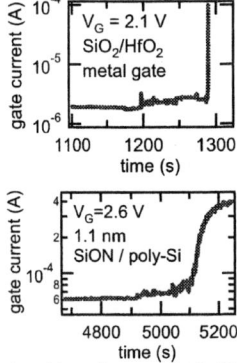

Fig. 14: In samples with small area, a single SBD leads to a current runaway after some wear-out time. When a metal gate is used, as in high-k stacks, the current-runaway is very fast and appears as a 'classical' HBD (left). With poly-Si gate, as for SiON, the runaway phase is slow (right).

In general, multiple SBDs occur in a sample (like for instance in Fig. 7) and the HBD is not necessarily caused by the wear-out of the first SBD. This competition effect needs to be accounted for in order to describe the correct shape of the HBD distribution. In recent work [20], we have constructed the analytical formula of this distribution using the 'noble art of area scaling'. It contains four parameters, the SBD Weibull shape and scale factors β_{SBD} and η_{SBD}, and the wear-out Weibull shape and scale factors, β_{WO} and η_{WO}.

INVITED PAPER

In [21], we demonstrate that for a carefully chosen voltage/area combination it is possible to extract all four parameters from a fit to the *hard breakdown distribution only*. This approach greatly simplifies the measurement of dielectric breakdown since HBD can be easily measured and it is no longer necessary to find an appropriate SBD detection method. Especially in large area samples, the relative current increase induced by a single SBD is very small and can be easily overlooked.

Once the four parameters of the SBD and wear out distributions are known, the TDDB reliability can be predicted at operating voltage by applying the voltage acceleration laws. A complete reliability prediction consists of three regions in the voltage/area domain, as shown in Fig. 15: in region 1 no SBD occurs before the specified lifetime of the device, in region 2 multiple SBDs will occur giving rise to an increased leakage current, but none of these SBDs will develop into a full HBD before the specified device lifetime, and in region 3 the HBD limits the device lifetime.

Because of the slow wear out phase, t_{HBD} is sufficiently high at low voltage for most Hf-based dielectrics. The main TDDB reliability problem of high-k/metal gate stacks is the gate leakage current increase due to multiple SBDs. In applications where leakage current increase cannot be tolerated, as for instance in memory devices, this will limit the usability of high-k gate materials.

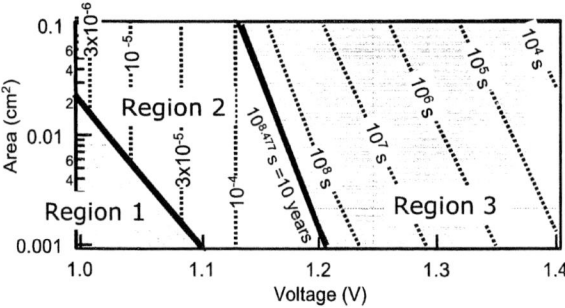

Fig. 15: TDDB reliability prediction for 0.9 nm EOT SiO_2/ALD-HfO_2/TaN stack. Three regions are indicated vs stress voltage and device area. In region 1 no SBD will occur after 10 years stress. In region 2, multiple SBDs will occur after 10 years stress. The contours show the average leakage current increase through the SBD spots divided by the tunnel current through an unstressed device. In region 3 HBD limits the reliability. Here the contours show the lifetime for 0.01% failures.

VI. CONCLUSIONS

An overview of our learning of reliability issues in high-k stacks has been presented. In general, most of the models and concepts that had been developed for SiO_2 and SiON reliability could be maintained on high-k stacks.

Degradation at the interface: NBTI

With the introduction of high-k materials, a thin layer of SiO_2 or SiON has been maintained in between the substrate and the high-k. Consequently, the substrate/dielectric interface did not undergo major changes. Since NBTI is a depassivation reaction of states at this interface, all NBTI characteristics and models are similar in high-k stacks as in single layer SiO_2 and SiON. NBTI in high-k stacks is dominated by the interface layer quality and shows nearly no dependence on the high-k composition, thickness or quality.

High-k bulk traps: hysteresis and PBTI

All Hf-based dielectrics have a defect band close to the conduction band edge. These bulk traps cause hysteresis, V_t-instability as well as PBTI. To control these reliability issues, one can either lower the impact of the bulk traps by thinning the dielectric, or reduce the bulk trap density by lowering the Hf-content and restrict to the use of metal gates.

Degradation of high-k bulk: SILC and TDDB

During degradation, additional bulk traps are generated and clusters of two or more traps form percolating paths that give rise to a considerable SILC. The percolation paths wear out with time leading to HBD, but at operating conditions this takes longer than the required lifetime. The main bulk trap-related reliability issue for high-k stacks is therefore the SILC, which has to be taken into account by circuit designers and will be a critical issue for memory applications.

REFERENCES

[1] M. L. Green, E. P. Gusev, R. Degraeve, and E. L. Garfunkel, *J. Appl. Phys.*, vol. 90, no. 5, pp. 2057-2121, 2001.
[2] V. Reddy *et al.*, *Proc. IRPS*, p. 248 (2002)
[3] G. Chen et al. *Proc. IRPS*, p. 196 (2003).
[4] V. Huard et al., *Microelectr. Reliab.* 46, p. 1 (2006).
[5] M. Denais et al., *IEDM Tech. Digest*, p. 109 (2004).
[6] T. Grasser et al., *IEDM Tech. Digest* (2007).
[7] M. Aoulaiche et al. , presented at IEEE SISC, San Diego, 2006.
[8] M. Aoulaiche et al., *Proc. ESSDERC*, pp. 197-200, 2005.
[9] M. Houssa et al., *Electrochem. Solid-State Lett.*, vol. 9, no. 1, pp. G10-G12, 2006.
[10] A. Kerber et al. *EDL*, vol.24 (2), 2003.
[11] M. Zahid et al., *Proc. IRPS*, 2007.
[12] R. Degraeve et al. , *IEDM Techn. Dig.*, 2005.
[13] L.-Å. Ragnarsson et al., *IEEE Trans. on Electron. Dev.*, vol. 53, no. 7, p. 1657, 2006.
[14] E.Y. Wu et al. *IEEE Trans. on Electron Dev.*, vol. 49, no. 12, p. 2244, 2002.
[15] R. Degraeve et al., *Proc. IRPS*, pp. 360-365, 2005.
[16] M.A. Alam et al., *Proc. IRPS*, pp. 406-411, 2003
[17] T. Kauerauf et al. , *IEEE Electron Dev. Lett.*, vol. 26, no. 10, p. 773, 2005.
[18] B. Kaczer et al. , *IEDM Techn. Dig.*, p. 713, 2004.
[19] R. Degraeve et al., *Proc. IRPS*, 2006.
[20] Ph. Roussel et al., presented at INFOS, Athens, Greece, 2007.
[21] S. Sahhaf et al., *IEDM Tech. Digest* , 2007.

978-1-4244-2039-1/08/$25.00 ©2008 IEEE

A New TDDB Lifetime Bi-Model for eDRAM MIM Capacitor with ZrO2 High-K Dielectrics

S.W. Chang, Chia Lin Chen,C.J. Wang, Kenneth Wu

Taiwan Semiconductor Manufacturing Company, Reliability Assurance Division. 121, Park Ave. 3, Hsinchu Science Park Hsinchu, Taiwan 300-77, R. O. C

Tel: 886-3-5672334, FAX: 886-3-5777507 e-mail: swchangd@tsmc.com

Abstract-A new TDDB lifetime bi-model on ZrO2 based capacitor is proposed as E-model for high field region (>2.7MV/cm) and power law model for low field region (<2.7MV/cm). The current conduction mechanism is identified as Poole-Frenkel tunneling. The mechanism of TDDB is explained as hole trapping enhanced ZrO2 breakdown in high field region and electron trapping dominated ZrO2 breakdown in low field region.

I. INTRODUCTION

Recently, high dielectric constant materials have been studied as a promising alternative for the dielectric material of eDRAM MIM capacitors because of its high dielectric constant [1]. Typically, TDDB lifetime model is extrapolated from accelerated high field to operation voltage by using conventional E-model [2], 1/E model [3], or power-law model [4]. However, there has been no clear answer so far as to which model is suitable for high K MIM capacitor, especially in low field region. In this study, the correlation among the current conduction characteristics, dielectrics breakdown behavior and TDDB lifetime model for ZrO2 MIM capacitors is well established. Besides, the mechanisms of current conduction and ZrO2 breakdown are included for discussion.

II. EXPERIMENTAL

The eDRAM MIM capacitors with 7.5-11.5nm ZrO2 high-K dielectric were fabricated. Time to breakdown (Tbd) was measured by applying a constant voltage under substrate injection at 1250C until gate oxide soft breakdown occurred.

III. RESULTS AND DISCUSSION

A. Current Conduction Characterization and Mechanism Discussion

Poole-Frenkel mechanism has been widely reported to dominate the high-K material conduction behavior [1]. Fig.1 shows the Jg/Eox versus $E^{\wedge}(1/2)$. Good linear fitting is found when Vg is smaller than 2.4V. Fig.2 shows the leakage current at various operation temperature. It is seen that the leakage current is strong dependent on temperature for the whole voltage range. Both the IV behavior and temperature dependence suggest the conduction current is dominated by Poole-Frenkel mechanism. However, the leakage current in Fig.1 fast increases and diverges from original trend when the voltage is larger than 2.4V. It is attributed that surface plasmon-induced-hole trapping enhances the leakage current increase in high voltage region.

B. ZrO2 Breakdown Behavior and Mechanism Discussion

Fig.3 shows typical time dependence of the leakage current under substrate injection. Jg decrease and abrupt soft breakdown are shown in low bias region (2.2V~1.9V). However, a gradual increase of leakage current is observed in high bias region (2.4~2.6V). It implies that electron trapping causes Ig decrease in low field region (1.9V~2.2V), while hole trapping causes Ig increase in high field region (2.4V~2.8V). Energy band diagrams are illustrated to explain the breakdown behavior shown in Fig.4.

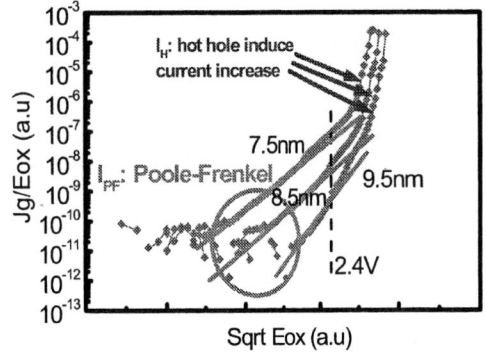

Fig.1 The J/Eox is plotted as a function of sqrt Eox for various ZrO2 thicknesses. Good linear fitting is found when Vg is smaller than 2.4V, indicating Poole-Frenkel dominates in low voltage region.

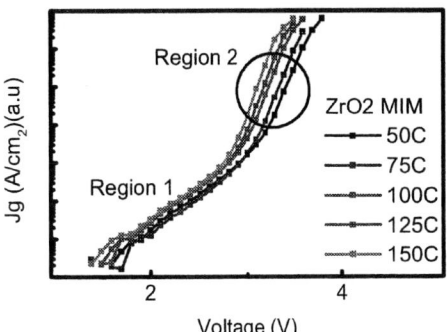

Fig.2 The temperature dependence of IV curve for ZrO2 MIM capacitor. Strong temperature dependence was found for the whole voltage range.

978-1-4244-2039-1/08/$25.00 ©2008 IEEE

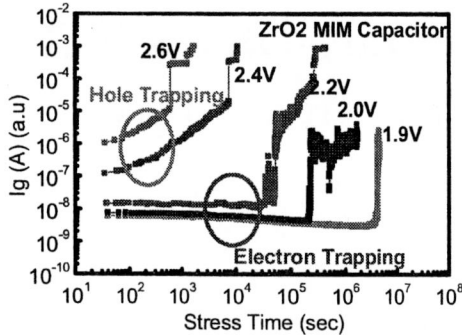

Fig.3 The time dependence of leakage current under substrate injection. Jg decreases in low-bias region (1.9V~2.2V) and Jg increases in high-bias region (>2.2V) before SBD occurs.

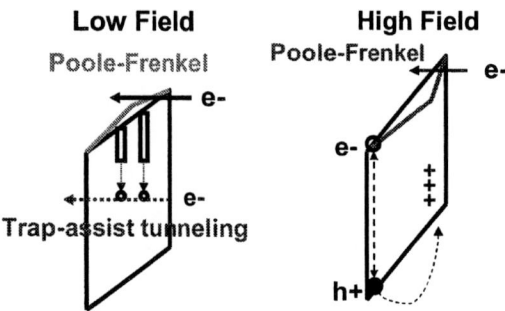

Fig.4 Schematic energy band diagrams explain the breakdown mechanism in high and low field regions respectively. In the high field region, the injected holes cause the current increase and ZrO2 breakdown, while in the low field region, trap-assisted electron traps dominate ZrO2 breakdown.

In high field region, electrons are injected from the cathod. The tunneling electrons generate holes by surface plasmons, and then holes inject into ZrO2 film from anode, consequently, resulting in leakage current increase and ZrO2 breakdown. On the contrary, the electron traps dominate the breakdown due to less holes being generated by surface plasmon in low field region. In order to further clarify the breakdown mechanism, the TDDB lifetime of ZrO2 with three different thicknesses is evaluated and shown in Fig.5. Takashi reported that thicker HfOx has longer Tbd lifetime and smaller voltage acceleration slope than thinner HfOx film due to larger grains and more defects in thicker HfOx film [6]. Similar results are observed in high field region, but inconsistent with our results in low field region. It is seen that leakage current of 7.5nm ZrO2 is around 2 orders higher than that of 9.5nm for the entire voltage range as shown in Fig. 5. However, the Tbd lifetime of 7.5nm ZO2 is almost comparable to that of 9.5nm ZrO2 in the low field region (2.2V~1.8V). It indicates that the Tbd lifetime well correlates to leakage current with hole trapping in high field region (>2.2V), but not in low field region. In low field region, the electrons hopping of Poole-Frenkel tunneling from traps to traps toward gate electrode does not create additional trap centers. However,

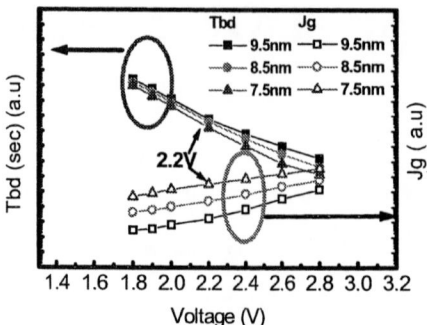

Fig.5 The Tbd and Jg of ZrO2 films are plotted as a function of voltage for various ZrO2 thicknesses. Lifetime correlates well with leakage current in the high voltage region (>2.2V) due to hole trapping while trap-assisted electron traps dominates in the low voltage region.

the thermally activated migration of trapping charges into the unoccupied state creates the new traps after long term stress and subsequently causes ZrO2 breakdown [1]. In other words, ZrO2 breakdown is dominated by trap-assisted electron traps rather than hole traps in the low field region.

C. ZrO2 High-K TDDB Characterization

Before encountering into TDDB lifetime model, the basic parameters are evaluated to assess the validity of our model. At least 3 voltages are applied on ZrO2 MIM capacitor as shown in Fig. 6, resulting in the Weibull cumulative failure distribution with the slope of 2, which is in good agreement with R. Degraeve et al. published in EDL 2005 [7]. The temperature dependence of ZrO2 is also executed on 100°C, 125°C, and 150°C at a fixed applied voltage. Fig. 7 shows the resulting activation energy of 1.2eV, reflecting stronger temperature dependence than silicon dioxide.

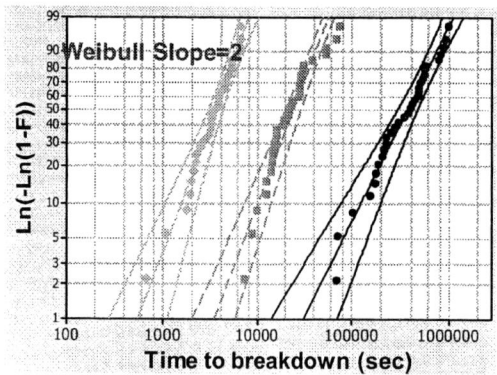

Fig.6 The plot describes the TDDB Weibull distribution for 3 voltages at 125°C.

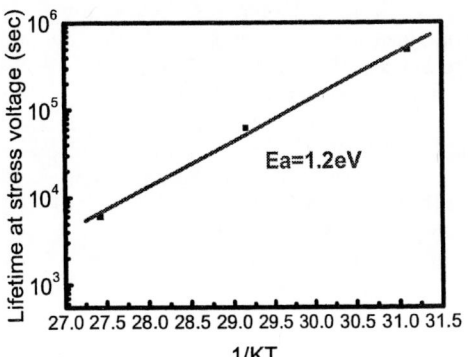

Fig.7 The logarithm of Tbd is plotted versus the reciprocal of temperature at 2.6V. Strong temperature dependence is found for ZrO2.

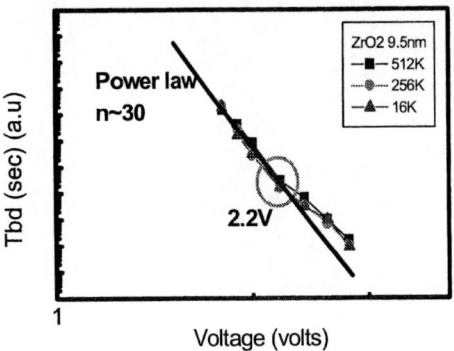

Fig.9 The logarithm of Tbd is plotted versus the logarithm of voltage. Lifetime fits very well in the low-field region, but it diverges in the high-field region.

D. TDDB Lifetime Model

E-model is used in the conventional lifetime prediction to extrapolate from high field to operation voltage for high-K film [8]. R. Degraeve proposed that the Tsbd has a power law dependence on Vg stress [9]. In order to clarify the lifetime model, the correlation between leakage current and lifetime model is discussed. Generally speaking, it is usually assumed that Tbd is inversely proportional to leakage current in thick dielectric, and that is Tbd~(1/Jg). Based on the above discussion, surface plasmon-induced-hole trapping in high field region and trap-assisted electron trapping in low field region dominate the ZrO2 breakdown. Therefore, a "bi-model" TDDB lifetime is proposed for ZrO2 MIM capacitor due to different current conduction and breakdown mechanisms in different voltage regions. Ln(T) is plotted as a function of V shown in Fig.8. The lifetime fits very well in high field region, but it is under-estimated by extrapolating to low field. Ln(T) is plotted as a function of ln(V) shown in Fig.9. Contrarily, the lifetime fits very well in low field region, but is divergent in high field region.

This means that E-model is not suitable for ZrO2 MIM capacitors in low field region. It needs to change the model from E-model to power-law model in the low field region. Long term stress at lower bias region is applied to demonstrate the power-law TDDB model, which is shown in Fig.9 with stress time >2000 hours. Experimental result shows that they are significantly located in the power law model line, not in the E model or sque E model line. Therefore, the Jg-Vg curve related to breakdown can be simulated as the surface plasmon-induced-hole traps in high field region and trap-assisted tunneling with electron traps in low field region, which are shown in Fig.10. Large Poole-Frenkel current masks trap-assisted tunneling in low field region.

E. Surface Roughness Effect on TDDB

ZrO2 MIM capacitors with smooth and rough interface are used to demonstrate TDDB lifetime "bi-model". Fig.11 shows the Tbd lifetime versus voltage plot. It is seen that Tbd lifetime of smooth interface sample is much better than that of rough interface sample, especially in low field region.

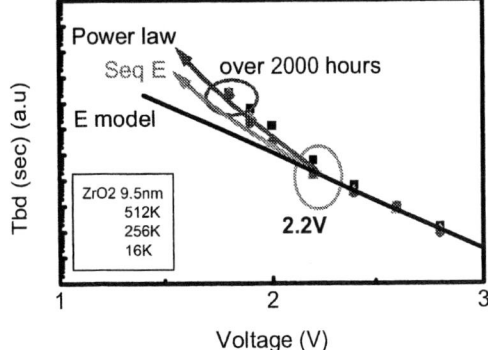

Fig.8 The logarithm of Tbd lifetime is plotted as a function of voltage. Lifetime fits very well in the high-field region, but it diverges in the low-field region.

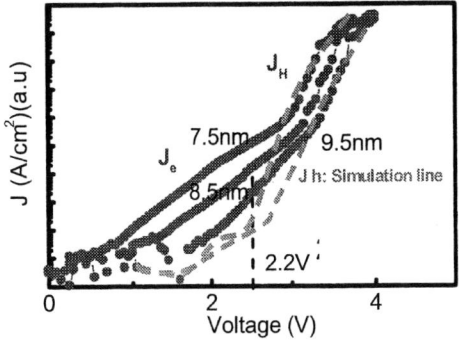

Fig.10 The Jg-Vg curve related to breakdown can be simulated as the Fowler-Nordheim tunneling with hole traps in high field region and traps-assisted tunneling with electron traps in low field region

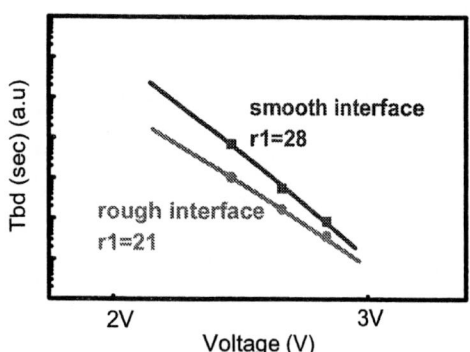

Fig.11 Tbd versus voltage for smooth interface and rough interface ZrO2 film, respectively.

However, I-V behavior of smooth interface samples is comparable to that of rough interface samples, which is shown in Fig.12. It implies that Poole-Frenkel tunneling can not be correlated to Tbd lifetime. Therefore, trap-assisted tunneling is proposed as the dominating lifetime mechanism in low field region. So, the electron tunneling currents can be reduced in smooth surface due to less interface traps, which results in larger voltage acceleration slope and better lifetime. Generally speaking, Tbd is inversely proportional to Jg. According to TDDB lifetime "Bi-model", the I-V curve can be simulated and shown in Fig.12 with S1=28 voltage acceleration slope for smooth interface sample and S2=21 for rough interface sample, respectively. Therefore, surface roughness effect on TDDB can be well explained by the TDDB lifetime "Bi-model". In the future study, the precise TDDB lifetime model for high-K material will be very critical.

IV. CONCLUSION

In this work, the current conduction mechanism is identified as Poole-Frenkel tunneling. TDDB lifetime model for ZrO2 MIM capacitor is proposed as E-model in high field region and power law model in low field region. The trap-assisted tunneling rather than Poole-Frenkle tunneling is proposed to correlate with Tbd in low field region. Surface roughness effect on TDDB can be well explained by the proposed TDDB lifetime "bi-model" with trap-assisted simulation current.

ACKNOWLEDGMENT

The authors would like to thank Dr. Rakesh Ranjan for the valuable technical discussions and suggestions for this paper.

REFERENCES

[1] G. Ribes., et al. IEEE Transactions on Device and Materials Reliability, p5, 2005
[2] Muhammad A. Alam, et al. IEEE IRPS, p.21, 2000.
[3] Chen-Ming Hu, et al. IEEE Transactions on Electron Devices, p2268, 1988.
[4] Ernest Y. Wu, et al. Microelectronics reliability , 2005.
[5] Wen-Jie Qi, et al. IEEE IRPS, p72, 2000.
[6] Takashi Ohtsuka, et al. SSDM, p132,2006
[7] R. Degraeve et al. EDL, 2005
[8] Young Hee Kim, et al. IEEE Electron Device letters, p.594, 2002.
[9] R. Degraeve, et al. IEEE, , 2005..

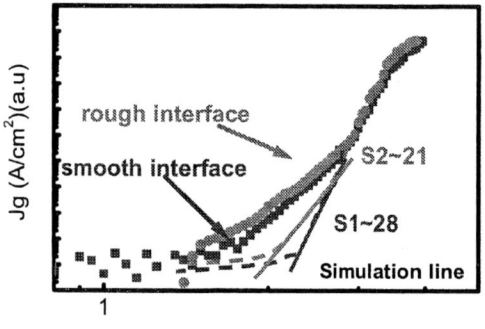

Fig.12 The I-V curves for rough and smooth ZrO2 interfaces, respectively. It can be simulated with S1=28 voltage acceleration slope for smooth interface sample and S2=21 for rough interface sample, respectively.

978-1-4244-2039-1/08/$25.00 ©2008 IEEE

A Rigorous Model for Trapping and Detrapping in Thin Gate Dielectrics

Wolfgang Goes°, Markus Karner°, Viktor Sverdlov[†], and Tibor Grasser°

°Christian Doppler Laboratory for TCAD at the Institute for Microelectronics, TU Wien
Gußhausstraße 27-29, 1040 Wien, Austria
[†]Institute for Microelectronics, TU Wien, Gußhausstraße 27-29, 1040 Wien, Austria

Abstract- **We rigorously model charge trapping and detrapping in ultrathin dielectrics. In addition to charge exchange with the substrate, the poly-gate interface is taken into account which gives rise to decreased charge trapping compared to conventional models designed for thicker gate dielectrics. Finally, an extension of this model also accounting for the shift of trap levels may possibly explain the large time scales experimentally observed during the recovery after application of an on-state voltage.**

I. INTRODUCTION

The success of semiconductor industry relies on continuously shrinking device dimensions which is accompanied by the emergence of quantum mechanical effects such as tunneling and trapping. Meanwhile research in this field encounters a point where these effects become increasingly important as it is the case for Hot Carrier Injection (HCI) and Stress-Induced Leakage Current (SILC) among others.

Even though considerable progress in oxide reliability has been made, the presence of traps in dielectric such as silicon oxynitrides and hafnium dioxide is experimentally confirmed by a series of measurement techniques. In particular Electron-Paramagnetic-Resonance (EPR) allows the identification of a large number of defects. Several variants of the E'$_\gamma$ center have been detected in silicon dioxide [1], while especially K centers in silicon oxynitrides have been discovered [2]. Some of these traps are suspected to be charge traps, however, their nature is not fully understood and no comprehensive model for charge trapping exists to date. It is well established that the trap behavior strongly depends on the spatial and energetical distribution of traps within the dielectric. Various sorts of traps have been proposed in literature, however, detailed knowledge is still vague. Zhang [3], for instance, proposed two types of traps: Cycling Positive Charges (CPC) which show an oscillating behavior for switching bias - in contrast to the so-called Anti-Neutralization Positive Charges (ANPC) which form a constant background charge. Annealing studies undertaken on irradiated samples reveal accelerated annealing behavior of traps at elevated temperatures [4]. Furthermore, traps featuring more than only two charge states, the so-called amphoteric traps, have been proposed in [5]. In short, several distinct properties for different sorts of traps may have to be taken into account when dealing with trapping in dielectrics.

An issue which has attracted some recent attention is the idea that charge trapping is suggested to be involved in Negative Bias Temperature Instability (NBTI). In literature, two kinds of explanations for NBTI are persued: First, the NBTI phenomenon is traced back to the creation and passivation of dangling bonds correlated to diffusion of hydrogen to and from the interface. Second, trapping and detrapping of charge carriers into the gate dielectric is regarded as the origin of NBTI or supposed to constitute an appreciable contribution to NBTI [6,7]. Tewksbury's trapping model [8] which addresses this issue in considerable detail has been used in this context [7]. However, our simulations extended by the impact of the gate contact indicate short storage times for trapped charges in thin dielectrics. This is in stark contrast to observations in experiments. As one possible explanation for these long storage times the impact of energy level shifts are investigated. They occur, when traps capture charge carriers followed by reformation of the trap and a shift of the corresponding trap level. Such shifts have been suggested in a speculative manner [3] and as a result of first principles calculations [9]. In this work, the focus is put on investigating the impact of level shifts in addition to charge trapping from the poly-gate interface. A rigorous model is suggested which incorporates the poly-gate interface as well as such level shifts.

II. THEORETICAL BACKGROUND

In the following, a brief overview of the models employed in this study is given. The present description of trapping and detrapping extends the approach followed by Tewksbury [8] - termed the fixed level model throughout this study - by allowing for charge injection from the poly-gate interface and incorporating trap level shifts. The rate equation (1) forms the basis of the fixed level model:

$$\partial_t f_t(E_t, x) = n(E_t) r_{in}(E_t, x)(1 - f_t(E_t, x)) \\ - p(E_t) r_{in}(E_t, x) f_t(E_t, x) \quad (1)$$

where $n(E_t)$ or $p(E_t)$ denote the number of occupied or empty states in the substrate and f_t stands for the occupancy of the traps located within the dielectric at a distance x from the substrate interface. Note that all quantities are evaluated at the same trap energy E_t. A derivation of the rates r_{in} and r_{out} based on Fermi's golden rule yields a WKB coefficient multiplied with a prefactor v_0:

$$r_{in/out} = v_0 \exp(-2 \int_{x_{if}}^{x_t} k \cdot x dx) \quad (2)$$

$$k^2 = \tfrac{2m}{\hbar^2}(E_{c/v} - E),$$

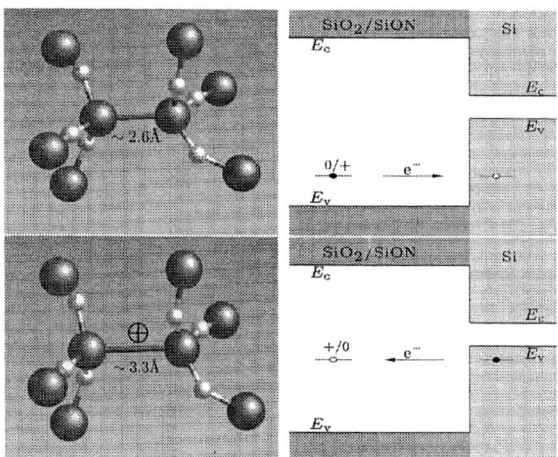

Fig. 1. Representation of a E'_δ [10] defect and its defect levels within the silicon dioxide bandgap. When a positive charge is introduced into this defect, the bond length increases, giving rise to a shift of the trap level upwards in the energy scale.

where x_t is the position of the trap, x_{if} the position of the interface. $E_{c/v}$ stands for the conduction or the valence band edge, respectively. Interface states may also be included in $n(E_t)$ and $p(E_t)$ but have to be specified by different prefactors due to their distinct nature compared to band states. The fixed level model, which was designed for thick gate dielectrics, neglects charge injection from the poly-gate interface - a shortcoming which has to be overcome for modeling modern semiconductor devices. This can be remedied by introducing additional rates on the right hand side of equation (3).

$$
\begin{aligned}
\partial_t f_t(E_t,x) = & \ n_s(E_t)r_{in}(E_t,x)(1-f_t(E_t,x)) \\
& - p_s(E_t)r_{in}(E_t,x)f_t(E_t,x) \\
& n_g(E_t)r_{in}(E_t,x)(1-f_t(E_t,x)) \\
& - p_g(E_t)r_{in}(E_t,x)f_t(E_t,x),
\end{aligned} \tag{3}
$$

where the subscript s stands for substrate quantities and g refers to poly-gate quantities. This refined model is also referred to as the extended fixed level model in the following.

Up to this point, it is assumed that energy levels do not depend on the charge state of the defect. This assumption has been proven to be unjustified. In fact, level shifts do occur after defects have been occupied by charge carriers. When a trapping process takes place, the defect configuration [9] rearranges itself and alters the nearby bond lengths (see Fig. 1), particularly in the amorphous layers used as dielectrics. This is combined with a change in the electrostatics due to the introduced charge. Both effects together cause a trap level shift which has been confirmed for a series of investigated defects suspected to be oxide traps [10,11]. In our level shift model, these level shifts are considered by introducing two types of energy levels, namely one for e⁻ capture (E_{in}) and another for e⁻ release (E_{out}):

$$
\begin{aligned}
\partial_t f_t(E_t,x) = & \ n(E_{in})r_{in}(E_{in},x)(1-f_t(E_{in},x)) \\
& - p(E_{out})r_{in}(E_{out},x)f_t(E_{out},x)
\end{aligned} \tag{4}
$$

Table I

VALUES USED FOR THE FIXED LEVEL MODEL WITH A NARROW TRAP DISTRIBUTION. THE TRAP LEVELS ARE RELATIVE TO THE CONDUCTION BAND EDGE OF SiO₂. $v_{0,if}$ AND $v_{0,band}$ RELATE TO THE PREFACTOR USED FOR TRAPPING FROM INTERFACE WITHIN THE BANDGAP OR FROM THE BANDS, RESPECTIVELY. SINCE THE THE DISTRIBUTION OF TRAP LEVELS IS ASSUMED TO BE UNIFORM, $\Delta_{in/out}$ RANGES FROM THE TOPMOST TO THE LOWERMOST TRAP LEVEL. m_t AND m_e DENOTES THE TUNNELING MASS AND THE ELECTRON MASS, RESPECTIVELY.

N_t	$3 \cdot 10^{18}$ cm⁻³
E_{in}	-4.8 eV
Δ_{in}	0.2 eV
E_{out}	-4.8 eV
Δ_{out}	0.2 eV
$v_{0,if}$	$6.3 \cdot 10^{-1}$ s⁻¹ cm² eV
$v_{0,band}$	$6.3 \cdot 10^{-12}$ s⁻¹ cm³ eV
m_t	$0.5 \cdot m_e$

Table II

VALUES USED FOR THE FIXED LEVEL MODEL WITH A BROAD TRAP DISTRIBUTION.

N_t	$3 \cdot 10^{18}$ cm⁻³
E_{in}	-4.8 eV
Δ_{in}	1.6 eV
E_{out}	-4.8 eV
Δ_{out}	1.6 eV
$v_{0,if}$	$6.3 \cdot 10^{-1}$ s⁻¹ cm² eV
$v_{0,band}$	$6.3 \cdot 10^{-12}$ s⁻¹ cm³ eV
m_t	$0.5 \cdot m_e$

where the level shift Δ is given by

$$
\Delta = E_{in} - E_{out} \tag{5}
$$

Solving these differential equations for each trap delivers trap occupancies which can finally be mapped to a measureable threshold voltage shift by making use of

$$
\Delta V_{th} = \frac{q_0}{C_{ox}} \int_0^{t_{ox}} (1-\frac{x}{t_{ox}})\rho_t \Delta f_t(E_t,x)dx, \tag{6}
$$

where C_{ox} denotes the capacitance of the dielectric, t_{ox} the thickness of the dielectric, and $\Delta f_t(E_t,x)$ the change in the trap occupancy.

In the following, a comparison between the individual models will be undertaken, thereby highlighting the differences observed in the timescales for the fixed level model and the level shift model. Numerical simulations based on the evaluation of the rate equations are performed coupled to a Poisson solver assuming Boltzmann statistics for the carrier concentrations [12].

III. THE FIXED LEVEL MODEL

Tewksbury's description of charge trapping is restricted to charge injection from the substrate and ignores the presence of the poly-gate interface as a source/sink of charge carriers. Recall that this assumption loses its justification when the gate dielectric shrinks into the nanometer range. At first, the fixed level model is recapitulated and on its basis the impact of the poly-gate interface will be discussed. Numerical simulations are performed on a pMOSFET with a 3 nm thick silicon dioxide

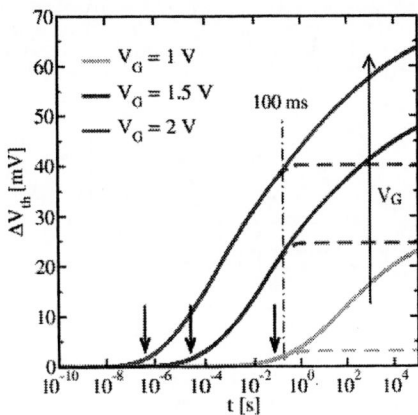

Fig. 2. Fixed level model: Time evolution of the threshold voltage during the on-state phase for different gate voltages. The solid lines denote charge trapping curves according to the conventional fixed level model, while the dashed lines belong to the extension of this model with a poly-gate interface. Within the extended fixed level model, only traps near the substrate interface (with short tunneling time constants) are capable of capturing h⁺. This reflects in an early saturation (indicated by the vertical arrows) during the on-state for the conventional fixed level model. The vertical line shows the beginning of the extended saturation behavior for all three gate voltages at approximately 100 ms.

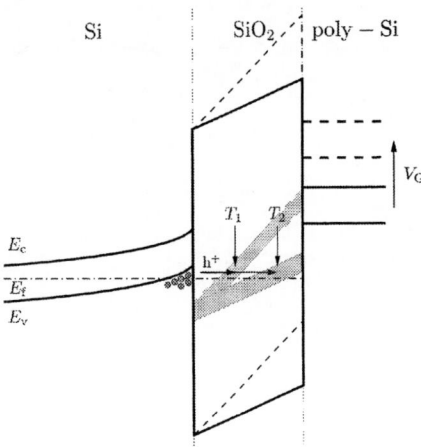

Fig. 4. Schematic of the band diagram for 2 different voltages. The crossing point between the Fermi level and the band of trap levels (grey regions) is linked to the earliest trapping events (T_1 and T_2) giving rise to the beginning of charge trapping. When the gate bias is increased, the crossing point is shifted closer to the substrate interface where traps with smaller tunneling time constants are situated. This leads to more earlier onset of charge trapping in T_1 compared to T_2, where the latter one corresponds to a higher gate voltage.

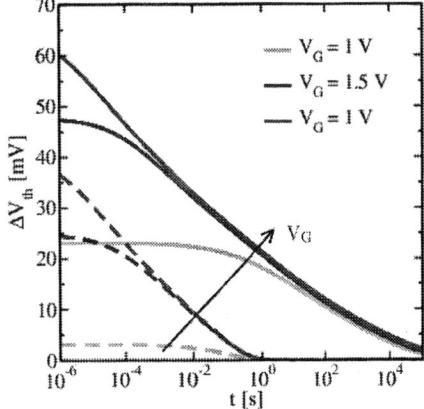

Fig. 3. The same as in Fig. 2 but for the off-state. Within the extended fixed level model, only traps with small tunneling times participate in charge trapping which is reflected in an early erase of trapped charges. They cause a slow decay of the threshold voltage shift during the off-state. In the case of the conventional fixed level model, also traps charged with long tunneling times are involved in detrapping again which results in slow decay in the V_{th}.

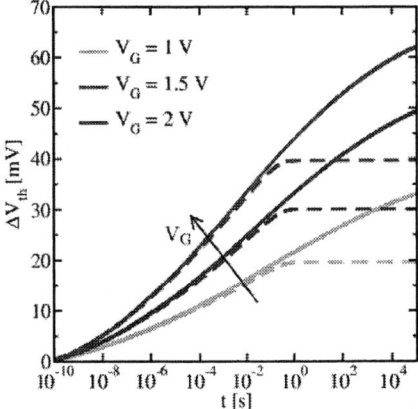

Fig. 5. The same as in Fig. 2 but for a wide spread of trap levels according to Table II. The onset of charge trapping occurs at the same time point which lies outside the measurable timescales. The different voltages are reflected in the different slopes of the curves.

layer and an n-poly-gate. The defect density N_t is assumed to amount to approximately $3 \cdot 10^{18}$ cm⁻³. The corresponding trap level is placed 0.55 eV below the substrate valence band edge with an uniform spread of 0.1 eV. Used quantities are listed in Table I. The first on-state phase is preceded by an equilibrium simulation. The off-phase directly continues after the end of the on-phase. For a proper analysis the tunneling time constants measured for thick oxides in [8] are converted into the respective v_0 applied for the present simulations.

In the conventional fixed level model, the e⁻ capture level coincides with the h⁺ capture level. Hence, the high h⁺

concentration in the valence band favors h⁺ emission into traps at energies close to and above the Fermi energy. The decay of the h⁺ concentration in energy scale, however, suppresses their capture in traps far below the Fermi level. The dependence of the WKB coefficient on the spatial depth of traps gives rise to increasing tunneling time constants from the substrate. At the beginning of the on-state, the negative gate bias decreases the Fermi level accompanied by a strong increase in the h⁺ concentration and consequently the onset of h⁺ trapping. Then the trapped h⁺ front penetrates into the gate dielectric with increasing time. The return to the off-state triggers the temporal refilling of traps which again proceeds from the substrate

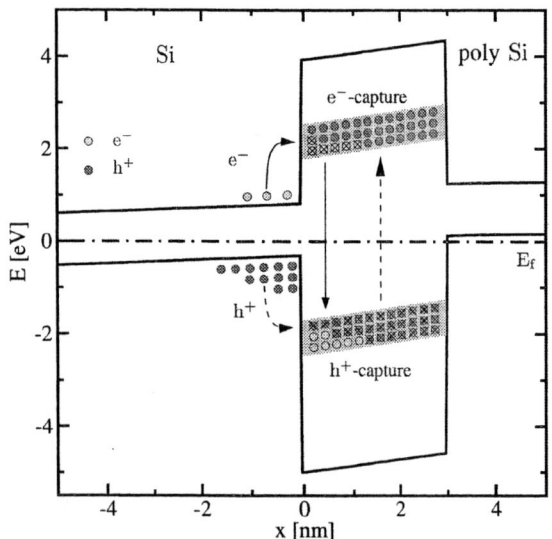

Fig. 6. Schematic band diagram including trap levels. The upper trap level indicated by the grey box enables only e⁻ (bright circles) capture. When this level is occupied, it is shifted down instantaneously and is capable of capturing h⁺ (dark circles) from the silicon valence band.

Table III
VALUES USED FOR THE LEVEL MODEL WITH A NARROW TRAP DISTRIBUTION. THE PREFACTORS $v_{0,in}$ AND $v_{0,out}$ REFER TO THE TRAP LEVEL AT AN ENERGY E_{in} OR E_{out}, RESPECTIVELY. NO PREFACTORS $v_{0,if}$ ARE GIVEN SINCE NO TRAPPING FROM STATES WITHIN THE BANDGAP IS ASSUMED.

N_t	$3 \cdot 10^{18}$ cm⁻³
E_{in}	-2.6 eV
Δ_{in}	0.2 eV
E_{out}	-5.0 eV
Δ_{out}	1.4 eV
$v_{0,in}$	$6.3 \cdot 10^{4}$ s⁻¹ cm³ eV
$v_{0,out}$	$6.3 \cdot 10^{-16}$ s⁻¹ cm³ eV
m_t	$0.5 \cdot m_e$

the substrate valence band edge with a wide spread are presented in Fig. 5. One can recognize that charge trapping sets in at the same time point for different gate voltages. However, a saturation of charge trapping is still not remedied by the broad distribution of traps and occurs at approximately 100 ms after beginning of the on-state again.

IV. THE LEVEL SHIFT MODEL

In the following, we discuss the implications emerging from the level shift. A band diagram indicating the trapping and detrapping processes is given in Fig. 6. Within the framework of this model h⁺ can only be injected into the e⁻ emission levels (lower trap levels), while e⁻ are only permitted to be captured by the e⁻ capture levels (upper trap levels). Hence a situation arises where two opposite processes, namely h⁺ capture and e⁻ capture, compete and determine whether trapping or detrapping occurs. The magnitude of their rates is governed by their corresponding concentrations $n(E_{in})$ and $p(E_{out})$ which are primarily affected by the position of the Fermi level at the interface.

For comparison between the fixed level model and the level shift model, the same device geometry, doping concentrations, and trap concentration have been used as in the case of the fixed level model. The e⁻ capture levels with only a small spread of 0.1 eV are assumed to be situated 0.5 eV above the silicon conduction band edge, while the h⁺ capture levels cover a wide energetical range of more than 1 eV (see Table II). In contrast to the fixed level model, the level shift model does not require to take charge trapping from interface states into account. The different nature of e⁻ capture levels and h⁺ capture levels is reflected in their respective prefactors v_0 (in r_{in} and r_{out} of equation (4)). However, the values of v_0 for the substrate and for the poly-gate must coincide since both describe trapping between band states of silicon bulk and the same sort of traps.

Data for the on-state and the off-state are shown in Fig. 7 and Fig. 8, respectively. When the device is operated in inversion, the Fermi level falls below the silicon valence band so that e⁻ injection into traps is impeded due to a lack of e⁻ above the conduction band edge. On the other hand, the high h⁺ concentration in the inversion layer gives rise to strong h⁺ trapping. During the on-phase, the level shift model predicts charge trapping in the long timescales as it is found for NBTI (cf. Fig. 7). When returning to the off-state (cf. Fig. 8), the Fermi

interface deep into the dielectric. Hence, positive charges built up during the on-state are swept out again.

Taking the poly-gate interface into account, traps with long tunneling time constants situated far from the substrate interface do not participate in charge trapping. For these traps, the capture rate of gate e⁻ outbalances the capture rate of substrate h⁺. Hence, the presence of the poly-gate interface establishes a spatial border to the penetrating h⁺ front. Only traps with small tunneling times are involved in charge trapping entailing an early saturation during the on-state and a fast erase of stored positive charges during the off-state.

Data of numerical simulations for the on-state and the off-state are depicted in Fig. 2 and 3. Fig. 2 clearly demonstrates the importance of accounting for the poly-gate interface in the on-state. As compared to the conventional Tewksbury model, the extended fixed level model yields an early saturation of the threshold voltage shift during the on-state. This saturation is reached at approximately 100 ms after the beginning of the on-state. While the off-state of the conventional Tewksbury's model covers many decades during the off-state, consideration of the poly-gate interface results in an early erase of trapped charges (see Fig. 3). A complete removal of trapped charges is already achieved after 1 s.

This behavior cannot explain NBTI measurements on thin oxide devices which yield long relaxation tails lasting for more than 10 s. Moreover, an earlier onset of trapping for higher voltages can be recognized in Fig. 2. This is originated in different regions of traps involved in charge trapping (see Fig. 4). For increasing voltage the trap levels are tilted upwards so that the ones situated at the same energy as the Fermi energy are located closer to the interface. These traps are associated with the earliest trapping events and the onset of charge trapping in the V_{th} transients. Simulations for traps centered 0.55 eV below

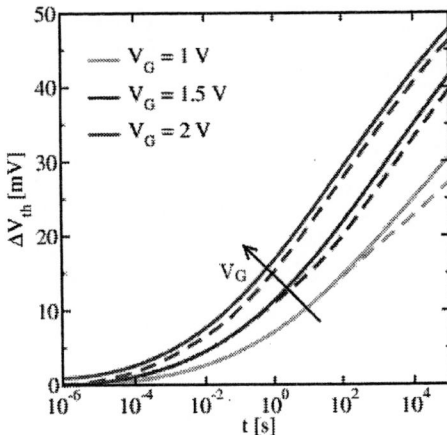

Fig. 7. The same as in Fig. 2 but for the level shift model. As in the fixed level model, the V_{th} transients shows that charge trapping sets in nearly simultaneously for different gate voltages. The simulations show that the influence of the poly-gate interface is of minor importance during the short-term part.

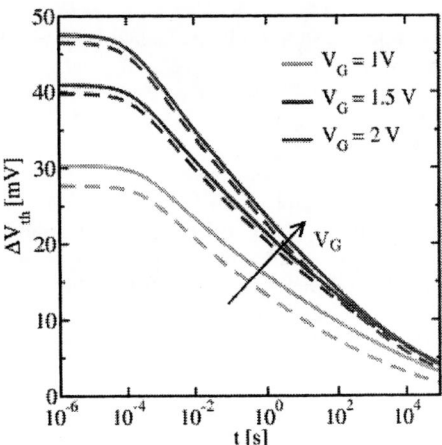

Fig. 8. The same as in Fig. 3 but for the level shift model. The V_{th} transients cover a wide range of decades in timescale.

level is situated above the silicon valence band accompanied by an increase of e$^-$ injection into e$^-$ capture levels and a vanishing h$^+$ injection into e$^-$ emission levels. In contrast to the fixed level model, the V_{th} transient plotted in Fig. 8 shows an overlap only at the tails of the curves for different voltages. For the case where the poly-gate interface is accounted for, no remarkable changes can be observed (see Fig. 7). This can be traced back to a small upwards shift of the gate Fermi level which prefers e$^-$ injection from poly-gate interface. The small amount of trapped e$^-$ from the poly-gate interface during the on-state is removed during the off-state again. This yields a similar detrapping behavior during the off-state as depicted in Fig. 8 for the level shift model without consideration the poly-gate interface.

V. CONCLUSION

We have shown that the extended fixed level model based on Tewksbury's approach shows a saturation behavior at 1 s due to the impact of the poly-gate interface [13]. This can be traced back to the fact that only traps with short tunneling times are involved in charge trapping. For the first time, trap level shifts have been rigorously incorporated into a new model and yield V_{th} transients covering a several decades for the on-state as well as the off-state. Even for the case where a second interface is taken into account, theoretical results still cover the timescales of interest.

REFERENCES

[1] J.F Conley, P.M. Lenahan, "Room Temperature Reactions Involving Silicon Dangling Bond Centers and Molecular Hydrogen in Amorphous SiO2Thin Films On Silicon", IEEE Trans. Nucl. Sci., vol. 39, no. 6, pp. 2186-2191 (1992).

[2] J.P. Campbell, P.M. Lenahan, A.T. Krishnan, S. Krishnan, Atomic-Scale Defects Involved in NBTI in Plasma-Nitrided pMOSFETs, 2007 IEEE IIRW Final Report, 12, (2007).

[3] J.F. Zhang, C.Z. Zhao, A.H. Chen, G. Groeseneken, R. Degraeve, "Hole Traps in Silicon Dioxides — Part I: Properties", IEEE Trans. Electr. Dev., vol. 51, no. 8, pp. 1267-1273 (2004).

[4] A.J. Lelis, H.E. Bosch, T.R. Oldham, F.B. McLean, "Reversibility of Trapped Hole Annealing", IEEE Trans. Nucl. Sci., vol. 35, no. 6, 1186-1191 (1988).

[5] Y.L. Yang, M.H. White, Charge Retention of Scaled SONOS Nonvotatile Memory Devices at Elevated Temperatures, Solid-State Electronics, vol. 44, 949 (2000).

[6] D.S. Ang, S. Wang, Recovery of the NBTI-Stressed Ultrathin Gate pMOSFET: The Role of Deep-Level Hole Traps, IEEE EDL, vol. 27, no. 11, 914 (2006).

[7] V. Huard, M. Denais, F. Perrier, N. Revil, C. Parthasarathy, A. Bravaix, E. Vicent, "A Thorough Investigation of MOSFETs NBTI Degradation", Microelectronics Reliability 45, 83-98 (2005).

[8] T.L. Tewksbury, "Relaxation Effects in MOS Devices due to Tunneling Exchange with Near-Interface Oxide Traps", doctor thesis, 1992, Massachusetts Institute of Technology.

[9] P.E. Blöchl, "First-Principles Calculations of Defects in Oxygen-Deficient Silica Exposed to Hydrogen", Phys. Rev. B, vol. 62, no. 10, pp. 6158-6179 (2000).

[10] W. Goes, T. Grasser, First-Principles Investigations on Oxide Trapping, SISPAD Proceedings 2007, 157-160 (2007).

[11] W. Goes, T. Grasser, Charging and Discharging of Oxide Defects in Reliability Issues, 2007 IEEE IIRW Final Report, 27, (2007).

[12] M. Karner, A. Gehring, S. Holzer, M. Pourfath, M. Wagner, W. Goes, M. Vasicek, O. Baumgartner, Ch. Kernstock, K. Schnass, G. Zeiler, T. Grasser, H. Kosina, S. Selberherr, "A Multi-Purpose Schrödinger-Poisson Solver for TCAD Applications", Journ. Comp. Electron., vol. 6, pp. 179-182 (2007)

[13] A.E. Islam, H. Kuflouglu, D. Varghese, S. Mahapatra, M.A. Alam, Recent Issues in Negative-Bias Temperature Instability: Initial Degradation, Field Dependence of Interface Trap Generation, Hole Trapping Effects, and Relaxation, IEEE TED, vol. 54, no. 9, 2143 (2007).

The Impact of Gate Dielectric Nitridation Methodology on NBTI of SiON p-MOSFETs As Studied by UF-OTF Technique

V. D. Maheta[1], C. Olsen[2], K. Ahmed[2], and S. Mahapatra[1]

[1]Department of Electrical Engineering, IIT Bombay, Mumbai, India
[2]Applied Materials, Santa Calara, CA, USA
Phone: 091-22-25764481. Fax: 091-22-25763707. Email: vrajesh@ee.iitb.ac.in

Abstract- **The impact of gate dielectric nitridation methodology on time, temperature and field dependence of NBTI in SiON p-MOSFETs is studied using Ultra-Fast On-The-Fly I_{DLIN} technique with 1μs resolution. It is shown that PNO devices with proper PNA show lower degradation magnitude, higher field dependence and therefore higher safe operating voltage compared to RTNO and PNO devices with improper PNA despite atomic N% is higher in PNO devices with proper PNA.**

I. INTRODUCTION

Negative Bias Temperature Instability (NBTI) of Silicon Oxy-nitride (SiON) p-MOSFET parameters (e.g. threshold voltage, V_T; linear drain current, I_{DLIN}; transconductance, G_M; etc) is a serious reliability issue as it limits the lifetime of modern CMOS IC's [1]-[3]. It is now well-known that NBTI increases with higher N content in the SiON gate dielectric [1]-[5] and recovers after the removal of stress [11], [12], [17]-[20]. It is also well-known that the degradation magnitude and its time evolution get influenced by recovery during stress interruption (for measurement) phase and therefore by measurement delay [11], [12], [19], [20]. It is now believed that NBTI degradation is not only due to interface trap generation (ΔN_{IT}) but trapping of inversion layer holes into (N related) pre-existing bulk traps (ΔN_h) in the SiON bulk as well [6], [19], [21]. However, the relative dominance of ΔN_{IT} or ΔN_h depends on N density and thickness of the gate insulator, which in-turn depends on SiON processing condition [11], [22], [23]. ΔN_{IT} is believed to show power law time dependence (t^n) with strong T activation and is presumed to be a relatively slower process [13], [23]. On the contrary, ΔN_h is believed to be a relatively faster mechanism with weak temperature (T) and logarithmic time dependence and get saturated in less than 1s [19], [22].

It has been shown that for a given total N dose, relatively lower N density at or near the Si/SiON interface results in lower NBTI and vice-versa [6],[7]. Plasma Nitrided Oxide (PNO) can show such N distribution in the gate oxide (with N peak at SiON/poly-Si interface), provided correct (and not any arbitrary) Post Nitridation Anneal (PNA) is used [8]-[10]. In this work, the time, temperature (T) and oxide field (E_{OX}) dependence of NBTI is studied in PNO devices subjected to different PNA, and compared to the results obtained for Rapid Thermal Nitrided Oxide (RTNO). It is shown that in spite of higher total atomic N%, PNO devices with proper PNA show lower degradation magnitude, higher E_{OX} acceleration factor, and as a result higher safe operating voltage when compared to RTNO and PNO devices with improper PNA. This work opens up interesting opportunities to optimize leakage and reliability by using higher atomic N% in deeply scaled SiON dielectrics.

II. DEVICE AND MEASUREMENT DETAILS

Table-I shows the details of SiON p-MOSFETs used in this work. D1 and D2 are PNO devices with proper two-step PNA as described in [9]. D4 and D5 are PNO devices with incorrect PNA (D5 has better PNA than D4), and D3 is the reference RTNO device. Note that total atomic N% for D1 is slightly higher, while for D2 is significantly higher compared to other devices. The atomic N% was calculated using XPS and equivalent oxide thickness (EOT) was calculated from CV measurements followed by Quantum Mechanical (QM) corrections for all the devices used in this work.

Table-I: Splits of SiON/poly-Si devices used in this work. PNO: Plasma Nitrided Oxide with proper PNA, RTNO: Rapid Thermal Nitrided Oxide, PNO': PNO with improper PNA1 and PNO": PNO with improper PNA2.

D#	Type	EOT(Å)	N%
1	PNO	17.70	19.5
2	PNO	15.6	34.6
3	RTNO	18.5	5.8
4	PNO'	22.2	11.9
5	PNO"	20.2	16.7

Recently developed Ultra-Fast On-The-Fly I_{DLIN} technique with minimum time-zero (t_0) delay (delay between application of stress V_G and measurement of first I-V point at stress V_G) of 1μs has been used [6] to avoid measurement delay and to capture the signature of fast degradation component (ΔN_h). The measured NBTI degradation is expressed as $\Delta V = -\Delta I_{DLIN}/I_{DLIN0}$ $*(V_{G,STRESS} - V_{T0})$ [12], where I_{DLIN0} is the initial drain current captured within 1 μs after the application of stress bias, $V_{G,STRESS}$ is the applied stress bias and V_{T0} is the pre-stress V_T. ΔV is related to but somewhat different (in magnitude) from V_T shift as mobility degradation due to ΔN_{IT} is not taken into account (see [24] for details of mobility correction).

III. RESULTS AND DISCUSSION

Degradation transients: Fig.1 and Fig.2 show the time evolution of ΔV measured using t_0 delay of 1µs for all devices (see Table-I) at different stress T but under identical stress E_{OX}. Both D1 and D2 (PNO with proper PNA) show clear T dependence for the entire stress duration, though D2 shows slightly higher overall degradation and lower T activation compared to D1 (see Fig.1). The relatively smaller increase in NBTI degradation in spite of significant increase in atomic N% can be attributed to small increase in N density at the Si/SiON interface for PNO with proper PNA [6]-[8].

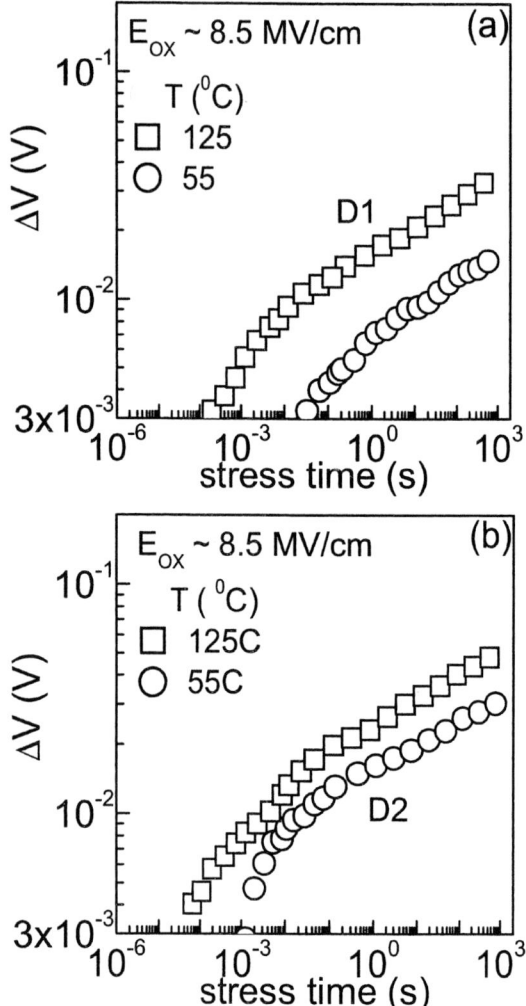

Fig. 1. Time evolution of ΔV degradation at different stress T for PNO (a) D1 and (b) D2 devices having different atomic N%.

RTNO (D3) shows very high short time degradation and much larger overall degradation compared to other devices in spite of lowest atomic N%, (see Fig.2 (a)). This can be attributed to higher N density at the Si/SiON interface for the RTNO

process [4], [6]. Though D4 and D5 are PNO, they were treated with improper PNA and therefore show higher short time degradation and larger overall degradation like D3 (RTNO) when compared to D1 and D2. This signifies higher N density at Si/SiON interface for PNO devices (D4 and D5) when not subjected to proper PNA.

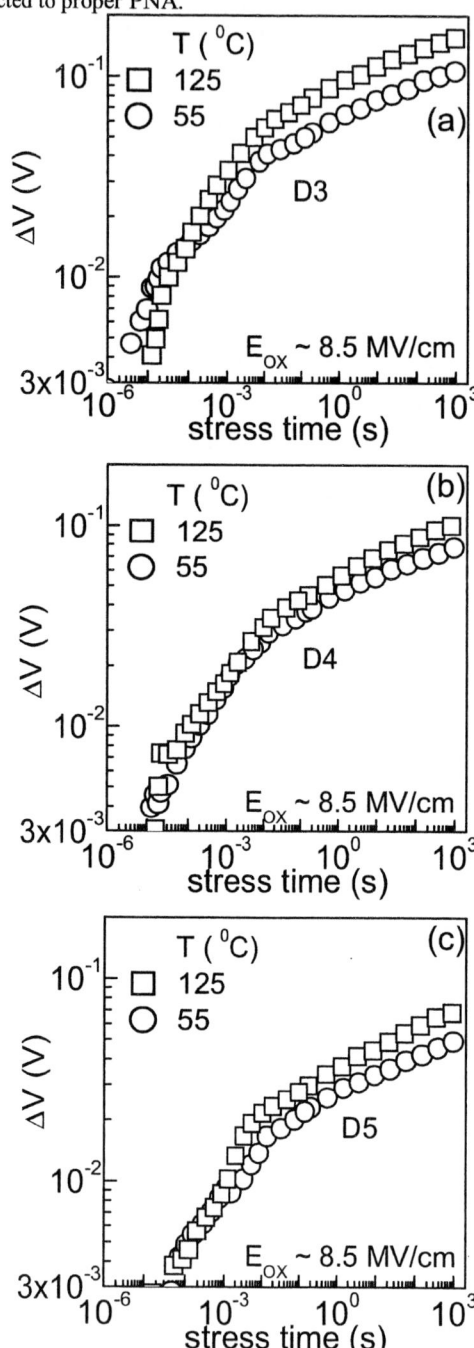

Fig. 2. Time evolution of ΔV degradation for (a) RTNO (D3), (b) PNO with improper PNA1 (D4) and (C) PNO with improper PNA2 (D5) devices at different stress T.

However, it is important to note that D5 (higher N%, relatively better PNA) shows relatively smaller degradation than D4 (lower N%, relatively worse PNA), implying larger Si/SiON N density for the latter device and both D4 and D5 show lower degradation compared to D3 (RTNO, lowest N %). Also note that (unlike D1, D2) D3, D4 and D5 show negligible T dependence during early stress time and weak T dependence for long stress time. The difference in overall degradation between D3 (or D4 or D5) and D1 (or D2) can be attributed to a large extent to this negligible T dependent degradation observed at early stress phase. It is obvious from the above discussion that NBTI degradation transients in PNO films depend significantly on the PNA step (which controls N density at the Si/SiON interface), and show no correlation to total atomic N% (or N dose) unless devices are subjected to proper PNA. Moreover, the difference in ΔV transients for different splits is due to dominance of one physical mechanism over another as reported in [6], [7]. Clear T dependence of ΔV at early stress time for PNO with proper PNA (D1 and D2) indicates ΔN_{IT} dominated NBTI degradation for these devices. On the other hand, negligible T dependence of ΔV at early stress time for RTNO (D3) and PNO with improper PNA (D4 and D5) indicates ΔN_H dominated NBTI, especially at early stress phase. Note that the negligible T dependent degradation observed at early stress time for PNO with improper PNA and RTNO can not be captured using other measurement methods such as conventional OTF [11], [12], [17], [23]; Fast-OTF[25], UF-SMS[20], and is only captured by UF-OTF technique as demonstrated in this paper.

Time exponent: Fig.3 shows extracted (using linear fit from 10s-10^3s of log-log ΔV versus time plot) time exponent (*n*) as a function of t_0 delay under identical stress T and E_{OX} for all devices.

All devices show reduction in n with reduction in t_0 delay. The saturation of n for $t_0 < 10\mu s$ is clearly observed for all devices, suggesting a faster version than the present UF-OTF would not give smaller n than reported in this work. PNO devices with proper PNA (D1, D2) show highest n (D2 shows lower n w.r.t D1), RTNO (D3) shows lowest n, while PNO devices with improper PNA show intermediate values. Slightly lower value of n for D2 compared to D1 can be attributed to slightly higher N at Si/SiON interface in spite of significant increase in total atomic N% for D2 [7]. Lowest value of n for RTNO (D3) among all other devices suggests highest N density at Si/SiON interface for RTNO. Since Si/SiON N density for D5 is lower compared to D4 owing to better PNA, D5 (higher N%) shows relatively higher n than D4 (lower N%). Note that devices with higher Si/SiON N density show lower n which can be attributed to increase in ΔN_h owing to increase in N related hole traps near the Si/SiON interface. Thus value of n is not unique but depends upon the SiON process condition for a given t_0 delay.

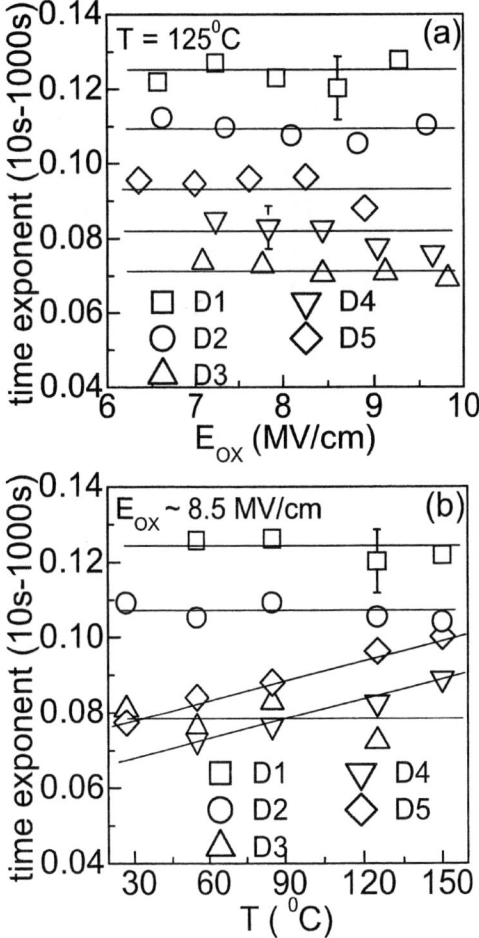

Fig. 4. Extracted n as a function of (a) stress E_{OX} at constant stress T (b) stress T at constant stress E_{OX} for t_0 delay of 1μs for PNO (D1, D2), RTNO (D3) and PNO with improper PNA (D4, D5) devices. Lines are guide to the eye. Maximum error in n due to noise induced scattering in I_{DLIN0} is ±0.007

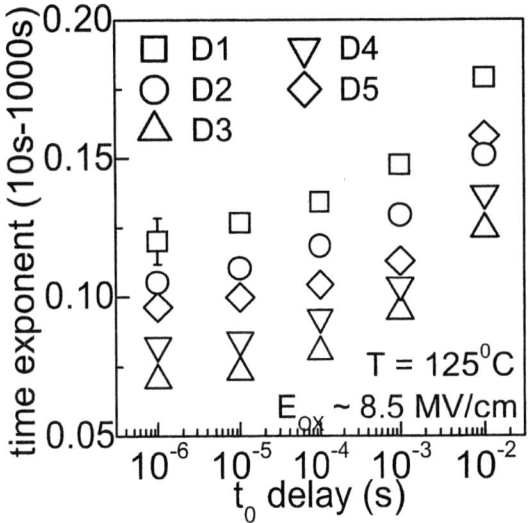

Fig. 3. Extracted n as a function of t_0 delay at constant stress T and EOX for PNO (D1, D2), RTNO (D3) and PNO with improper PNA (D4, D5) devices.

Fig.4 shows the E_{OX} and T dependence of n (t_0=1μs, fitted in range of t-stress=10s-10^3s) for all devices. Note that n is independent of stress E_{OX} (constant T) for all splits (see Fig.4 (a)), and small variation in n (due to noise in measured I_{DLIN0}) is well within the error bar. The E_{OX} independence of n implies no bulk insulator trap generation in the range of stress V_G used in this work [14]. Moreover, constant n for varying E_{OX} validates the usage of power-law extrapolation of extracted ΔV in time to the end-of-life for determining the safe gate over drive voltage. Stress T independence of n is observed for D1, D2 and D3 devices suggesting Arrhenius T activated NBTI degradation for long stress time (t-stress ≥ 10s). In contrast, linear T dependence of n is observed for D4 and D5 devices. Though n for a particular T and E_{OX} is higher for D5, the slope of n versus T is identical for D4 and D5 devices. The T dependence of long-time n for improper PNA devices suggests different physical mechanism than that for devices having proper PNA (n independent of T) [15], [16].

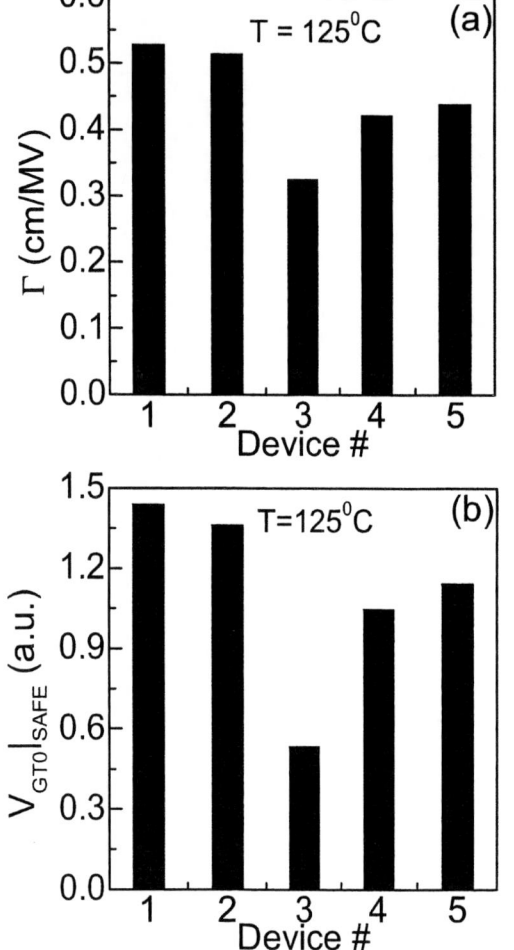

Fig. 5. (a) Slope of field dependence (Γ) of NBTI degradation and (b) calculated safe gate overdrive voltage ($V_{GT0|SAFE}$ = (V_G-V_{T0})|$_{SAFE}$) for 10 years lifetime to reach ΔV=60 mV for PNO (D1, D2), RTNO (D3) and PNO with improper PNA (D4, D5) devices.

E_{OX} dependence: Figs.5 (a) and (b) respectively show the slope (Γ) of E_{OX} dependence of ΔV and extrapolated safe gate overdrive voltage ($V_{GT0|SAFE}$) for 10 years lifetime to reach ΔV=60 mV, under identical stress T for all devices (t_0=1μs). Note that Γ has been shown to be a good indicator of Si/SiON interfacial N density [4], [5]. Much lower Γ and $V_{GT0|SAFE}$ are observed for RTNO (D3) due to much higher N density at the Si/SiON interface in spite of having lowest total N%. Small reduction in Γ and $V_{GT0|SAFE}$ is observed for D2 compared to D1 due to slight increase in N density at the Si/SiON interface despite large increase in N%, as these devices were subjected to proper PNA. D1 and D2 (PNO with proper PNA) show higher Γ and $V_{GT0|SAFE}$ compared to D4 and D5 (PNO with improper PNA), as improper PNA results in higher N density at the Si/SiON interface. D4 shows lower Γ and $V_{GT0|SAFE}$ compared to D5 though atomic N% is higher for D5, which can be attributed to inferior PNA for D4. Thus, higher N density at Si/SiON interface reduces Γ which can be attributed to reduction in hole tunneling barrier [13]. Such reduction in Γ consequences in higher degradation and lower lifetime as stress data is extrapolated to operating condition.

IV. CONCLUSION

To summarize, PNO devices with incorrect PNA but lower atomic N% show larger NBTI compared to PNO devices with proper PNA having higher atomic N%. The improper PNA devices show lower E_{OX} dependence and larger T independent degradation at early stress time, similar to that observed for RTNO devices. A proper, two-step PNA reduces N density at the Si/SiON interface and hence overall NBTI. Finally, mere atomic N% can not be used to predict NBTI in SiON devices unless proper PNA is done.

ACKNOWLEDGMENT

The Authors would like to thank E. N. Kumar, S. Purawat for development of measurement setup.

REFERENCES

[1] N. Kimizuka, T. Yamamoto, T. Mogami, K. Yamaguchi, K. Imai, and T. Horiuchi, "The impact of bias temperature instability for direct tunneling ultra-thin gate oxide on MOSFET scaling," in proc., *VLSI Tech. Symp.*, pp.73-74, 1999.
[2] G. Chen, K.Y. Chuah, M. F. Li, C. H. Ang, J. Z. Zhen, and D. L. Kwong," Dynamic NBTI of PMOS transistors and its impact on device lifetime," in proc., *Int. Reliability Physics Symp.*, p.196, 2003.
[3] G. La Rosa, F. Guarin, A. Acovic, J. Lukatis, and E. Crabbe," NBTI-Channel hot carrier effects in PMOSFETs in advanced CMOS Technology," in proc., *Int. Reliability Physics Symp.*, p. 282, 1997.
[4] C.H. Liu, M.T. Lee, Lin Chih-Yung, J. Chen, K. Schruefer, J. Brighten, N. Rovedo, T.B. Hook, M.V. Khare, Shih-Fen Huang, C. Wann, Tze-Chiang Chen, and T.H. Ning, "Mechanism and process dependence of negative bias temperature instability (NBTI) for pMOSFETs with ultrathin

gate dielectrics," in proc., *Int. Electron Device Meet.*, p. 861, 2001.

[5] Y. Mitani, M. Nagamine, H. Satake and A. Toriumi, "NBTI mechanism in ultra-thin gate dielectric - nitrogen-originated mechanism in SiON," in proc., *Int. Electron Device Meet.*, pp.509-512, 2002.

[6] E. N. Kumar, V. D. Maheta, S. Purawat, A. E. Islam, C. Olsen, K. Ahmed, M. Alam and S. Mahapatra, "Material dependence of NBTI physical mechanism in silicon oxynitride (SiON) p-MOSFETs: A comprehensive study by ultra-fast On-the-fly (UF-OTF) I_{DLIN} technique," in proc., *Int. Electron Device Meet.*, pp. 809-812, 2007. [7] V. D. Maheta, C. Olsen, K. Ahmed and S. Mahapatra, "The impact of nitrogen engineering in silicon oxynitride gate dielectric on negative bias temperature instability of p_MOSFETs: A study by Ultra-fast On-the-fly I_{DLIN} technique," *IEEE Trans. Electron Devices, 2008* (under review).

[8] Philip A. Kraus, Khaled Z. Ahmed, Chris S. Olsen and Faran Nouri, "Physical Models for predicting plasma nitrided SiON gate dielectric properties from physical metrology," *IEEE Electron Device Letters,* Vol. 24, No. 9, pp. 559–561, September 2003.

[9] C. Olsen, "Two-step post nitridation annealing for lower EOT plasma nitrided gate dielectrics," U.S. patent No.0175961A1, Sep.9, 2004.

[10] A. Veloso, F.N. Cubaynes, A. Rothschild, S. Mertens, R. Degraeve, R. O'Connor, C. Olsen, L. Date, M. Schaekers, C. Dachs, and M. Jurczak, "Ultra-thin oxynitride gate dielectrics by pulsed-RF DPN for 65 nm general purpose CMOS applications," *ESSDERC*, pp.239-242, 2002.

[11] S. Rangan, N. Mielke and E. C. C. Yeh, "Universal recovery behavior of negative bias temperature instability," in proc., *Int. Electron Device Meet.*, pp. 341–344, 2003.

[12] D. Varghese, D. Saha, S. Mahapatra, K. Ahmed, F. Nouri and M. Alam, "On the dispersive versus arrhenius temperature activation of NBTI time evolution in plasma nitrided gate oxides: measurements, theory, and implications," in proc., *Int. Electron Device Meet.*, p.684, 2005.

[13] A. E. Islam, H. Kufluoglu, D. Varghese, S. Mahapatra and M. A. Alam, "Recent issues in negative bias temperature instability: Initial degradation, field dependence of interface trap generation, hole trapping effects and relaxation," *IEEE Trans. Electron Devices*, p.2143, v.54, September, 2007.

[14] S. Mahapatra, P. Bharath Kumar and M. A. Alam, "Investigation and modeling of interface and bulk trap generation during negative bias temperature instability of p-MOSFETs," *IEEE Trans. Electron Devices*, Vol. 51, No. 9, pp. 1371–1379, Sept. 2004.

[15] B. Kaczer, V. Arkhipov, R. Degraeve, N. Collaert, G. Groeseneken and M. Goodwin, "Disorder controlled kinetics model for negative bias temperature instability and its experimental verification," in proc., *Int. Reliability Physics Symp.*, pp. 381–387, 2005.

[16] V. Huard, C.R. Parthasarath, C. Guerin, and M. Denais," Physical Modeling of Negative Bias Temperature Instabilities for Predictive Extrapolation," in proc., *Int. Reliability Physics Symp.*, pp. 733-734, 2006

[17] A. T. Krishnan, C. Chancellor, S. Chakravarthi, P. E. Nicollian, V. Reddy, A. Varghese, R. B. Khamankar, and S. Krishnan, "Material dependence of hydrogen diffusion: implications for NBTI degradation," in proc., *Int. Electron Device Meet.*, p.688, 2005.

[18] M. Ershov, S. Saxena, H. Karbasi, S. Winters, S. Minehane, J. Babcock, R. Lindley, P. Clifton, M. Redford and A. Shibkov, "Dynamic recovery of negative bias temperature instability in p-type metal–oxide–semiconductor field-effect transistors," *Applied Physics Letters*, vol. 83, no. 8, pp. 1647-1649, Aug 2003.

[19] H. Reisinger, O. Blank, W. Heinrigs, A. Muhlhoff, W. Gustin and C. Schlunder, "Analysis of NBTI degradation- and recovery- behaviour based on ultra fast V_T measurements," in proc., *Int. Reliability Physics Symp.*, p.448, 2006.

[20] C. Shen, M. F. Li, C. E. Foo, T. Yang, D. M. Huang, A.Yap, G. S. Samudra and Y. C. Yeo, "Characterization and physical origin of fast Vth transient in NBTI of pMOSFETs with SiON dielectric," in proc., *Int. Electron Device Meet.*, s.12-p5, 2006.

[21] K. Sakuma, D. Matsushita, K. Muraoka and Y. Mitani, "Investigation of nitrogen originated NBTI mechanism in SiON with high nitrogen concentration," in proc., *Int. Reliability Physics Symp.*, p.454, 2006.

[22] S. Mahapatra, K. Ahmed, D. Varghese, A. E. Islam, G. Gupta, L. Madhav, D. Saha and M. A. Alam, "On the Physical Mechanism of NBTI in Silicon Oxynitride p-MOSFETs: Can Differences in Insulator Processing Conditions Resolve the Interface Trap Generation versus Hole Trapping Controversy?," in proc., *Int. Reliability Physics Symp.*, p1, 2007.

[23] S. Mahapatra and M.A.Alam, "Defect Generation in p-MOSFETs under Negative Bias Stress: An Experimental Perspective," *IEEE Tran. Device and Materials Reliability*, pp. 35-46, vol. 8, March, 2008.

[24] A. E. Islam, E. N. Kumar, H. Das, S. Purawat, V. Maheta, H. Aono, E. Murakami, S. Mahapatra and M. A. Alam, "Theory and practice of ultra fast measurements for NBTI degradation: Challenges and opportunities," in proc., *Int. Electron Device Meeting*, pp. 805-808, 2007.

[25] M. Denais, C. Parthasarathy, G. Ribes, Y. Rey-Tauriac, N. Revil, A. Bravaix, V. Huard and F. Perrier, "On-the-fly characterization of NBTI in ultra-thin gate oxide PMOSFET's," in proc., *Int. Electron Device Meet.*, pp. 109–112, 2004.

Trapping and De-trapping Characteristics in PBTI and Dynamic PBTI between HfO$_2$ and HfSiON Gate Dielectrics

Wei-Liang Lin[1], Yao-Jen Lee[2], Wen-Cheng Lo[3], King-Sheng Chen[2], Y. T. Hou[4], K. C. Lin[4], and Tien-Sheng Chao[1]

[1] Department of Electrophysics, NCTU, Hsinchu, Taiwan
[2] National Nano Device Laboratories, Hsinchu, Taiwan
No. 26, Prosperity Road 1, Science-based industrial park, Hsin-Chu, Taiwan
Tel: +886-3-5726100-7793, FAX: +886-3-5722715, e-mail: yjlee@ndl.org.tw
[3] Department of Electronic Engineer, NCTU, Hsinchu, Taiwan
[4] Taiwan Semiconductor Manufacturing Company

Abstract - **PBTI degradation for HfO$_2$ and HfSiON NMOSFETs has been demonstrated. The generated oxide trap dominated the PBTI characteristics for Hf-based gate dielectrics. In addition, the reduction of ΔV_{TH} and oxide trap generation under PBTI indicates that the HfSiON is better than HfO$_2$. On the other hand, the electron trapping/de-trapping effect has been investigated. As compared to HfO$_2$ dielectrics, the HfSiON has shallower charge trapping level due to elimination of deep dielectric vacancies, and the temperature effects are quite different between the HfSiON and HfO$_2$ gate dielectrics.**

I. INTRODUCTION

High-k dielectrics are especially advantageous for low-power application. Among high-k gate dielectric materials, Hf-based gate dielectric including HfO$_2$ and Hf-silicate are the attractive materials because it has good device characteristics. However, before Hf-based gate dielectrics being successfully integrated into future technologies, their reliability characteristics still need to be better identified. NMOS positive bias temperature instability (PBTI) could be a potential scaling limit of CMOS technology with Hf-based gate dielectrics[1]. Most of the previous studies showed a significant positive threshold voltage shift for the high-k gate stack under PBTI stress, which was attributed to the preexisting traps in the high-k layer or the holes induced oxygen vacancy traps[2-3]. In addition, one of main issues for high-k gate dielectrics is the charge trapping/de-trapping characteristics during reliability test. Initial observation of instability was current–voltage (IV) in V_{th} change. Since electrons can be trapped and de-trapped in the high-k dielectrics with a minimal residual damage to its atomic structure, a V_{th} instability associated with electron trapping/detrapping in high-k layer can significantly affect the transistor performance[4].

II. EXPERIMENT RESULTS

The high-K dielectric including HfO$_2$ and Hf-silicate were deposited by atomic-layer deposition (ALD). Chemical oxide was used as the interfacial layer unless it is specifically mentioned. The nitridation of HfSiO with Hf/(Hf+Si) ratio of 50% was carried by NH$_3$ annealing in the ambient. The fabrication of a heavily doped source/drain junction by implantation was followed by a rapid thermal annealing (RTA) of 1000 ºC for 5s for S/D activation and thermal stability of HfSiON.

III. RESULTS & DISCUSSIONS

At a higher field (>1 MV/cm), the mobility of HfSiON gate dielectric was large than HfO$_2$ gate dielectrics as shown in Fig. 1. We speculate that the Si-O and Si-N bodings were formed for the HfSiON gate dielectrics resulting in annihilation of oxygen vacancies to offer mobility enhancement at high electric field. In order to understand the mechanism of PBTI in our high-k dielectrics, Figure 2 shows the threshold voltage degradation (ΔV_{th}) of HfO$_2$ and HfSiON gate dielectrics under PBTI stress at room temperature. Figures 3 show the charge pumping current (I_{CP}) before and after PBTI stressing. HfSiON dielectrics had larger initial I_{CP} than HfO$_2$, but the increase in I_{CP} during PBTI stress was almost identical for HfO$_2$ and HfSiON gate dielectrics.

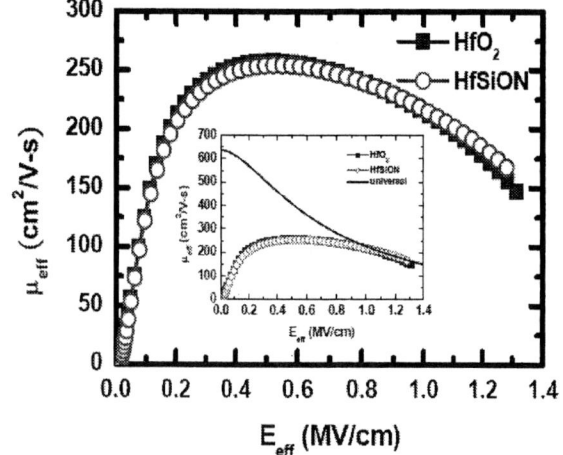

Fig. 1 The effective electron mobility measured on HfO$_2$ and HfSiON gate dielectrics using split CV method.

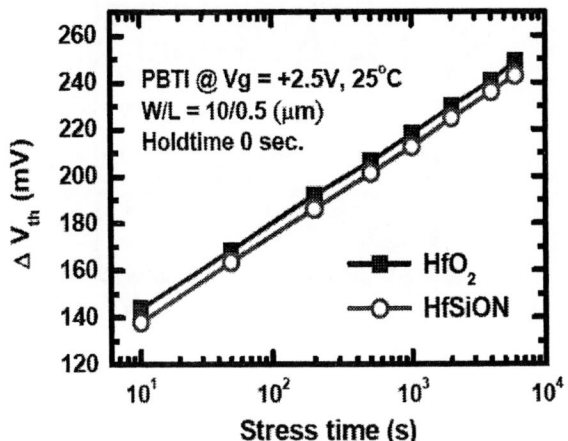

Fig. 2 ΔV_{th} of HfO$_2$ and HfSiON gate dielectrics under the same PBTI stress (Vg = +2.5 V) at room temperature

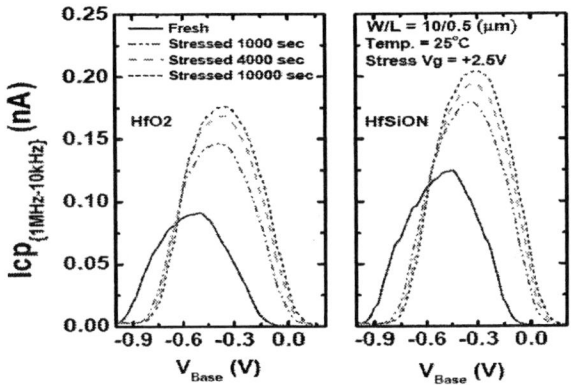

Fig. 3 Charge pumping current (I$_{CP}$) before and after PBTI stressing at +2.5V for HfO$_2$ and HfSiON dielectrics, respectively.

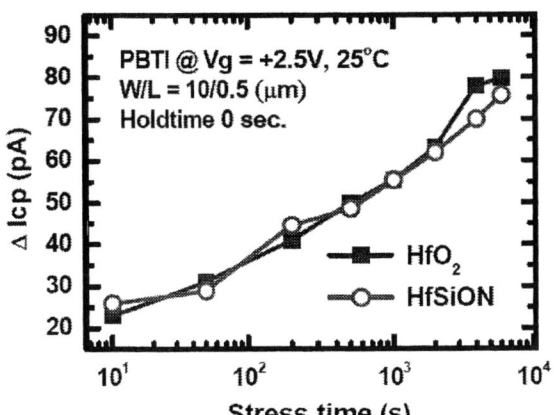

Fig. 4 Charging pumping increase (ΔI_{CP}) of HfO$_2$ and HfSiON gate dielectrics during the same PBTI stress bias (Vg = +2.5 V) at room temperature.

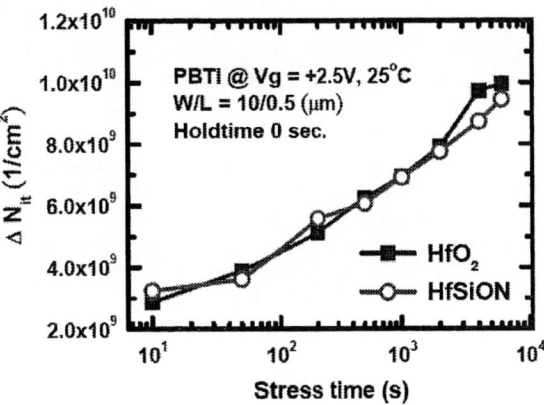

Fig. 5 Interface trap increase (ΔN_{it}) which extracted from ΔI_{CP} of HfO$_2$ and HfSiON gate dielectrics during the same PBTI stress bias (Vg = +2.5 V) at room temperature.

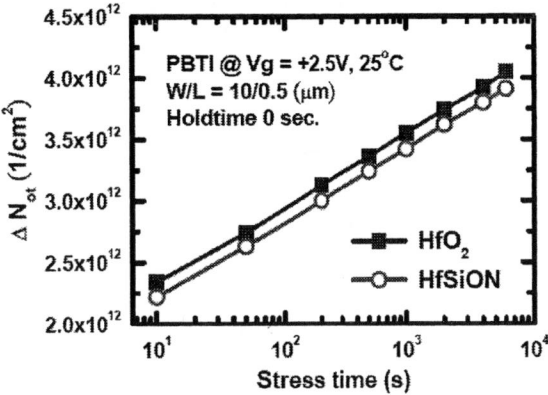

Fig. 6 Generated oxide trap (ΔN_{ot}) in the bulk of higk-k film during PBTI stress for both of HfO$_2$ and HfSiON dielectrics, respectively

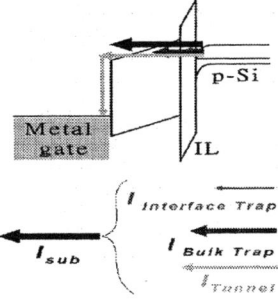

Fig. 7 Illustration of electron trapping model for Hf-based gate dielectrics under PBTI stress.

From Figs. 4 and 5, it is noted that the N_{it} increase is quite low (< 2×10^9 cm^{-2}) in this work. Figure 6 shows the N_{ot} increase for both of HfO$_2$ and HfSiON dielectrics during PBTI stress, respectively. The charge trapping model for Hf-based gate dielectrics under PBTI stress was illustrated in Fig.7. We also investigate the temperature-dependent trapping and de-trapping characteristics for both HfO$_2$ and HfSiON splits. In Figs. 8, we compare the HfO$_2$ and HfSiON gate dielectrics under PBTI stress (Vg = +2.5 V) with and without hold time at different temperatures, including 25, 75 and 100 °C. The obvious improvement in PBTI characteristics was observed for HfSiON gate dielectrics while the hold time is 100 seconds as indicated in Fig. 8(b). This result indicates that the electron de-trapping will be easily occurred in HfSiON gate dielectrics. It means that the extra Si-O and Si-N bodings effectively removed the dielectric vacancies to have a lower trapping cross section and a lower concentration of generated traps, and reduce some trapping levels. This implies that some deep electron traps were effectively eliminated for HfSiON gate dielectrics, resulting in the characteristics as shown in Figs. 9(a) & (b). In Fig.9 (a), it indicates that the HfO$_2$ gate dielectric generated deeper charge trap during PBTI stress and trapped electrons need higher thermal energy to de-trap from HfO$_2$ film. In Fig. 9(b), because of some deeper trap were effectively eliminated; the ΔV_{TH} in PBTI stress with 0 s hold time is larger than ΔV_{TH} with 100 s hold time at different temperatures. Figures 10(a) & (b) illustrate the charge trapping/de-trapping models for HfO$_2$ and HfSiON gate dielectrics under PBTI stress respectively. As compared to HfO$_2$ dielectrics, the HfSiON has deep charge trapping level under PBTI stress as illustrated in Fig. 10(b).

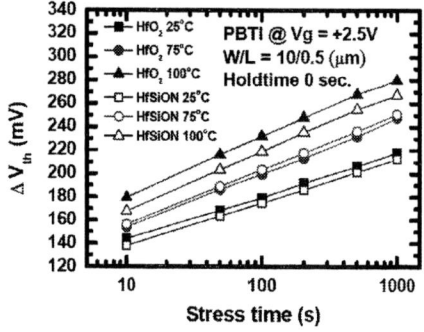

Fig. 8(a) ΔV_{th} of HfO$_2$ and HfSiON gate dielectrics during PBTI stress (Vg = +2.5 V) with no hold time at different temperatures, including 25, 75 and 100 °C.

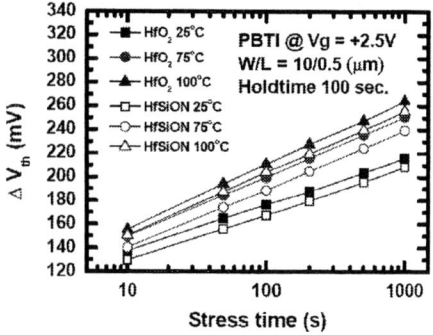

Fig. 8(b) ΔV_{th} of HfO$_2$ and HfSiON gate dielectrics during PBTI stress (Vg = +2.5 V) with hold time 100 s at different temperatures, including 25, 75 and 100 °C

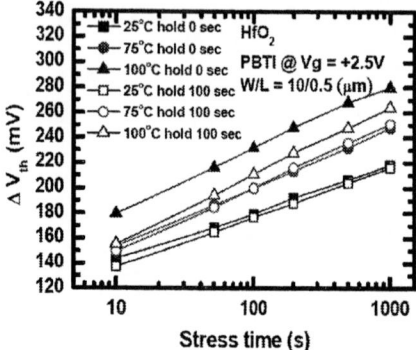

Fig. 9(a) Comparison of different hold time (0 & 100 s) under PBTI stress (Vg = +2.5V) at different temperatures, including 25, 75 and 100 °C for HfO$_2$ gate

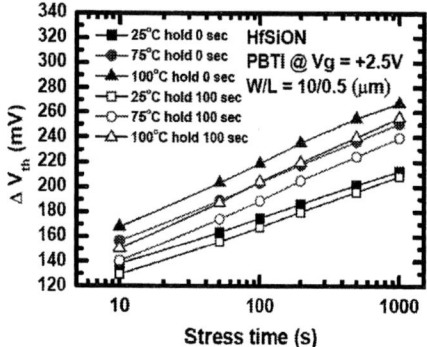

Fig. 9(b) Comparison of different hold time (0 & 100 s) under PBTI stress (Vg = +2.5V) at different temperatures, including 25, 75 and 100 °C for HfSiON gate dielectrics.

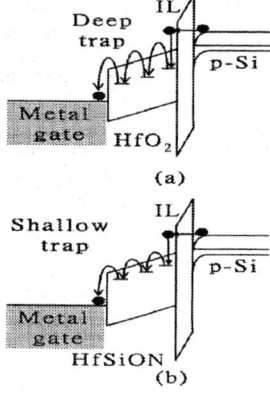

Fig. 10 Illustration of the charge trapping/de-trapping models for (a) HfO$_2$ and (b) HfSiON gate dielectrics under PBTI stress respectively.

In Figs. 11, we also compare the HfO$_2$ and HfSiON gate dielectrics under dynamic positive bias temperature instability (DPBTI) stress (Vg = +2.5V under stress, while Vg=0V under passivation) at 25°C, 75°C, 100°C. In Fig. 11(a), the ΔV_{TH} of HfSiON is lower than that of HfO$_2$ at 25°C due to the annihilation of oxygen vacancies by the Si-O and Si-N bonding. From the results under higher temperature DPBTI stress in Figs. 11(b) and (c), we speculate that electrons need

978-1-4244-2039-1/08/$25.00 ©2008 IEEE

higher thermal energy to trap in deep level in HfSiON gate dielectric. In addition, the recovery amplitudes at different stress temperature during passivation periods are almost identical for both HfO$_2$ and HfSiON splits. Finally, we extracted the coefficient n and K from DPBTI stress and summarized in Table I, where n stands for trapping /detrapping ability and K is the amplitude which is function of stress voltage Vg, temperature, dimension size, and process conditions. From the n and K values in Table□, the temperature effects are quite different between the HfSiON and HfO$_2$ gate dielectrics.

Table I
Extracted Coefficients n and k after stress/passivation: $\Delta V_{th} = K * t^n$

		1st stress		1st passivation		2nd stress		2nd passivation	
		HfSiON	HfO2	HfSiON	HfO2	HfSiON	HfO2	HfSiON	HfO2
25°C	K	68.429	116.75	152.95	214.69	139.69	203.27	164.79	226.59
	n	0.113	0.089	-0.0194	-0.0146	0.0197	0.0138	-0.0167	-0.0116
75°C	k	121.63	125.94	250.11	253.7	236.32	238.26	263.96	276.88
	n	0.1065	0.1016	-0.016	-0.0148	0.0144	0.0137	-0.014	-0.0132
100°C	k	134.98	131.85	265.17	266.38	253.18	251.33	275.16	280.46
	n	0.0998	0.1039	-0.0147	-0.0154	0.0098	0.0134	-0.0116	-0.0132

IV SUMMARY

The generated oxide trap during PBTI stress will dominate the PBTI characteristics for Hf-based gate dielectrics. In addition, the reduction of threshold voltage degradation and oxide trap generation under PBTI stress indicates that the HfSiON thin film quality is better than HfO$_2$ attributed to HfSiON gate dielectrics had the extra Si-O and Si-N bodings resulting in annihilation of oxygen vacancies. As compared to HfO$_2$ dielectrics, the HfSiON has shallower charge trapping level under PBTI stress due to elimination of deep dielectric vacancies. Finally, the temperature effects are quite different between the HfSiON and HfO$_2$ gate dielectrics.

REFERENCES

[1] K. Onishi, in *Proc. IRPS*, 2002, pp. 419-420.
[2] G. Ribes, J. *IEEE Trans. Dev. Mat. Reliability.* 2005, pp. 5–19
[3] R. Degraeve, in *IEDM Tech. Dig.*, 2003, pp. 935–938.
[4] S. Zafar, A *J. Appl. Phys.*, vol. 93, pp. 9298-9303, 2003

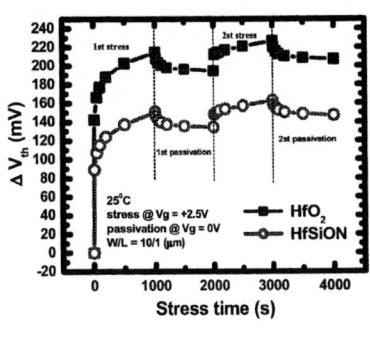

(a)

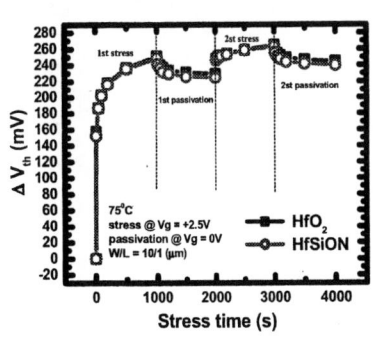

(b)

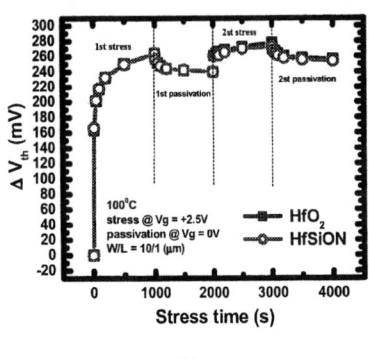

(c)

Fig. 11 ΔV_{th} of DPBTI stress versus stress/passivation at different temperature.(a) 25 °C(b) 75 °C(c) 100 °C

A Study of NBTI in HfSiON/TiN p-MOSFETs Using Ultra-Fast On-The-Fly (UF-OTF) I_{DLIN} Technique

Shweta Deora and Souvik Mahapatra
Department of Electrical Engineering, IIT Bombay, India
Phone: 091-22-25764481. Fax: 091-22-25763707. Email:shwetad@ee.iitb.ac.in

Abstract- **Negative Bias Temperature Instability (NBTI) is studied in HfSiON/TiN p-MOSFETS having thin (2nm) and thick (3nm) HfSiON layer on top of 1nm SiO$_2$ interfacial layer. By using ultra fast on the fly I_{DLIN} technique, the impact of stress temperature (T) and oxide field (E$_{OX}$) on NBTI time evolution is studied. The thickness of the HfSiON layer is shown to have negligible impact on time, T and E$_{OX}$ dependence of NBTI. The impact of time-zero (t_0) delay on power law time exponent (n), E$_{OX}$ acceleration (Γ) of degradation and E$_{OX}$ acceleration (β) of time to fail (tt_F) is also studied. The t_0 does not impact Γ but strongly impacts n, β and hence extracted safe operating voltage (V$_{GSAFE}$).**

I. INTRODUCTION

The introduction of metal gate and high-k gate dielectric is a necessity to enable the EOT scaling of high performance logic transistors below 1.2nm to meet the requirements of 45nm technology node and beyond [1]. Hf based dielectrics are promising candidates in terms of thermal stability and gate leakage, and are being used on top of a SiO$_2$(N) interfacial layer (IL) to preserve interface quality, crucial for meeting mobility requirements to achieve high drive current [2,3,4].

Degradation of p-MOSFET parameters due to NBTI is a serious reliability concern [5]. Though exhaustively studied in SiON dielectric films [6-9], not much has been done to study the effect of NBT stress on Hf-based dielectrics [10,11], especially using ultra-fast techniques [12-14] that are required to eliminate recovery related artifacts [6,7], which are well-known to cause significant impact on measured time evolution and extracted device lifetime.

In this work, the impact of stress oxide field (E$_{OX}$) and temperature (T) on NBTI time evolution is studied in HfSiO$_X$ p-MOSFETs using the recently developed Ultra Fast On-The-Fly (UF-OTF) I_{DLIN} technique [14]. The impact of time-zero (t_0) delay on power-law time exponent (n), E$_{OX}$ acceleration (Γ) of degradation and E$_{OX}$ acceleration (β) of time-to-fail (tt_F) and extrapolated safe operation voltage (V$_{GSAFE}$) are studied. It is shown that HfSiO$_X$ layer thickness has negligible impact on n, Γ and T activation energy (E$_A$), suggesting NBTI in these devices is dominated by Si/SiO$_2$(N) interface or IL degradation.

II. RESULTS AND DISCUSSION

A. *Device and measurement details*:

TiN/HfSiOx p-MOSFETs were made using a standard CMOS flow at SEMATECH. Gate stack consists of 2nm and 3nm HfSiO$_X$ were deposited using ALD on thermally grown SiO$_2$ (1nm), followed by anneal in NH$_3$. The nitrogen content of films is ~8% with an EOT of 1.1nm and 1.3nm for the thin and thick stack [7]. Measurements were performed by using UF-OTF I_{DLIN} with minimum t_0 delay of 1µs as described in [14]. Measured stress induced I_{DLIN} degradation is expressed as $\Delta V = -\Delta I_{DLIN}/I_{DLIN0}*(V_G - V_{T0})$ [17], where I_{DLIN0} is measured I_{DLIN} at the initiation of stress, V_G is stress gate voltage and V_{T0} is pre-stress threshold voltage. Note that ΔV is proportional to but somewhat different (in magnitude) to ΔV_T as mobility degradation is not taken into account (see [15] for details of mobility correction).

B. *Time dependence:*

Figure 1 plots (in a log-log scale) the time evolution of ΔV for different t_0 delay under identical stress E$_{OX}$ and T.

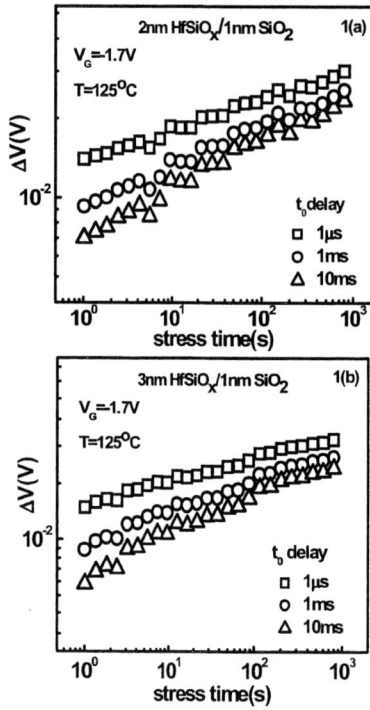

Figure .1 Time evolution of ΔV from ultra fast on the fly I_{DLIN} technique for different t_0 delay for (a)2nm HfSiOx and (b)3nm HfSiOx under identical stress field and temperature.

Power-law dependence is observed for both thin and thick stacks with similar n for a given t_0. The increase in n and reduction in ΔV as t_0 is increased are due to lower measured I_{DLIN0} and the impact of t_0 on ΔV is more at short stress time as explained in [16].

Fig.2 shows the t_0 delay impact on n extracted for long stress time. The reduction in n with reduction in t_0 saturates (variation within the error bar caused by noise in measured I_{DLIN0}) for $t_0 < 10\mu s$. This implies that a faster OTF implementation will not yield lower n values than that reported in this work. This saturation behavior is observed independent of stack thickness, stress bias (as shown) and stress T (not shown). It is also important to note that measured n is independent of the HfSiO$_x$ layer thickness. Finally, note that while UF-OTF shows a saturated slope of $n\sim 0.11$, conventional OTF [17] with $t_0\sim 1ms$ in these stacks would yield $n\sim 0.13$-0.14, identical to previous reports [18]. Therefore, the lower value of n obtained from UF-OTF gives a larger extrapolated lifetime compared to conventional OTF technique as shown later in this paper.

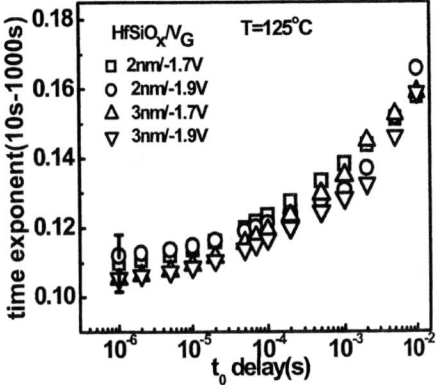

Fig. 2. Power law time exponent (10s-1000s) as a function of t_0 delay at different V_G and constant T.

C. E_{OX} and T dependence:

Figure 3 and 4 respectively show the stress bias and stress T dependence of n for thin and thick stacks. Within measurement error, n is independent of stress V_G and T for a wide range of t_0 delay. For all stress V_G and T and for a given t_0, n is also independent of stack thickness. The stress V_G independence of n

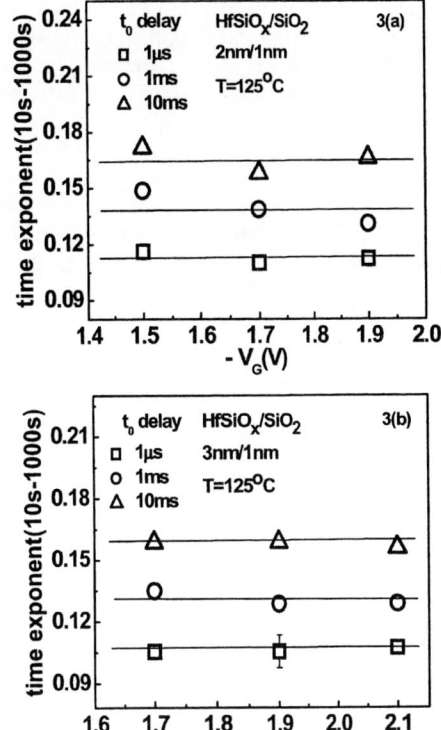

Fig. 3 Power law time exponent (10s-1000s) dependence on V_G stress and t_0 delay for (a) 2nm HfSiOx and (b)3nm HfSiOx oxide for constant temperature stress.

shows the absence of bulk trap generation in the range of V_G studied [5], which is consistent with stack thickness independence of n. As stack thickness variation comes from variation of the HfSiO$_X$ layer, similar n for thin and thick stacks signifies absence of any significant charge generation (hole trapping) in the high-k layer. However, the saturated n ($t_0\sim 1\mu s$) is smaller than that predicted by the Reaction-Diffusion(R-D) model for interface trap generation with molecular H_2 diffusion [19], and therefore some small amount of hole trapping in the IL bulk cannot be ruled out [11].

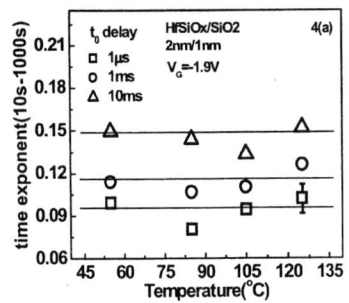 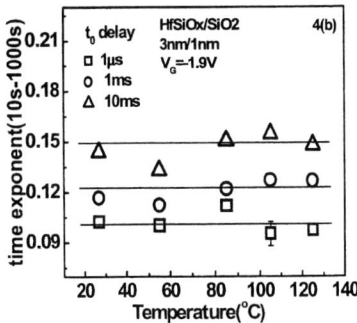

Figure 4 Power law time exponent (10s-1000s) dependence on temperature for various t_0 delay is shown for (a) 2nm and (b) 3nm HfSiOx for constant V_G.

On the other hand, T independence of n is similar to that observed for SiON films [14] and suggests non-dispersive H_2 diffusion controlled mechanism for long stress time [6].

Fig.5 shows ΔV as a function of stress E_{OX} for different t_0 delay for thin and thick stacks. The E_{OX} dependent slope (Γ) is independent of t_0 for both stacks, similar to that obtained for SiON films [14]. However Γ is slightly lower for the thinner stack. As Γ shows strong dependence on N density at the $Si/SiO_2(N)$ interface (reduces with higher N) [14], the slightly lower Γ implies slightly higher N incorporation in the IL during

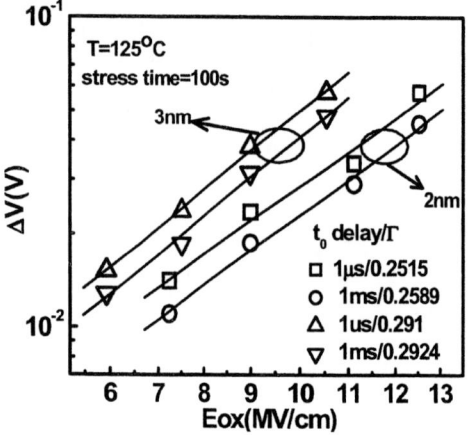

Figure.5 Eox dependence of ΔV degradation measured at 100s stress time at constant T for 2nm and 3nm HfSiOx oxide at two t_0 delays showing slope gamma(Γ).

NH$_3$ PDA for thinner stack (which is expected). Fig.6 shows ΔV as a function of stress T for thin and thick stack at t_0=1µs. High activation energy of E_A~0.11eV implies dominant contribution from interface trap generation for both stacks (the small difference between the two stacks is within the error bar).

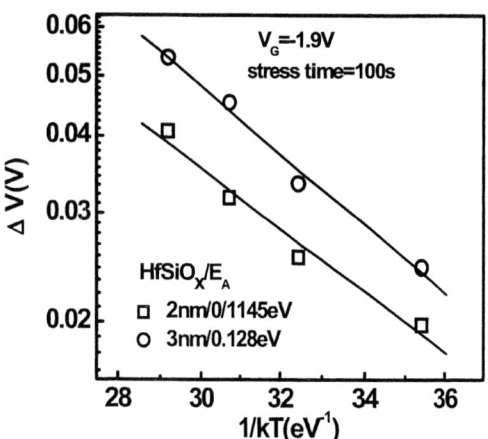

Figure.6 Temperature activation of ΔV measured at 100s stress for 2nm and 3nm HfSiOx at t_0 delay=1µs for constant V_G showing activation energy (E_A).

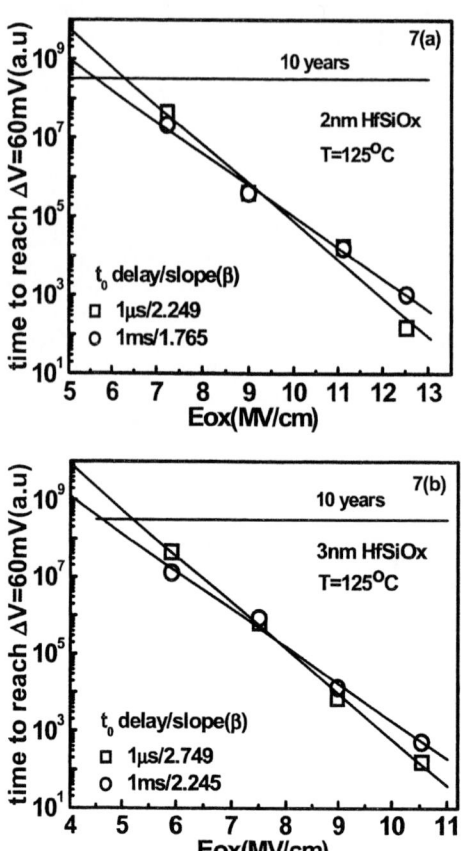

Figure.7 Time to reach ΔV=60mV(a.u) as a function of Eox for (a)2nm HfSiOx and (b)3nm HfSiOx at different t_0 delays and identical temperature stress of 125°C. Slope(β) is shown for different t_0 delays.

D. *Lifetime determination:*

Fig.7 plots the time (tt$_F$) to reach a particular failure criterion (ΔV=60mV in this work) for thin and thick stacks as a function of E_{OX} for different t_0 delay. Note that the E_{OX} acceleration (β) reduces with higher t_0 and yield lower safe E_{OX} for a given lifetime criterion (10 years in the present case). At high stress E_{OX} failure is met within the measurement window, and higher t_0 results in lower ΔV and higher tt$_F$. For lower stress E_{OX}, tt$_F$ was obtained by extrapolation of measured ΔV by assuming power law dependence with n extracted from 10s to 10^3s. Therefore, higher t_0 causes higher n and lower tt$_F$ for lower E_{OX}, and explains lower β obtained for higher t_0 delay. Fig.8 shows β and V_{GSAFE} (normalized) as a function of t_0 delay. Note that increase in t_0 reduces β and extracted V_{GSAFE}. For example, more than 10% reduction in V_{GSAFE} is observed when conventional OTF data is compared to UF-OTF.

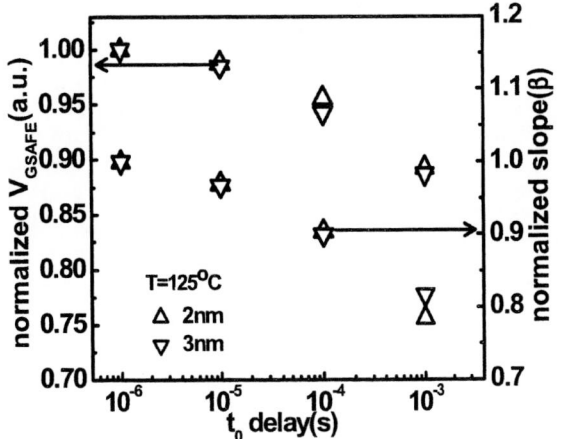

Figure .8 Effect of t_0 delay on safe overdrive voltage ($V_{GSAFE}=V_G-V_{T0}$) and slope(β) for 2nm and 3nm HfSiOx at 125°C.

III. CONCLUSIONS

To summarize, NBTI in HfSiON p-MOSFETs is studied using UF-OTF I_{DLIN} technique having minimum t_0 resolution of 1µs. For a given t_0, the time dependence at long stress time is independent of stress E_{OX} and T, showing the robustness of the underlying physical process. The time, T and E_{OX} dependence of NBTI is independent of thickness of the HfSiON layer, which suggests Si/SiO$_2$(N) interface or IL dominated physical process (interface trap generation and small hole trapping in the IL layer). Control of N density in the IL layer can improve NBTI in these films. The t_0 does not impact E_{OX} dependence (at fixed t-stress), but strongly impacts n, and hence the E_{OX} dependence of time to fail and extracted V_{GSAFE}. Implementation of ultra-fast techniques is a must to determine reliable safe operating condition.

ACKNOWLEDGMENT

The authors would like to thank SEMATECH for providing devices and process details and SRC (GRC) for financial support.

REFERENCES

[1] http://www.itrs.net/Links/2006Update/2006UpdateFinal.htm
[2] K.Mistry et al., "A 45nm Logic Technology with High-k+Metal Gate Transistors, Strained Silicon, 9 Cu Interconnect Layers, 193nm Dry Patterning, and 100% Pb-free Packaging" in *IEDM Tech. Dig.*,pp. 247-250,2007.
[3] E.P.Gusev et al , "Charge Trapping in Aggressively Scaled Metal Gate/High-k Stacks", *IEDM Tech. Dig*, pp.729-732, 2004.
[4] A.Shanware et al., "Reliability Evaluation of HfSiON Gate Dielectric Film with 12.8Å SiO$_2$ equivalent thickess" in *IEDM Tech. Dig.*, pp. 137-140, 2001.

[5] S.Mahapatra and M.A.Alam, "Defect generation in p-MOSFETs Under Negative Bias Stress: An Experimental Perspective" in *IEEE Trans. Device Mater Rel.*, pp. 35-46, 2008
[6] D.Varghese et al., "On the Dispersive versus Arrhenius Temperature Activation of NBTI Time Evolution in Plasma Nitrided Gate Oxides: Measurements, Theory and Implications" in *IEDM Tech. Dig.*, pp. 684-687, 2005.
[7] A.T.Krishnan et al.,"Material Dependence of Hydrogen Diffusion: Implications for NBTI degradation", *IEDM Tech. Dig.*, p. 4, 2005.
[8] Y. Mitani, H.Satake and A.Toriumi "Influence of Nitrogen on Negative Bias Tempertaure Instability in Ultrathin SiON", in *IEEE Trans. Device Mater. Rel.* pp.6-13., 2008.
[9] K. Sakuma, D. Matsushita, K. Muraoka and Y. Mitani, Investigation of Nitrogen-Originated NBTI Mechanism In SiON With High-Nitrogen Concentration", *Proc. Int. Reliab Phys. Symp.*, pp.454-460, 2006.
[10] S. Zafar et al. "A Comparative Study of NBTI and PBTI (Charge Trapping) in SiO$_2$/HfO$_2$ Stacks with FUSI, TiN, Re Gates" in *Symposium on VLSI tech.*, pp. 23-25, 2006.
[11] A.Neugroschel, "An Accurate Lifetime Analysis Methodology Incorporating Governing NBTI Mechanism In High-k/SiO$_2$ Gate Stacks" in *IEDM Tech. Dig.*, pp.1-4, 2006
[12] H.Reisinger, O.Blank, W. Heinrigs, W.Gustin and C.Schlünder " A Comparison of Very Fast to Very Slow Components in Degradation and Recovery Due to NBTI and Bulk Hole Trapping to Existing Physical Models" in *IEEE Trans. Device Mater. Rel.*,pp.119-129,2007.
[13] C.Shen et al., "Characterization and Physical Origin of Fast V_{th} Transient in NBTI of p-MOSFETs with SiON Dielectric" *IEDM Tech. Dig.*, pp.1-4, 2006.
[14] E.N.Kumar et al., " Material Dependence of NBTI Physical Mechanism in Silicon Oxynitride (SiON) p-MOSFETs: A Comprehensive study by Ultra-Fast On-The-Fly (UF-OTF) I_{DLIN} Technique" in *IEDM Tech. Dig.*,pp.809-812, 2007.
[15] A.E.Islam et al., "Theory and Practice of On-the-fly and Ultra-Fast V_T Measurements for NBTI Degradation: Challenges and Opportunities", in *IEDM Tech. Dig.*, pp.805-808, 2007.
[16] A.E.Islam, H.Kufluoglu, D.Varghese and M.A. Alam, "Critical analysis of short-term negative bias temperature instability measurements: Explaining the effect of time-zero delay for on-the-fly measurements", in *Appl. Phys. Lett.* 90, 083505, 2007.
[17] S.Rangan, N.Mielke and E.C.C.Yeh, "Universal Recovery Behaviour of Negative Bias Temperature Instability", in *IEDM Tech. Dig.*,pp. 341-344, 2003.
[18] V.D.Maheta, S.Purwat and G.Gupta, " Comparison of Negative Bias Temperature Instability in HfSiO(N)/TaN and SiO(N)/poly-Si p-MOSFET" in *IPFA*,p.91,2007.
[19] S.Chakrvarti, A. T. Krishnan, V. Reddy, C. F. Machala and S.Krishnan, "A Comprehensive Framework For Predicitive Modeling of Negative Bias Temperature Instability", in *proc. Int. Reliability Physics Symp.*,pp.273-282, 2004.

978-1-4244-2039-1/08/$25.00 ©2008 IEEE

Session 10:

Physical & Chemical Characterization

INVITED PAPER

Nanoanalysis of High-k Dielectrics on Semiconductors

A. J. Craven, M. MacKenzie and D W McComb[1]

Department of Physics & Astronomy, University of Glasgow, Glasgow, G12 8QQ, Scotland, UK.
[1]Department of Materials, Imperial College London London, SW7 2AZ, UK.
Phone: (+44 141) 330 5892 Fax: (+44 141) 330 4464 Email: a.craven@physics.gla.ac.uk

Abstract- **Replacement high-k dielectrics for Si(O,N) in MOSFETs undergo many physical and chemical changes during deposition and processing.   The situation is even more complicated when a metal electrode is inserted into the gate stack.   Investigation of such systems using advanced nanoanalytical techniques in the transmission electron microscope is discussed.**

I. INTRODUCTION

Introducing a replacement for the Si(O,N) gate dielectric currently used in Si MOSFETs requires a high-k dielectric layer and a metallic gate electrode. These systems must meet a large number of stringent conditions to give the performance and reliability essential for large scale integration in devices. The deposition and processing of such systems lead to many physical and chemical changes in the deposited layers. Thus it is crucial to understand and be able control the changes if the required performance is to be achieved.

Hafnia based dielectrics will be in the first generation of devices incorporating such high-k dielectrics.   One issue is crystallization of HfO_2 during processing and the addition of SiO_2 raises the crystallization temperature but leads to phase separation [1-3].   Since the aim of the high-k dielectric is to increase the capacitance without decreasing the thickness, sources of smaller series capacitance must also be removed. One of these is the depletion capacitance associated with the traditional doped poly-Si gate electrode.   To remove this contribution, a metallic gate electrode is required [4].   Such a metallic electrode has the additional benefit of removing the high concentration of dopant atoms adjacent to the channel, to which the dopant atoms can diffuse during processing with adverse effects on performance [5].

The effective work function of the gate electrode must be chosen to match the device (p- or n-type). With poly-Si, this can be done by a change of dopant but, for a metallic electrode, the material itself must be changed.   For n-type devices TaN_x and Ta(C,N) are candidate materials.   One of the problems encountered has been a change of effective work function with processing so that the Fermi level moves from close to the conduction band edge towards mid-gap during processing [6]. The reason for such changes is ill-understood but one possibility is interface reactions or reconstructions that occur during processing.   Nanoanalytical transmission electron microscopy

(TEM), used in conjunction with electron energy loss spectroscopy (EELS), is an excellent tool for investigating such reactions, with the energy loss near edge structure (ELNES) providing critical information on local chemistry [7-9].   An earlier study of a TiN inserted gate stack showed a definite Si(O,N) interface layer forming at the TiN/poly-Si interface with a reaction at the dielectric/TiN interface that depended on whether the dielectric was HfO_2 [10-12] or HfSiO [13]. Below, an investigation of a TaN_x inserted HfSiO gate stack is reported. This is a much more challenging system for nanoanalytical TEM because of the high and adjacent atomic numbers of Ta and Hf. This study illustrates what can be achieved currently and what developments are needed to take the analysis further.

II. EXPERIMENTAL DETAILS

A. Deposition and Processing of the Gate Stack

The native oxide was removed from a Si (100) wafer by etching in $HF:H_2O$ (1:100) solution. The wafer was then put through an O_3/DI water cleaning sequence resulting in the growth of ~1 nm chemical oxide.   Metal organic chemical vapour deposition (MOCVD) was used to deposit approximately 4 nm of HfSiO (co-deposition of 70% HfO_2 and 30% SiO_2).   The wafer was then given a standard degas treatment at 330 °C for 40 s and 10 nm of TaN_x was deposited by physical vapour deposition (PVD). Here x is ~0.5. 100 nm of amorphous Si was then deposited and the stack annealed at 1000 °C for 10 s to simulate dopant activation. During this step, the amorphous Si crystallised into poly-Si. The wafer surface was exposed to the clean room atmosphere as it was moved between process tools.   In particular, this occurred after the HfSiO and TaN_x deposition steps.   However, there was no vacuum break between the degas stage and the TaN_x deposition.

B. TEM Specimen Preparation and Nanoanalysis

Standard grinding, polishing, dimpling and ion milling were used to prepare cross-sectional TEM specimens. The sample was oriented so that the growth direction was normal to the electron beam using the diffraction pattern from the single crystal Si substrate. Commercial monoclinic HfO_2 powder was used to give an ELNES reference shape for the O K-edge in HfO_2.

INVITED PAPER

The specimens were examined in an FEI Tecnai F20 TEM/STEM operated at 200kV and equipped with a field emission gun and a Gatan ENFINA electron spectrometer. Bright field conventional TEM (CTEM) images and high angle annular dark field (HAADF) scanning TEM (STEM) images were recorded.

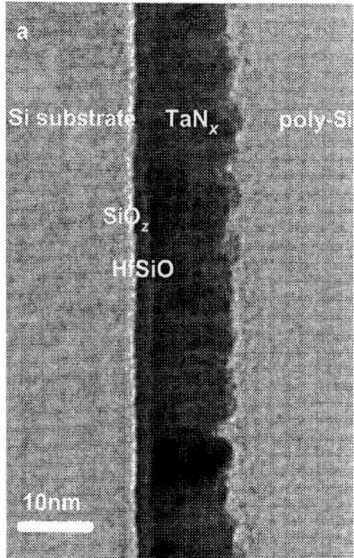

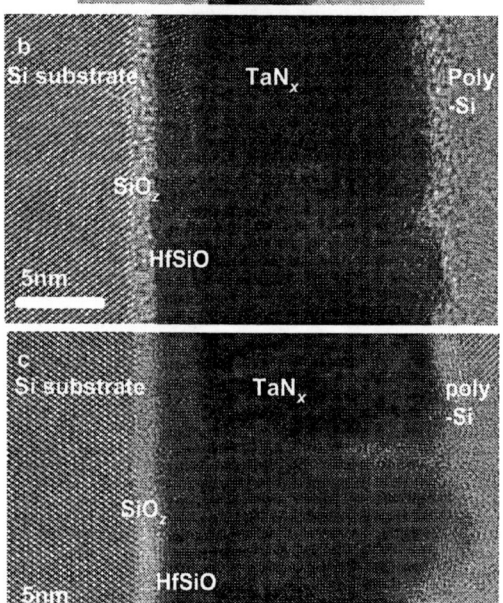

Fig. 1 Bright field CTEM images of the TaN$_x$ inserted HfSiO gate stack.

The scanning was performed using Gatan DigiScanII hardware controlled by Digiscan and DigitalMicrograph software. Spectrum images (SIs), where one or more spectra are collected at each pixel, were recorded from key regions identified in the HAADF images. In some cases, both core loss and low loss regions of the EELS spectrum were recorded at the same pixel, using the Glasgow dual-EELS system [14]. Here a fast beam switch is used to limit the integration time for the low loss region. The dual-EELS acquisition is controlled by special scripts written for DigitalMicrograph. SIs were recorded over the energy ranges of the low loss spectra and those appropriate to the core loss edges from Si, Hf, Ta, N and O. A ~0.5 nm diameter probe with a convergence semi-angle of 9 mrad was used for the data presented here. A dispersion of 0.3eV/ch, a spectrometer collection semi-angle of 27 mrad and an integration time of 5 sec per pixel were used for most of the core losses but the recording conditions were changed to 1eV/ch, 45 mrad and 10 sec per pixel for acquisition of the Hf M$_{4,5}$- TaM$_{4,5}$- and Si K-edges in the 1600 – 2000eV energy loss region.

III. RESULTS AND DISCUSSION

A. Bright Field Images

Fig. 1a shows a bright field CTEM image of the gate stack. Between the substrate and the HfSiO there is a layer of SiO$_z$ which is somewhat wider that the 1nm chemical oxide formed on the wafer before deposition. The TaN$_x$ is polycrystalline with a significant number of grains not running through the thickness of the layer. The interface between the HfSiO and the TaN$_x$ is rough on a scale similar to the TaN$_x$ grain size and this is particularly clear at the top of the image. The interface between the TaN$_x$ and the poly-Si is very rough and controlled by the grain growth. Figs. 1b and 1c are higher magnification images. Fig. 1b shows a clear amorphous layer between the TaN$_x$ and the poly-Si while Fig. 1c shows amorphous material penetrating down the boundary between two TaN$_x$ grains.

Fig. 2 shows a HAADF image of the gate stack. Here, to a first approximation, the signal is proportional the sum of the Rutherford scattering from each of the atoms along the beam path. The Rutherford scattering cross-section is proportional to Z^2 where Z is the atomic number. The signal from the boundaries is clearly weaker than from the grains. Thus, either there are fewer atoms at the boundaries because the grains do not touch or lower Z elements have penetrated down the boundaries.

B. EELS Spectrum Imaging

To investigate further, line SIs were recorded across the gate stack, one avoiding a boundary and one following a boundary. The two lines shown on Fig.2 show the typical positions of the two types of SIs. The spectra in these SIs were core loss only and covered the range including the O and N K-edges and some Hf and Ta N-edges. Figs. 3a and b show the O and N K-edge intensities across the gate stack. It is difficult to extract the intensities of the Hf and Ta N-edges because they are overlapping and have low signal to background ratios due to their slow onsets. However, a good indication of the combined distribution of these high Z elements can be obtained from the HAADF signal. To give a better indication of the Hf+Ta distribution, the HAADF signals in Fig. 3 have had constants subtracted so that they are zero in the poly-Si. The O and N

978-1-4244-2039-1/08/$25.00 ©2008 IEEE

INVITED PAPER

intensities are normalised to the same maximum value and the HAADF signal scaled to match the N in the TaN_x layer.

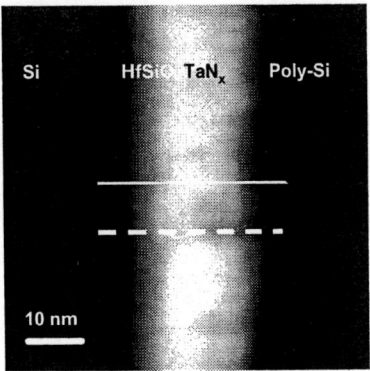

Fig. 2 HAADF image of the gate stack. The solid white line is a typical position for a line SI through a grain while the dashed one is a typical position for a line SI through a boundary.

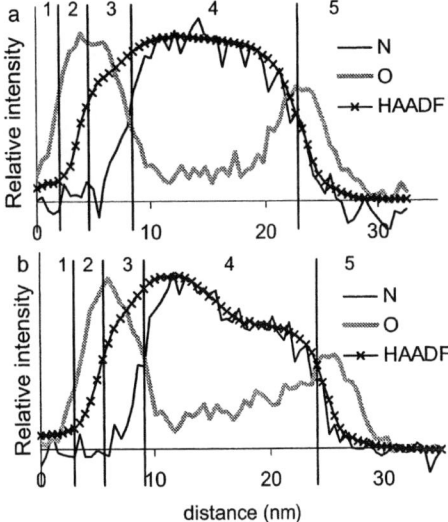

Fig. 3 Intensity profiles across a) a grain and b) a grain boundary showing the N and O K-edge and the HAADF signals, the last with a constant subtracted to make it zero in the poly-Si. The O and N signals are scaled to the same maximum intensity and the HAADF signal is scaled to the N in the TaN_x.

Fig. 3 shows several things very clearly. In the SiO_z region marked 2, the O intensity rises more rapidly than the HAADF signal, as expected. In the HfSiO (region 3) both the O and the HAADF signal are high but the N is low as expected. In the TaN_x (region 4), the N and HAADF signals track each other closely but the O:N ratio varies in different ways in the two cases suggesting possible penetration of O down the boundary. At the interface between the TaN_x and the poly-Si, the O signal peaks after both the N and the HAADF signals have dropped, suggesting that the amorphous layer at the interface could be SiO_z.

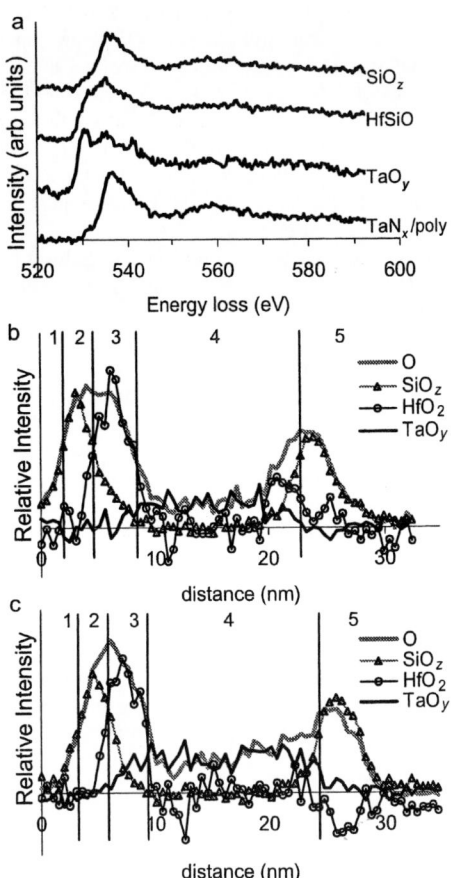

Fig. 4 a) The O K-edge shapes from SiO_z, HfSiO, oxidised TaN_x (TaO_y) and the TaN_x/poly-Si interface; b) a plot of the MLLS fit coefficients to the spectra in the SI across the grain; c) a plot of the MLLS fit coefficients across the grain boundary. The corresponding O K-edge intensities are shown

Fig. 4a shows the O K-edge shapes obtained from different parts of the stack: SiO_z, HfSiO, TaO_y (from oxidation of the TaN_x) and the TaN_x/poly-Si interface. To show the distribution of the phases, each O K-edge shape in the SI can be modelled as a linear combination of the edge shapes from SiO_z, TaO_y and the HfO_2 standard using a multiple linear least squares (MLLS) fit. Figs. 4b and c show plots of the fit coefficients along with the total O K-edge intensities, which have been scaled to give the best match. It is clear that the amorphous interface layer between the TaN_x and the poly-Si is indeed SiO_z. Also, in both cases, the HfSiO has effectively phase separated with the SiO_z moving towards the substrate and the HfO_2 moving towards the TaN_x. In Fig. 4b, the TaO_y coefficient follows the O K-intensity until the boundary with region 5 is approached when a contribution from HfO_2 appears. In this part of the SI, there was a pre-peak in the O K-edge data, possibly resulting from beam damage [15]. The MLLS fit adds a contribution from the HfO_2 shape in this region because of this pre-peak. In Fig. 4c, there is no artefact like this but the TaO_y seems constant across the layer while the O K-edge intensity increases steadily. One possible

INVITED PAPER

explanation is that another element has penetrated into the boundary and combined with the extra O. However, there is nothing in the spectra to suggest this but such an element could have edges that are difficult to detect or lie outside the energy range recorded. The presence of such an additional element might be expected to show up in the current MLLS fits as an increase in the residual where it is present. Here, there is no obvious pattern in the residuals that would suggest a missing element but it cannot be ruled out.

Another possibility is that the incomplete analysis is leading to a false interpretation. For instance, the degree of oxidation of the Ta could change with position. During preparation, the surface of the TaN_x was exposed to the clean room atmosphere during transfer between tools. The resulting degree of oxidation of the surface and the grain boundaries may differ. Amorphous Si is then deposited and annealed and, again, there could be differences in the interaction at the interface and in the grain boundaries. Finally a TEM sample is made and its surfaces are exposed to atmosphere resulting in oxidation. Such surface oxidation of TEM specimens is why O tends to be seen at some level at all points in an SI.

C. Developments Required

Table I
Relevant Ionization Edge Thresholds

	Edge	Energy	Comment
Si	$L_{2,3}$	100	Sensitive to bonding but on a high background whose shape is perturbed by multiple scattering, especially in the presence of higher Z elements.
Si	L_1	149	Very weak.
Hf	$N_{4,5}$	214	Slow onset and poor signal to background makes it difficult to extract.
Ta	$N_{4,5}$	229	Slow onset and poor signal to background makes it difficult to extract.
Hf	$N_{2,3}$	380	Slow onset and poor signal to background makes it difficult to extract.
N	K	399	Good signal to background and sensitive to bonding.
Ta	$N_{2,3}$	404	Slow onset and poor signal to background makes it difficult to extract.
O	K	532	Good signal to background and sensitive to bonding.
Hf	N_1	538	Very weak.
Ta	N_1	565	Very weak.
Hf	M_5	1662	Slow onset and low signal level but can be extracted.
Hf	M_4	1716	Normally take with the M_5.
Ta	M_5	1735	Slow onset and low signal level but can be extracted and separated from the Hf $M_{4,5}$ using the edge shapes.
Ta	M_4	1793	Normally take with the M_5.
Si	K	1840	Sensitive to bonding with good signal to background ratio but low signal level.

Table I lists the core ionisation edges of relevance to this gate stack. In the presence of high Z elements, the Si $L_{2,3}$-edge is difficult, but not impossible, to work with. As noted above, the Hf and Ta N-edges are difficult to extract. Moreover, there are overlaps with the N and O K-edges. However, the Hf and Ta edges that overlap have smaller cross-sections than the O and N K-edges in the regions used to integrate the N and O K-edges. Thus the N and O intensities can be extracted with reasonable accuracy provided they are not too low, at which point background subtraction needs careful attention. The Hf and Ta $M_{4,5}$-edges and the Si K-edge can be separated and extracted but

the signal level is low. In addition to the core-loss edges, the low loss region is required to correct for the image contrast and to normalise out thickness variations as described below.

Thus it is necessary to record three regions of the spectrum, each with a very different signal level: the low loss region; the N and O K-edge region; and the Hf $M_{4,5}$, Ta $M_{4,5}$ and Si K edge region. The three spectra in Fig. 5 show the intensity range present in the spectrum from TaN_x over the required energy loss range.

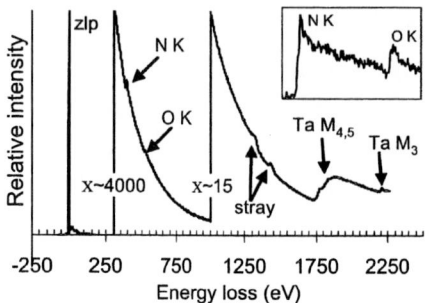

Fig. 5 Three spectra from TaN_x covering the energy range -250eV to 2250eV and showing the intensity range present. The features marked stray are the result of scattering in the electron gun. Inset are the background subtracted N and O K-edges.

To convert the edge intensities to atomic ratios, cross-sections are needed. Cross-section models exist with those for K and L cross-sections giving an absolute accuracy of 10-20% [16]. Where ratios of these cross-sections are taken, the approximations will tend to cancel giving improved accuracy. Use of suitable standards gives better accuracy. Here the aim is to refer everything back to one edge (or possibly two) whose cross-section is known e.g. the O K-edge. This can be very successful e.g. [17].

Normalisation by the low loss intensity removes the image contrast very effectively and, with the cross-section, gives the number of atoms per unit area of specimen for each species [14]. The low loss spectrum also gives the local value of $t/\Box$ where t is the thickness and $\Box$ is the inelastic mean free path. If the composition is known, a parameterised Kramers-Kronig model can give the local $\Box$ [18] so that the local thickness can be determined. Hence the number of atoms per unit volume can be determined.

This has been demonstrated successfully on a steel sample using the Fe $L_{2,3}$-edges and the Glasgow dual EELS system [14]. However, this can currently only use two integration times at a pixel and the problem here requires three. The hardware for this is currently being developed.

D. The Hf and Ta $M_{4,5}$-edges

Fig. 6a shows the HAADF image of the TaN_x inserted gate stack and Fig. 6b shows the line SI in the energy range ~1600 to ~2000eV after background subtraction. With only two variables, a 1-D SI can be presented as an intensity map with position in one direction and energy loss in the other. Fig 6b clearly shows the Ha and Ta $M_{4,5}$-edges and the Si K-edge. Between the substrate and the HfSiO, the chemical shift of the

INVITED PAPER

Si K-edge in SiO$_z$ can be seen and this shift is also present at the TaN$_x$/poly-Si interface, as expected from the analysis of the O K-edge shapes in Fig. 4. Fig. 6c shows the distribution of crystalline Si (c-Si), SiO$_z$, Hf and Ta using an MLLS fit of the edge shapes, demonstrating that it is possible to extract intensities from these Hf and Ta edges.

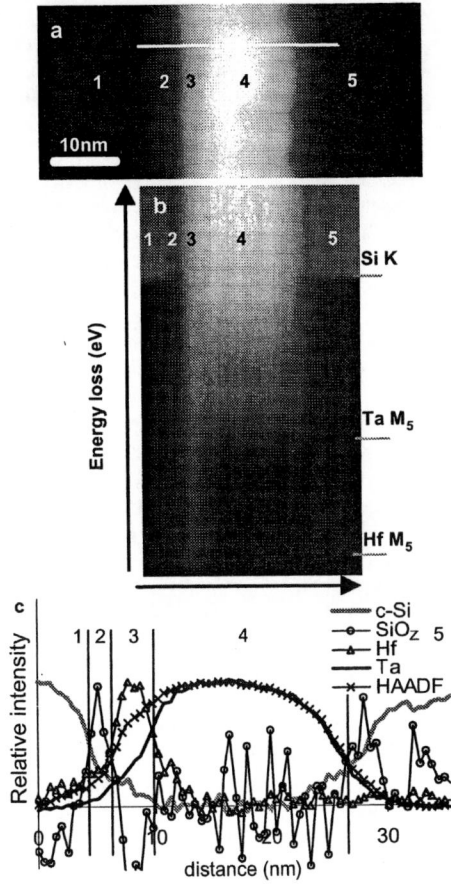

Fig. 6 a) HAADF image showing the position of the SI; b) 1-D SI showing the Hf M$_{4,5}$-, Ta M$_{4,5}$- and Si K-edges with the chemical shift on the Si K-edge showing regions of crystalline Si and regions of SiO$_z$; c) plot of MLLS fitting coefficients and the HAADF signal minus a constant.

E. Dual EELS

Fig. 7a shows the region from which a 2-D spectrum image was recorded using the dual EELS system. Both the low loss region and the core loss region were taken covering the range from -20 to 735eV. Fig 7b shows the extracted Si L$_{2,3}$-, N K- and O K-edge intensity profiles from the bottom part of the SI region indicated on Fig. 7a. The HAADF signal is added after subtraction of a constant to make it zero in the poly-Si. Once again, the N signal appears to follow the HAADF signal but the O signal rises as the poly-Si is approached.

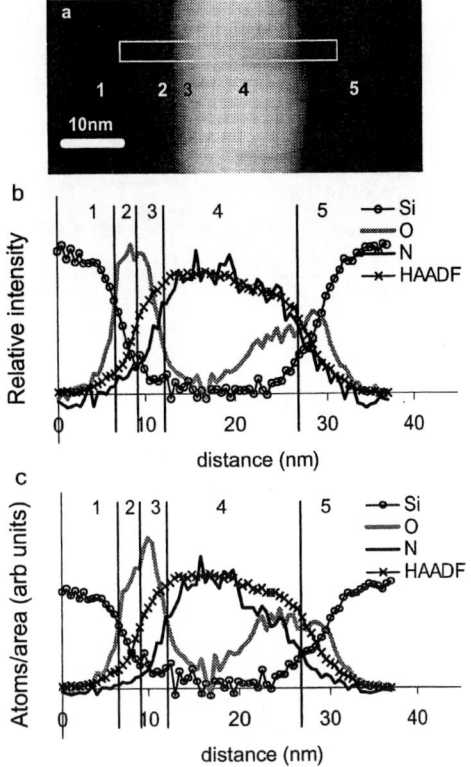

Fig. 7 a) HAADF showing the region of the 2-D SI; b) Si L$_{2,3}$-, N K- and O K-edge intensities extracted from the bottom part of the SI; c) atoms/area after normalizing the edge intensities by the low-loss intensity and the cross-sections. The HAADF signal with a constant subtracted is also shown.

However, if the Si, N and O intensities are normalised by the cross-sections and the low-loss intensity to give the number of atoms per unit area, a different picture emerges, as shown in Fig. 7c. Now the number of N atoms per area is seen to drop and the number of O atoms per unit area to rise at approximately the 20nm position on the profile. Here the HAADF signal also drops but only slightly, corresponding to the slightly darker region along the grain boundary seen at the bottom of the SI region in Fig. 7a. At the boundary between regions 4 and 5, there is still a significant HAADF signal where the N has dropped to low level. These observations are consistent with the TaN$_x$ having oxidised at the surface (~28nm) and down the grain boundary to ~20nm. The region around 28 - 30nm, where the O concentration remains high and the HAADF signal falls, corresponds to the amorphous SiO$_z$ layer that forms at the interface.

Without the corresponding Hf and Ta distributions together with suitable experimental cross-sections, it is not possible to determine, x, y and z but this will be possible in the future. It is, however, worth noting that the dual EELS system allows accurate calibration of the absolute energy loss. Figure 8 shows the Si L$_{2,3}$–edge shapes displayed with absolute energy scales from the substrate (c-Si), the poly-Si, the SiO$_z$ layer on the substrate and the amorphous layer at the TaN$_x$/poly-Si interface.

INVITED PAPER

It is clear that the shapes from the c-Si and the poly-Si have identical threshold energies and shapes. The edge shapes from the SiO$_z$ and the amorphous layer are also extremely similar, confirming that the amorphous layer is SiO$_z$ with z being very similar in the two cases.

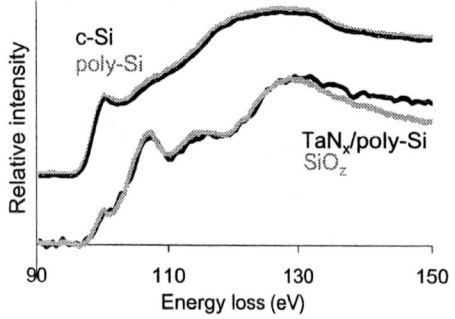

Fig. 8 Si L$_{2,3}$-edge shapes with absolute energy calibration taken from the dual EELS SI data in Fig. 7 to compare the shape from the substrate (c-Si) with that from the poly-Si and the shape in the silica adjacent to the substrate (SiO$_z$) with that from the TaN$_x$/Poly-Si interface.

IV. CONCLUSIONS

It is clear from the results presented that the physical and chemical processes occuring in metal inserted gate stacks can be quite complex and that care must be taken to avoid undesirable effects such as exposure to atmosphere when changing tools. Ideally a cluster tool should be used.

Understanding the details of such physical and chemical changes is a challenge because of the fine layers, the rough interfaces and the wide range of Z present. Nanoanalytical techniques to provide this understanding are continuing to improve, giving information on composition and chemistry over a wide range of Z. Information is needed on all elements present. In EELS, it is important that corrections are made for the underlying contrast in the signal entering the spectrometer and for thickness variations in the specimen. Thus the ability to collect the whole spectral range with accurate spatial registration and adequate signal to noise ratio is crucial. Continued development of such systems is essential.

ACKNOWLEDGMENTS

The authors thank: Prof. S. de Gendt of IMEC for providing the gate stack; Dr. S. McFadzean and Mr B. Miller for technical support and specimen preparation; EPSRC for financial support under grant GR/S44280/01.

REFERENCES

[1] J.P. Maria, D. Wickaksana, J. Parrette, and A.I. Kingon, "Crystallization in SiO2-metal oxide alloys," *J. Mat. Res.*, vol. 17, pp. 1571-1579, 2002.

[2] S. Stemmer, et al., "Application of metastable phase diagrams to silicate thin films for alternative gate dielectrics," *Jap. J. App. Phys. Pt 1-Regular Papers Short Notes & Review Papers*, vol. 42, pp. 3593-3597, 2003.

[3] D.W. McComb, A.J. Craven, D.A. Hamilton, and M. MacKenzie, "Probing local coordination in high-k materials for gate stack applications," *App Phys Lett* vol. 84, pp. 4523-4525, 2004.

[4] H.J. Cho, et al., "The effects of TaN thickness and strained substrate on the performance and PBTI characteristics of poly-Si/TaN/HfSiON MOSFETs," *Tech. Dig.- Int. Electron Devices Meet.*, pp. 503-506, 2004.

[5] L.-Å. Ragnarsson, et al., "The impact of sub monolayers of HfO2 on the device performance of high-K based transistors," *IEDM 2003 Tech. Digest*, pp. 87-90, 2003.

[6] S.B. Samavedam, et al., "Fermi level pinning with sub-monolayer MeO$_x$ and metal gates," *IEDM 2003 Tech. Digest*, pp. 307-310, 2003.

[7] A.J. Craven, M. MacKenzie, D.W. McComb, and F.T. Docherty, "Investigating physical and chemical changes in high-k gate stacks using nanoanalytical electron microscopy," *Microelectronic Eng.*, vol. 80, pp. 90-97, 2005.

[8] B. Foran, J. Barnett, P.S. Lysaght, M.P. Agustin, and S. Stemmer, "Characterization of advanced gate stacks for Si-CMOS by electron energy-loss spectroscopy in scanning transmission electron microscopy," *J. Electron Spectrosc.*, vol. 143, pp. 149-158, 2005.

[9] M. MacKenzie, A.J. Craven, D.W. McComb, and S. de Gendt, "Interfacial reactions in a HfO2/TiN/Poly-Si gate stack," *App. Phys. Lett.*, vol. 88, 192112, 2006.

[10] M. MacKenzie, A.J. Craven, D.W. McComb, and S. de Gendt, "Advanced nano-analysis of high-k dielectric stacks," *J. Electrochem. Soc.*, vol. 153, pp. F215-F218, 2006.

[11] M. MacKenzie, et al., "Advanced nano-analysis of a high-k dielectric gate stack prior to activation," *Electrochemical and Solid-State Letters*, vol. 10, pp. G33-G35, 2007.

[12] F.T. Docherty, et al., "A nanoanalytical investigation of elemental distributions in high-k dielectric gate stacks on silicon," *Microelectronic Eng.*, vol. 85 pp. 61-64, 2008.

[13] M. MacKenzie, A.J. Craven, D.W. McComb, S. McFadzean, and S. de Gendt, unpublished.

[14] J. Scott, et al., "Near-simultaneous dual energy range EELS spectrum imaging," unpublished.

[15] M. MacKenzie et al., "Electron energy-loss spectrum imaging of an HfSiO high-*k* dielectric stack with a TaN metal gate," J. Phys: Conf. Ser., in press.

[16] R.F. Egerton, *Electron Energy Loss Spectroscopy in the Electron Microscope*, 2nd ed., New York: Plenum, 1996, pp. 420-427.

[17] M. MacKenzie, P. Harkins, A.J. Craven, and D.W. McComb, "Quantitative electron energy-loss spectroscopy (EELS) analyses of lead zirconate titanate," *Micron* doi:10.1016/j.micron.2007.10.016, 2007.

[18] T. Malis, S.C. Cheng, and R.F. Egerton, "EELS log-ratio technique for specimen thickness measurement in the TEM," *J. Electron Microsc. Tech.*, vol. 8, pp. 193-200, 1988.

Physical Characterization Challenges in 45 nm Technology Node

K. Li, P. Liu, Q. Wang, I. Tee, and J. Teong
Chartered Semiconductor Manufacturing Ltd.
60 Woodlands Industrial Park D, Street 2, Singapore 738406
Phone: (65) 63604617 Fax: (65) 63604592. Email: likun@charteredsemi.com

Abstract-The advent of 45 nm technology poses a real challenge to device physical characterization. The shrinkage in dimension makes the characterization of some critical structures very difficult or impossible. The adoption of ultra low K materials even worsens the situation. In this paper, an attempt is made to address some of the challenging characterization issues and some solutions are provided with the aim to facilitate 45 nm process development and optimization.

I. INTRODUCTION

The stringent demand on speed and portability of electronic devices leads to rapid development of IC technology. The speed requirement makes the gate length in MOS transistors shrink rapidly from 130 nm to 90nm, 65 nm, and 45 nm. The portability requirement makes packing density increases rapidly, and as a result the spacing between metal lines reduces significantly. To reduce RC delay, Cu has been used to replace Al due to its lower resistivity and better electromigration resistance, and low K (LK) and ultra low K (ULK) materials (k value 3.0 to 2.4) have been used to replace the normal silicon dioxide (k value 4.0).

The shrinkage of IC device dimension and the incorporation LK and ULK materials pose significant challenges in physical characterization and failure analysis. With IC devices scaling down to 65nm and 45 nm, there is a quantum leap in the difficulty of characterization as compared with the linear increase in difficulty with the technology migrating from 130 nm to 90nm.

One of the main reasons is that the critical device feature dimension becomes smaller than the thickness of a normal TEM sample, which is about 80 to 100nm. For example, the poly length of a typical 45 nm transistor is around 30 nm, and the active width is about 50 nm. To make thing worse, the surrounding structure such as STI is normally not at the same level as the active (STI elevation). As a result, an along/cross-poly sample of normal thickness will show shadowing effect, making important information such as gate oxide thickness and gate oxide corner thinning not easily available. The key is that the sample has to be made as thin as possible, while amorphization of the sample has to be minimized [1-3].

Another challenging issue is beam damage of low K/ultra low K materials, which causes the distortion of the structure being characterized. How to set up the electron beam and ion beam condition and how to do imaging are critical.

This paper studies these issues systematically and proposes a couple of solutions.

II. EXPERIMENTAL

The samples used in this study are from 45 nm technology node. To illustrate the issue, the narrowest poly line is not chosen. The ultra low K materials used is with a k value smaller than 3.

Three types of focused ion beam (FIB) machines, i.e., single beam FIB (FEI FIB TEM205), dual beam FIB (FEI Helois 400S), and triple beam FIB (Seiko XVision 200TB), are used for sample preparation to compare the effect of different experimental conditions.

TEM used in this study is a 200 kV FEI Tecnai. The purpose of imaging after different period of time of illumination is to study the effect of high KV electron beam on the degradation of low k materials.

III. RESULTS AND DISSCUSSIONS

A. Low dimension gate oxide sample preparation

At 65/45 nm technology nodes, one big challenge for characterization is TEM sample preparation of gate oxide samples along or cross poly, as poly length and active width are significantly smaller than the thickness of a normal TEM sample, which is about 80~100 nm thick. Due to STI divot and elevation, the surrounding oxide is normally not at the same level as the active. As a result, shadowing effect will appear in the image.

Fig. 1(a) shows the layout of a typical 45 nm transistor. The width of the active is 70 nm, and the length of the poly is 50nm. For across poly sample preparation, if the sample is made thicker than 70nm, or if FIB milling does not stop accurately within the active, shadowing effect will appear, as shown in Fig. 1 (b), where two layers of nitride and poly can be seen, and no gate oxide information can be obtained. The occurrence of double layer imaging is due to the sample containing the poly running on both active and STI, which are at different levels. It is more challenging to prepare an along ploy sample, as the requirement for sample thickness is more stringent. When the sample is thicker than 50 nm or when the end pointing is not controlled well to be within the poly, the adjacent silicide will be seen, as shown in Fig. 1 (c). The information about gate oxide cannot be obtained either.

978-1-4244-2039-1/08/$25.00 ©2008 IEEE

The challenge here is that the sample has to be made thinner than 50 nm and the end point for both sides has to be accurately within the poly. To get a high resolution image to reveal the gate oxide, the amorphization has to be well controlled.

To make a sample thinner than 50nm, a high resolution ion beam is necessary. A well-tuned high-resolution ion beam also helps to reduce amorphization. To minimize amorphization and maintain good crystallinity of a less than 50 nm thick sample, high quality low kV ion beam cleaning (down to 3 to 2 kV) is also required. An even better way is to use low energy argon mill after FIB milling, either in-situ or ex-situ.

In this paper, two methods are tried. The first method employs a Helios 400S dual beam FIB system with very good ion beam profile (5 nm resolution) and low kV (5, 3, and 2 kV) performance. The sample is first milled by a 30 kV gallium ion beam to a thickness of about 60 nm. Then 5kV cleaning is

applied followed by 2 kV cleaning. As the system has an SEM column of very good resolution at both high and low kV, the end point detection becomes relatively easy. Fig. 2 (a) is the low magnification photo of the along poly profile. It is much improved as compared with Fig. 1(c). There is no silicide shadow at all and the poly profile is very clear. Fig. 2 (b) is the high resolution of the gate oxide region. The lattice fringe can be seen easily in both the active and poly, indicating that the sample is of good crystallinity and amorphization is minimum. Normally without good low kV cleaning, the sample got amorphization severely, and poly lattice is difficult to see, increasing the difficulty in differentiating poly and gate oxide.

The second method employs an SIIT XVsion 200TB triple beam system. In addition to the ion beam and electron beam equipped in a normal dual beam system, this system has an additional argon ion gun operating at 500 v to 1 kV. The sample is first prepared by normal FIB Ga milling at 30 kV to a thickness of about 60~70 nm, followed by 5kV low energy Ga ion beam cleaning. Finally the sample is cleaned with the 1kV Argon beam. When doing Ar milling, the SEM can be turned on and used as an in-situ monitoring tool. As Ar atom is lighter and smaller than Ga atom, it causes less damage to silicon substrate than Ga ions at the same energy. Therefore post FIB Argon milling helps to further reduce the amorphization. The results are shown in Fig. 2 (c) and (d). Fig. 2 (c) is the low magnification photo, where the poly and active are clearly resolved, and no shadowing happens. The high resolution image of the gate oxide (Fig. 2 (d)) shows the lattice fringe very clearly. The lattice fringe of poly in Fig.2(d) is even clearer than that in Fig.2(b), which demonstrates the benefit of Ar low kV

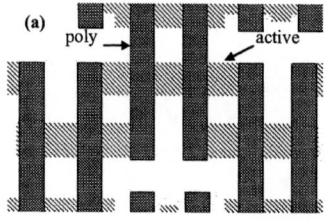

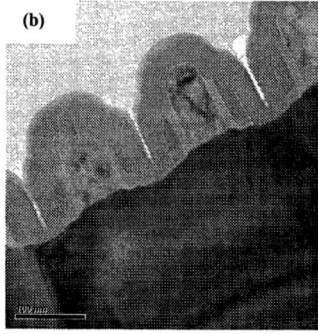

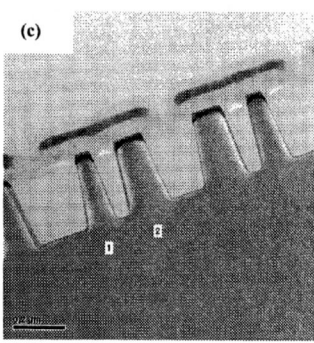

Fig. 1(a) A schematic layout of poly, active and STI structure of field effect transistors. The active width in TEM sample thickness direction is 70nm and the poly length is about 50 nm; (b) the sample is cut slightly off the center or slightly thicker than 709 nm, the shadowing effect can be clearly seen and the image is not clear; (c) along poly length direction cross-section image; as the sample is thicker than 50 nm, silicide is visible, and gate oxide information is not available.

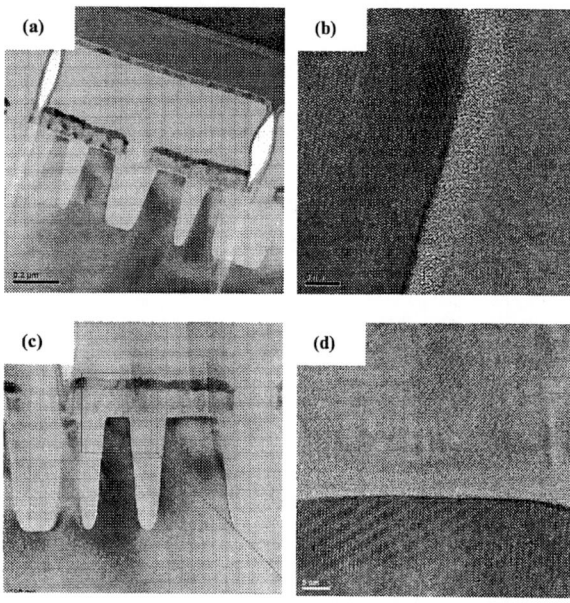

Fig. 2 Along poly gate oxide images. (a) Low magnification image of 2kV gallium cleaned sample; (b) high magnification image from the same sample of (a); (c) low magnification image of post-FIB Ar cleaned sample, and (d) high magnification image from the same sample of (c).

cleaning. Obviously, this technique combines the advantages of accurate positioning from FIB milling and low lattice damage from Ar milling, and is recommended for low dimension gate oxide sample preparation.

B. Low K/Ultra low K materials damage

With the value of dielectric constant K decreases to be lower than 3 and reaches 2.7 to 2.4, the distortion of low k materials in TEM images starts to appear and becomes a concern for process characterization. It sometimes gives rise to the ambiguity in the origin of some abnormal profiles; is it due to sample preparation/imaging or really due to processing condition.

This paper studies the effects of both sample preparation and imaging on low K/ultra low K materials distortion, as the both sample preparation and imaging can cause the degradation of the low K materials.

We notice that in SEM imaging different imaging conditions can cause the distortion of lowk/ultra low K materials to different extent. Therefore, we focus on the different SEM imaging/monitoring condition in FIB TEM sample preparation when we study the effects of TEM sample preparation. Three conditions are selected for comparison: dual beam FIB milling with live electron beam monitoring for end point detection, dual beam FIB milling with alternating SEM imaging/FIB milling, and FIB milling without SEM monitoring, i.e., single beam FIB. The results are shown in Fig. 3 (a), (b), and (c) respectively. When using live SEM monitoring, delamination between Ta/TaN and Cu (indicated by arrow) as well as some distortion are observed in the profile (Fig. 3(a)). In comparison, the samples prepared with alternative SEM monitoring (Fig. 3 (b)) and single beam (Fig. 3 (c))do not show delamination and obvious profile distortion. The results suggest that (i) the ultra low k materials are sensitive to electron illumination and (ii) the ultra low K materials are less sensitive to ion beam illumination. The first deduction is believed to be related to electron dose. Compared with alternative FIB milling/ SEM imaging, live SEM imaging while FIB milling uses higher beam current to achieve reasonable signal to noise ratio due to the interference from the FIB ion current. As a result, more electrons are pumped into the ultra low materials when doing live SEM imaging, causing it to deform more. The second deduction is related to the impinging angles of electron and ion beams with respect to the sample surface. The ion beam is parallel to the final sample surface, and most of the ions dislodge the sample atoms and are not left within the sample, while the electrons impinge the sample surface at an angle of 52° to 54°; significant portion of the electron will be trapped inside the samples, causing Joel heating and low k materials degradation. It should be noted, however, that the success rate for single beam sample preparation is low as end point detection is difficult when dealing small features in 45 nm technology node.

Another example of low K distortion is shown in Fig. 4 (a), where Metal A and Metal B are supposed to be at the same level and aligned horizontally. But the vertical distortion of the low materials causes the whole metal line profile distorted.

To study the effect of TEM imaging (high energy electron beam) on ultra low K materials distortion, a single beam FIB

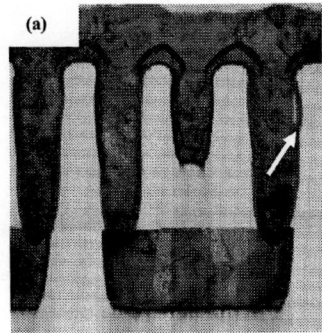

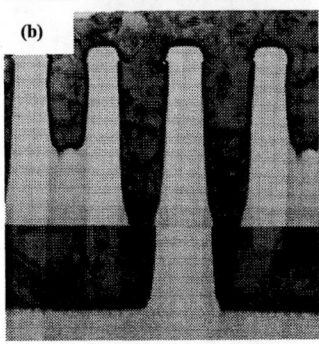

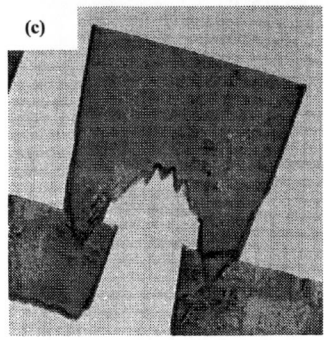

Fig. 3 Comparison of the effect of different FIB conditions on Ultra low K materials distortion; (a) dual beam FIB with live SEM monitoring; (b) dual beam FIB with alternative DEM monitoring; and (c) single beam FIB milling.

prepared TEM sample is illuminated with electron beam (200 kV) for different period of time, namely, 0, 1, and 60 minutes. A TEM image is taken after each period of time of illumination and the results are shown in Fig. 4 (b), (c) and (d) respectively. The results show that even after 60 minutes of illumination, there is no obvious distortion in low K materials. Obviously low K materials are not sensitive to high energy electron beam illumination.

There are two beam damage mechanisms in TEM, knock-on and Joule heating [4]. Knock-on damaging increases with electron beam energy, while Joule heating damage decreases with increased electron beam energy. Normally high atomic number elements such as metals are more prone to knock-on damage, while light element materials such as polymer is more prone to Joel heating. The results indicate that the dominant

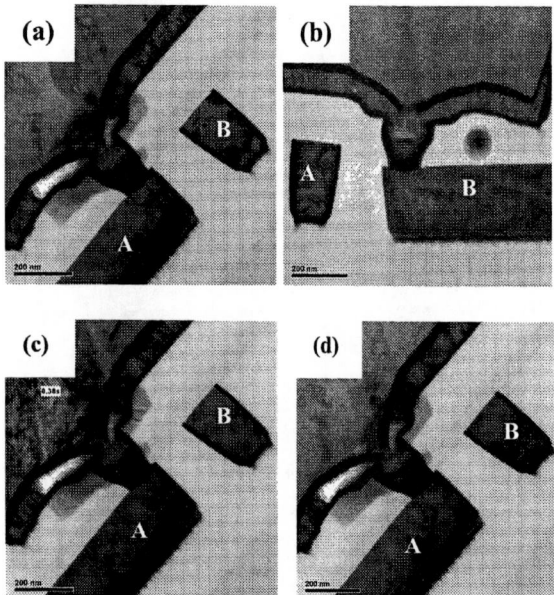

Fig.4 Effect of sample preparation and imaging conditions on sample distortion. Sample (a) is prepared by dual-beam FIB with live SEM. Samples (b), (c) and (d) are prepared with single beam FIB. The images (b), (c), and (d) were taken after 0, 1 and 60 min 200 kV electron beam illumination respectively.

beam damage mechanism for low k materials should be Joel heating.

IV. CONCLUSIONS

Physical characterization of 45 nm IC devices is very challenging due to the shrinkage of critical feature sizes to a level smaller than the thickness of a normal TEM sample and the incorporation of the ultra low K materials. To prepare site-specific ultra-thin gate oxide sample with minimum amorphization high-resolution ion beam and excellent low kV performance has to be used. To further improve the sample quality, post-FIB low energy argon ion milling is recommended. Low K materials distortion mainly happens during FIB TEM sample preparation. To minimize the distortion, proper SEM imaging protocol has to be applied. The key is to minimize the electron dose during sample preparation.

ACKNOWLEDGMENT

The authors would like to thank the TEM Sample Preparation Team of Physical Failure Group of Chartered Semiconductor Manufacturing Ltd. for their contributions in carrying out the sample preparations.

REFERENCES

[1] L.A. Giannuzzi and F.A. Stevie, "A review of focused ion beam milling techniques for TEM specimen preparation," *Micron*, Vol 30, pp.197-204, 1999.

[2] Q.Gao, M. Zhang, C. Niou, M. Li, W.T. K. Chien, "Experiment study on Crystal/Amorphous Structure of TEM Samples Prepared by FIB Milling, " in *Proceedings of the 32nd International Symposium for Testing and Failure Analysis (ISTFA-2006),* pp.76-78, November 2006.

[3] Z. Wanga,, T. Katoa, T. Hirayamaa, N. Katob, K. Sasakic, and H. Saka, "Surface damage induced by focused-ion-beam milling in a Si/Si p–n junction cross-sectional specimen", Applied *Surface Science*, Vol.241, pp. 80-86, 2005.

[4] D.B. Williams and C.B. Carter, **Transmission Electron Microscopy**, Plenum Press, New York and London, P63, 1996.

Understanding Soft Defect Localization Set-Points for Reducing Cause-Not-Founds in Integrated Circuits

V.K. Ravikumar, S.L. Phoa, V. Narang, J.M. Chin
Advanced Micro Devices Singapore Pte Ltd
508, Chai Chee Lane, Singapore 469032
Phone: (65) 67969888 Fax: (65) 62339080 Email: venkat-krishnan.ravikumar@amd.com

Abstract- **Soft Defect Localization (SDL) is an established fault isolation process for localizing soft defects using laser heating. This paper consolidates interesting analysis cases to identify good relation between SDL set-points and defect types that enables better approach in exposing such defects during Physical Analysis and reduce Cause-Not-Founds.**

I. INTRODUCTION

As devices shrink and speeds increase, subtle defects also known as soft defects gain prominence and result in reduced sort yield and reliability. Soft defects also result in reduced speeds and increased power envelops. Usually characterised to pass and fail at certain conditions of voltage, temperature and clock frequecies within operating specifications, these defects are extremely hard to detect through conventional physical failure analysis (PFA) techniques and go unobserved in a majority of cases, leading to high rates of Cause Not Found (CNF).

Soft Defect Localization (SDL) introduced early this decade by Bruce et al [1] has become an established Fault Isolation technique for Advanced Failure Analysis. With capabilities to detect sensitive circuits from design marginalities and defects induced by processing, the SDL technique has immense applications in dynamic Fault Isolation.

The technique is widely used world-wide along with the similar derivatives Resistive interconnection localization (RIL) [2], Laser Assisted device Alteration (LADA) [3], Dynamic Laser Stimulation (DLS) [4]. Seah et al [5] had published several interesting case-studies using SDL to successfully isolate and expose soft defects in AMD SOI microprocessor at the 90 nm node.

With the use of SDL, it is possible to reduce the area of interest to a small location. However, it is still dependent on the PFA technique chosen to be able to find the physical defect. The defects being subtle in nature often go undetected even after SDL based fault isolation has been obtained.

The aim of this work is to obtain additional information that benefits PFA from the electrical behavior of the DUT during SDL with a focus on processing defects and interconnect anomalies. With the help of a few case studies we elucidate the significance of gathering additional information from the die and combining the data with background information that can potentially reduce the cause not founds and explain the failure better.

II. SETUP AND REQUIRMMENTS

Hardware

SDL is primarily performed using an automated tester equipment (ATE) in tandem with a Scanning Optical Microscope (SOM). The Tester keeps the unit biased and the failing test pattern looping continuously. The pass/ fail information is passed to the SOM using a digital signal. A low signal indicates that the unit has passed that particular instance of the loop, and a high may indicate a fail. This information is mapped using the SOM to create an SDL image as it scans the exposed portion of the die.

The DUT is thinned down, mirror polished and loaded into the ATE. It is held in place by the cooling solution used. The cooling solution ensures that the DUT does not exceed temperatures beyond the operating specifications while being optically transparent, thus allowing imaging of the die.

The SOM consists of a scanhead that is equipped with 1340nm NIR laser. The advantages of using 1340nm laser in terms of its low absorption in lightly doped silicon and its ability to deliver the heat to interconnects is well known [3]. 1340nm NIR laser is used for both imaging and intrusive heating. Fig. 1 summarizes the hardware used in SDL. The SOM stores an optical image of the suspect location of the failure, also known as laser scanning microscope (LSM) image formed by the reflection of laser from the active circuitries. This image is overlaid with that obtained from the SDL analysis

Other requirements may involve a vibration platform which is useful when taking high magnification images and anti-reflective coating to improve the intensity of laser stimulation and imaging.

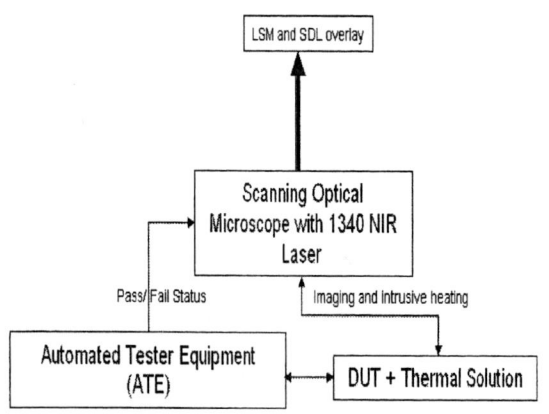

Fig. 1: Block diagram of hardware interactions in SDL analysis.

SDL Setup

In this paper we present the working of SDL and reasons to choose certain set-points for SDL. As mentioned before, soft defects are those that have a passing and a failing region in a shmoo plot as shown in Fig 2. The discussed set-points include parameters such as supply voltage and clock frequencies, which are common variables for speed characterization in any integrated circuit. More often than not, the unit has a failing region at lower voltages and higher frequencies and a strong passing region at higher voltages.

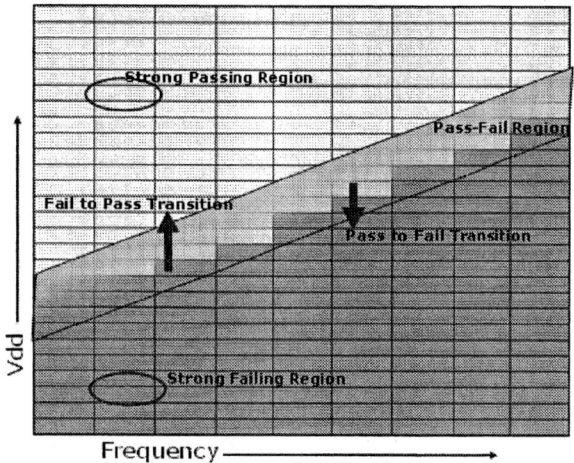

Fig. 2: Example of Shmoo indicating pass-fail transitions that occur during SDL analysis.

The failing test which is mostly functional in nature is exercised continuously at speed and the unit is biased at set-points close to the edge of the pass/fail region while under laser stimulation and controlled thermal envelope. The laser stimulation causes localized heating and results in a signal alteration sufficient to drive the DUT to a strong passing or a strong failing region. At every instance, the pass/fail information is sent to the SOM which creates the SDL image.

By controlling the laser power delivered, we can obtain a precise SDL location.

III CONSIDERATIONS FOR SDL BASED FAILURE ANALYSIS

Physical analyses on DUTs with soft defects are quite difficult because these defects are quite subtle and may go unnoticed if slightly different methods are undertaken to expose the defect. As a consequence, the cause-not-founds are extremely high even though precise SDL site has been obtained and the PFA direction should be addressed after considering background information and clues from the set-points.

Type of reject: A yield reject with known or suspected process issues like metal planarization or via underetch could have a different failing mechanism from a qualification reject which has been stressed to fail and both different from design related marginalities.

Temperature sensitivity: Temperature sensitivity is usually considered very important information for SDL analysis and for PFA. SDL uses laser to heat the exposed portions of the unit, and temperature sensitivity increases the confidence of an SDL site since heat causes circuits to change speed.

Simulator results: ATE testing with DFT features has enabled major portions of the die circuitry to be tested and the failing net instances accurately determined. Though the callouts might refer to multiple circuitries, they are usually the starting point for SDL. The instances have been simulated to cause the failure and hence any defect in that location can more easily be proven to cause the reported failure.

Intrinsic sensitivities and Technology nodes: As technology advances, new failure mechanisms (ex: Ni-silicide piping), and reliability issues (NBTI) are introduced, and SDL data should be interpreted rightly to expose the defect.

Information from set-points during SDL: As will be evident in the next section, there is some information that can be obtained from the electrical set-points. Defects can be classified as those that trigger the unit to become predominantly passing when the laser intrusively heats the location as compared to those that cause the unit to predominantly fail under similar conditions.

IV ELECTRICAL BEHAVIOR OF DUT DURING SDL

Set-Points are values of voltages and clock frequencies in the pass fail interface region where an SDL is performed and when no external stimuli is exerted the unit has an equal chance of passing or failing. When the laser hits the location causing the failure, the unit may pass or fail more depending on the result of interaction between heat and the electrical behavior of the defect. The result of this interaction can also be represented in terms of an increase or a reduction in speed of the bottleneck circuitry. Characterization of this interaction can lead the analyst to better predict the type of failure, and in turn select the best PFA approach and expose the defect.

We observe some dependence in the PFA approach to different SDL set-points especially when the defect lies in the Interconnects. It is clear that there are two types of transitions possible, namely pass-to-fail transition and the fail-to-pass transition.

Units that show pass-to-fail transitions are observed to typically fail from a reasonably strong passing state with the application of low power (3-10 mW) of 1340nm laser but are unable to pass from a reasonably failing state even with 50-70mW power of laser. The same scenario is observed in typical fail-to-pass units. This also means that the defect can either cause an increase or a decrease in speed of the circuit in contention, but not both.

Throughout the discussion that follows, we will focus on defects that lie in the interconnects. Models relating to interconnects are easier to assimilate since there are just two possible electrical defects, one being a resistive short and the other being a resistive open. A totally open or a dead short is more likely to create hard failure than a soft defect.

Heating with 1340nm laser is generally known to slow down active circuits of the 90 and 65nm nodes at typical operating voltages and frequencies. This generally translates to the shmoo shifting to the left, implying a larger chance to fail at that particular tested voltage and frequency.

Pass-to-Fail transition: A pass-to-fail transition is observed when heating the defect causes the speed of the failing circuitry to reduce. Assuming a metal short is the cause of the failing shmoo in Fig 2, SDL performed on this defect causes the device to fail at frequencies lower than the set-point. It can be better understood from the following case where SDL was performed on a built-in-self-test (BIST) reject.

The device had a pass-fail region at room and hot, with more fails at hot. Fig. 3 shows a pass-to-fail transition spot during SDL and physical analysis exposed shorts between the bit-line and ground within the array. This stringer defect could have caused a reduced speed when heating the location. Assuming metal expands on heating, the stringer initially having high resistance would temporarily have its resistance lowered, thereby increasing the leakage and causing the device to fail at a lower speed.

Fig. 4 shows another case of a pass-to-fail SDL localization and the corresponding PFA result. This particular sample from a 65nm qualification failed scan tests with a pass-fail region at room temperature and hotter temperatures. SDL was performed at around 1 V of supply voltage and 100MHZ core frequency

which set the unit in a good passing region. When it was heated with the laser near the callout circuitry, a good Pass-to-fail SDL site was observed.

In both cases, a top-down deprocessing approach was adopted which exposed shorts between two metal lines (dual damascence) in Fig. 3 and contact shorts in Fig. 4. Using a cross-sectional approach on the other hand would have caused a long FIB (focused ion beam) time and the defect could have been milled away within a few cuts.

There is a possibility that heat on the other hand, causes an increase in resistance of the shunt circuitry and in which case we might observe a fail-to-pass transition. In our observations hitherto however we have primarily observed pass-to-fail set-points characteristics and hence propose the mechanism described above

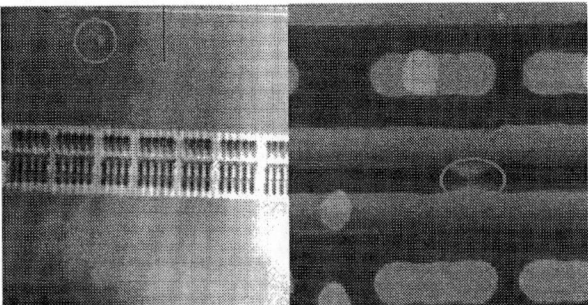

Fig. 3: SDL showing pass-to-fail transition at the failing bit location and the image on the right showing a Top-Down deprocessed image of a short between two metal lines.

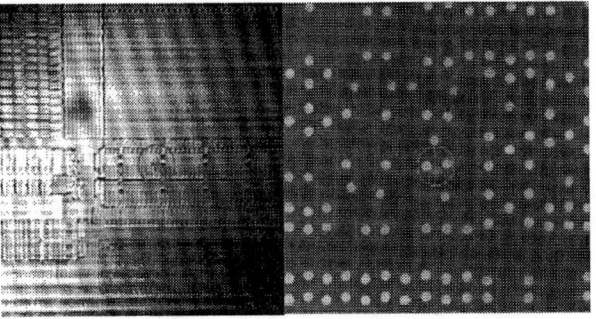

Fig. 4: SDL showing Pass-to-Fail transition at the callout, and the right showing the top-down deprocessed image of a short between two contacts.

Fail-to-pass transition: A fail-to-pass transition is on the other hand observed when heating the defect causes the speed of the failing circuitry to improve. A metal void for example, has higher than normal resistance. When the laser heats up the void, the void closes up partially causing the metal line to conduct better with a lower resistance than before. This would cause the unit to start passing as the void was the high-resistance bottleneck of the circuitry.

From our analysis, we have observed that defects typically having fail-to-pass signatures are more likely to be open or partially open conductors. This observation is useful in deciding the PFA direction. A cross section approach would then be recommended as compared to a top down delayering approach which might miss the open conductor.

Fig. 5 and Fig. 6 show cases of fail-to-pass SDL localization and the corresponding PFA results. These units were also from similar qualifications as the unit in Fig. 3 and had a pass/fail region in room temperature. The device in Fig. 5 had a good pass shmoo plot at hot while Fig. 6 was still marginal at hot. The SDL conditions set the unit to a stable failing region and the laser heating switched the state to a passing region. PFA on the SDL location showed an open conductor through FIB cross-sectioning in Fig. 5 and missing silicide in Fig. 6.

Fig.5: SDL showing fail-to-pass transition at the callout location, and the image on the right showing a FIB cross-section showing a broken contact.

Fig.6: SDL showing fail-to-pass transition at the callout location, and the image on the right showing a FIB cross-section exposing missing silicide.

V. CONCLUSIONS

This work is published with an aim of understanding better the electrical behavior of defects during Soft Defect Localization. Cases put forward here indicate that, along with test and characterization knowledge, there is a definite dependence between the SDL set-points and the defect characteristics. Though the focus of this work is interconnect oriented, it is understandable that transistor related defects too would show relations between defect type and electrical behavior during SDL. Realization of the possible nature of defect and therefore selecting the correct approach before actual physical analysis greatly increases the chances of finding these defects.

ACKNOWLEDGEMENT

We would like to thank the Device Analysis laboratory of Advanced Micro Devices Singapore for sharing their analysis work. We would also like to thank Mike Bruce of AMD Austin for his useful suggestions which made this paper more complete.

REFERENCES

1. M. Bruce, V. Bruce, D. Eppes, J. Wilcox, E. Cole, P. Tangyuanyong, C. Hawkins, "Soft Defect Localization (SDL) on ICs". Proc.28th International Symposium for Testing & Failure Analysis, 2002, pp. 21-27

2. E. Cole, P. Tangyuanyong, C. Hawkins, M. Bruce, V. Bruce, R. Ring, WL Chong, "Resistive Interconnection Localization" Proc.27th International Symposium for Testing & Failure Analysis, 2001, pp. 11-15

3. J. Rowlette, T Eiles, "Critical Timing Analysis in Microprocessors Using Near-IR Laser Assisted Device Alteration (LADA)", International Test Conference 2003 Proceedings, pp. 264-73

4. F. Beaudoin, K. Sanchez, P. Perdu, "Dynamic laser stimulation techniques for advanced failure analysis and design debug applications". Microelectronics Reliability Vol 47 (2007), Pg. 1517–1522

5. Seah, Y.X. Palaniappan, M. Chin, J. M. Applications of Soft Defect Localization (SDL) on AMD Advanced SOI Microprocessors, International Symposium for Testing & Failure Analysis 2006, VOL 32, pages 311-31

Gate Oxide Integrity Failure Caused by Molybdenum Contamination Introduced in the Ion Implantation

D. Gui, Y.H. Huang, G.B. Ang, Z.X. Xing, Z.Q. Mo, Y.N. Hua and J. Teong

Chartered Semiconductor Mfg Ltd
Woodlands Industrial Park D, Street 2, Singapore 738406
Phone: (+65) 63604378.   Fax: (+65) 63624592.   Email: guid@charteredsemi.com

Abstract- **The gate oxide is the most fragile element of metal-oxide-semiconductor (MOS) transistor. Metal contamination is fatal to gate oxide integrity because metallic contamination in the gate oxide leads to high leak current and even gate oxide breakdown. In this paper, a case of gate oxide integrity failure was investigated. The fact that the emission spot locates in the bulk area instead of poly edge excludes the possibility of failure cause of poly/spacer etch, gate oxide thinning at the edge or gate oxide thinning at STI corner issue. Molybdenum (Mo) contamination was detected in the gate oxide of NMOS using magnetic sector secondary ion mass spectrometry because of its excellent sensitivity. The results showed that Mo contamination is introduced in the process of germanium pre-amorphization implantation mainly in the form of $(^{98}Mo^{12}C^{19}F_2)^{++}$, which has the same nominal mass to charge ratio as $^{74}Ge^+$. The different properties of poly-Si resulted in that NMOS but not PMOS was affected by the Mo contamination. Unlike iron (Fe) contamination, Mo contamination is rarely reported in the GOI failure. On the basis of that, suggestions have been proposed to greatly suppress the Mo contamination.**

I. INTRODUCTION

Complementary metal-oxide-semiconductor (CMOS)-based silicon process has been the mainstream of ultra-large scale integration (ULSI) technology due to its low power consumption [1]. As its name states, MOS has a unique element, 'oxide' namely 'gate oxide' here, distinguishing itself from other types of transistor structures, e.g. bipolar transistor. The gate oxide is the most fragile element of a transistor because it is an extremely thin layer of material biased during normal operation. As the transistor size continuously shrinking, the reducing of the gate oxide thickness causes the oxide to be more susceptible to gate current leakage, defect creation, and ultimately dielectric breakdown. Among the root causes of the gate oxide failure, metal contamination is the most common one and fatal to the integrity of gate oxide. The presence of metallic contamination has been well known to have a negative impact on the performance of MOS device, e.g. increasing the leak current and reducing the minority carrier lifetime as well as gate oxide quality [2].   The transition metals are one major contamination in the wafer fabrication because of their high diffusion coefficient [3] and the fact that transition metals, e.g. molybdenum (Mo) and tungsten (W) etc., are commonly employed in the process equipment. To improve the line yield and device reliability, it is necessary to monitor and determine the metallic contamination including the transition metals in the starting materials and the processes, which requires a diverse array of process characterization tools. Among them, secondary ion mass spectrometry (SIMS) is a critical tool of characterizing the dopant profile and contamination due to its excellent sensitivity and high depth resolution [4]. However, SIMS faces a big challenge in small area analysis for IC failure analysis. The test area is usually the test pad located in the scribe line or other structure with smaller size. To scan the primary beam over the small test pad, the spot size of the primary beam should be much smaller than the test area, which will limit the primary beam current. Furthermore, the small volume of material limits the total available atoms of trace elements. These two factors result in poor sensitivity in small area SIMS analysis, compared to large area analysis in blanket wafers [5].

In this paper, we will present the investigation on a failure case of early breakdown of NMOS gate oxide.   Emission microscope was used to isolate the GOI defect. SIMS depth profiling was performed at electrical test (E-test) pad located in the scribe line using Cameca Wf SIMS machine. The root cause has been identified as Mo contamination in the gate oxide. Further study showed that the Mo was introduced in the process of germanium pre-amorphization (PAI) implantation. The mechanism of Mo contamination has been discussed.   The measures to improve the process have been proposed on the basis of the SIMS results. Unlike the other metal contaminants, e.g. iron (Fe) etc., Mo contamination is rarely reported in the GOI failures.

II. EXPERIMENTAL

The product wafers were sorted as 'good' or 'bad' wafers by the end of line (EOL) E-test. Emission microscope SEMICAPS SPEMS 1350 with a C-CCD detector was employed for isolating the GOI defects. SIMS depth profiling was carried out on CAMECA IMS Wf system. Before SIMS experiment, the wafers were delayered to expose poly Si layer using diluted hydrofluoric acid (HF). A primary O_2^+ beam with 5 keV net impact energy was used to obtain metal contaminant profiles from poly Si till Si substrate.   To avoid any mass interference to Mo, a technique of energy window shift was employed to reduce the Mo background. The SIMS depth profiling was carried out in the in-line measurement pad structures located in the scribe line. Therefore, the primary beam was confined to scan over an

978-1-4244-2039-1/08/$25.00 ©2008 IEEE

area within 50μm x 50μm. The secondary ions were detected at the centre part of the scan area with a diameter of 12 μm to avoid the crater effect.

Short loop blanket wafers were depth profiled to investigate the effect of arc current of implanter ion source on the Mo concentration. The scan area and detection area are 200μm x 200μm and 62μm in diameter, respectively. The depth scale was determined using surface profilometer KLA Alpha Step IQ. And the relative sensitive factor for Mo was calibrated using a Mo implanted standard sample with a known dose.

III. RESULTS AND DISCUSSION

The product wafers were sorted as 'good' or 'failed' by at the end of line (EOL) electrical test (E-test), in terms of gate oxide breakdown voltage. All failed wafers have a signature that the breakdown voltage of NMOS gate oxide is very low and PMOS gate oxide works well. This phenomenon implies that the failure is unlikely due to the gate oxide growth process. IV curve tracing performed on the ET GOI N-bulk capacitor showed early breakdown on these failed sites, as shown in Fig.1.

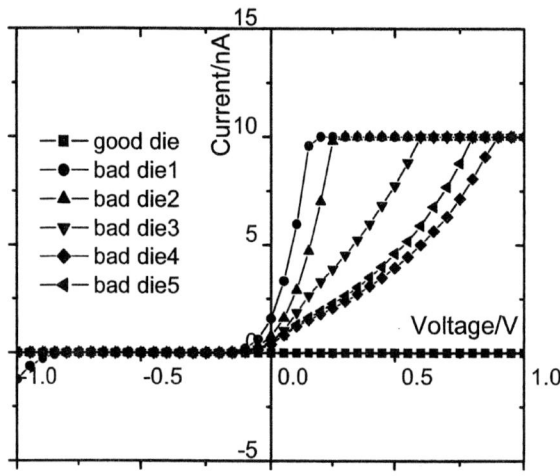

Fig. 1. IV curves of GOI N-bulk capacitor.

Fig. 2. Backside emission at GOI N-bulk capacitor. The emission spot is pointed by the arrow.

To isolate the GOI defect, emission microscope was employed to analyze the GOI N-bulk capacitor. Due to the fact that emission was not able to penetrate through the metal bondpads on top of the poly capacitor plate, emission analysis was performed from backside. As seen from the Fig. 2, emission spot located at the center of the capacitor structure, which indicates that the cause of the GOI failure was not due to poly/spacer etch, gate oxide thinning at the edge or gate oxide thinning at STI corner issue. In other words, the failure may be caused by contamination.

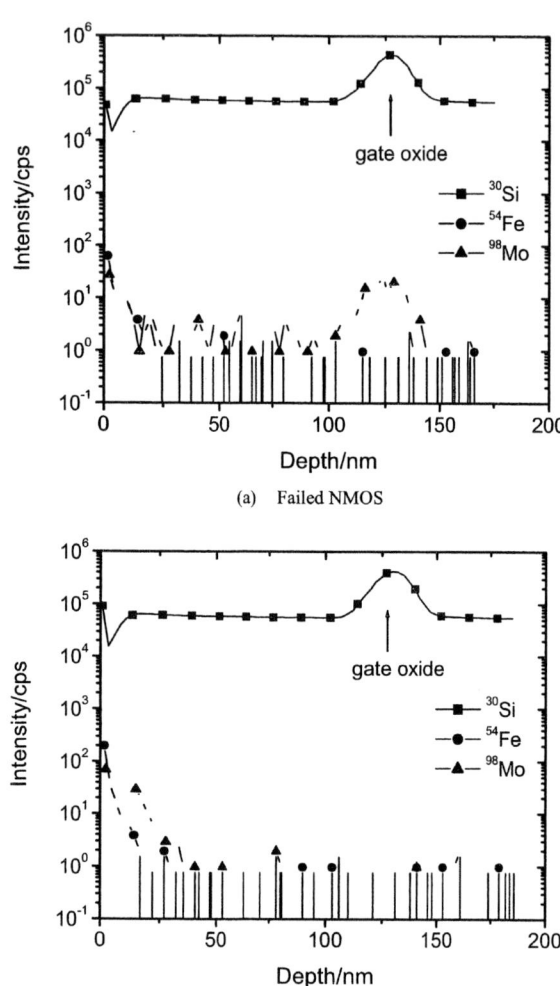

(a) Failed NMOS

(b) Failed PMOS

Fig. 3. SIMS depth profiles of failed sample. Mo contamination was detected in gate oxide of failed NMOS.

To confirm whether the contamination is the root cause, SIMS was used to depth profile the GOI bulk capacitors of both good and failed samples. Before SIMS experiment, the product wafers were delayered to expose poly crystalline silicon (poly-Si) layer using diluted hydrofluoric acid (HF). The SIMS depth profiles of good and failed wafers are shown in Fig. 3 and

4. Thanks to the enhance effect of oxygen on positive ion yield, Si peak can serve as the marker of gate oxide. It is evident that there is Mo contamination in the gate oxide region of failed NMOS, whereas no obvious Mo contamination was detected in PMOS of failed die. Fe concentration is below the detection limit for both categories. It is notable that $^{54}Fe^+$ instead of $^{56}Fe^+$ was detected for better detection limit because $^{56}Fe^+$ is interfered by $^{28}Si_2^+$, which has the same nominal mass to charge ratio.

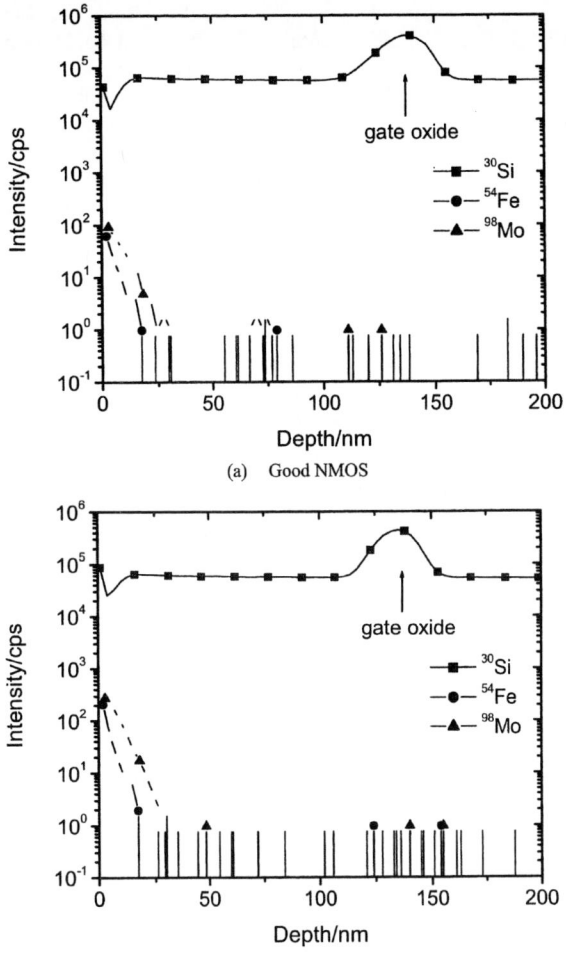

(a) Good NMOS

(b) Good PMOS

Fig. 4. SIMS depth profiles of good sample. No obvious Mo and Fe contamination were detected in gate oxide.

The tool commonality study showed that all failures have gone through one implanter, which runs germanium pre-amorphization implantation (PAI). Ge source is GeF$_4$ and the source arc chamber is made of Mo. The implanted specie was selected as ^{74}Ge, which is the most abundant isotope of Ge. Mo can be co-implanted into Si wafer in the form of $(^{92}Mo^{19}F_3)^{++}$ and $(^{98}Mo^{12}C^{19}F_2)^{++}$, with the mass to charge ratio of 74.5 and 74, respectively, which are very close to or the same as ^{74}Ge. It is possible that Mo can be co-implanted into the wafer because the resolution of implanter's mass filter is not high enough [6].

Further investigation of the effect of arc current in Ge PAI on Mo contamination was carried out to confirm whether the Mo contamination is introduced in the implantation process rather than gate oxide growth. The Ge PAI was made with different arc current intentionally. The typical Mo profile is shown in the Fig. 5. The Mo doses are summarized in the Table 1. It can be seen that the Mo dose increases with the arc current. The total dose of ^{92}Mo and ^{98}Mo increases from 2E10 at/cm^3 to 7.5E11 at/cm^3. The high arc current results in more free F radicals, which enhances the erosion of Mo arc chamber. The results support the conclusion that Mo is indeed introduced in the Ge PAI process.

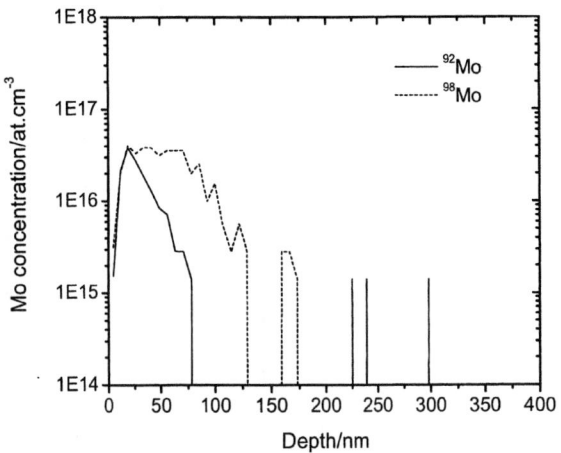

Fig. 5. SIMS depth profiles of simulation wafer.

Table I
MO CONTAMINATION INTRODUCED WITH DIFFERENT ARC CURRENT

arc current	Dose/ats.cm^{-2}	
	^{92}Mo	^{98}Mo
low	8.72E+09	1.20E+10
medium	9.81E+10	1.35E+11
high	3.42E+11	4.12E+11

The unaffectedness of PMOS can be explained by the different poly-Si properties. For NMOS, the Ge PAI was made before n-type lightly doped drain (NLDD) annealing, whereas the Ge PAI was made after NLDD annealing. The grain size of un-annealed poly Si was smaller, which is favorable for Mo to diffuse at subsequent thermal process and to segregate at the gate oxide because the grain boundary provides fast diffusion path. On the contrary, Mo diffusion is less in P-poly because the grain size became greater after annealing. Furthermore, Boron gettering effect in P-poly is helpful to block Mo from segregating to gate oxide region.

Based on the mechanism of Mo contamination, several measures can be taken to reduce the risk of Mo contamination. Firstly, use W arc chamber instead of Mo Chamber to eliminate the Mo contamination source. Secondly, change the implant specie from ^{74}Ge to ^{72}Ge to avoid the mass interference of Mo contained clusters. Thirdly, change the source from the gas source of GeF$_4$ to solid Ge source to avoid forming the interfering cluster of $(^{92}\text{Mo}^{19}\text{F}_3)^{++}$ and $(^{98}\text{Mo}^{12}\text{C}^{19}\text{F}_2)^{++}$. Last but not least, reduce the arc current to lower the corrosion of Mo chamber, if aforementioned measures cannot be taken.

IV. CONCLUSION

In this paper, we have investigated a case of GOI failure, with a signature that the breakdown voltage of NMOS gate oxide is very low and PMOS gate oxide works well. The fact that the emission spot locates in the bulk area of GOI test structure excludes the possibility of poly/spacer etch issue, gate oxide thinning at the edge or gate oxide thinning at STI corner. With SIMS analysis in GOI test pad in scribe line, the root cause of the failure has been concluded as Mo contamination introduced in Ge PAI process, which is rarely discussed in previous literatures. The mechanism of Mo contamination has been further discussed. The different properties of poly-Si resulted in that NMOS but not PMOS was affected by the Mo contamination. On the basis of that, measures have been proposed to suppress the Mo contamination, i.e. switching from Mo arc chamber to W arc chamber, from ^{74}Ge to ^{72}Ge, from the gas source of GeF$_4$ to solid Ge source, and/or reducing the arc current.

ACKNOWLEDGMENT

The authors would like to thank Mr. Jun Zhang and Mr. Cheng Kit Pang with Chartered Semiconductor Manufacturing Ltd for providing the samples and useful discussion.

REFERENCES

[1] P. Chatterjee, "ULSI market opportunities and manufacturing challenges", *IEDM Tech. Dig. 11* (1991)

[2] S.Q. Hong, T. Wetterroth, H. Shin, S.R. Wilson, D. Werho, T.C. Lee and D.K. Schroder, " Improvement in gate oxide integrity on thin-film silicon-on-insulater substrates by lateral gettering", *App. Phys. Lett.* Vol. 71(23), p. 3397 (1997)

[3] D.L. Kwong, "Si device processing", in R. W. Cahn (ed.) Materials science and technology: a comprehensive treatment, Vol. 16 Processing of semiconductors / Vol. ed.: K.A. Jackson, VCH Verlagsgesellschaft, Weinheim, 1996

[4] F.A. Stevie, R.G. Wilson, D.S. Simons, M.I. Current and P. C. Zalm, "A Review of SIMS Characterization of Contamination Associated with Ion Implantation," *J. Vac. Sci. Technol. B,* vol. 12, 2263-79 (1994)

[5] H. Ogata, "Gigabit devices and technologies: SIMS challenges of the near future, in *Proceedings of SIMS X*, 1997, p. 3

[6] D. Gui, Y.N. Hua, Z.X. Xing and C.W. Tan, "SIMS Analysis of Boron Cross Contamination in Al Implantation Process of Wafer Fabrication", Proceedings of IEEE NSM 2005, Kuching, Malaysia, November 21st- 24th, 2005, p.116

Tool Cleanliness Characterization for Improving Productivity and Yields

Victor K.F. Chia

Balazs NanoAnalysis - Singapore
15 Gul Drive, Singapore 629466
Phone: (65) 9236 6258   Email: victor.chia@balazs.com

Abstract-**Tool cleanliness is a prerequisite for increased production ramp, reduced tool down-time and high process yields. Any materials used in the build of material (BOM) must be verified to be clean both in the bulk of the material and on its surface after machining and cleaning. Bulk analyses are destructive due to the nature of the test and the information required. In contrast, surface analyses should be non-destructive so the part may be re-used after surface cleanliness testing or re-cleaned if the test indicates the part does not pass its cleanliness specification. Wafer tool specifications are in place for particles and metals by the original equipment manufacturers (OEMs). Particle specification depends on the method of wafer clamping and wafer metal specification for tool acceptance is typically in the range of $1\text{-}5\text{x}10^{10}$ atoms/cm^2. No generally accepted organic specifications are in place. Currently, there are no accepted tool parts particle, metal and organic specifications. Very few machine shops, cleaning houses, OEMs and fabs have developed baseline tool parts cleanliness specifications. This paper describes key analytical techniques for bulk and surface characterization of tool parts.**

I. INTRODUCTION

In the sub-100 nm technology node even irreducible differences in the components of identical tool chambers can influence yield and mean time between failures (MTBF). Advanced process control is required to minimize systematic and random variability in 100s of active tool parts or build of material (BOM). Tool cleanliness is a prerequisite for clean processing; it is an invisible condition that can change the integrity of the wafer surface during processing. The overall equipment productivity in fabs is only ~60%. Tool downtime relating to contamination issues includes unscheduled equipment stops, wafer tests, equipment PM, equipment set up time and equipment start-up standby time. The current focus is to increasing profit margin that requires fabs to maximize yield and increase overall equipment productivity and wafer throughput. This places greater emphasis on having cleaner tools and deliberate selection of BOM is essential because any surface and bulk contamination is a contamination source. In addition, a systematic approach using advanced characterization techniques must be applied when an escalation occurs to quickly identify and resolve the tool contamination.

II. MATERIAL SELECTION AND BULK CLEANLINESS

A complete characterization of starting materials used to machine tool parts is necessary. Bulk contamination can migrate

to the surface during thermal treatment or after many cleaning cycles that involves the removal of the material. Materials such as ceramic, quartz and O-ring vary in bulk cleanliness by vendors and by lot-to-lot. Traditional O-rings with inorganic fillers such as SiO_2, $BaSO_4$, ZnO, Carbon or TiO_2 will shed metallic particles onto the wafers after ~6,000 wafer counts. Laser ablation ICP-MS (LA ICP-MS) can determine bulk metal composition so a low metallic concentration filler O-ring can be selected for use. Alternatively, an O-ring using organic filled material can be used to extend MTBF (mean time between failures) to 20,000 wafer counts. Ceramic and quartz parts must be bulk characterized to ensure the bulk contaminants are present at low concentrations. The reason is these contaminants will eventually become near or at the surface of the tool part after many cleaning cycles or after extended plasma etching and onto the wafer surface during processing. Figures 1 and 2 show LA ICP-MS profiles of ceramic and quartz analyzed to several microns in depth. Quality differences of materials provided by vendors have been revealed by LA ICP-MS and demonstrated to be the root cause of wafer contamination during processing.

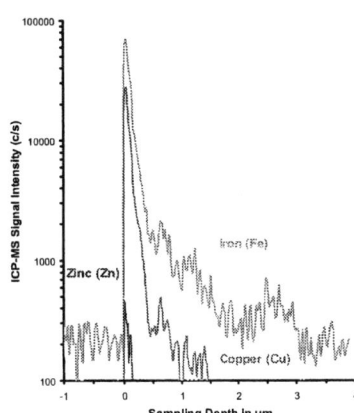

Fig. 1. Laser ablation ICP-MS depth profile of ceramic

Machined parts are likely to have major surface and sub-surface contamination from machine lubricant oil, metal cross-contamination from drill bits, water and solvent residues from rinsing and contamination from the oven during thermal treatment to relieve stress. Contamination on machined parts may be ranked in importance as Organic > Particle > Metal >

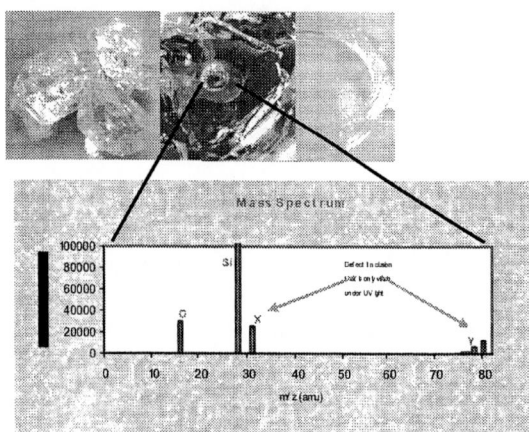

Fig. 2. Laser ablation ICP-MS bulk elemental survey of quartz

Anion. These machined parts will be precision cleaned resulting in minor surface contamination remaining typically from handling, cleanroom environment and packaging. The contaminants of concern are Metal > Particle > Organic > Anion. It is important when making a decision for material selection and component design that both functionality and its cleanliness requirement are taken into account. A simple material contamination cycle is shown in Figure 3.

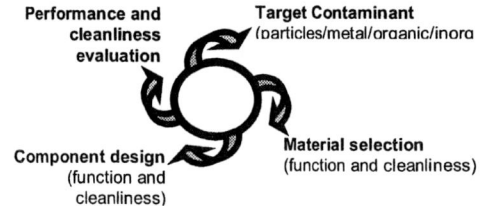

Fig. 3. Material contamination reduction cycle

III. TOOL PARTS CHARACTERIZATION

Table I shows the Chemical Surface Test Methods that are non-destructive so the part may be used in the tool after it has passed cleanliness qualification without additional cleaning. Common test methods used are summarized in the table.

Table I
CHEMICAL SURFACE TEST METHODS
(NON-DESTRUCTIVE)

SEMICONDUCTOR PROCESS		Acid extraction & ICP-MS	1
	Metal	UPW extraction & ICP-MS	2
Wafer Production Thermal Oxidation/Film Photolithography Etch Doping/Ion Implant Dielectric Deposition CMP		Drop scan etch & ICP-MS	3
	Organic	Solvent extraction & GC-MS	4
		Solvent extraction & NVR/FTIR	5
	Ionic	UPW extraction & Ion Chromatography	6
	Particle	UPW extraction & LPC (SEM-EDS)	7

1. Metal - whole surface extraction
2. Metal - extraction efficiency less than acid
3. Metal - localized surface acid extraction
4. Organic - solvents to extract organic residue and UPW/TOC
5. Organic - weight of NVR and organic identification
6. Ionic - whole surface extraction
7. Particle - whole surface particle counting and identification

Table II shows the Physical Surface Techniques that are destructive and used primarily on coupons in R&D of cleaning recipes and material compatibility studies, first articles and if absolutely necessary on real parts. Common test methods used are summarized in the table.

Table II
PHYSICAL SURFACE TEST METHODS
(DESTRUCTIVE)

SEMICONDUCTOR PROCESS		AES	1
	Metal	TXRF	2
		VPD ICP-MS	3
Wafer Production Thermal Oxidation/Film Photolithography Etch Doping/Ion Implant Dielectric Deposition CMP		SIMS	4
		TOF-SIMS	5
	Organic	Full Wafer Outgassing TD-GCMS	4
		TOF-SIMS	5
		XPS	6
	Ionic	XPS	7
	Particle	FE-AES	8

1. AES: 30-50 Å, at% DL, elemental survey, conducting surface
2. TXRF: 30-50 Å, 109-1015 at/cm^2, elemental survey, flat surface
3. VPD ICP-MS: SiO_2, 10^7-10^{15} at/cm^2, elemental survey
4. SIMS: any depth, 10^8-10^{15} at/cm^2, elemental specific
5. TOF-SIMS: monolayer, 10^7-10^{15} at/cm^2, elemental survey, any surface

6. Full Wafer Outgassing: ng/cm^2, organic survey on selected wafer surface
7. TOF-SIMS: monolayer, ng/cm^2, organic survey, any surface
8. XPS: 30-50 Å, at% DL, elemental/chemical state survey, non-conducting surface
9. XPS: 30-50 Å, at% DL, elemental/chemical state survey, non-conducting surface
10. FE-AES: 10 nm spatial resolution for elemental characterization

Additional surface techniques frequently used include:

- UV (black) light: visual inspection for residue polymer on the surface
- Profilometry: surface roughness and surface layer thickness

IV. CASE STUDY

After a weekend PM the base pressure increased and Cl was detected at $5x10^{13}$ at/cm^2 instead of $5x10^{10}$ at/cm^2 by TXRF. All metal concentrations measured by TXRF were at or below $5x10^{10}$ at/cm^2 that was the wafer cleanliness specification. Chlorine on the wafer was determined to be from an organo-chloride compound using Full Wafer TD GC-MS. Chlorine was not from insufficient rinsing after acid cleaning that included the use of HCl since no residue ionic Cl was detected on the wafer surface from UPW extraction and ion chromatography of the extract aliquot.

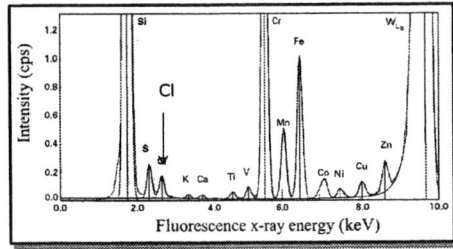

Fig. 4. TXRF spectra of wafer processed in the tool

Interestingly, a static wafer that was left on the ESC for 1h in the tool showed no Cl by TXRF whilst a dynamic test with a wafer cycled through the tool showed surface Cl at $5x10^{13}$ at/cm^2. This experimental observation will reveal its significance once we identify the contamination source.

The organo-chloride compound was identified as a common flame retardant. Flame retardants are often used on foam cushions, sofas and beds to prevent them from catching on fire. After reviewing the BOM, the root source of the organo-chloride compound was eventually traced to a vibration isolation pad that was blue in colour. The BOM specified a black vibration isolation pad that does not outgas. This was confirmed by outgassing the blue and black isolation pads by

ATD GC-MS using the industry standard method, IEST WG-CC031. The outgassed organic compounds from the blue isolation pad matched the wafer outgassed organic signature.

The static and dynamic wafer test results now become clear. No water leaks if you hold a wet sponge. However, if you squeeze the sponge lightly or even shake the sponge with your hand some water will leak out. This is the case with the vibration isolation pads. When the wafer handler is static the pads are not active and do not outgas. In contrast, when the wafer handler is moving and transporting the wafer the pads are adsorbing any vibrations produced and in the process they will compress and outgass. So, even though the design specification for vibration insulation was met using the blue pad its bulk properties was not investigated resulting in a contamination escalation.

V. CONCLUSION

All tool components and parts must be designed using materials that are compatible to both its function and cleanliness. This means individual tool parts in the completed build tool must have cleanliness specifications for its technology node. The smaller the technology node the cleaner the tool must be. One way of establishing a parts cleanliness baseline is to select a tool that passes all particles, metal, ionic and organic wafer testing. If the tool passes theses wafer acceptance tests then the individual part cleanliness is likely to be acceptable too. This paper describes the test methods for surface and bulk material characterization. Most importantly the surface test methods are non-destructive and when carried out with meticulous care the part may be packaged with a Certificate of Analysis (CoA) and returned to the end user for installing into the tool. The case study illustrates the consequence of not having a part cleanliness specification. OEM, fabs and their supply chain operating with parts cleanliness specifications will maximize their yield and increase overall equipment productivity and wafer throughput which in turn will increase their profit margin.

Studies and Applications of Standardless EDX Quantification Method in Failure Analysis of Wafer Fabrication

Hua Younan, Liu Binghai, Mo Zhiqiang & Teong Jennifer
Chartered Semiconductor Mfg Ltd
Woodlands Industrial Park D, Street 2, Singapore 738406
Tel: 65-63604154, Fax: 65-63622935 Email: huayounan@charteredsemi.com

Abstract-Energy-dispersive X-ray microanalysis technique has been widely used in failure analysis of wafer fabrication. However, we still face some common problems. In this study, we will introduce standardless element coefficients to improve accuracy of quantitative results, propose an estimating method to select beam acceleration voltage & demonstrate application cases and discuss identification methods of the spectra overlapped. The accuracy of EDX results have been improved, for example, the relative error Si3N4 layer is reduced from (-67.5% ~ +101%) to (-2.83% ~ +3.93%).

I. INTRODUCTION

Energy-dispersive X-ray microanalysis technique (EDX) has been widely used to determine the compositions of particles, contamination and thin film layers in wafer fabrication. However, we still face some common problems such as how to further improve accuracy of quantitative EDX results, especially for thin films; how to select electron beam acceleration voltage to make a balance between exiting the characteristic X-ray lines and controlling beam penetration depth to identify thin film layers and how to identify the spectra overlapped etc. In this paper, we will study and standardize EDX analysis parameters & methods and introduce standardless EDX quantification method with SEC (Standardless Element Coefficient) factors to improve accuracy of EDX results; introduce an estimating method of electron beam acceleration voltage for different characteristic X-ray lines of elements and demonstrate application cases, and discuss identification methods of the spectra overlapped. The application results show that after performing the calibrations and corrections, EDX results have been greatly improved, for example, the relative error of the passivation layer of Si_3N_4 is reduced from (-67.5% ~ +101%) to (-2.83% ~ +3.93%).

II SELECTION OF ELECTRON BEAM ACCELERATING VOLTAGE

During EDX analysis of wafer samples, which are with more layers, we have to select suitable electron beam acceleration voltage so as to identify elemental information of the surface layer from the underneath layers. For example, it is difficult using EDX to determine TiN residue on Al bondpads with Al/TiW/Ti metallization (top is Al layer, underneath is TiW & Ti barrier layers) due to the limitations of EDX technique.

When we determine TiN by detecting Ti Kα peak (4.510 keV), Ti Kα peak cannot be excited by using a 5 kV beam accelerating voltage. A higher kV beam acceleration voltage should be used to excite Ti Kα peak (4.510 keV). However, when a higher beam accelerating voltage is used, another problem raised is that the penetration depth of the electron beam will be greatly increased. As results, the signal from the underneath TiW & Ti barrier metal layers will be also detected. This may make confusion in identification of Ti Kα peak signal detected. In the other word, it is unable to identify whether Ti Kα peak signal is from the TiN residue on surface of bondpad, or from the underneath TiW/Ti barrier metal layers (Figure 1). Therefore, in failure analysis, it is necessary for us to select the beam accelerating voltage used and estimate the penetration depth of the electron beam under this accelerating voltage before performing EDX analysis.

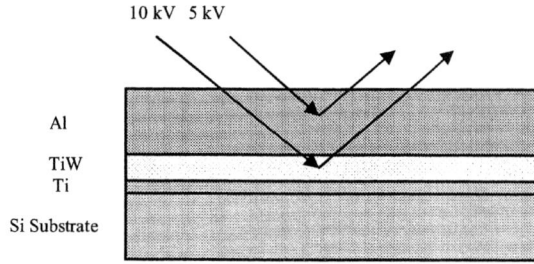

Figure 1. Diagrammatic sketch showed cross section of a microchip Al bondpad with Al/TiW/Ti metallization and penetration of the electron beam on Al bondpad using different beam acceleration voltage. When using 5 kV, the beam only penetrates in Al layer. However, if using 10 kV, the beam will penetrate to underneath TiW/Ti layer.

From EDX theory, we can obtain a simple method to estimate electron beam accelerating voltage. In EDX analysis, we have to approximately estimate the beam accelerating voltage used for an interesting element. In this study, Ti Kα line is taken as an example. From theory of EDX, the energy of the characteristic X-ray line Ti Kα is 4.510 keV, and its critical excitation energy is 4.965 keV and that corresponds to a critical wavelength (λ):

$$\lambda = c/\nu = hc/E = 12.398/E \qquad (1)$$

where λ is in Å and E is in keV.

In this case, the critical excitation wavelength of the K shell Ti Kα lines is 2.497 Å. This means that the excitation wavelength from the excitation source must be shorter than 2.497 Å. On other hand, we can also obtain the short wavelength limit of an excitation source. According to Duane-Hune relation [1]:

$$h\nu_0 = hc/\lambda_0 = eV_0 \qquad (2)$$

This expression signifies that the energy of any electromagnetic radiation in the emission spectrum can never be higher than the kinetic energy of the electrons. Equation (2) is more conveniently written into:

$$V_0 = 12.398/\lambda_0 \qquad (3)$$

where λ_0 is in Å anV_0 is in kV. For convenience, if we use the energy of the characterization X-ray line (E_i) to replace $12.398/\lambda_0$ in Equation (3) and add a k as the constant of the over-voltage factor, then we can obtain a simple estimation formula:

$$V_i = k\,E_i \qquad (4)$$

In Eqn (4), k is the over-voltage constant. The unit of Vi is in kV and Ei is in keV. If we find the value of the k, we can quickly estimate the actual beam accelerating voltage (Vi) used for a characteristic X-ray line interested (Ei).

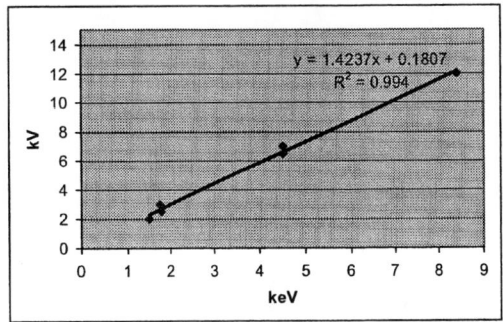

Figure 2. The graph of the beam accelerating voltage vs. the energy of the characteristic X-ray line of elements interested. It is a liner relation and the gradient of the graph (over-voltage constant k in Eqn (4) is about 1.42.

We have done some experiments using thin film layers of TiN & TiW and Al bondpad samples. Using different beam accelerating voltages from 1 to 20 kV, N Ka, Al Ka, Si Ka, W Ma, Ti Ka and W La peaks were detected, and then their over-voltage constant, k, was obtained, which was about 1.42 (Figure 2). Considering to easily to use, we recommend to set it at k=1.5. Therefore,

$$V_i \,(kV) = 1.5\,E_i\,(keV) \qquad (5)$$

where V_i is the beam accelerating voltage in kV used in the SEM/EDX, E_i is the energy of the characteristic X-ray line of the element interested in keV.

Using Eqn (5), one can quickly to estimate the beam accelerating voltage for detecting a characteristic X-ray line of an element interested. For example, for detecting Ti Ka (4.510 keV) peak, we should use accelerating voltage larger than 6.75 kV. Similar, for detecting W La (8.396 keV) peak, we should set it larger than 12.5 kV, we often set it at 15 kV.

III USING THE SAME BEAM ACCELERATING VOLTAGE FOR BOTH SEM AND EDX

For EDX microanalysis, generally different beam voltages are chosen for different elements. High voltage is needed for high-Z elements in order to excite high-energy X-ray characteristic lines, while low beam voltage is chosen for the low-energy lines. However, for some samples with charging problems, it should be cautious when performing EDX analysis due to changing effects. To obtain accurate EDX results, EDX analysis should be performed with properly focused imaging conditions at a specified voltage. For example, to analyze contamination on the surface of microchip Al bondpads, a low voltage such as 5 kV is preferred for EDX analysis. This is to prevent deep penetration of electron beam so as to obtain true signals from Al bondpad surface. Before EDX analysis, imaging focus should be properly done at 5 kV. It means that both SEM image focusing and EDX analysis should be done at the same voltage. It is not recommended to focus SEM images at 15 kV, but do EDX at 5 kV without re-focusing after changing voltage. The reason lies in the fact that the analysis area focused at the high kV may be shifted at the low kV due to charging & alignment problems, and thus EDX results may not be from the area focused/interested. The following case may be a good example to elucidate the importance of the correlation between imaging focusing and EDX analysis at different voltages.

There is one case in which the bondpad under-etch issue was reported as EDX analysis detected high Si & N peaks. After discussing with the customer, we understood that they focused the SEM image at 15kV as they faced changing problem at 5 kV, and then they decreased the voltage from 15 kV to 5kV and did EDX analysis without re-focusing. The EDX results showed the strong N and Si peaks, and thus they reported the bondpad under etch issue. As such, the EDX results & conclusion needed to be confirmed, as SEM imaging and EDX analysis were not done at the same voltage. For the purpose of verification, one unit was sent back for further FA. In this study, Auger, surface SEM/EDX analysis were done to determine the presence of Si & N on bondpad surface.

First, Auger analysis was performed on the unit and the results showed that no Si & N peaks were detected on bondpad surface, which was contrary to the results from the customer. In order to understand the reason behind it, EDX analysis was done on the unit using the same experimental conditions as those used by the customer. SEM imaging was focused at a bondpad area at 15 kV (Fig. 3(a)). And then the beam voltage decreased to 5 kV without moving sample position & re-focusing and the SEM imaging is shown in Fig. 3(b), which clearly indicated that the position of the interested area was shifted a lot. The EDX analysis in this area detected strong Si and Ni peaks. However, one can clearly see that the signals were actually from the passivation layer, as shown in Fig.4 (b).

To get accurate EDX results on the bondpad, the sample was then repositioned and focused at 5kV, and then EDX was done at 5kV on the area near the right bottom edge of the bondpad. EDX results (5kV) showed that only Al and O were detected and no N & Si peaks were detected (Fig.4(c)). For confirmation, we did EDX at 5kV on a few bondpads in the units and similar results were obtained, which are well consistent with Auger analysis results.

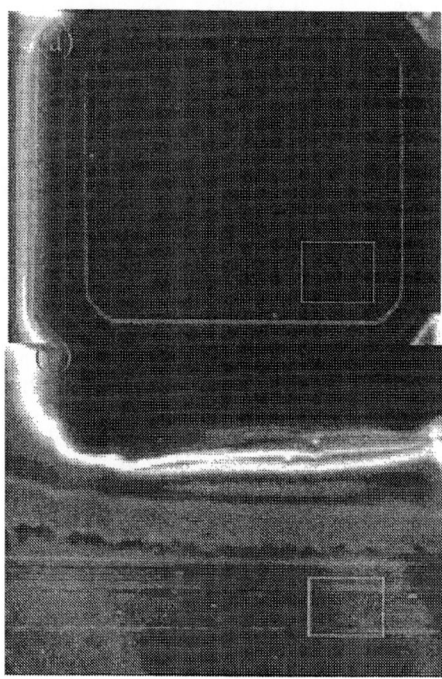

Figure 3 SEM images of a bondpad taken (a) at 15 kV (b) at 5kV (decreased from 15 kV) without imaging re-focus and sample re-positioning

Such a large image shift at different voltages can be understood in terms of the different configurations of the electron optics at different voltages. At the different working voltages, the excitation levels and thus focus lengths of condenser and objective lenses will be different. Therefore the shift of the imaging is expected when the working voltage is changed, especially for the samples with charging problem. To avoid such a large image shift, good alignments of electron optics are required. However, usually, small image shift will still exist due to the magnetic hysteresis effects of magnetic lens system. Therefore, it is always a necessary practice to adjust imaging focus and analytical areas/positions before EDX analysis whenever changing the working voltages. The issue can be very critical for those samples with serious surface charging. In case of surface charging, accurate sample positioning and EDX analysis could be very difficult and the results of EDX analysis may not be reliable. To achieve the accurate EDX analysis of such charging samples, it is

recommended to coat the samples with the anti-charging layers such as Pt or Au and use the same beam voltage for focusing of SEM imaging & EDX analysis.

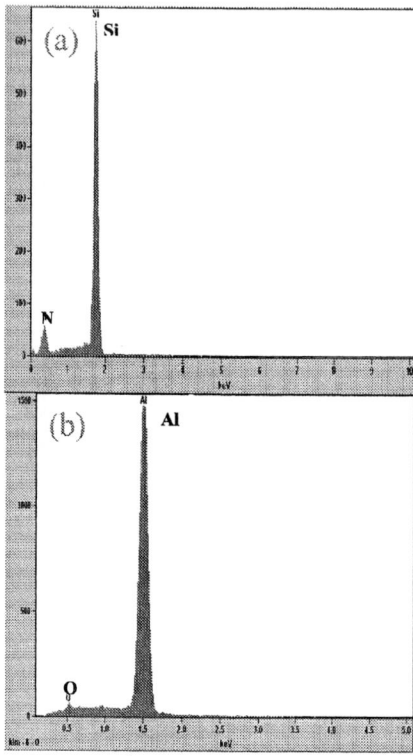

Figure 4 EDX spectra taken (a) at 5kV with voltage down from 15kV and without readjustment of imaging focus and sample re-positioning; (b) at 5kV with proper readjustment of focus and sample re-positioning

IV IDENTIFICATION METHODS OF ELEMENTAL PEAKS OVERLAPS

A result of peak broadening is overlap occurring for X-rays of different energies, both for complex peaks from a single element such as L-lines and for peaks from different elements. For example, Si Ka (1.740 keV) peak is overlay with W Ma (1.775 keV) peak. F Ka (0.677 keV) is overlay with Fe La (0.705 keV. In order to provide accurate elemental information, we have to do some identification and confirmation process.

A. Identification using the Peak Fitting Technique

When we determine Si and W peaks in WSix sample, Si (1.740 keV) peak is overlay with W Ma (1.775 keV) peak. We can confirm the present of W Ma peak using the peak fitting technique. EDX results from 5 kV are shown in Figure 5. From Figure 5 (left), one can see that the fitting peak of Si Ka (blue line) is not fully overlap with real peak (red color peak), but it is shifted to left of the real peak. This indicates that this peak (red color peak) is not only Si Ka peak, but also the possibility of the present of W Ma peak. In Figure 5 (middle), the fitting peak of W Ma peak (blue line) is also not fully overlap with real peak (red color peak), but it is shifted to right of the real peak. This also indicates that the peak (red color peak) is not only W Ma peak, but also the possibility of the present of Si Ka peak. However, if we fit the result using both Si Ka and W Ma peaks, the fitting peak (blue line) is fully overlapped with the real peak (red color peak). This indicates that the real peak is contributed by both Si Ka and W Ma peaks and both Si and W elements are present in WSix material.

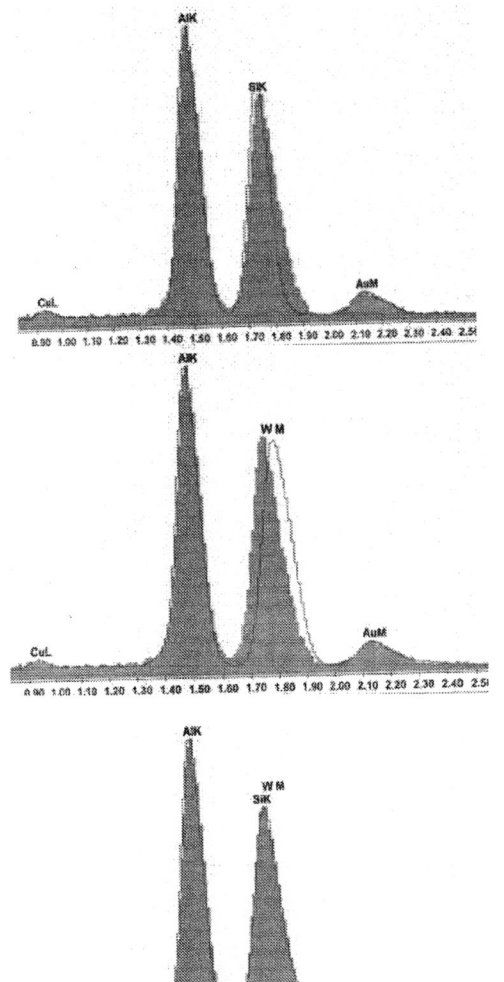

Figure 5. The fitting peak of Si Ka (blue line) is not fully overlap with real peak (red color peak) (up); the fitting peak of W Ma (blue line) is also not fully overlap with real peak (red color peak) (middle); but if put both Si and W together, the fitting peak (blue line) is fully overlap with real peak (red color peak) (bottom). This indicates the presence of both Si and W elements in WSix material.

From the above case, one can know that the fitting method can help us to identify elements. However, we must think all possibility of the present of elements.

B. Identification using Different Beam Voltage Method

Another method is to use different beam voltage to identify elements overlapped. For example, for the above WSix case. If using 5 kV, we can only detect Si Ka and W Ma peaks, which are overlapped. However, if using 15 kV, we can also detect W La peak. Therefore, if W La (8.396 keV) peak is to appear, then we can confirm the present of W.

Similarly, Mg Ka peak (1.254 keV) is overlapped with As La peak (1.282 keV). As such, if Mg Ka peak is detected at 1.254 keV using 5 kV, we have to confirm it using 15-20 kV. If no As Ka peak (10.542 keV) is detected, the peak at 1.254 keV is confirmed to be Mg Ka peak. Otherwise, if As Ka peak (10.542 keV) is detected, then the peak at 1.254 keV may not be Mg Ka peak, but it may be As La peak.

Another example is that F Ka peak (0.677 keV) is overlapped with Fe La (0.705 keV). If F Ka peak is detected at 0.677 keV using 5 kV, we have to confirm it using 15 kV. If no Fe Ka peak at 6.403 keV is detected using 15 kV, the peak at 0.677 keV is confirmed to be F Ka peak. If Fe Ka peak at 6.403 keV is detected, then the peak at 0.677 keV may not be F Ka peak, but it may be Fe La peak. The below is real previous case study.

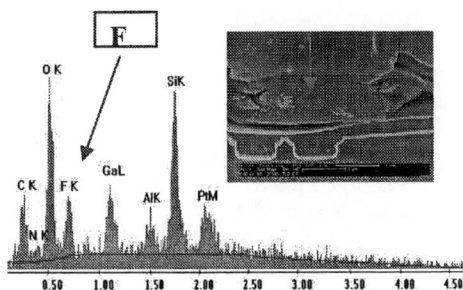

Figure 6. EDX (5kV) detected F peak besides C, O, Ga (from FIB), Al, Si and Pt (from Pt coating) peaks on the particle.

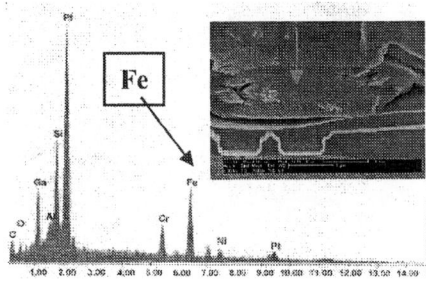

Figure 7. EDX (15kV) detected Fe Ka peak besides C, O, Ga (from FIB), Al, Si, Pt (from Pt coating); Cr and Ni on particle. This indicated that the particle was not a F related particle, but a stainless steel particle.

In failure analysis, it was reported that the unit failed functional test during multi port testing. After decapsulation, EMMI was performed on the failure unit and found the hot spot. The unit was performed delayering process. Visual inspection found a particle after removal of passivation (Si3N4 and SiO2) and metal 3 layers. After that, cross sectional FIB/SEM and EDX were carried out on the particle. FIB/SEM results showed that the particle was the above metal 1 layer

(Fig 6, SEM). EDX analysis was also done on the particle using 5kV beam acceleration voltage, the result showed that C, O, F, Ga, Al, Si and Pt peaks (Fig. 6, EDX), in which Ga was introduced by FIB sample preparation and Pt was from sample Pt coating.

Based on the EDX (5kV) results (Fig. 6, EDX), one can see that besides C, O, Al and Si, higher F peak (F Ka at 0.677keV) was detected on the particle. Thus, it was identified into a fluorine contamination related particle. The Fab team tried to find the possible root cause of F particle. However, we found that between metal 1 and metal 2 layers, the process was only related IMD layer process and no F source could be introduced. In this situation, the team could not find the root cause and further failure analysis was requested.

From the above section, we have understood that the F Ka peak (0.677 keV) may be overlapped by Fe La peak at 0.705 keV as EDX cannot identify F Ka peak from Fe La peak due to its poor energy resolution. In other words, it is possible that the F peak detected may be Fe La peak. Therefore, to confirm the suspicion, we have to do EDX using high beam acceleration voltage. If Fe Ka peak at 6.403 keV is detected, then the peak at 0.677 keV may not be F Ka peak, but Fe La peak. We did EDX again on the particle using 15 kV and the results are shown in Figure 7. One can see that Fe Ka peak (6.403 keV) was detected. Moreover, besides Fe Ka peak, Cr Ka and Ni Ka peak were also detected. Thus, based on EDX (15kV) results, we confirmed that the peak detected at 0.677 keV was not F Ka peak, but was Fe La peak. Therefore, the particle was not a fluorine contamination related, but most likely to be a stainless steel particle as Cr, Fe and Ni were detected together. With this new information using high kV, the Fab team found the root cause and concluded that the stainless steel particle was from the thin film machine. From this case study, we can understand that it is very important to identify the overlap of elemental peaks.

V. IMPROVEMENT OF ACCURACY OF QUANTIFICATION EDX ANALYSIS WITH SEC CORRECTIONS

In wafer fab, the standardless EDAX ZAF quantification method is widely used to perform elemental microanalysis [2]:

$$W_{i(wt\%)} = \frac{K_i}{Z_i A_i F_i} \times 100\%$$
(6)

$$K_i = \frac{I_i^{Meas}}{I_{i,Calc}^{Std}}$$
(7)

where Wi(wt%) is the weight fraction of the analyzed element i, Z is the atomic number correction factor, A is the absorption correction factor, F is the fluorescence correction factor. I^{Meas} and I^{Std} are the measured and standard intensities of the

Element involved.

For standardless analysis, we have no I^{Std} available and can chose to use calculated standards. Therefore, in the ZAF method, the theoretical intensities in the standard sample can be calculated by [3-4]:

$$I_{i,Calc}^{Std} = n^{e-} \cdot \frac{\Omega}{4\pi} \cdot \varepsilon_d \cdot \omega_j \cdot p_{jl} \cdot (1+f_c) \cdot (1+g_{ck}) \cdot f(x) \cdot \frac{N^o}{A} \cdot R \cdot \int_{E_o}^{E_k} \frac{Q_j(E)}{dE/d(\rho s)} \cdot dE.$$
(8)

Since n is unknown, and thus set to unity, the calculated intensity might be in a totally different order of magnitude as the measured intensity. Normalizing the weight percentage to 100% solves this problem.

A. SEC (Standardless Element Coefficient) Factors

The standardless EDAX ZAF quantification method (Eqn. (6)) is used to do a good and reliable compositional analysis for particles and contamination in wafer fab. However, in some cases, EDX results may not be accurate, especially for the purpose of thin film layer identification.

For thin film analysis, SEC factor correction is recommended, as it is very important. For example, using the standardless EDAX ZAF quantification method without SEC factors, the relative error of EDX results was very high up to (-67.5% ~ +101%) for Si_3N_4 of passivation layer analysis although the data have performed ZAF correction using Eqn. (6). This is because the theoretical intensities for standard sample have been used and one disadvantage is that the detector efficiency (ε_d) cannot be predicted with sufficient accuracy for X-ray lines below 1 keV. Small variations in detector parameters such as window thickness, Si dead layer, etc can cause variations in measured intensity. Therefore, the SEC (standardless element coefficient) factors have to use so as to make a correction to obtain accurate results:

$$W_{i(wt\%)} = \frac{K_i}{Z_i A_i F_i (SEC)} \times 100\%$$
(9)

In Eqn. (9), (SEC) are the SEC (standardless element coefficient) factors, which can be calculated by the EDAX ZAF software.

However, it should be noticed that the SEC factors are relative numbers of course, again due to the normalization to 100%. So a set of SEC factors will normally contain a value of 1.00 for one element, while the SEC factors for all other elements are scaled relative to it [2]. Moreover, the SEC factors are independent of the accelerating voltage. As they only correct for variations in detector parameters, the accelerating voltage with which the X-ray was generated is irrelevant. The SEC factors are also matrix independent. Therefore, the SEC factors remain unchanged for a long time as detector parameters hardly change.

978-1-4244-2039-1/08/$25.00 ©2008 IEEE

B. Calculation of the SEC Factors and Application

The SEC factors can be calculated by the EDAX ZAF software. In this study, we use Si_3N_4 analysis as an example. The SEC factor of Si was set to 1.00 and the SEC factor of N calculated is scaled relative to that of Si and it was 10.88 (ratio is 1: 10.88). Moreover, we also set the SEC factor of Si to 58.04, and then the SEC factor of N was 631.6 (the ratio is still 1:10.88) so as to compare the EDX results using two sets of SEC factors. The EDX results of Si_3N_4 are summarized in Table 1. From Table 1, one can see that if without SEC factors correction (Si = 1 & N = 1), the relative error of EDX results of Si_3N_4 is very high up to (-67.5% ~ +101%) although the data

have performed ZAF correction. However, if with SEC factors correction, the relative error of Si_3N has been greatly improved and is reduced to (-2.83% ~ +3.93%) using the set of the SEC factors (Si = 1.00 and N = 10.88). If using another set of the SEC factors (Si = 58.04 and N = 631.60), the results are almost the same and the relative error is also reduced from (-67.5% ~ +101%) to (-2.83% ~ +3.91%).

We also did similar experiments for passivation SiO_2 layer and barrier metal TiW layer. The EDX results also showed that the accuracy has been greatly improved. For example, the relative error of SiO_2 layer is reduced from (-62.3% ~ +54.9%) to (-2.76% ~ + 3.15%) and that of the barrier metal TiW layer is reduced from (-76.9% ~ +20.06%) to (-10.9% ~ +11.6%).

Table 1. Comparison of EDX Analysis Results of the Passivation Layer of Si_3N_4 Without and With the SEC Factor Correction (Theoretical Values: Si = 60.06 wt% and N = 39.94 wt% in the Passivation Layer Si_3N_4)

Sample No.	Without the SEC Factor Correction (Si = 1, N = 1)			With the SEC Factor Correction (Si = 1, N = 10.88)			With the SEC Factor Correction (Si = 58.04, N = 631.6)		
	Ele.	Wt (%)	Diff. (%)	Wt (%)	Diff. (%)	Diff. (%)	Ele.	Wt (%)	Diff. (%)
SN (SEC)	N	79.44	+ 98.9	40.08	+ 0.35		N	40.07	+ 0.33
	Si	20.56	- 65.8	59.92	- 0.23		Si	59.93	- 0.22
SN #1	N	79.00	+ 97.8	39.43	- 1.28		N	39.42	- 1.30
	Si	21.00	- 65.0	60.57	+ 0.85		Si	60.58	+ 0.87
SN #2	N	78.57	+ 96.7	38.81	- 2.83		N	38.81	- 2.83
	Si	21.43	- 64.3	61.19	+ 1.88		Si	61.19	+ 1.88
SN #3	N	79.76	+ 99.7	40.54	+ 1.50		N	40.53	+ 1.48
	Si	20.24	- 66.3	59.46	- 1.00		Si	59.47	- 0.98
SN #4	N	80.41	+ 101	41.51	+ 3.93		N	41.50	+ 3.91
	Si	19.59	- 67.5	58.49	- 2.61		Si	58.50	- 2.60

ACKNOWLEDGEMENT

The author would like to thank QRA-FA Dept MFA personal for providing technical advice and making contributions to this technical paper.

REFERENCES

[1] Tertian, R. Claisse, F. *Principles of Quantitative X-Ray Fluorescence Analysis*, Heyden, London, pp.35-35. 1982.

[2] Hua Younan, Z. R. Guo & K. W. Chau, "Studies of ZAF Standardless EDX Quantification Method and Application in Failure Analysis of Semiconductor". *Journal of Trace and Microprobe Techniques*, vol. 15 (1), pp.13-31, 1997.

[3] S. Steinbrecher, *Microscopy and Micoanalysis*, Vol.9, pp.160-161, 2003.

[4] JI Goldstein, DE Newbury, P Echlin, DC Joy, AD Romig, CE Lyman, C Fiori and E Lifshin, *Scanning Electron Microscopy and X-ray Microanalysis*, 1992.

978-1-4244-2039-1/08/$25.00 ©2008 IEEE

SESSION 11:

BEOL II - LOW-K & ADVANCED INTERCONNECTS

INVITED PAPER

New Models for Interconnect Failure in Advanced IC Technology

J.R. Lloyd
IBM T.J. Watson Research Center
Yorktown Heights NY 10598
jrlloyd@us.ibm.com

Abstract- **The two major interconnect failure modes most commonly considered in IC technology are electromigration (EM) and Time Dependent Dielectric Breakdown (TDDB) in interlevel dielectrics. In order to evaluate reliability, tests must be performed under conditions that are much more severe than those expected under normal use in order to obtain failure statistics in reasonable times. These tests only make sense if the extrapolated lifetimes and failure distributions can be extrapolated back to use conditions, so that assessments of the reliability under normal operation can be made. The simple use of the traditional Black's Law for EM and either the E or 1/E model for TDDB can be shown not to provide accurate projections of product lifetime in the light of recent knowledge of the failure mechanisms. In this paper, the most recently developed techniques for extrapolating test data to use conditions will be described with suggestions for determining operational reliability performance.**

ELECTROMIGRATION

More than 40 years ago, the appearance of electromigration as a failure mechanism in the then newly developed integrated circuit technology shocked the industry and nearly brought it to its knees. Solid state devices were supposed offer major improvements in reliability compared to vacuum tubes (valves) and discrete devices as well as having promise for cost savings and complexity. In these new devices, thin metal films replaced copper wires for interconnecting circuit elements. Unfortunately, when these new devices entered the market, mysterious open circuit failures occurred in depressingly short times, threatening the entire semiconductor industry with extinction. The large companies involved in this industry spent a total of nearly a billion dollars (when this was a very large number) trying to find a solution to this impending disaster. The problem was identified as electromigration, where, in simple terms, friction between conducting electrons and the atoms in the conductor produced metal migration in the direction of current flow. Wherever there was a mass flux divergence, and there are necessarily many in any practical circuit, either a void leading to an open or an extrusion leading to a short would eventually appear.

The first reasonably consistent failure model was proposed by Jim Black of Mototola in 1966 (1) where he empirically determined that the lifetime of a metal conductor was inversely proportional to the square of the applied current density, j which has become known as Black's Law.

$$t_{50} = \frac{A}{j^2}\exp\left(\frac{\Delta H}{kT}\right) \qquad (1)$$

Subsequent research had shown that eqn (1) was not always strictly obeyed. The current density exponent, n, was not always two. To accommodate this, Black's Law was then rewritten as;

$$t_{50} = \frac{A}{j^n}\exp\left(\frac{\Delta H}{kT}\right) \qquad (2)$$

Often n could be found to vary substantially from 2, ranging from as low as 1, but in fact rising without limit in extreme cases. Very high values of n can often be attributed to overstressing with too high a current density, rendering the test results meaningless. The failure mode in these cases would be due to the presence of temperature gradients that would, presumably, not be present in properly designed products. There are, however, other reasons that n may be high, approaching infinity, unrelated to temperature gradients that need to be considered. Most importantly is the onset of the Blech Length Effect. (2)

The major misconception about the current exponent is that it is a material specific property. There is no n value for Cu or Al or any other metal for that matter. n will vary from experiment to experiment, depending as much on the geometry of the test structure as on the test conditions. Nowhere is this more evident than in the treatment of the Blech Length Effect. If the test structure has diffusion barriers on either end, which could be contacts to Si or interlevel vias with a barrier layer, there will be a mechanical stress gradient generated in response to the electromigration driving force that, if the conductor is sufficiently short, or the current density sufficiently low, will effectively stop electromigration in its tracks. Therefore, as we approach the Blech Condition, where the product of the current density and the length of the conductor is below a critical value,

$$jl \leq \chi_B \qquad (3)$$

the current exponent rises without bound. If the Blech Product is less than χ_B, failure is impossible and if in a test at least one of the experimental conditions exceeds χ_B, the apparent current exponent becomes infinity. If the test conditions approach the Blech Condition but do not meet it, the end effects will be "felt"

INVITED PAPER

by the electromigrating atoms and large values of n will be obtained. For a given current density, the length of the conductor that must be exceeded to permit electromigration failure, called the Blech Length (l_B), must be exceeded by at least a factor of three to minimize contributions to the lifetime. Design of test structures must take this into consideration lest we make optimistic lifetime projections using unrealistically high n values obtained from the flawed experiments. Note that if in operational use the Blech Condition is satisfied, the device, as far as electromigration is concerned, can be considered immortal. This can be used by designers to improve reliability.

This is one reason that the unconditional use of Black's Law is problematic. It is not the only reason. The stress gradient that is generated by electromigration is hydrostatic in nature. Damage in the form of voids, however, is not determined by the attainment of a tensile hydrostatic stress that leads to void nucleation, but most probably by reaching a normal stress that produces a delamination at an interface. (3) When this happens, the stress at the delamination surface vanishes, changing the boundary conditions dramatically and promoting rapid void growth which leads to open circuit failure. It has been shown theoretically that electromigration induced void nucleation should depend on the inverse of the square of the current density, like the classical Black's Law. (4,5) This has also been seen experimentally. (6) If void growth to failure is short compared to the nucleation time, n will be close to but will not exceed 2. If n values greater than 2 are observed, and we are not in a regime where the Blech Length Effect is operating, it is probably due to the effects of Joule Heating which would make the data irrelevant for extrapolation to use conditions.

Looking at electromigration, then, we see that it should be recognized that failure would be composed of stages of nucleation and growth. (7) Nucleation of voids (or extrusions) would be inversely dependent on the square of the current density, whereas growth would be more related to drift velocity which depends directly on the current density. Thus, for nucleation and growth the lifetime due to electromigration is best described by

$$ t_f = t_{nucleation} + t_{growth} = \left(\frac{A}{j} + \frac{B}{j^2} \right) \exp\left(\frac{\Delta H}{kT} \right) \quad (4) $$

where it is assumed here that the activation energy for nucleation and growth are the same. This is a reasonable since they should both be controlled by the same mass transport mechanism. The first term in the pre-exponential brackets represents void growth and the second term, nucleation.

Although eqn (4) represents something quite different from what is commonly used today, it is no more difficult to apply than the modified Black Equation (n variable) since there are the same number, two, of parameters, but is more reasonable physically. In eqn. (2) we have to determine A and n and in eqn. (4) we need to determine A and B. In cases where we fit both

models to lifetime data, eqn (4) fits somewhat better, and is consistent with failure analysis, where early fails were characterized by failure locations requiring smaller voids and consequently less time for growth. The early fails and the late fails had similar values for the nucleation time, but the later fails showed significantly longer growth times.

This way of looking at electromigration suggests that things are actually quite a bit more complicated than we generally assume. A and B each can be derived from first principles and are expected to be temperature dependent and contribute subtle errors in calculating the apparent activation energy if this is not considered.

Realizing that void growth would be equivalent to the edge motion in finite "island" samples first observed by Blech (2) we can re-write eqn (4) as

$$ t_{50} = t_{nuc} + t_g = \left(\frac{AkT}{j} + \frac{B(T)}{j^2} \right) \exp\left(\frac{\Delta H}{kT} \right) \quad (5) $$

where B(T) has a temperature dependence corresponding to the failure model chosen. (1,4,5,7,8)

Clearly, if we were to apply Black's original equation to failure that really followed eqn. (5) we would calculate a value for n that would lie between 1 and 2. It is also clear that the value of n would not be constant, but would vary according to the range of current densities used in the experiment.

It is also interesting that the proportion of the lifetime dedicated to nucleation and growth would vary as a function of current density. If a test were performed that would, if eqn (2) were applied, result in n = 1.7, over the current density range between 1 and 2×10^6 A/cm^2, at the higher current density, the lifetime would be 1/3 nucleation and 2/3 growth, but at the lower current density it would be roughly split evenly between nucleation and growth. At a use condition of 5×10^5 A/cm^2, the proportions reverse with 2/3 of the lifetime devoted to nucleation and 1/3 to growth.

This can have serious consequences when considering the effects of stress voiding on electromigration lifetime. If stress voiding occurs, nucleation time vanishes, leaving only growth of the damage, presumably a void, to determine the lifetime. Void growth obeys n = 1 kinetics. In addition, stress voiding probably will not occur at accelerated stress conditions, but may occur at use conditions where the thermal stresses are substantially higher. Thus, eqn (5) may be appropriate at test conditions, but if voids form due to induced tensile stresses, the coefficient B in eqn (5) becomes zero, and all lifetime extrapolations will only be via the first term, reducing the problem to the use of eqn (2) with n = 1, after, unfortunately, you subtract the time to nucleation from the recorded failure times at test.

INVITED PAPER

Therefore, the prudent thing to do, since stress voiding can never really be eliminated with 100% certainty, is to determine at the accelerated conditions what the relative proportions of nucleation and growth are by fitting the data at constant temperature to eqn (4), then finding out what proportion of the lifetime is devoted to damage nucleation at a reference condition. This then should be subtracted from the total lifetime, and the remainder, which is devoted to growth, be decelerated to use currents with n = 1.

One might argue that observations of void nucleation and migration might make this model invalid. However, it must be realized that both void motion as well as void growth will depend on the drift velocity. A void may nucleate, then grow to a fatal size, which may be a function of the location where the failure will occur. The void growth will continue as long as there is a flux divergence, but, unless the flux divergence is located at a strongly blocking boundary, such as a contact, it is likely that the divergence may disappear. If this were to occur, the void will no longer grow but, since mass motion is still taking place, may then act as a "marker" and be transported at the drift velocity to the location where the failure ultimately takes place. The post-nucleation actions both follow n=1 kinetics. Therefore, the model holds.

The activation energy for failure is also not what it classically appears to be. Since electromigration is essentially a diffusion process, it is expected that the activation energy for electromigration failure should be that of the predominating diffusion pathway. This is undoubtedly true, but the measured apparent activation energy will be affected by the fact that there are processes active that are not expressed as an Arrhenius relation, necessary for the simple extraction of an activation energy by plotting the log of the failure time against reciprocal temperature. In reality, what we do in classical activation energy determinations is to calculate

$$\Delta H_{app}^{g} = k \frac{\partial \ln t_{50}}{\partial \frac{1}{T}} \qquad (6)$$

The most important contribution to error in the determination of the activation energy is the presence of thermal stresses. Due to differences in the thermal expansion of the metal conductors and the surrounding Si wafer and oxide passivation, at use conditions the metal lines are in a state of hydrostatic tension. Therefore, under these conditions, the amount of stress that must be built up to nucleate a void is less than it might be at elevated test temperatures. If we apply the model of Korhonen et. al. (5) to void nucleation, we find that the apparent activation energy for void nucleation becomes

$$\Delta H_{app}^{nuc} = \Delta H - kT \left(1 + \frac{2E\Delta\alpha T}{\sigma_n - E\Delta\alpha(T_0 - T)} \right) \qquad (7)$$

where ΔH is the activation for diffusion, E is the elastic

modulus of the conductor, $\Delta\alpha$ is the difference in the thermal coefficient of expansion between the metal conductor and the passivation constraint, σ_n is the stress required to nucleate the void, and T_0 is the elevated temperature where there conductor is stress free, presumably the temperature at which the passivation layer is applied, or, in the case of low-k dielectrics, the curing temperature.

Eqn. (7) predicts significant differences between the apparent activation energy and the activation energy for diffusion, especially at the temperature where the denominator in the brackets vanishes. It is at this temperature void nucleation (delamination) occurs via thermal stresses alone, generally called stress voiding. The activation becomes negative infinity, meaning no temperature dependence on void nucleation and the time to nucleation is essentially zero.

If we assume reasonable values for the activation energy for grain boundary diffusion (1.1 eV) (9), the modulus (120 GPa), the difference in thermal expansion between the metal and the surrounding dielectric (15 ppm/K), and the required stress to form a void through delamination (500 MPa), we see that the apparent activation energy in the regime where we perform tests is on the order of 0.8 to 0.9 eV. This compares favorably with the activation energy determined by many for electromigration failure. Note that this figure should only be valid for nucleation dominated failure. It should also be noted that small changes in any of the parameters may make relatively significant changes in the measured activation energy. This may account for the notable differences from one experiment to another that is commonly seen in electromigration testing

The measured activation energy for void growth will also not be exactly correct since the Einstein equation for drift velocity will yield,

$$\Delta H_{app}^{g} = k \frac{\partial \ln t_{50}}{\partial \frac{1}{T}} = \Delta H - kT \qquad (8)$$

representing a significantly smaller error than what we get for nucleation, but present nonetheless. Therefore, the calculated activation energy will always be somewhat less than the actual activation energy for diffusion. It also implies that over large temperature ranges, the Arrhenius plot will not be linear.

For the apparent activation energy for both nucleation and growth together, the rather formidable equation appears.

$$\Delta H_{app}^{f} = \Delta H - kT - \frac{2\gamma E\Delta\alpha kT^2 [\sigma_n - E\Delta\alpha(T_0 - T)]}{\gamma[\sigma_n - E\Delta\alpha(T_0 - T)]^2 + Aj} \qquad (9)$$

This is virtually impossible to use for practical purposes.

INVITED PAPER

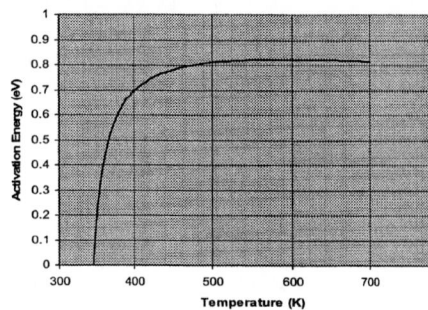

Figure 1 Apparent activation energy for void nucleation in Cu as a function of temperature for the parameters in the text. The activation energy for diffusion is assumed to be 1.1 eV.

The model suggests, however, that the activation energy obtained from electromigration kinetic studies using eqn. (6) would not be the correct activation energy for the mass transport. In all cases the calculated value would be less. However, since the error would be substantially greater for nucleation than for growth, one would predict that higher activation energies should be calculated from studies that had lower values of the current exponent. The effect would be relatively small, however, and may not be clearly visible in the data.

WHAT TO DO?

The arguments above make it unclear exactly what one should do to safely estimate operational lifetimes from test data. A somewhat conservative scheme is here described. Since it is implied by Figure 1 and Eqn (7) that void nucleation is a near certainty at use conditions, we would be better off assuming that it's going to happen rapidly at use conditions, and that failure will be determined solely by the growth rate. (The closer a calculated n is to 1, the better this assumption is.) Since nucleation at use temperatures is not a factor, the portion of the lifetime at test that is devoted to nucleation should be determined, then subtracted from the test lifetime. This can be accomplished by fitting the data to Eqn (4). The remaining growth time can then be extrapolated using n = 1 kinetics and the activation energy for diffusion obtained by an alternate method than electromigration failure. Obtaining the activation energy from void growth or drift velocity experiments would be a reasonable way to get this number.

Let us imagine that we performed an electromigration test at 300C and when we doubled the current density from 1 to 2 x 10^6 A/cm^2, we arrived at a current exponent of -1.2 using Black's Law. Further assume that by varying the temperature we obtained an apparent activation energy of 0.8 eV. If we extrapolate via the modified Blech Equation, eqn (2), to a potential use condition of j = 5 x 10^5 A/cm^2 and T = 100C, we get an acceleration factor of ~32,500 from 2 MA/cm^2.

However, if we we use the scheme described above we get ~225,000, even though we essentially threw out 30% of our lifetime. If this analysis is correct, we actually have, in this particular case, almost 7 times the lifetime we think we do. If the apparent activation were the same, (which it may not be) but the current exponent was calculated at -1.8, Black's Law would provide an acceleration of ~75,000 whereas the above scheme would predict an acceleration of ~98,000. The error is then a function of the use condition and the testing condition as well as the calculated value of n and not a constant that can be used for every test indiscriminately.

It is recognized that this practice requires a leap of faith that many engineers may be loathe to make without considerable experimental justification. However, it must be pointed out that it is easy to misinterpret the results of many of these experiments if the issues that have been presented here are not considered.

BIMODAL ELECTROMIGRATION FAILURE

One of the most intriguing observations with electromigration testing of Cu metallization was that in almost all cases Cu possessed a bimodal failure distribution when Al alloys generally did not. (10,11) In addition, the measured Blech Product for Cu was significantly smaller than that for Al alloys, even though Cu is a much stronger material than Al. (12) These problems can be accounted for by considering the anisotropic nature of the Cu elastic constants, and a somewhat different way of modeling the void nucleation. (3)

It has been argued that classical nucleation theory cannot account for the generation of electromigration induced voids. (13) However, if void nucleation is actually due to a delamination of the copper from the liner or the cap, then things start to make sense. First, it should be recognized that delamination will occur if a sufficiently high normal stress is achieved, whereas classical void nucleation will require a hydrostatic stress. It also must be realized that the driving force opposing electromigration is a gradient in the hydrostatic stress. For Cu, with highly anisotropic elastic constants, this means that although there may be a smooth hydrostatic stress profile, the stresses normal to an interface where delamination might occur may change markedly from grain to grain.

Therefore, if delamination is the event that presages electromigration failure, the unique properties of Cu are what makes both of these observations expected. (9) The normal stress required for a delamination to occur is

$$\sigma_{delam} = \sqrt{\frac{2E_n G_{ad}}{h}} \qquad (10)$$

where E_n is the elastic modulus normal to the interface where the delamination is taking place, h is the thickness of the metal conductor, and G_{ad} is the specific strain energy released when

INVITED PAPER

the delamination occurs. The ratio of the Blech Products for Al and Cu can be shown to be given by

$$\frac{(jl)_{Al}}{(jl)_{Cu}} = \frac{\Omega_{Al}}{\Omega_{Cu}} \sqrt{\frac{E_{Al}G_{Al}}{E_{Cu}G_{Cu}}} \qquad (11)$$

where Ω is the atomic volume. Cu does not adhere as well to most oxides and dielectrics as does Al, which, because of its remarkably stable oxide, adheres to almost anything. So although the modulus of Cu is significantly greater than Al, eqn (11) predicts that the ratio of the Blech Products favors Al by a factor of about 2.5, which is in the range of that observed.

Furthermore, it has been shown that due to the anisotropy of the elastic moduli in Cu, that, given identical hydrostatic stress, the normal stresses can vary up to 40%. The bimodal nature of the failure distribution arises from this variation. Early failures occur when the electromigration induced voiding occurs just under a via, where the hydrostatic stress is at its highest and a much smaller void is required to be fatal. This would occur whenever the grain orientation favored a maximum of the normal stress in the via. If the orientation were not favorable under the via, but were such that failures would occur somewhat away, failure would take longer to happen because the fatal void size is significantly larger. Thus a distribution of early failures will be observed from those examples where the crystal orientation produces the highest normal stresses at a via, and more "mainstream" failures will develop a failure distribution when the maximum normal stress is developed away from the via. Therefore, it seen that the bimodal character of Cu metallization failure distributions is not necessarily caused by the presence of manufacturing defects leading to early failure, but are an unavoidable function of the mechanical properties of Cu. It is not likely that improved process control will have any effect on this.

The bottom line for modeling purposes is that electromigration testing for Cu is expected to have a bimodal failure distribution. With relatively small sample sizes, say a dozen or so samples, the bimodal character may not be unambiguous and fitting the data to a single lognormal distribution will produce a lognormal standard deviation (σ) that is much larger than is justified and this can lead to very pessimistic predictions for the reliability. The expression for the failure rate is very sigma sensitive and even if the lifetime of a weak mode is shorter than what one would estimate assuming a single mode, the erroneous sigma will predict failure rates at early failure probabilities orders of magnitude higher than what is justified. It is, therefore, important to understand that the bimodal nature of Cu interconnect failure is expected and extrapolations should be calculated accordingly.

TDDB IN INTERLEVEL DIELECTRICS

Time Dependent Dielectric Breakdown (TDDB) is the failure of

a dielectric by breakdown at applied fields below the characteristic breakdown field of the dielectric material. It is the electrical analogy to mechanical creep, where loads below the yield strength are applied to metals, but the material slowly distorts and eventually fails. Previously, TDDB had only been a problem with very thin gate oxides and was never an issue in interlevel dielectrics. The thickness of the dielectric was much thicker than at a gate, and TDDB was unheard of. With the introduction of low-k dielectrics, the picture has changed.

As the dielectric constant of a material becomes smaller, so does the breakdown field and the performance under TDDB conditions. This, coupled with the very small dimensions that are now a feature of the newest semiconductor technologies, has created a failure mode that has become a matter of great concern. It should be pointed out that the intralevel spacing in the latest designs approach the gate oxide thickness in the early days of semiconductor technology when the problem of TDDB was first identified.

Throughout the many years that TDDB had been studied in gate oxides, it has been the subject of some controversy. The most important has been the "E vs 1/E" battle that is continually fought. The E model, (14) where failure is exponentially dependent on the negative of the applied field, and the 1/E model (15) where failure is proportional to the exponential of the reciprocal of the applied field are virtually indistinguishable at the test conditions generally used for testing. Unfortunately, the extrapolations to use condition can differ by tens of orders of magnitude. Given the same data, one can predict a disaster with the E model and at the same time predict there is no conceivable reliability problem with the 1/E model. This is a distinctly undesirable state of affairs for a reliability engineer. The controversy has not disappeared when applied to the interlevel dielectrics, but in fact it has deepened with the introduction of a class of new models that depened on the exponential of the negative of the square root of the applied field, now called "root-E" models.

There are now at least three distinct versions of the root-E TDDB failure model (16-18) for interlevel dielectrics and another based on trap creation that doesn't lend itself to a simple formula. (19) All assume that the rate of damage accumulation will be proportional to the number of electrons being pumped into the dielectric, hence the dependence on root-E. In one model (16) TDDB is caused by the diffusion of Cu either through the dielectric or along a dielectric/cap interface. It is assumed that conducting electrons create Cu ions from the metal atoms and after they are injected into the dielectric they are neutralized by capturing electrons and then diffuse by Fick's Law. The Cu serves to increase the conductivity of the dielectric interface leading to local magnification of the electric field and eventual breakdown. In a second similar model (17) it is assumed that the Cu remains as ions and drift from anode to cathode and build up a space charge near the cathode, again magnifying the electric field and causing damage. In the other two models (18,19) Cu does not play a direct role, but the

INVITED PAPER

electrons create damage in the dielectric eventually leading to failure.

Ref	Model Formula	Acceleration (3 - .3 MV/cm)	Comments
(14)	$Exp(-E)$	7×10^5	Thermochemical Model
(16) (17)	$Exp(-E^{1/2})$	2×10^{10}	Cu ionization and diffusion
(18)	$Exp(-E^{1/2} + 1/E)$	$> 10^{31}$	Impact damage

Table 1 Comparison of the root-E models to the E model

As we can see from Table 1 the projections from test to use conditions differs greatly depending upon which of the models you choose. Haase's model (19) must be solved numerically, so it was not included in the table, but it is expected to be approximately like the Lloyd et. al. model (18). Since all these models are almost indistinguishable in a normal testing range of, say, 3 to 5 MV/cm, it is critically important to the reliability engineer to have a way of choosing the correct model. Nothing in your everyday testing can signal which model should be used, but a more than 25 order of magnitude difference in predicted lifetime makes a difference.

Each of the models have problems that make it uncertain which, if any, of the models should be used if we are prepared to reject the most conservative E model. Two of the models (16,17) share a conceptual problem with the E model in that when E = 0, the equations governing failure predict there is still a finite lifetime. In ref (16) it is not clear why the Cu ions that were apparently created by the impact of the conducting electrons should neutralize and diffuse randomly. In ref (17) where the Cu is assumed to remain as ions, it is not clear why they should not follow kinetics that would be linear in the electric field instead of root-E in the exponent. (20) The models of Ref (18) and Ref (19) do not have the E = 0 paradox, but do not consider Cu as a contributor to failure, when experiments have shown that the absence of Cu results in substantially longer lifetimes. (21)

However, the industry consensus is gravitating towards a root-E formalism. Since the majority of the testing is done under conditions of extreme acceleration and yet over relatively small ranges of lifetime, making it difficult to prove one model's efficacy over another. In one case where over 7 orders of magnitude in lifetime ,it was found the data supported a model that deviated from root-E in the direction of a second 1/E term in the exponent. (18) More importantly, the coefficients could be derived from the model and the best fit to the data provided perfectly reasonable values for these coefficients. Convincingly, a threshold energy for causing damage of on the order of 1 eV and a mean free path for the electron in the dielectric of ~ 1 nm were suggested by the data.

The consequences of accepting a model with a second 1/E term in the exponent are significant. If incorporating 1/E is correct, it

is almost inconceivable that TDDB could be a real reliability problem. It also suggests that there would be a threshold voltage below which TDDB would be impossible. It stands to reason, and there is some supporting experimental evidence that this energy would be on the order of 1 eV. Therefore, since the maximum energy that could be gained by an electron in the field would be the applied voltage, if under operational conditions, this approaches 1 volt, we will not have to worry about TDDB. TDDB failure in real systems would be highly unlikely, regardless of the results of accelerated testing. Such a position is very risky, with the consequences of being wrong potentially disastrous. Thus, even though these models might even allow more relaxed use conditions, it has become accepted that the root-E model can be used with some relaxation of requirements over the E model, but still be conservative enough to engender confidence in the product.

SUMMARY

Recently, insight into BEOL failure mechanisms has enabled advances in reliability modelling for both electromigration and TDDB that more accurately predicts the lifetime under use conditions from accelerated data. Generally, the news is good in that it appears that we have been a bit more conservative than we need to be in our designs.

ACKNOWLEDGEMENT

I would like to thank my colleagues at IBM Research, especially Eric Liniger, Bob Rosenberg and Tom Shaw for useful discussions and support in the work that lead up to the modeling outlined in this paper.

REFERENCES

1) J.R. Black, Proc. 6th Ann. Reliab. Phys. Symp., 148 (1966)
2) IA Blech. J. Appl. Phys. **47**, 1203 (1976)
3) J. R. Lloyd, C. E. Murray, T. M. Shaw, M. W. Lane, X.-H. Liu, and E. G. Liniger, 8th International Workshop on Stress-Induced Phenomena in Metallization, AIP Conf. Proc.Vol. 817, 23 (2005)
4) M. Shatzkes and J.R. Lloyd, J. Appl. Phys., **59**, 3890 (1986)
5) M.A. Korhonen, P. Borgesen, K.N. Tu and C.-Y. Li, J. Appl. Phys. **82**, 3790 (1993)
6) S. Yokogawa and H. Tsuchiya, Jap. J. Appl. Phys., 44, 1717 (2005)
7) J.R. Lloyd, Microelectronics Reliab., **47**, 1468 (2007)
8) J.J. Clement, and C.V. Thompson J. Appl. Phys. **78**, 900 (1995)
9) I. Kaur, W. Gust and L. Kozma, HANDBOOK OF GRAIN AND INTERPHASE BOUNDARY DIFFUSION DATA, Ziegler Press Stuttgart 1989
10) C.-K. Hu, L. Gignac, E. Liniger, R. Rosenberg, and A. Stamper, Proc. 2002 IEEE International Interconnect Technology Conference, 133 (200)

INVITED PAPER

11) M. H. Lin, Y. L. Lin, K. P. Chang, K. C. Su and Tahui Wang, Jpn. J. Appl. Phys. **45**, 700 (2006)

12) L. Arnaud, Proc. 40th Ann. Intl. Reliab. Phys. Symp., 433 (2002)

13) W.D. Nix and E. Arzt, Metallurgical and Materials Transactions A, 2007 (1992)

14) J.W. McPherson, J. Appl. Phys., 95, 8101 (2004)

15) I.C. Chen, S. Holland and C. Hu, IEEE Trans. On Elec. Dev., ED-32, 413 (1985)

16) F. Chen, O. Bravo, K. Chanda, P. McLaughlin, T. Sullivan, J. Gill, J.R. Lloyd, E. Wu, R. Kontra and J. Aitken, Proc. Proc 44th Ann. Int. Reliab. Phys. Symp.54, (2006)

17) N. Suzumara, S. Yamamoto, D. Kodama, K. Makabe, J. Komari, E. Murukami, S. Maegawa and K. Kubota, Proc. Proc 44th Ann. Int. Reliab. Phys. Symp. 484 (2006)

18) J.R. Lloyd, E. Liniger and T.M. Shaw, J. Appl. Phys., **98**, 084109 (2005)

19) G. Haase, Proc 46th Ann. Int. Reliab. Phys. Symp., 556 (2008)

20) J. R. Lloyd, C. E. Murray, S. Ponoth, S. Cohen, and E. G. Liniger, Microelectron. Reliab. **46**, 1643 (2006)

21) J.R. Lloyd, S. Ponoth, E. Liniger and S. Cohen, Proc 45th Ann. Int. Reliab. Phys. Symp., 410 (2007)

Interfacial Characterization of ultra low-k film (kappa=2.55) with Time of Flight Secondary Ion Mass Spectrometry (TOF-SIMS)

J. Widodo*, Z. X. Xing^, Z. Q. Mo^, T. Ouyang*, D. Gui^, Y.N. Hua^, H. Liu*, W. Lu*
*Technology and Development Department
^ Quality and Reliability Assurance Department
Chartered Semiconductor Manufacturing. Ltd
Woodlands Industrial Park D
Street 2, Singapore 738406
Phone: 65-64137614. Fax: 65-64137506 Email: *johnnywidodo@charteredsemi.com*

Abstract- **Interfacial characterization of ultra low-k film with the layer underneath is very important for reliable manufacturing of ultra low-k film. In this study, we proposed a new application of TOF-SIMS to predict the interfacial behavior of the ultra low-k with the layer underneath. Strong correlation between carbon intensity at the bottom interface with the adhesion energy of the ultra low-k with the layer underneath is observed. We also observed a linear correlation between the quantified carbon peak intensity and the adhesion energy.**

I. INTRODUCTION

Continuous improvement in device performance and density has significantly impacted the feature size and complexity of the wiring structure for on-chip interconnects. As the critical dimension of integrated circuits is scaled down, resistive and capacitive delay of interconnects, power dissipation, and cross talk noise of the interconnect structure become one of the major limiting factors for ultra-large-scale-integration of integrated circuits. Therefore, there is a need to replace aluminum as the metallization and SiO_2 as the interlayer dielectric with copper and low-k, respectively, to reduce the RC delay.

Numerous studies on various low-k (LK) and ultra low-k (ULK) have been reported over the years[1, 2]. Reduction in the dielectric constant can be brought about by introducing pores into the films or by reducing the polarization[3]. In this study, we have chosen an ultra low-k film that has its reduction in the dielectric constant by introducing pores. Unlike SiO_2, carbon-doped oxide interlevel dielectrics (ILD) are known to have fundamentally weak interfacial strength to underlying films. The interfacial strength of ULK to an underlying film such as SiCN can typically withstand routine processing steps. However, more severe stress involving chip packaging can induce interfacial failures[4]. Simple PECVD deposition of bulk ULK directly on SiCN does not provide adequate interfacial strength based upon 4-point bending measurement[5, 6]. It was found that the interfacial failure was mostly a-near interface cohesive failure in the ULK.

This paper describes the correlation found between the composition profiles with the adhesion energy quantified by the 4-point bending measurement. Modifications on the general film deposition conditions have improved the adhesion energy

that is accompanied with the changes in the composition profiles. Finally, the correlation between the adhesion energy and carbon peak quantification obtained from the composition profile is discussed.

II. EXPERIMENTAL PROCEDURE

By means of plasma-enhanced chemical-vapor deposition (PECVD) methods using Applied Materials[TM] Producer, ultra low-k films (kappa 2.55) were deposited on 300mm Si. The ultra low-k film was cured in Applied Materials Nanocure[TM]. The adhesion energy was measured using 4-point bending method while the composition profile was characterized using Time of Flight Secondary Ion Mass Spectrometry (TOF-SIMS).

The typical specimen setting for 4-point bending test is shown in figure 1[7]. Details the derivation of the Gc equation is described elsewhere[7, 8].

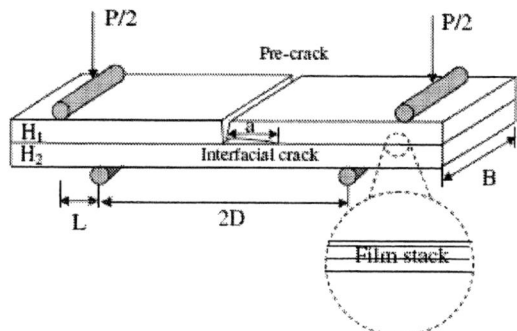

Fig. 1. 4-point bending test specimen

In the 4-point bending method, the adhesion energy was calculated using equation (1).

$$Gc = \frac{2L(1-v^2)M^2}{4Eh^3} \quad (1)$$

Where,
Gc= adhesion energy in J/m²
M = PL/2B is the bending moment per unit width
E is the elastic modulus

v is the Poisson's ratio
h is the thickness of the beam
P is the force
L is the pin spacing as shown in figure 1
B is the specimen width as shown in figure 1

In the TOF-SIMS experiment, Bi^+ Liquid Metal Ion Gun (LMIG) was used as primary source for analysis gun, and operated at 25keV with 1pA current with analysis area set at 100μm x 100μm. Cs source was used as sputtering gun for depth profiling, applied low beam energy and beam current to achieve a good depth resolution. The Cs ion gun was operated at 500eV, and beam current is 30nA. To avoid crater effect, the raster size is set at 300μm x 300μm. To compensate the possible charging issue, E-gun flooding at 500eV was also applied to the experiment.

III. RESULT AND DISCUSSION

In this study, the initial un-optimized film has poor adhesion energy (G_c), at <4J/m², characterized by four-point bending technique. Composition profile analysis identified the association of poor interfacial strength with a high concentration of carbon near the bottom interface. The best approach to consistently avoid any initial carbon spike is to deposit a thin oxide layer, followed by a continuously graded transition region to the bulk ULK composition. The initial oxide layer is deposited using the same chemicals as used for bulk ULK, but with higher O_2 and lower ULK precursors flow rates. Optimization of the graded transition layer is crucial to maintain good interfacial strength[9]. TOF-SIMS sputter depth profiling is used to analyse the composition profile and is achieved using 500eV Cs sputtering with profiles being collected in the negative ion mode.

Initial high concentration of carbon near the bottom interface for un-optimized ULK film is shown in figure 2. High carbon concentration ("C" spikes) near the bottom interface is accompanied with low oxygen concentration ("O" dips). The "O" dip was not clearly observed in figure 2 as logarithm scale is used. Various hardware improvements and process modifications have been performed to remove the "C" spikes to improve the adhesion energy.

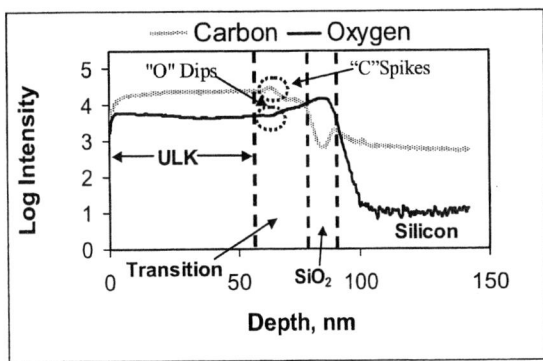

Fig. 2. Carbon composition profile for un-optimized ULK film

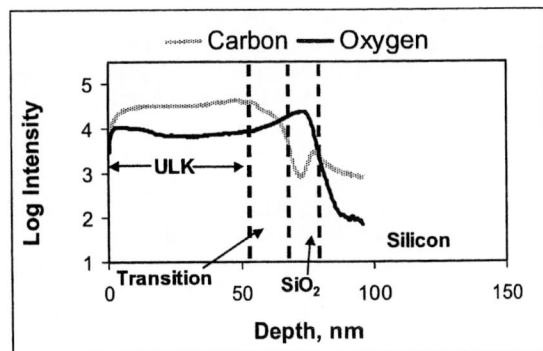

Fig. 3. Carbon composition profile for optimized ULK film

Figure 3 shows the carbon composition profile for optimized ULK film. The optimized ULK film does not exhibit any high initial carbon concentration or any low initial oxygen concentration near the bottom interface. The G_c measured for optimized ULK film is 5.5J/m². Figure 4 showed the comparison of oxygen compositional profile between optimized and un-optimized ultra low-k film in linear scale. "O" dips was clearly shown accompanying the "C" spikes in the un-optimized ultra low-k film that resulted in low adhesion energy.

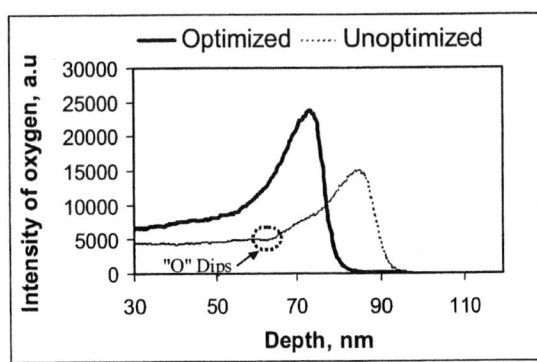

Fig. 4. Carbon composition profile for optimized ULK film

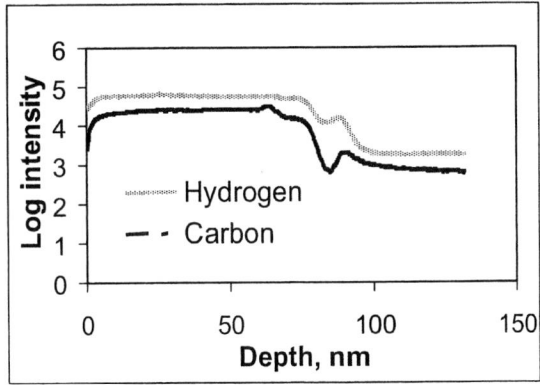

Fig. 5. C and H composition profile for un-optimized ULK film

However as shown in figure5, we observed that the hydrogen concentration at the transition region remains unchanged. This observation has led us to a hypothesis whereby the ratios of carbon, hydrogen, oxygen and silicon that bond to each other at the transition region of the ultra low-k film is extremely crucial to avoid the low adhesion energy. Therefore, there is a possibility of having part of the transition layer not properly bonded when we observed an extra carbon peak in the transition region as shown by the un-optimized ultra low-k film.

Figure 6 shows the G_c comparison between optimized ULK film with un-optimized ULK film. Significant improvement on the G_c value was observed. The G_c was improved by almost 50% of the original value. The significant improvement was achieved by suppressing the initial high carbon concentration at near bottom interface. A minor modification in the transition region was done to suppress the initial high carbon concentration as shown in the optimized film.

In order to obtain an optimised transition region composition profile in ULK film, several factors must be considered for process equipment and recipe. The ULK precursor ramp rate and its carrier gas for the liquid flow controller should be adjusted to an optimum value. If the ramp rate is set too low, then the transition region will widen up, thus resulting in the increment of the effective dielectric constant for the entire ULK film. If set too high, then carbon spikes is likely to happen. Another important factor is to grade the transition region without turning off the plasma throughout the entire oxide and graded layer deposition sequence. Deposition of SiO_2 layer followed by a separate deposition of ULK did improve only slightly the interfacial strength9.

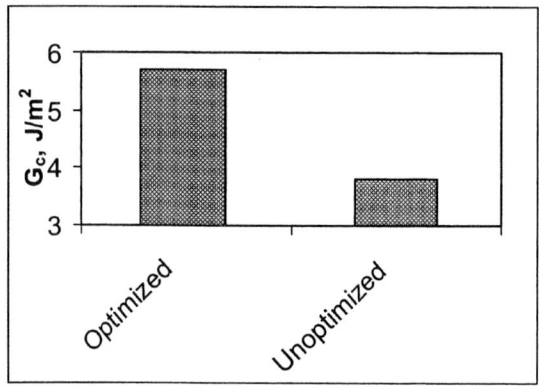

Fig. 6.   Comparison of G_c for optimized and un-optimized ULK film.

As described elsewhere[9], defect performance was monitored as well to ensure that no particle formed due to gas phase nucleation. We monitored the defect performance for several splits of the optimized film by varying the precursors ramp up rate. All the splits showed not only clean defect performance as shown in figure 7 but also Gc value >5 J/m² as shown in figure 8. Those adhesion energy data indicate that the optimised ULK film described in this work is very robust.

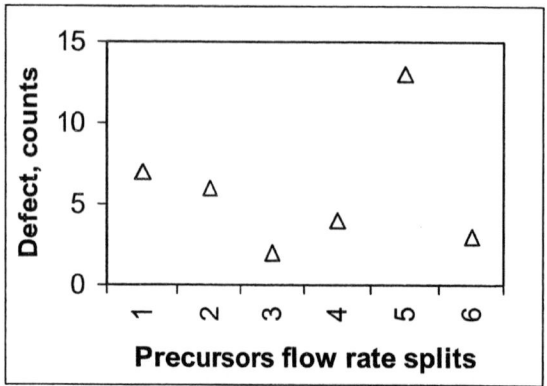

Fig. 7.   Defect performance of the precursors ramp up splits.

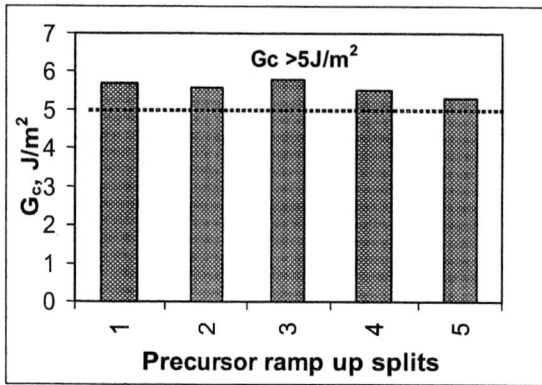

Fig. 8. G_c performance of the precursors ramp up splits

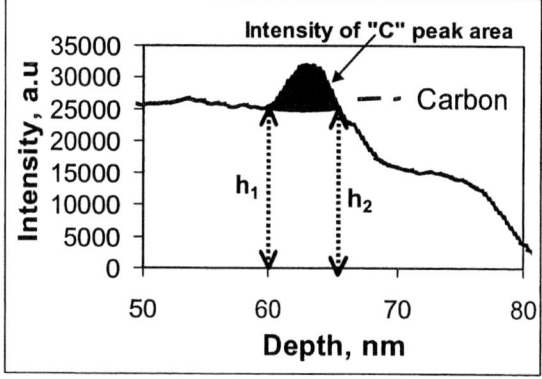

Fig. 9.   Typical example of C peak area calculation

Figure 9 showed a typical example of compositional profile for un-optimized film whereby we observed a very distinct "C" spike that can be quantified with equation (2).

$$C = \frac{\text{Intensity of "C" peak area}}{(h_1 + h_2)/2} \qquad (2)$$

Where,

C: Carbon peak intensity, a.u

h_1: Intensity number of "C" spike starting point

h_2: Intensity number of "C" spike ending point

In table 1, we listed the quantification of carbon peak intensity derived from the TOF-SIMS compositional profile for various Gc value. We observed a strong linear correlation between the carbon peak intensity and the adhesion energy as shown in figure 10 although a typical 4-point bending and carbon intensity data will give about 10% standard deviation.

Table.1 List of carbon peak intensity over various Gc value

Split	Carbon Peak Intensity, a.u	Gc, J/m^2	Remarks
1	2.27	5.7	Optimized
2	2.15	5.6	
3	2.03	5.8	
4	2.15	5.5	
5	2.26	5.3	
6	3.58	3.9	Unoptimized

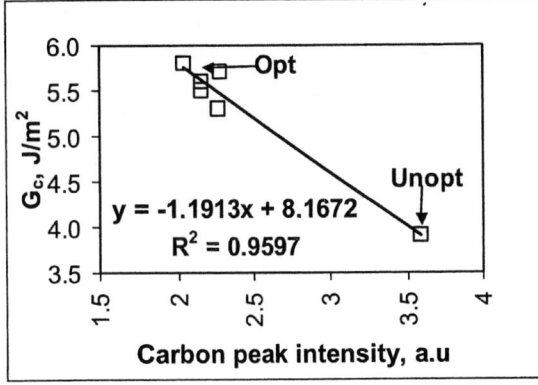

Fig.10. Correlation of Gc with the quantification of carbon peak intensity

IV. CONCLUSION

In this paper, we have shown that we have a strong correlation between "C" spikes and "O" dips intensity characterized by TOF-SIMS with the adhesion energy, characterized by 4-point bending method, of the ultra low-k film with the layer underneath. Strong linear correlation was observed as well for different range of adhesion energy value. Current optimized ultra low-k film showed a wide process window for carbon composition profile, adhesion energy and defects performance.

ACKNOWLEDGEMENTS

The author would like to thank S. Reiter from Applied Materials. Inc and A. Jain, R. Leong, and H. S. Tang from Applied Materials. South East Asia. Pte. Ltd. for the support on 4-point bending measurement.

REFERENCES

[1] J. Widodo et al., "*Comparisons on 3MS and 1,3,5,7 TMCTS based Low-k films: Dielectric and Mechanical Properties*", Journal of The Electrochemical Society 152(4), G246-G251.

[2] M. Damayanti et al, "*Effect of porosity and adhesion promoter layer on adhesion energy of nanoporous inorganic low-kappa*", Journal of Thin Solid Films, 504/1-2, 2006, 213-217.

[3] M. Morgen, E.Todd Ryan, Jie-Hua Zhao, Chuan Hu, Taiheui, and Paul S. Ho, Annu. Rev. Mater. Sci. 30, 645 (2000)

[4] W. Landers et al., "*Chip-to-package interaction for a 90nm Cu/PECVD Low-k technology*", Proceedings IITC 2004, pp.108-110

[5] M. Lane et al, J. Mater. Res., 15, 203(2000)

[6] M. Damayanti et al., "*Adhesion study of low-k/Si system using 4-point bending and Nanoscratch Test*", Journal Materials Science and Engineering B, 121 (2005) 193-198.

[7] Zhenyu Huang, Z. Suo, Guanghai Xu, Jun He, J.H.Prèvost, N. Sukumar, "*Initiation and arrest of a interfacial crack in a four-point bend test*", Engineering Fracture Mechanics, 72 (2005), 2584-2601.

[8] Sassan Roham, Kedar Hardikar, and Peter Woytowitz, "*Stress Analysis of 4-Point Bend Test for Thin Film Adhesion*", Mat. Res. Soc. Symp. Proc. Vol 778, 2003.

[9] D. Restaino et.al.," *Optimized Interfacial Strength for Dense and Porous SiCOH*", Invited paper in 24th Advance Metallization Conference 2007, USA.

Etching of Copper in Deionized Water Rinse

J. Gambino, J. Robbins, T. Rutkowski, C. Johnson, K. DeVries*, D. Rath**, P. Vereecken**,
E. Walton*, B. Porth, M, Wenner, T. McDevitt, J. Chapple-Sokol, S. Luce
IBM Microelectronics, 1000 River Street, Essex Junction, VT, 05452
*IBM Microelectronics, Rt 52, Hopewell Junction, NY
**IBM T.J. Watson Reseach Center, Yorktown Heights, NY 10598

Abstract – **A new yield loss mechanism is described that is related to the etching of Cu in deionized water. Water that contains high concentrations of dissolved oxygen can etch Cu at the bottom of vias during pre-metallization wet cleans. The etching creates voids in the Cu which remain after metallization, resulting in high resistance and functional fails in the affected array circuits. The dissolved oxygen concentration in the deionized water must be minimized to prevent etching of Cu.**

I. INTRODUCTION

Copper interconnects have gained wide acceptance in the microelectronics industry due to improved resistivity and reliability compared to Al interconnects [1]. However, there are many yield issues associated with Cu metallization, especially in terms of stability during wet chemical cleans. Corrosion or etching of Cu is often observed after wet cleans associated with chemical mechanical polishing (CMP) [2,3,4] or via etch (i.e. prior to etch stop removal)[5].

Three different types of corrosion have been reported for Cu interconnects; photo-corrosion, galvanic corrosion, and chemical corrosion.   For corrosion to occur, at least two reactions are required; an anode reaction and a cathode reaction [6].   At the anode, the metal surface is oxidized, forming metal ions and electrons;

$$Cu(s) \delta Cu^{2+} + 2e^- \qquad (1)$$

At the cathode, electrons are consumed by several possible reactions:

$$2H^+ (aq) + 2e^- \delta H_2 (g) \qquad (2a)$$
$$O_2 (g) + 2H_2O (\textit{l}) + 4e^- \delta 4OH^- (aq) \qquad (2b)$$
$$Cu^{2+} + 2e^- \delta Cu(s) \qquad (2c)$$

In addition, there must be an electrical connection between the anode and the cathode, and there must be an electrolyte in contact with both the anode and the cathode.

Corrosion of copper interconnects often depends on the pattern. Photocorrosion is observed in the presence of light when Cu is electrically connected to a p-type Si region in the substrate [2,3]. The Cu connected to p-type Si (anode) is at a more positive potential than Cu connected to n-type regions (cathode), making it more susceptible to corrosion. Light allows current to flow through the Si and complete the electrochemical cell with the electrolyte, providing the charge carriers required for corrosion to occur (i.e. in equations (1) and (2)).

Etching of Cu is often associated with DI H₂O rinses [4.5]. At first, this appears to be a surprising result, because DI H₂O is often considered to be an innocuous cleaning treatment.

However, in the electrochemistry literature, it is well known that Cu is thermodynamically unstable in H_2O in the presence of oxygen (Fig. 1)[7,8]. If oxygen is present in DI H_2O, it can consume the excess electrons produced by dissolution of Cu (equations 2b and 1, respectively), allowing Cu corrosion to proceed. The corrosion rate of Cu in DI H_2O depends on dissolved oxygen concentration, temperature, and pH, with a maximum corrosion rate for a dissolved oxygen concentration of 200 to 300 ppb [8]. The use of dissolved oxygen in DI H_2O has been proposed as a way to improve the effectiveness of post-CMP cleans (i.e. for more effective removal of Cu contamination on the surface) [9]. However, high oxygen concentrations in the DI H_2O can cause corrosion during post-CMP cleans [4]. The etching of Cu in DI H_2O rinses after via etch has also been associated with the presence of "oxygen from the cleaning solution" [5].

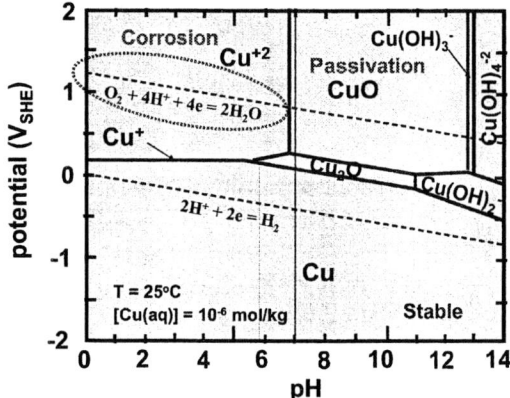

Fig. 1. Pourbaix diagram for copper at 25°C (ref. 7). Copper is thermodynamically unstable in deionized H_2O (with dissolved oxygen) at pH below 7.

Although the etching of Cu after DI H_2O rinses has been reported after CMP and after via etch (prior to etch stop removal), there are no reports in the literature on Cu corrosion associated with pre-metallization cleans. In this study, we show that localized Cu etching can in fact occur due to deionized water rinses associated with pre-metallization cleans. We propose that this is due to a corrosion reaction that is accelerated by the presence of dissolved O_2 gas in the water.

978-1-4244-2039-1/08/$25.00 ©2008 IEEE

II. EXPERIMENT

Samples were fabricated using a 0.13 μm CMOS process with via-first dual damascene Cu in an FSG dielectric [10]. The preclean prior to metallization consists of dilute HF and deionized water rinses [11].

Electrical fails were detected on fully integrated logic chips. The electrical location of fails in the product chip were localized using a scan chain method. Scan chains are embedded in logic circuits to test small subsets of the circuit [12]. Additional test pins are added to provide inputs and outputs to critical portions of the circuit. When fails are detected in scan chains, the failing area is relatively small, and therefore can be easily detected by delayering techniques.

The failing chips were delayered and were then analyzed by scanning electron microscopy (SEM), transmission electron microscopy (TEM), and scanning transmission electron microscopy (STEM). The chemical composition of the failing structures was analyzed in the STEM (JEOL 2010F) using energy dispersive spectroscopy (EDS). The TEM sample preparation consisted of delayering to V1, capping with SiO₂ to protect the surface, then extracting a thin sample (150 nm) from the region of interest using focused ion beam (FIB) milling.

III. RESULTS

During the early ramp of this technology in manufacturing, it was observed that the yield was abnormally low for scan chains that contained a "special" array circuit. The yield for the scan chains generally decreased as the percentage of array circuits in the scan chain increased (Fig. 2). The area of these scan chains is very small, so the yield loss was much higher than that expected based on random defects.

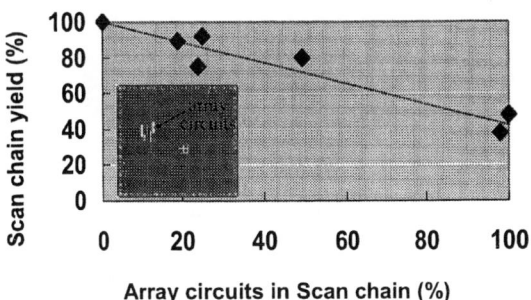

Fig. 2. Scan chain yield vs percentage of array circuits in the scan chain.

Initial failure analysis indicated that the fails were due to voids in the M1 metal, near vias (Fig. 3). The Ta liner is visible adjacent to the void, indicating that the problem is not related to the M1 liner deposition. TEM images (Fig. 4 and 5) show that the SiN layer on top of the void is smooth, indicating that the void formed after the SiN deposition. The

TEM also shows that the Ta liner at the bottom of the via is missing (Fig. 5), suggesting that the void formed prior to metal deposition in the via. EDS analysis confirms that the most of the Ta liner at the bottom of the via is missing (Fig. 6). Hence, the Cu in M1 directly underneath the via is from the V1/M2 metal deposition, rather than the M1 deposition (i.e. the metal from the via deposition partially fills in the void in M1).

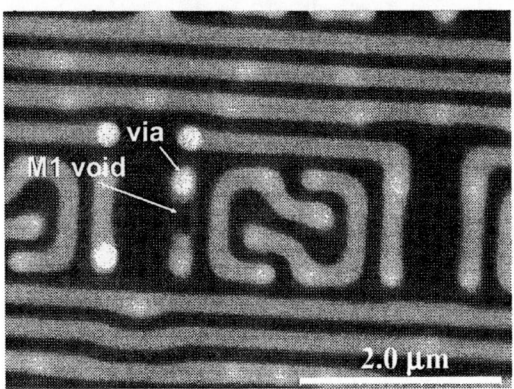

Fig. 3. Top-down SEM of void in M1 copper.

The failure analysis indicated that the voids in the M1 metal occurred between the SiN deposition on top of M1 and the metal deposition into the via. Based on this information, automated bright field optical microscope inspections of the array were used to determine which process was causing the voids. Based on the inspections, the voids were found to form as a result of the DHF preclean and deionized water rinse prior to V1 metallization (Fig. 7). Note that the number of fails detected by the automated optical microscope inspection is only a fraction of those detected electrically.

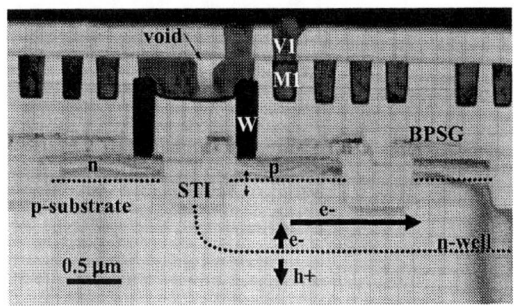

Fig. 4. TEM micrograph of void in M1 copper.

IV. DISCUSSION

The etching of the Cu during the wet clean is unique to the special array circuit. Other circuits without this array have high yield (Fig. 2). Hence, the underlying Si in the special array contributes to the corrosion. This was confirmed by running some wafers where the contacts were intentionally underetched. An automated optical microscope inspection showed no voids in M1, indicating that the etching of Cu does not occur if the M1 is floating.

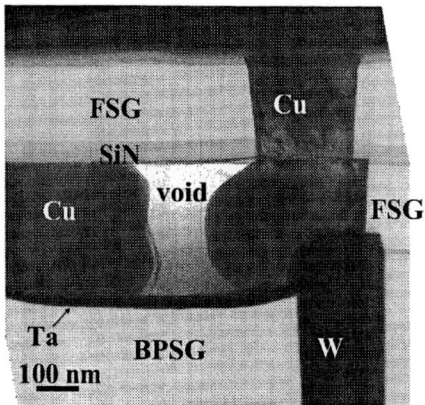

Fig. 5. TEM micrograph of void in M1 copper.

The Cu that is etched is connected to the substrate at a p-type junction. This suggested that photocorrosion might be contributing to the etching of the M1 Cu [3,4]. Therefore, experiments were run modulating the amount of light incident on the wafers in the wet etch tool. However, the automated optical microscope inspection showed a similar amount of voids in the M1 Cu, with or without the presence of light. This suggests that the etching of Cu is not due to photocorrosion.

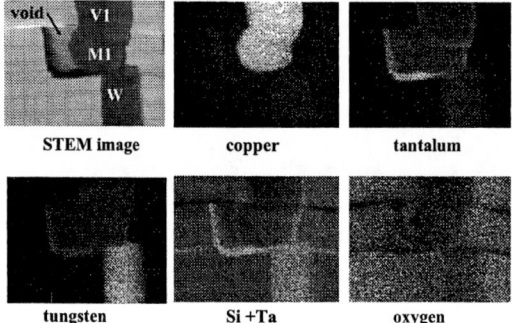

Fig. 6. EDS analysis of defect region.

Dissolved O_2 gas in water is known to enhance the corrosion of Cu (Fig. 1)[4-9]. Therefore, the formation of voids in the M1 Cu was studied as a function of the O_2 concentration in the deionized water rinse. The automated optical microscope inspections showed that the voids could be eliminated by minimizing the O_2 content in the deionized water. This was confirmed by electrical measurements on scan chains with a high percentage of the special arrays (Fig. 8). It was also determined that a deionized water rinse by

itself (i.e., with high O_2 content, but no DHF) could cause voids in Cu.

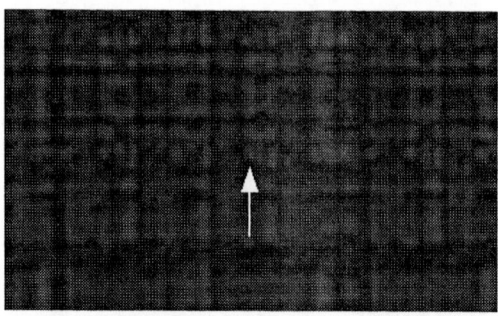

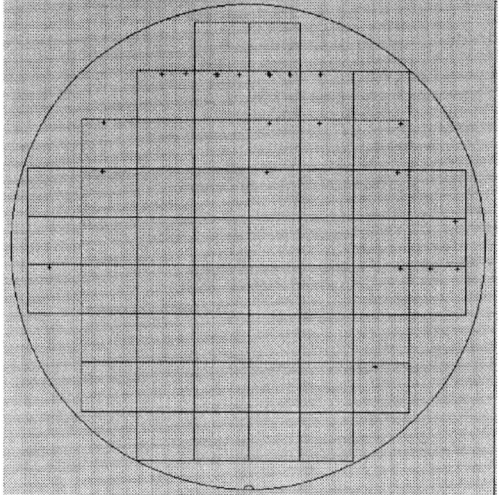

Fig. 7. Optical micrograph and defect map of void in M1 after M2 dilute HF preclean.

The etching of M1 Cu in the special arrays is clearly related to having a high O_2 concentration in the water rinse. However, that exact mechanism is still unclear. There are at least two possibilities; galvanic corrosion and junction-enhanced corrosion (Fig. 9).

Enhanced galvanic corrosion of Cu with respect to the Ta liner is expected due to the high O_2 concentration in the water. All of the M1 patterns that exhibit corrosion have minimum width lines, so Ta as well as Cu is exposed to preclean at the bottom of the via. For this mechanism, the Cu is the anode and the Ta is the cathode. The role of O_2 is to consume the electrons at the Ta cathode (equation (2b) and reaction (a) in Fig. 9). However, there are many other partially landed vias where the M1 metal doesn't corrode, so the underlying junctions in the Si must also be playing a role.

The role of the underlying Si junctions could be similar to role of the junctions in photocorrosion. The Cu connected to the p-type junctions is at a lower potential than the Cu connected to the n-type regions (and therefore more susceptible to corrosion), if a bias is applied to the junction.

One way to apply a bias to the junction is with light. However, etching of M1 Cu was observed even in the absence of light, suggesting that photocorrosion is not the mechanism. A more likely mechanism is that the O_2 is consuming electrons at the Cu connected to n-type regions (equation (2b) and reaction (b) in Fig. 9), and thereby biasing the junctions.

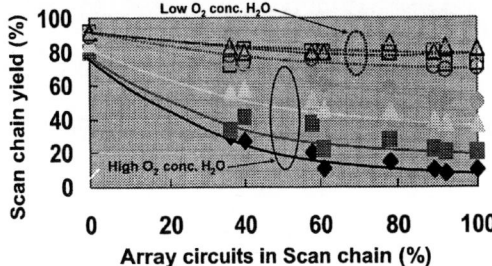

Fig. 8. Effect of deionized H_2O rinse (as a function of dissolved oxygen) on scan chain yield.

V. CONCLUSION

A new yield loss mechanism is described that is related to the etching of Cu in deionized water. Water that contains high concentrations of dissolved oxygen can etch Cu at the bottom of vias during pre-metallization wet cleans. The etching creates voids in the Cu which remain after metallization, resulting in high resistance and functional fails in the affected array circuits. The dissolved oxygen concentration in the deionized water must be minimized to prevent etching of Cu.

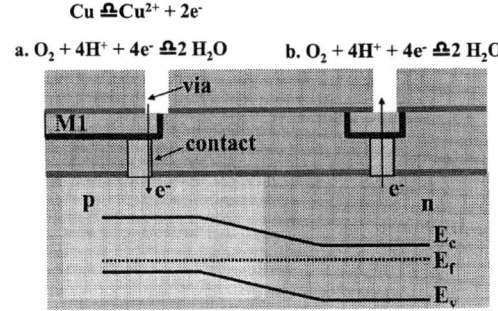

Fig. 9. Schematic of etching of copper during deionized H_2O rinse.

ACKNOWLEDGEMENT

The authors acknowledge C. Farris, P. Charron, D. Forcier, G. Tidman, and the staff of the IBM Burlington manufacturing facility for help with processing the wafers. The authors also acknowledge helpful discussions with T. Vogel, W. Lepuschenko, R. Wisnieff, D. Kneble, L. Gignac, D. Miura, B. Clark, J. Crafts, G. Shelley, A. Stamper, M. Tiersch, and O. Krom.

REFERENCES

1. D. Edelstein, J. Heidenreich, R. Goldblatt, W. Cote, C. Uzoh, N. Lustig, P. Roper, T. McDevitt, W. Motsiff, A. Simon, J. Dukovic, R. Wachnik, H. Rathore, R. Schulz, L. Su, S. Luce, J. Slattery, "Full Copper Wiring in a Sub-0.25µm CMOS ULSI Technology", *IEDM Proc.*, 1997, p. 773.
2. A. Beverina, H. Bernard, J. Palleau, J. Torres, F. Tardiff, "Copper Photocorrosion Phenomenon during Post CMP Cleaning", *Electrochem. Sol. St. Lett.*, **3**, 156 (2000).
3. Y. Homma, S. Kondo, N. Sakuma, K. Hinode, J. Noguchi, N. Ohashi, H. Yamaguchi, N. Owada, "Control of Photocorrosion in Copper Damascene Process", *J. Electrochem. Soc.*, **147**, 1193 (2000).
4. M. Kodera, Y. Matsui, H. Kosukegawa, N. Miyashita, M. Kamezawa, K. Ito, "Corrosion Control Technique in Copper Metallization Using Gas Dissolved Water", *IITC. Proc.*, 2002, p. 105.
5. Y.C. Ee, C. Perera, J.B. Tan, B.C. Zhang, Y.K. Siew, B.M. Seah, R. Joy, C.H. Low, H. Liu, S.T. Chua, F.C.H. Lim, T. Fu, L.C. Hsia, "Charging Induced Missing Pattern on Metal Layers in 65nm Technology Node", *AMC 2006 Proc.*, MRS, 2007, p. 557.
6. J.C. Kotz, K.F. Purcell, *Chemistry and Chemical Reactivity*, Saunders College Pub., N.Y., 1987, Chap. 19.
7. B. Beverskog, I. Puigdomenech, "Revised Pourbaix Diagram for Copper at 25 to 300°C", *J. Electrochem. Soc.*, **144**, 3476 (1997).
8. R. Dortwegt, E.V. Maughan, "The Chemistry of Copper in Water and Related Studies Planned at the Advanced Photon Source", *Proc. IEEE Part. Accel. Conf.*, 2001, p. 1456.
9. K.K. Christenson, C. Pizetti, "Use of Dilute HF with Controlled Oxygen for Post Cu CMP Cleans", in *Chemical Mechanical Polishing – Fundamentals and Challenges*, ed. S.V. Babu, S. Danyluk, M. Krishnan, M. Tsujimura, MRS, vol. 566, 2000, p. 267.
10. A. Stamper, C. Adams, X. Chen, C. Christiansen, E. Cooney, W. Cote, J. Gambino, J. Gill, S. Luce, T. McDevitt, B. Porth, T. Spooner, A. Winslow, R. Wistrom, "0.13 mm Generation Integration and Manufacturing of Dual Damascene Copper in FSG", *AMC 2002 Proc.*, ed. B.M. Melnick, T.S. Cale, S. Zaima, T. Ohta, MRS, 2003, p. 485.
11. J. Gambino, E. Cooney, S. Barkyoumb, J. Robbins, A. Rutkowski, A. Piper, M. Moon, C. Benson, E. Walton, C. Johnson, B. Laughlin, M. Gibson, J. Coffin, H. Wildman, "Precleans for Copper Vias in an FSG Process", *AMC Proc. 2001*, ed. A.J. McKerrow, Y. Shacham-Diamond, S. Zaima, T. Ohba, MRS, 2002, p. 49.
12. R.B. Norwood, E.J. McCluskey, "Synthesis-for-scan and scan chain ordering", *IEEE VLSI Test Symp.*, 1996, p. 87

Current-Induced Breakdown of Carbon Nanofiber Interconnects

Hirohiko Kitsuki, Tsutomu Saito, Toshishige Yamada, Drazen Fabris,
Patrick Wilhite, Makoto Suzuki, and Cary Y. Yang
Center for Nanostructures, Santa Clara University
Santa Clara, California 95053-0569, USA
Phone +1-408-554-6817, Fax: +1-408-554-5474, E-mail: kituki@mac.com

Abstract- **Current-induced breakdown phenomena of carbon nanofibers (CNFs) for future on-chip interconnect applications are presented. The effect of heat dissipation via the underlying substrate is studied using different experimental configurations. Scanning electron microscopy (SEM) techniques are utilized to study the structural damage by current stress. While the measured maximum current density in the suspended CNF in air is inversely proportional to nanofiber length and independent of diameter, SiO_2-supported CNFs improves their current capacity, which implies effective heat dissipation to the oxide. The correlation between maximum current density and electrical resistivity confirms the importance of local Joule heating, showing strong coupling between electrical and thermal transport in CNFs.**

I. INTRODUCTION

Copper interconnect is rapidly proceeding toward its minimization limit as a result of material failure due to electromigration. Carbon-based nanostructures such as carbon nanotubes (CNTs) [1-5] and carbon nanofibers (CNFs) [6,7] are being investigated for high-performance device and interconnect applications, because of their high electrical and thermal conductivities as well as current capacity. The growth of CNFs consistently yields high conductivity and high directionality, which are attractive for realistic interconnect fabrication processes [6]. In our previous study [7], CNF vias embedded in SiO_2 demonstrated high degree of reliability while being subjected to a stressing current of 1×10^7 A/cm^2. Thus such a structure is expected to achieve the current density target set by the International Technology Roadmap for Semiconductors (ITRS) [8] for the year 2015.

In CNT systems, breakdown phenomena have been observed under high electric fields, including nonlinear transport in single-walled CNTs [9] and successive graphitic wall breakdown in multi-walled CNTs [10]. In recent studies [7,11], proof of concept of the high-current reliability of CNFs for on-chip interconnects and the high-field transport properties of CNFs have been demonstrated. These results indicate that CNF breakdown mainly depends on resistive heating [12], but details of the failure mode due to high current and the accompanying physical breakdown mechanisms have yet to be investigated. While current annealing has been reported to drastically reduce the overall resistance and attributable to significant lowering of contact resistances in CNT devices [13][14], the significance of Joule heating in current-induced breakdown [11] implies that thermal contact coupling between CNF, electrodes, and surrounding materials (e.g., SiO_2) affects the current capacity of

the CNF. In this work, the effect of heat dissipation via the electrodes and underlying substrate is studied for CNFs under high-current stress using different experimental configurations.

II. EXPERIMENT

The CNF samples are grown by plasma-enhanced chemical vapor deposition (PECVD) with a Ni catalyst layer on Si substrate. The detailed growth conditions have been described elsewhere [14]. Figure 1(a) shows a CNF sample suspended between gold electrodes, with a SEM image at 75° tilted-angle view. This planar geometry is a model of horizontal on-chip interconnect configuration, where the CNF sidewall is in contact with the electrodes [15]. A DC current source connected to these electrodes is also shown in Fig. 1(a). Constant-current stress (equivalent to current annealing) is then carried out to monitor the electrical resistance prior to breakdown in air. In order to study the failure mode of the CNFs fabricated in a more realistic device structure on a Si substrate, CNFs supported by SiO_2 are also examined, as shown in Fig. 1(b). Results for twenty devices, with CNFs ranging from 100 to 200 nm in diameter and 1.5 to 6 μm in length, are presented in this paper.

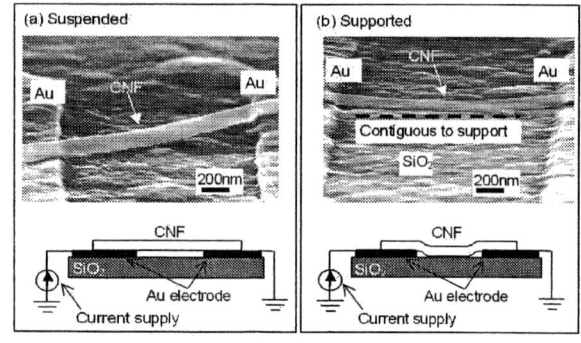

Fig. 1 Set-up for current-stressing experiments. (a) CNF suspended by gold electrodes. (b) CNF supported by SiO_2 substrate. Upper figures show SEM image of a CNF sample at 75° tilted-angle view, lower figures illustrate schematic of electrical measurement.

The progression of constant-current stress cycles (at 180 sec. each) is illustrated in Fig. 2(a). At the end of each cycle, *I-V* characteristics are obtained around $V = 0$. Figure 2(b) shows the resistance of a suspended CNF device after each annealing cycle versus annealing current. Increasing the annealing current results in a gradual decrease in resistance before the nanofiber breaks down, at 700 μA for this particular device. The measured resistance consists of CNF bulk and contact resistance between the fiber and electrodes, and the CNF consistently breaks near the middle. This result implies that current annealing reduces the contact resistances significantly, while the device approaches breakdown due to resistive Joule heating in the bulk of the CNF.

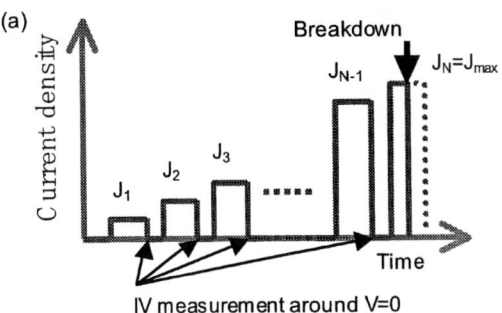

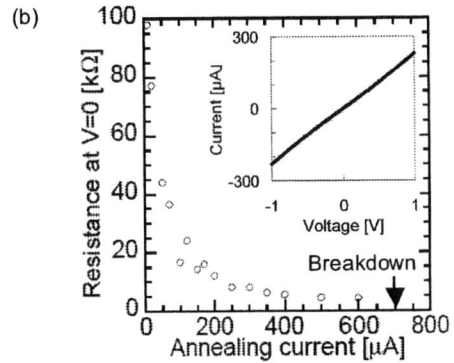

Fig.2 Resistance reduction of CNF device due to current annealing. (a) Schematic of successive current annealing cycles using stepwise increment of stressing current. (b) Resistance of the suspended CNF device at $V=0$ obtained after each annealing cycle. The inset shows the current-voltage behavior at the end of one of the anneal cycles.

III. RESULTS AND DISCUSSION

SEM image of a CNF suspended by gold electrodes before and after current-induced breakdown is shown in Fig. 3(a). In all experiments for suspended CNFs, we have confirmed that breakdown always occurs near the middle of the nanofiber. This is consistent with diffusive heat transport observed in CNTs at high bias [10], suggesting that resistive heating [12] is critical to CNF breakdown.

For CNFs supported by SiO$_2$, result of current-induced breakdown is shown in Fig. 3(b). The interface between CNF and the substrate is imaged using SEM contrast [16]. A dark-contrast region along the CNF indicates a section of the nanofiber contiguous to the substrate, while the bright section is not in contact or suspended. Breakdown occurs in the segment where CNF is suspended above the substrate. It is expected that

there is more heat transfer from CNF to substrate in the supported segment, thus resulting in higher current capacity than the suspended CNF. Meanwhile, the suspended segment is prone to failure because of poor thermal coupling to surrounding materials.

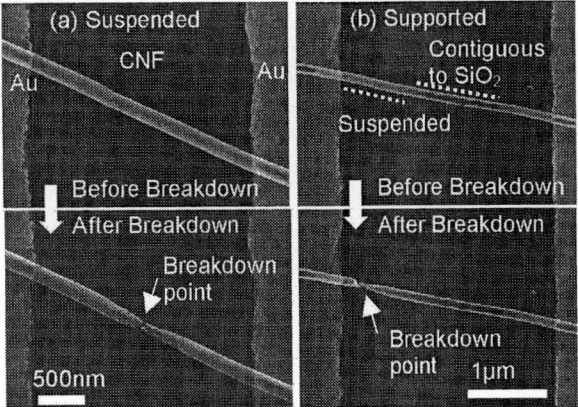

Fig. 3 SEM images of CNFs before and after current stressing at the top view. (a) CNF suspended by gold electrode. (b) CNF supported by SiO$_2$ substrate.

Figure 4 shows the relationship between the maximum current density (J_{max}) and reciprocal CNF length. Data obtained for the suspended CNF are given by solid circles. The decrease of J_{max} with increasing length is consistent with current-induced breakdown of single-walled carbon nanotubes (SWNT) [17] and gold nanowires fabricated using conventional lithography [18]. It should be noticed that a rapid current sweep of a few seconds led to breakdown near the electrodes, far from the middle of the fiber, suggesting substantial Joule heating due to contact resistance. Consequently, the measured J_{max} deviates significantly from the J_{max} versus length behavior as shown in Fig. 4. The entire annealing sequence as illustrated in Fig 2(b) typically takes about 45 minutes, consistently resulting in breakdown near the middle of the nanofiber as shown in Fig 3(a). The successive current annealing process effectively reduces the rise in temperature at or near the electrodes due to gradual decrease of contact resistance and more time for heat dissipation via the contacts, making the Joule heating along the CNF length the primary contribution to breakdown. This observed behavior is useful for predicting the current capacity of CNFs with different lengths when designing interconnects.

A one-dimensional thermal transport model [12,19] is used to analyze correlation between the observed current capacity J_{max} and length L of the CNFs, and the results is given by [20]

$$J_{max} = [(T_{max} - T_\infty)\sigma\gamma w / A]^{1/2}[1 - 1/\cosh(aL/2)]^{-1/2}. \quad (1)$$

Here $a^2 = w\gamma/A\kappa$, where σ is the electrical conductivity, κ the CNF thermal conductivity, w the effective contact line width, and γ a coupling coefficient to account for the efficiency of heat transport through the CNF surroundings. T_{max} is the temperature at the breakdown point and T_∞ the ambient temperature. In the suspended case, heat

978-1-4244-2039-1/08/$25.00 ©2008 IEEE

dissipation is expected to be negligible, or $aL \ll 1$. In this limit [20],

$$J_{max} \approx 2\sqrt{2(T_{max} - T_\infty)\sigma\kappa} \, / \, L \qquad (2)$$

In Fig. 4, the measured J_{max} is plotted as a function of $1/L$. It is seen that the J_{max} versus $1/L$ behavior can be fitted to a straight line, as predicted by Eq. (2). For breakdown of SWNTs, a similar development using heat transport equation yielded the same result as here [9]. This agreement confirms that for a suspended CNF, heat dissipation to the surroundings (air) is small, and heat conduction along the length of the CNF causes the highest temperature to occur at the middle of CNF.

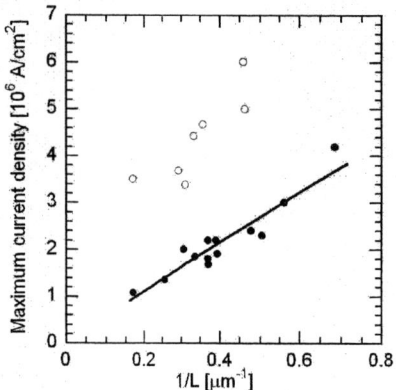

Fig. 4 Dependence of maximum current density on CNF length obtained using 20 devices. The solid and open circles show the results for suspended and supported CNFs, respectively. The straight line is a linear fit for suspended CNFs, as predicted by the heat transport model.

The maximum current density observed for the supported configuration is plotted with open circles in Fig. 4. The supported CNFs show significantly improved current capacity in comparison to the suspended CNFs. However, the dependency of maximum current density on length is not as clear. While J_{max} increases with with decrease in CNF length as in the suspended case, its behavior is largely affected by the area in contact with the support material, which contributes significantly to the heat dissipation process. This finding indicates that heat dissipation via its immediate surroundings is a critical determining factor for high-current transport in CNF. While the contrast in SEM images is useful to estimate where the CNF is contiguous to the supporting SiO_2, thermal coupling between the nanofiber and the substrate is yet to be investigated for analyzing the current capacity versus the length behavior.

Combining the experimentally fitted line for J_{max} versus $1/L$ in Fig. 4 with Eq. (2), $T_{max} = 1260$ K for the suspended CNFs. Here, a CNF thermal conductivity of 12 W/m K [21] is used for the calculation of critical temperature. While the thermal contact resistance was reduced by platinum coating on the CNFs [21], the breakdown near the middle of the fiber as shown in Fig. 3(a) seems to suggest that successive current annealing improves thermal conductivity as well. Regarding σ, we use the maximum electrical conductivity prior to the breakdown from

the present data. Despite of the CNFs grown using the same process, the measured conductivity varies within a factor of two after the annealing process. For MWNTs under high current stress, the conductivity drastically decreases due to successive graphitic wall removal [10]. In addition, current-voltage characteristics of CNTs saturate in the high-current region, being attributed to electron-phonon scattering [22,23]. Meanwhile, the resistance of CNF is monotonically reduced during the annealing steps as shown in Fig 2(b). It can be assumed that, the improvement in conductance is dominated by changing the coupling between CNFs and the electrodes, and significant contact resistance still remains even in late annealing cycles, resulting in variation of the measured conductivity. Thus the maximum conductivity obtained with two-point current stress measurements can be assumed to approach the CNF conductivity.

This result for $T_{max} = 1260$ K is comparable to the CNF synthesis temperature, estimated to be in the 1000 K range [24]. For current-induced breakdown of CNTs, critical temperature under high current stress has been experimentally obtained as above 800 K [9,25]. Though having no radiative heat transfer and in the limit of no coupling with the substrate, the temperature obtained is only an indication of CNF durability, it is a reasonable estimation of critical temperature for CNF breakdown and points to the need for systematic local temperature measurement.

IV. CONCLUSION

We have investigated experimentally the CNF breakdown caused by high-current stress using two distinct geometrical configurations. With reduced contact resistance due to current annealing, it is found that the maximum current density monotonically decreases with increasing CNF length for both suspended and supported fibers. In the suspended case, a simple heat transport model confirms the observed linear J_{max} versus $1/L$ behavior. The effective heat dissipation by the surrounding materials and ambient is shown to improve the nanofiber's current-carrying capacity, and the present study represents an important first step toward the understanding of the reliability of CNF for potential interconnect applications.

ACKNOWLEDGMENT

The authors are grateful to John R. Jameson of Santa Clara University, Jun Li of Kansas State University, and Alan M. Cassell of NASA Ames Research Center for their helpful advice. They also thank Hitachi High-technologies America for its assistance in electron microscopy. This work was supported by the United States Army Space and Missile Defense Command (SMDC) and carries Distribution Statement A, approved for public release, distribution unlimited.

REFERENCES

[1] See, e.g., *"Carbon Nanotubes: Synthesis, Structure, Properties, and Applications,"* ed. by M. S. Dresselhaus, G. Dresselhaus, and Ph. Avouris, Boston:, Springer, 2001.

[2] R. Martel, T. Schmidt, H.R. Shea, T. Hertel, P. Avouris, *"Single- and multi-wall carbon nanotube field-effect transistors ,"* *Appl. Phys. Lett.* Vol. 73, pp. 2447-2449, October 1998

[3] A. Javey, J. Guo, Q. Wang, M. Lundstrom, H. Dai, *"Ballistic carbon nanotube field-effect transistors,"* *Nature*, Vol. 424, pp. 654-657, August 2003.

[4] T. Yamada, *"Modeling of carbon nanotube Schottky barrier modulation under oxidizing conditions,"* *Phys. Rev. B*, Vol. 69, 125408, June 2004.

[5] M. Nihei, M. Horibe, A. Kawabata, Y. Awano, *"Simultaneous Formation of Multiwall Carbon Nanotubes and their End-Bonded Ohmic Contacts to Ti Electrodes for Future ULSI Interconnects,"* *Japan. J. Appl. Phys.*, Vol. 43, pp. 1856-1859, April 2004.

[6] J. Li, Q. Ye, A.M. Cassell, H.T. Ng, R. Stevens, J. Han, M. Meyyappan, *"Bottom-up approach for carbon nanotube interconnects,"* *Appl. Phys. Lett.*, Vol. 82, pp. 2491-2493, April 2003.

[7] Q. Ngo, A.M. Cassell, A.J. Austin, J. Li, S. Krishnan, M. Meyyappan, C.Y. Yang, *"Characteristics of aligned carbon nanofibers for interconnect via applications,"* *IEEE Electron Device Lett.*, Vol. 27, pp. 221-224, April 2006.

[8] International Technology Roadmap for Semiconductors (ITRS) 2006 Update. [Online]. Available: http://public.itrs.net

[9] E. Pop, D.A. Mann, J. Cao, K.E. Goodson, H. Dai, *"Electrical and thermal transport in metallic single-wall carbon nanotubes on insulating substrates,"* *J. Appl. Phys.*, Vol. 101, 093710, May 2007.

[10] J.Y. Huang, S. Chen, S.H. Jo, Z. Wang, D.X. Han, G. Chen, M.S. Dresselhaus, Z.F. Ren, *"Atomic-scale imaging of wall-by-wall breakdown and concurrent transport measurements in multiwall carbon nanotubes,"* *Phys. Rev. Lett.*, Vol. 94, 236802, June 2005.

[11] M. Suzuki, Y. Ominami, Q. Ngo, C.Y. Yang, J. Li, A.M. Cassell, *"Current-induced breakdown of carbon nanofibers,"* *J. Appl. Phys.*, Vol. 101, 114307, June 2007.

[12] M.A. Kuroda, A. Cangellaris, J.-P. Leburton, *"Nonlinear transport and heat dissipation in Metallic carbon nanotubes,"* *Phys. Rev. Lett.*, Vol. 95, 266803, December 2005.

[13] J.-O Lee, C. Park, J.-J. Kim, J. Kim, J.W. Park, K.-H. Yoo, *"Formation of low-resistance ohmic contacts between carbon nanotube and metal electrodes by a rapid thermal annealing method,"* *J. Phys. D.*, Vol. 33, pp. 1953-1956, August 2000.

[14] B.A. Cruden, A.M. Cassell, Q. Ye, M. Meyyappan, *"Reactor design considerations in the hot filament/direct current plasma synthesis of carbon nanofibers,"* *J. Appl. Phys.*, Vol. 94, pp. 4070-4078, June 2003.

[15] L. Zhang, D. Austin, V.I. Merkulov, A.V. Meleshko, K.L. Klein, M.A. Guillorn, D.H. Lowndes, M.L. Simpson, *"Four-probe charge transport measurements on individual vertically aligned carbon nanofibers,"* *Appl. Phys. Lett.* Vol. 84, pp. 3972-3974, May 2004.

[16] M. Suzuki, Y. Ominami, Q. Ngo, C.Y. Yang, T. Yamada, J. Li, A.M. Cassell, *"Bright contrast imaging of carbon nanofiber-substrate interface,"* *J. Appl. Phys.*, 100, 104305, November 2006.

[17] E. Pop, D. Mann, J. Cao, Q. Wang, K. Goodson, H. Dai, *"Negative Differential Conductance and Hot Phonons in Suspended Nanotube Molecular Wires,"* *Phys. Rev. Lett.*, Vol. 95, 155505, October 2005.

[18] C. Durkan, M.A. Schneider, M.E. Welland, *"Analysis of failuar mechanisms in electrically stressed Au nanowires,"* *J. Appl. Phys.*, Vol. 86, pp. 1280-1286, April 1999.

[19] H.S. Carslaw, J.C. Jaeger, *Conduction of Heat in Solids Second Edition*; Oxford University Press, Oxford, 1986.

[20] H. Kitsuki, T. Yamada, D. Fabris, J.R. Jameson, P.Wilhite, M. Suzuki, C.Y Yang, *"Length dependence of current-induced breakdown in carbon nanofiber interconnects,"* *Appl. Phys. Lett.*, Vol. 92, 173110, May 2008.

[21] C. Yu, S. Saha, J. Zhou, L. Shi, A.M. Cassell, B.A. Cruden, Q. Ngo, J. Li., *"Thermal Contact Resistance and Thermal Conductivity of a Carbon Nanofiber,"* *J. Heat Transfer*, Vol. 128, pp. 234-239, March 2006

[22] Z. Yao, C. L. Kane, and C. Dekker, *"High-Field Electrical Transport in Single-Wall Carbon Nanotubes,"* *Phys. Rev. Lett.* Vol. 84, pp. 2941-2944, March 2000.

[23] P.G. Collins, M. Hersam, M. Arnold, R. Martel, P. Avouris, *"Current Saturation and Electrical Breakdown in Multiwalled Carbon Nanotubes,"* *Phys. Rev. Lett.*, Vol. 86, pp. 3128-3131, April 2001.

[24] K.B.K. Teo, D.B. Hash, R.G. Lacerda, N.L. Rupesinghe, M.S. Bell, S.H. Dalal, D. Bose, T.R. Govindan, B.A. Cruden, M. Chhowalla, G.A.J. Amaratunga, M. Meyyappan, W.I. Milne, *"The Significance of Plasma Heating in Carbon Nanotube and Nanofiber Growth,"* *Nano Lett.*, Vol. 4, pp. 921-926, April 2004.

[25] T. D. Yuzvinsky, W. Mickelson, S. Aloni, S. L. Konsek, A. M. Fennimore, G. E. Begtrup, A. Kis, B. C. Regan, and A. Zettl., *"Imaging the life story of nanotube devices,"* *Appl. Phys. Lett.* Vol. 87, 083103, August 2005.

Solderability and Reliability of Printed Electronics

B. Salam*, B.K. Lok

Singapore Institute of Manufacturing Technology, Multi-functional Substrate Technology (MST)
71 Nanyang Drive, Singapore – 638075
*Corresponding Author: Tel:+65- 67938526, Fax:+65- 67916377, Email: budimans@simtech.a-star.edu.sg

Abstract- **Two types of printed conductors were studied for their effects on the solderability and reliability of printed electronics. The metal particles of the studied printed conductors were copper and silver and the bonder was phenol resin. The test vehicle was a 6 x 25mm FR4 coupon, where the printed conductors were formed on it. Dip and look test results indicated that printed silver heavily leached and hence had poor solderability. Wetting balance test showed the studied printed copper had poor solderability. Surface roughness measurement and microstructure observation confirmed that the poor solderability of printed copper was due to its surface roughness and heterogeneity (mixed of copper and void).**

The reliability of printed interconnects were also studied by analyzing their microstructures. The study found that the solder formed the interconnection by bonding the outer metal particles. However, the outer metal particles did not closely packed and hence the bonding was not continuously formed. The discontinued bonding confirms the heterogeneity of the printed conductor surface, which causes the poor solderability of the printed copper and affects the strength of the printed interconnection. The interface intermetallic layers of the aged printed interconnects were found to be thicker than those of the as-soldered printed interconnects.

I. INTRODUCTION

Printed electronics refer to the printing of electronic circuitry on a common media such as paper, plastic, or textile. The conductive materials are available in liquid and paste forms, and can be applied using ink jetting and screen printing respectively. Printed electronics are portable, thin, tiny and suitable for customized applications such as smart labels, animated posters, and active clothing.

Printed electronics is being regarded as the second coming of the semiconductor industry because it can significantly reduce their cycle times and cost structure. New products will have a much faster time to market due to the use of high volume manufacturing methods such as inkjet printing. Printed electronics also lead the way for simplified materials processing and large area electronics manufacturing because of the elimination of lithographic masks and economic custom design for small orders and low volume production. Moreover, the cost of setting up a printed electronics factory is likely to be less than $10 million whereas that of a silicon chip manufacturing facility exceeds $1 billion currently [1]. This is because printed electronics eliminate the need for equipment intensive clean room processing and reduce material and chemical wastage in the process.

One important condition for achieving a good solderability is

that the surface must have a good wettability. Good wettability results in good solderability, which means the formation of a uniform, smooth, unbroken, adherent coat of solder on the base metal [2]. Wettability is determined by contact angle. The contact angle that is below 90° indicates wetting ($\theta = 0°$ corresponding to perfect wetting) and the opposite the contact angle that is above 90° is a sign of poor wetting ($\theta = 180°$ corresponding to non-wetting). Similarly, soldering process on printed electronics also depends on wetting for the formation of solder-to-printed base metal contact.

Measuring solderability can be conducted with wetting balance technique. It is versatile, popular and provides reliable data for evaluating solderability [2]. The technique involves measuring the force values after dipping specimens into a solder bath. The measured force can be illustrated in the following equation:

$$F = \gamma \cdot \cos\theta \cdot c - g \cdot p \cdot v \ \dots\dots\dots\dots (1)$$

Where: γ is surface tension of molten solder. c is specimen perimeter. g is gravitation. p is solder density and v is specimen immersed volume.

Theoretically, the interfacial surface forces dictate the shape of the molten solder. The resulting surface tensions and their effect on wetting is best illustrated by the Young' relation [2], which defines the contact angle, given by:

$$\frac{\gamma_{SV} - \gamma_{SL}}{\gamma_{LV}} = Cos\theta \ \dots\dots\dots\dots\dots\dots (2)$$

Where the symbols γ_{SV}, γ_{SL}, & γ_{LV} refer to the solid-vapour, solid-liquid and liquid-vapour surface tension, respectively. For optimum wetting, the solid-vapour surface tension must be maximised, and the solid-liquid and liquid-vapour surface tension must be kept minimised. For the solid-vapour surface tension, the value can be increased by applying flux to remove the oxides and contaminants on the substrate. For the liquid-vapour surface tension, the value is relatively constant because the liquid-vapour surface tension is referred to as the surface tension of the solder, which the value is a function of the solder composition, and the flux. The solid-liquid surface tension is influenced by the chemical reaction between the liquid and the solid substrate such as the formation of the intermetallic compound layer.

978-1-4244-2039-1/08/$25.00 ©2008 IEEE

Printed conductors consist of metal powder and resin binder. Commonly used metal powders are silver and copper. Silver has good conductivity but dissolves easily in molten solder, which causes leaching of silver conductor during soldering. Suggested solution of soldering silver conductor is to use a lower-melting point solder alloy but it is unknown whether this method will work on the printed conductor [3]. Copper has average conductivity compared to silver, and does not leach but soldering on printed copper can be challenging if the organic binder can not withstand the temperature shock of molten solder. Currently high-temperature binder has been developed and printed conductors using the high-temperature binder have also been formulated. However, soldering the printed conductors is still challenging. Therefore the objective of this study is to investigate the solderability and reliability of printed conductors.

II. EXPERIMENTAL PROCEDURE

Two types of printed conductors were studied. Their metal particles were copper and silver, and the bonder was phenol resin. The test vehicle was a 6 x 25mm FR4 coupon with the studied printed conductors formed on it.

The solderability of the printed conductors was evaluated by dipping the test vehicles into molten Sn-Ag-Cu and Sn-Pb solders using wetting balance machine. Before conducted the solderability test, the test coupons were cleaned with 5%wt NaOH and 5%wt HNO3 diluted in distilled water. The used flux was liquid no-clean rosin mildly activated (RMA). The wetting balance test was conducted at the temperatures of 230°, 240° and 250°C. The wetting balance tester had pre-immersion time of 10s, immersion time of 10s, immersion speed of 20 mm/s and immersion depth of 5mm.

The dipped coupons were also cross-sectioned and polished to reveal their microstructures. Afterward, an in-situ observation equipment was used to analyze the microstructures. The in-situ observation system was developed to conduct various tests e.g. thermal and electromigration tests. In this study, the test vehicles were isothermally aged at 150°C for 66 hours.

III. RESULTS AND DISCUSSION

The surface of the printed copper was fully covered with the studied solder after dipping, which indicated the molten solder could wet and react with the printed copper. Meanwhile, the surface of the printed silver was distorted after dipping into the molten solder. The results after dip and look test can be seen in Fig.1. It showed the leached printed silver and indicated the printed silver could not withstand the temperature shock of the molten solder. The results suggested that the solderability of the printed copper was much better than that of the printed silver. Printed silver obviously requires alternative approach to form interconnection. Details of how to form interconnects on printed silver will be reported in the future.

Fig.1 Printed Silver and Copper Test Coupons, after Dip and Look Test

Wetting balance test was started by dipping the test vehicle into the liquid no-clean RMA flux. Then the test coupon was dipped into molten solder at the previously mentioned parameters. The forces exerted on the test coupon were then plotted as a function of time, as seen in Fig.2. The shape of the wetting balance curve can be explained as follows: At the beginning, the buoyancy forces (due to the interfacial tension) were predominant, and so the wetting force was negative. Gradually, the solder began to wet, and the wetting force increased with time, overcoming the buoyancy force. When the measured wetting force is zero, the two forces are balanced. This point is called the "zero cross time" (t0), and indicated the transition from no-wetting (overall force less than zero) to wetting (overall force greater than zero). For printed copper, the force never reached a flat zone after dwelling time of 10s. It is an indication of poor wetting.

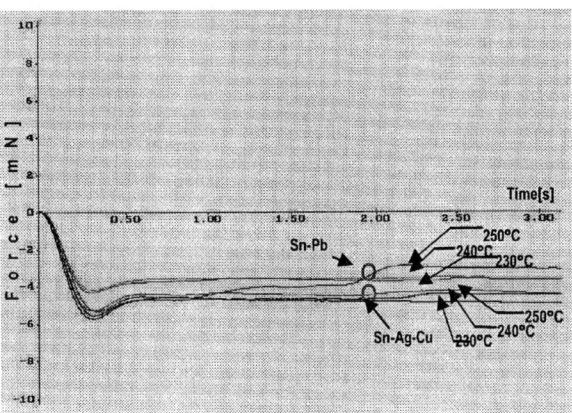

Fig.2 Wetting Force Plotted against Time, for Printed Copper

The forces exerted on the printed copper are affected by different solder alloys and test temperature, as shown in Fig.2. The wetting forces at 2 seconds from start of test for Sn-Pb solder are higher than those for Sn-Ag-Cu solder. Increasing the heating temperature from 230° to 250°C slightly raised the wetting forces at 2 seconds from start of test for both solders.

Solderability is a complex phenomenon sensitive to large number of factors such as materials of solder and conductor, surface roughness, heterogeneity, atmospheric condition and flux usage [2]. The main factors that could deteriorate the solderability of printed copper are its surface roughness and heterogeneity. Fig.3, 4 and 5 present the top and side views of the printed copper and demonstrate that the surface of the studied printed copper is coarse and heterogeneous (mixed of copper and void). It surface roughness (Ra) measured using profilometer was about 14.84μm. It was relatively high compared to that of laminated copper (Ra=0.57μm). The printed copper might be solderable when its surface roughness and heterogeneity could be improved.

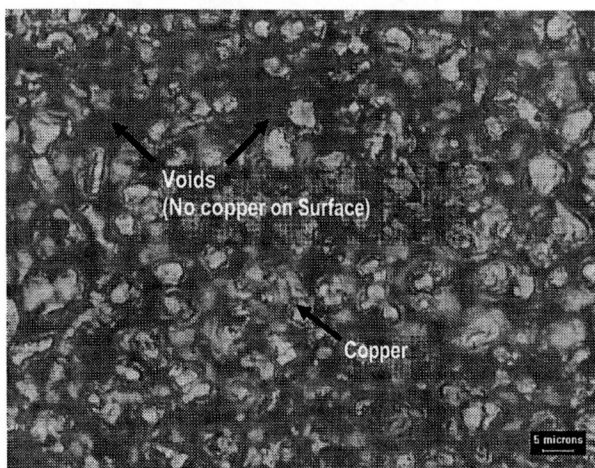

Fig.3 Surface of Printed Copper (Top View)

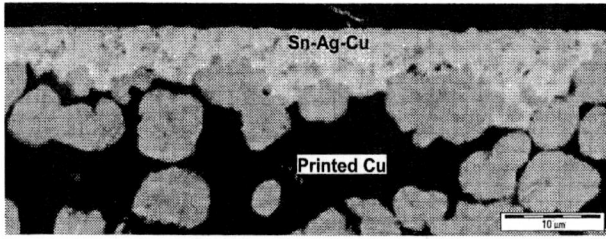

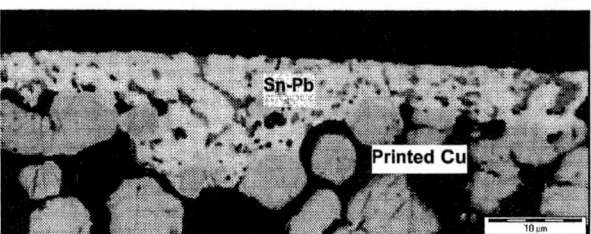

Fig.4 Microstructures of As-soldered Printed Interconnects

The microstructures of the as-soldered printed interconnects consists of metal particles and cured phenol resin, as seen in Fig.4. The phenol resin was seen to bind well the metal particles. The solder formed the interconnection by bonding the outer metal particles. However the outer metal particles on the surface did not closely packed and hence the bonding was not continuously formed. The discontinued bonding confirms the heterogeneity of the printed conductor surface, which may

affects the solderability of the printed copper and the strength of the printed interconnection.

Fig.5 presents the microstructures of the aged printed interconnects. EDX analysis confirms that the formed intermetallic layers are Cu-Sn intermetallics. Cu_6Sn_5 and Cu_3Sn were generally found intermetallics at the interface between aged Sn-Pb and Sn-Ag-Cu solders and copper [6, 7]. It was also expected that the interface intermetallic layers of the aged printed interconnects grow thicker than those of the as-soldered printed interconnects.

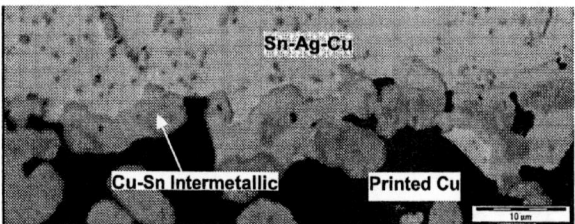

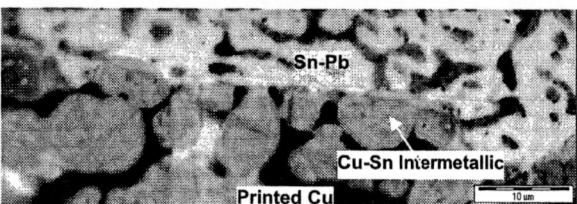

Fig.5 Microstructures of Aged Printed Interconnects

IV. SUMMARY

Solderability of printed silver and copper has been studied using wetting balance method. Dip and look test results indicated that printed silver heavily leached and hence had poor solderability. Wetting balance test showed the studied printed copper had poor solderability. Surface roughness measurement and microstructure observation confirmed that the poor solderability of printed copper was due to its surface roughness and heterogeneity (mixed of copper and void).

The reliability of printed interconnects were also studied by analyzing their microstructures. The study found that the solder formed the interconnection by bonding the outer metal particles. However, the outer metal particles did not closely packed and hence the bonding was not continuously formed. The discontinued bonding confirms the heterogeneity of the printed conductor surface, which could cause the poor solderability of the printed copper and may affect the strength of the printed interconnection. The interface intermetallic layers of the aged printed interconnects were found to be thicker than those of the as-soldered printed interconnects.

ACKNOWLEDGMENT

We would like to express our gratitude to all those who gave us the possibility to complete this study. We want to thank our colleagues Chua Cheng Hwee, Gan Hiong Yap and Cheng Chek Kweng for all their help, support, interest and valuable hints. We also thank our manager, Dr Albert Lu, for his stimulating suggestions, motivation and encouragement.

REFERENCES

1. Frost & Sullivan, "Printed Electronics--Technologies and Applications – 2007," www.frost.com, assessed on 31 March 2008.
2. G. Kumar and K.N. Prabhu, "Review of Non-reactive and Reactive Wetting of Liquids on Surfaces", Advances in Colloid and Interface Science, 133, 2007, pp.61–89.
3. H.H., Manko, "Solders and Soldering" McGraw-Hill, 4th Edn., 2001.
4. K. Gilleo "Polymer Thick Film" Von Nostrand Reinhold, 1996, pp.202-230.
5. M.G. Varadarajan, K.J. Lee, S.K. Bhattacharya, A. Bhattacharjee, L. Wan, R. Pucha, R.R. Tummala, S. Sitaraman, "Studies on design, fabrication and reliability assessment of embedded passives on a high-density interconnect (HDI) organic substrate using a sequential build-up process" 2006 Conference on High Density Microsystem Design and Packaging and Component Failure Analysis, 2006, pp. 188-198.
6. D. Frear, H. Morgan, S. Burchett and J. Lau (ed.), The Mechanics of Solder Alloy Interconnects, Van Nostrand Reinhold, New York, 1994, pp. 42-86.
7. D.R. Frear, Solder Mechanics: A State of the Art Assessment, TMS, Minerals Metals Materials, Pennsylvania, USA, 1991, pp. 29-104.

978-1-4244-2039-1/08/$25.00 ©2008 IEEE

SESSION 12:

NOVEL DEVICES II

The Device Characteristics of Partially Undoped Poly-Silicon Gate P-LDMOS Power Transistors

R.Y. Su, P.Y. Chiang, J. Gong, J.L. Tsai*, T.Y. Huang*, Mingo Liu*and C.C. Choub*

Dept. of Electronics Engineering, National Tsing Hua University, HsinChu, Taiwan.
*High Voltage Department, TSMC, Hsin-Chu, Taiwan
101, Sec2, Kuang Fu Road, Hsin-Chu, Taiwan, 30055.
Tel: (+886)-3-5715131 ext.34041   Fax: (+886)-3-5752120   Email: d945038@oz.nthu.edu.tw

Abstract –The effect of partially undoped poly-silicon gate above the drift region in P-Lateral Double-Diffused MOS (P-LDMOS) Transistors is investigated. Experiment results show that it can improve the off-state leakage current and reduce the on-state resistance. For hot carrier performance, this structure induces a higher initial current shift due to less vertical field. The long-term hot carrier degradation behavior of this device is the same as that of the standard devices.

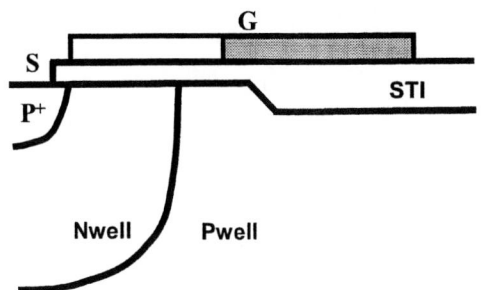

Fig. 1b: The cross section structure of partially undoped poly-silicon gate (shaded area) above the drift region

I. INTRODUCTION

The development of smart-power technologies is mainly based on the introduction of lateral DMOS transistors in a standard CMOS process flow. Such DMOS devices are designed to withstand medium to high voltages (20-100V) on the drain [1]. However, since submicrometer LDMOS transistors present unique features that are able to substantially affect the hot-carrier phenomena, a specific investigation of the degradation mechanisms occurring in these devices is required [2][3].

In this paper, we designed a different-to-normal structure on ordinary LDMOSFET to observe the difference with standard lateral DMOS transistors. The experimental result shows better off-state leakage current and smaller on-state resistance for the DC characteristics. In the hot carrier performance, this structure induces a higher initial current shift due to the smaller vertical field.

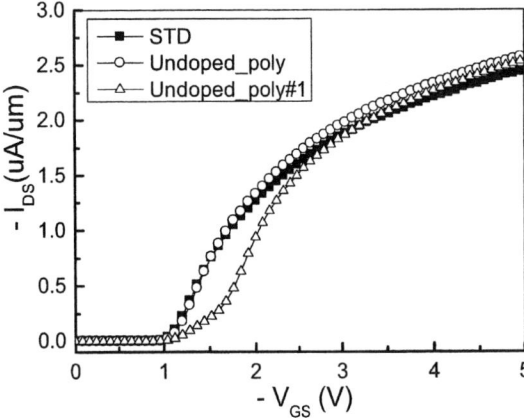

Fig. 2: The $I_{DS}\_V_{GS}$ characteristics (measured at V_{DS}=-0.1V) of the standard and partially undoped poly-gate transistors. The undoped poly #1 sample has its undoped poly-Si across the channel region.

II. RESULTS AND DISCUSSION

A) DC characteristics

Fig. 1a is the schematic cross section of the standard lateral DMOS transistors, and Fig. 1b demonstrates the structure of the partially undoped poly-gate (PUPG) transistor. The undoped poly region is undesirable to across the channel region because it may affect the $I_{DS}\_V_{GS}$ characteristics.

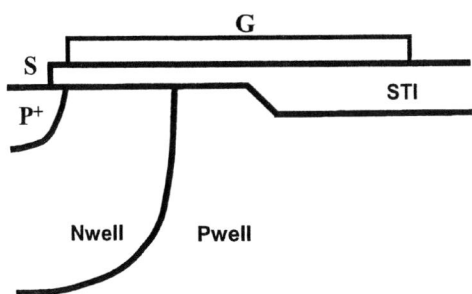

Fig. 1a: The schematic cross section structure of standard lateral DMOS transistors.

978-1-4244-2039-1/08/$25.00 ©2008 IEEE

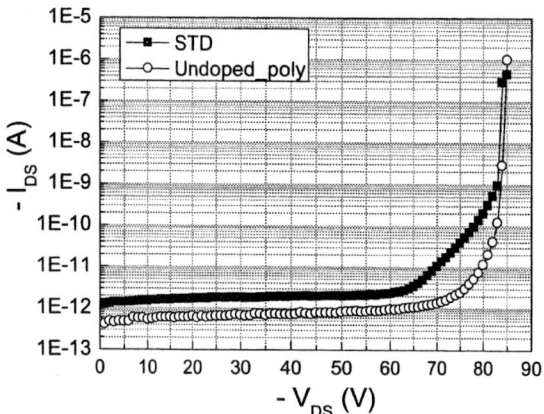

Fig. 3: The off-state characteristics. The PUPG structure has lower leakage current and sharper breakdown curve.

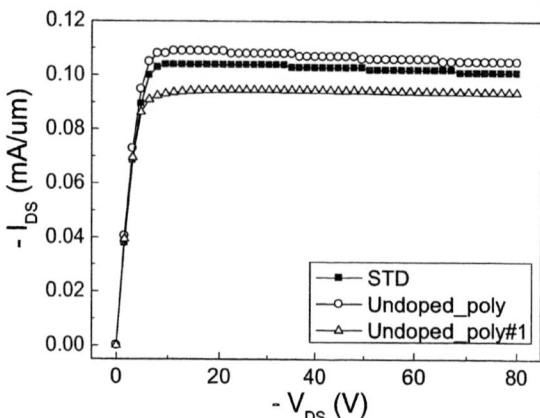

Fig. 5: The I_{DS}-V_{DS} characteristics of different structured transistors for V_{GS}=-5V.

Fig. 2 is the I_{DS}_V_{GS} characteristics (measure at V_{DS}=-0.1V) of the standard and the PUPG devices. The threshold voltage remains the same between the standard device and regular PUPG devices. Significant difference is observed in the PUPG devices with the undoped poly-Si across the channel region (undoped poly #1).

Fig. 3 shows the off-state performance of the transistors. The standard device has an I-V property similar to a soft breakdown. The PUPG structure has lower leakage current and sharper breakdown curve. According to reference [4], when the gate is biased to form an accumulation layer at the silicon surface, due to field crowding and peak field increase, it forms a path for the GIDL (gate induced drain leakage) current. In order to identify the leakage source, we sweep the gate voltage from negative to positive at a constant drain voltage. Experiment results are shown in Fig. 4. There is a significant difference between the standard and the PUPG devices at high drain voltage.

The I_{DS}-V_{DS} characteristics of different structure transistors for V_{GS}=-5V is shown in Fig. 5. The regular PUPG transistor has the largest drain current. The specific resistance, R_{dson}, for the standard and the regular PUPG transistor is 250 $m\Omega mm^2$ and 235 $m\Omega mm^2$, respectively. A smaller vertical electric field in the undoped poly-Si region reduces 6% on-resistance because the current will flow with a wider cross-section area and/or the carrier may have larger mobility. In the undoped poly #1 samples, due to the larger threshold voltage that it demonstrates the lowest current drive in Fig. 5.

B)   Hot carrier performance

Recently, it has reported that hot electron injection and trapping into the thick gate oxide and field oxide result in the Ron decreases with the stress time [5].

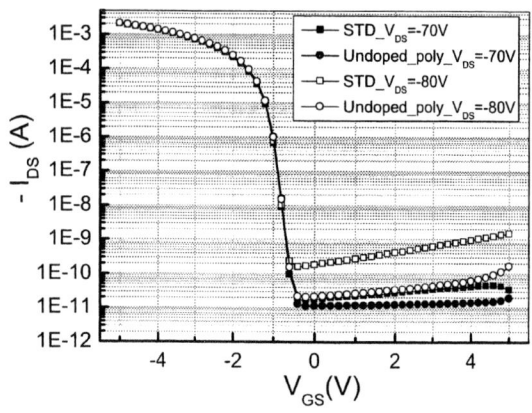

Fig. 4: GIDL current for the standard and the PUPG devices.

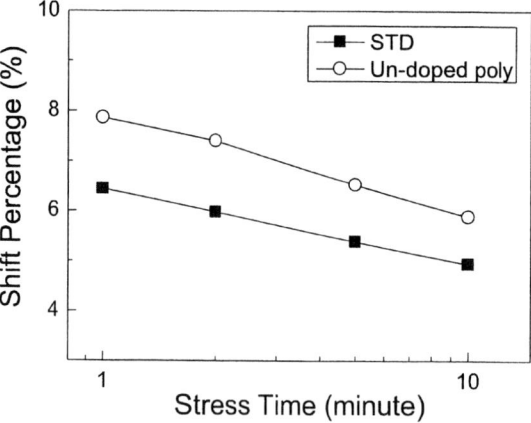

Fig. 6: I_{DLIN} shift percentage in PUPG and STD transistors.

978-1-4244-2039-1/08/$25.00 ©2008 IEEE

Similar experiments were performed. The stress bias condition is the maximum I_{SUB} point, V_{GS} = -2.9V and V_{DS}=-66V, therefore, the gate voltage is more positive with respect to the drift region during the stress period. Fig. 6 shows the hot carrier degradation in standard and PUPG devices during a stress period of 10 minutes. It demonstrates that PUPG transistor has a higher initial current shift ΔI_{DLIN} in the first minute of the stress period and then it exhibits normal hot carrier degradation. Fig.7 is the measured results of I_{SUB} for these two kinds of transistors. Again, the PUPG transistor has a higher I_{SUB}. Fig. 8 shows the energy band diagram under stress condition (negative gate bias) for both device structures and the PUPG transistor has less vertical electric field.

To further study this phenomenon, two-dimensional simulator is used to analyze the experiment results. Fig. 9 presents the simulated result of less surface electric field at PUPG transistor. In addition, fig. 10 is the simulation result of impact ionization rate contour at stress condition: V_{GS}=-2.9V and V_{DS}=-66V.

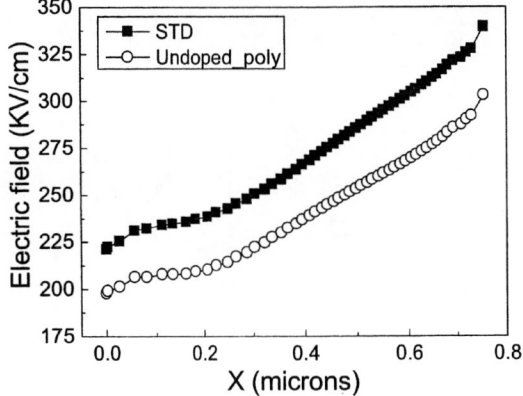

Fig. 9: Surface electric field in STD and PUPG transistors.

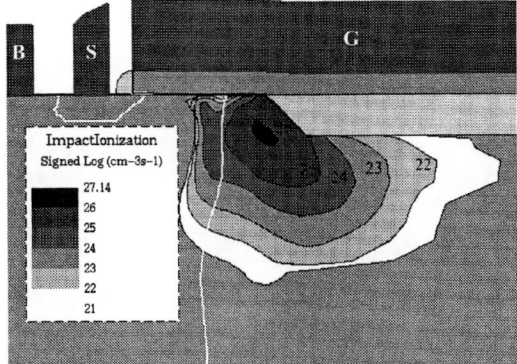

Fig. 10: Maximum impact ionization rate at stress condition locates in the accumulation region under poly-Si gate. It is favorable for electrons to inject into STI.

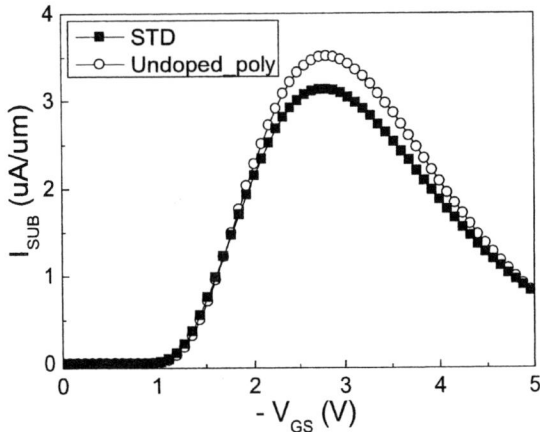

Fig. 7: I_{SUB} in PUPG and STD transistors.

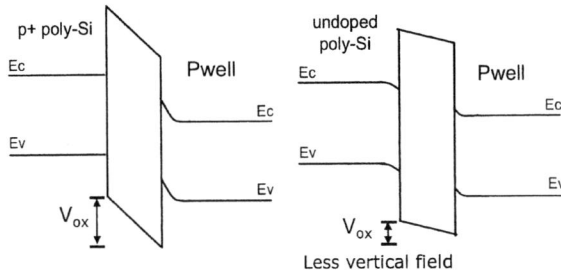

Fig. 8: Energy-band diagram under stress condition (negative gate bias).

As can be seen, the maximum impact ionization generation rate (IIGR) in the accumulation region generated electron-hole pairs, which is favorable for electrons to inject into the STI edge and the trapped charges will change the electrical field distribution in the device. This may be the reason of the initial current shift. The maximum impact ionization generation rate for standard and PUPG devices are shown in Fig. 11. It was found that in both cases the distance between the maximum IIGR position and STI is strongly correlated to the amount of the initial current increase during the hot-carrier stress period.

Shorter distance from the maximum IIGR position to STI (along the electric field direction) causes higher initial current shift in Fig. 6. On the other hand, the IIGR in PUPG transistor is larger at the STI edge compared to the STD case. Both phenomena confirm that the initial current shift is severe in partially undoped poly-silicon gate transistor.

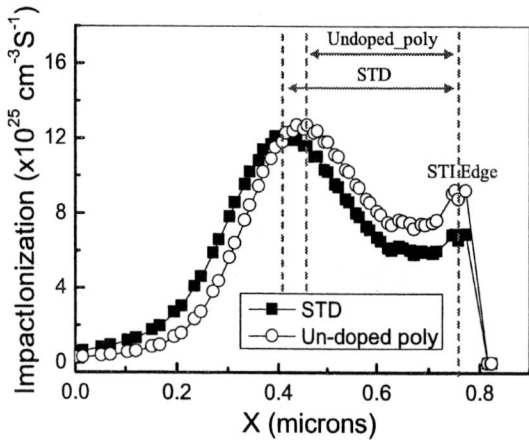

Fig. 11: The distance of the maximum impact ionization generation rate (IIGR) point to STI edge.

III. CONCLUSION

Partially undoped poly-Si gate above the drift region in P-Lateral Double-Diffused MOS (LDMOS) Transistors is discussed for the first time. Experiment results show that it has better DC characteristics. As to hot carrier performance, hot electrons are shown to inject into the STI edge that causes an initial drain current increase in the stress period. It was found that the unusual initial current shift is strongly correlated to the distance between the maximum IIGR position and the STI edge.

ACKNOWLEDGMENT

This work is also supported by National Science Council, R.O.C., under the contract of NSC 95-2221-E-007-259-MY2

REFERENCES

[1] P. Moens, M. Tack, R. Degraeve and G. Groeseneken, "A Novel Hot-Hole Injection Degradation Model for Lateral NDMOS Transistors," IEDM, 2001.

[2] R. Versari et al. "Experimental Study of Hot-Carrier Effects in LDMOS Transistors" IEEE Trans. Electron Dev, p.1228, 1999

[3] Chin-Chang Cheng, J. W. Wu, C. C. Lee, J. H. Shao and T. Wang "Hot Carrier Degradation in LDMOS Power Transistors" IEEE Proceedings of 11th IPFA 2004.

[4] Kaushik Roy, Saibal Mukhopadhyay, and Hamid Mahmoodi-meimand "Leakage Current Mechanisms and Leakage Reduction Techniques in Deep-Submicrometer CMOS Circuits" Proceedings of THE IEEE, Vol. 91, NO. 2, February 2003.

[5] W. Sun, H. Wu, L. Shi, Y. Yi, and H. Li et al. "On-Resistance Degradations for Different Stress Conditions in High-Voltage PLEDMOS Transistor With Thick Gate Oxide" IEEE Electron Devices Letters. Vol. 28, no. 7, pp631-633 2007.

Hot Carrier Degradation in Nanowire (NW) FinFETs

T. K. Maiti, M. K. Bera, S. S. Mahato, P. Chakraborty, C. Mahata, M. Sengupta, A. Chakraborty, and C. K. Maiti

Dept. of Electronics & ECE, Indian Institute of Technology Kharagpur, 721302, India

Phone: +91 3222 281475. Fax: +91 3222 255303. Email: tkm.iitkgp@yahoo.com

Abstract- **Hot carrier reliability of a nanowire Ω-FinFET is investigated for the first time. Hot holes injected into the gate oxide via hot-carrier injection (HCI) at the silicon (Si) - silicon dioxide (SiO$_2$) interface of Ω-FinFETs results in the formation of dangling silicon bonds due to the breaking of silicon-hydrogen bonds and lead to high interface traps generation. The trapping and/or bond breaking creates oxide charge and interface traps affect the Coulomb mobility. A quasi-two dimensional (quasi-2D) physics-based screening Coulomb scattering mobility model has been developed and implemented in Synopsys Sentaurus Device simulator.**

I. INTRODUCTION

Hot-carrier induced phenomena are of great interest due to their important role in device reliability [1]. High energy carriers (also known as hot carriers) are generated in MOSFETs by high electric field near the drain region. Hot carriers transfer energy to the lattice through phonon emission and break bonds at the Si/SiO$_2$ interface. The trapping or bond breaking creates oxide charge and interface traps that affect the channel carrier mobility and the effective channel potential. Interface traps and oxide charge also affect the transistor parameters, such as, the threshold voltage and drive currents.

Several workers have reported the results of their investigation on hot-carrier effects on the performance of PMOS transistors [2, 3]. It has been shown that the degradation of PMOS transistors is caused by the interface state generation and hole trapping in the gate oxide from the hot-carrier injection. Reliability assurance of analog circuits requires a largely different approach than for the digital case. It is generally accepted that injected and trapped electrons dominate the degradation behavior. In this work, we describe a physics based Coulomb mobility model developed to describe Coulomb scattering at the Si-SiO$_2$ interface and implement in device simulator. Hot-carrier induced current and subsequent degradation in nanowire (NW) Ω-FinFETs are investigated using simulation and validation with reported experimental data. The influence of the hot carriers on the threshold voltage and drive currents is examined in detail for nanowire Ω-FinFETs.

II. PROCESS CONSIDERATION

The process flow used in this work is similar to the process flow reported by Singh et al. [4, 5]. The starting material is an SOI wafer with a top silicon layer (p-type, ~10^{15}/cm^3) thickness of 200-nm over a 150-nm BOX. All lithographic patterning is based on a line width of 50-nm. To define the mask for the

source/drain area, the lift-off technique was used. The S/D implantation was carried out after gate formation. To obtain a more uniform doping profile at the left and right sides of the fin, the implantation was performed in two steps. S/D implant activation anneal (1 s at 1025 K, with a 3 s ramp up and a 2 s ramp down) was used to ensure dopant diffusion throughout the gate electrode and thick S/D (and extension) regions beneath the gate. The contact areas were etched (free from the screening oxide) and electrical contacts were defined for device simulation. Fig. 1 presents tilted view of a 100-nm long and 10 nm thick Si NW fin.

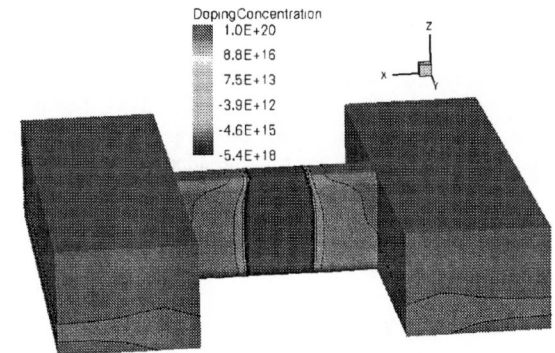

Fig. 1. Simulated net dopant distribution. For a better view, only the silicon layer is shown.

III. QUASI-2D COULOMB MOBILITY MODEL

The silicon (Si)–silicon dioxide (SiO$_2$) interface in nanowire (NW) Ω-FinFETs shows a very large number of trap states. These traps become filled during inversion causing a change of conduction charge in the inversion layer and increase the Coulomb scattering of mobile charges. Owing to the large number of occupied interface traps, Coulomb interaction is likely to be an important scattering mechanism in nanowire (NW) Ω-FinFET device operation, resulting in very low surface mobilities and may be described by a quasi-2D scattering model. The Coulomb potential due to the occupied traps and fixed charges decreases with distance away from the interface. So, mobile charges in the inversion layer that are close to the interface are scattered more than those further away from the interface; therefore, the Coulomb scattering mobility model is required to be depth dependent. We assume that the electron gas can move in the x-y plane and is confined in the z direction. Electrons are considered confined or quantized if their

978-1-4244-2039-1/08/$25.00 ©2008 IEEE

deBroglie wavelength is larger than or comparable to the width of the confining potential. The deBroglie wavelength of electrons, given by $\lambda = \hbar / \sqrt{2m^* k_B T}$, is approximately 150Å at room temperature, where as the thickness of the inversion layer is typically around 50Å to 100Å. Thus, one may justify treating the inversion layer as a two dimensional electron gas.

The scattering from charged centers in the electric quantum limit has been formulated by Stern and Howard (1967). We consider only the p-channel inversion layer on Si (100) surface where the Fermi line is isotropic and calculate the potential of a charged center located at (r_i, z_i). Using the image method, we get

$$V_i(r,z) = \frac{e^2}{4\pi\varepsilon_0 \tilde{k}\sqrt{(r-r_i)^2 + (z-z_i)^2}} \tag{1}$$

where $r^2 = x^2 + y^2$, $z = 0$ corresponds the Si/SiO$_2$ interface. $z > 0$ is in silicon whereas $z < 0$ is in the oxide. Where $\tilde{k} = (k_{Si} + k_{ox})/2$ for $z < 0$, and ε_o is the permittivity of free space. We assume parabolic sub bands with the same effective heavy-hole mass, m^*. Since inversion layer electrons are restricted to move in the x-y plane, they would only scatter off potential perturbations that they see in the x-y plane. Therefore, we are only interested in determining the potential variations along that plane. To do so, one needs to calculate the two dimensional Fourier transforms of the potential appearing in Eqn. (1). The hole wave functions are then given by

$$\psi_{i,k}(r,z) = \frac{1}{\sqrt{A}}\xi(z)e^{ik.r} \tag{2}$$

where i represent the subband index and $k = (k_x, k_y)$ is the two-dimensional wavevector parallel to the interface. $\xi(z)$ is the quantized wave function in the direction perpendicular to the interface, E_i its corresponding energy and $r = (x, y)$. We denote the area of the interface by A. The effective unscreened quantum potential for holes in the inversion layer in the electric quantum limit in terms of the 2D Fourier transform is given by

$$v(q,z_i) = \frac{e^2}{2\tilde{k}\varepsilon_0 q}\iint \xi_i(z)\xi_j(z)e^{-q|z-z_i|}dz \tag{3}$$

We now consider the effect of screening due to inversion layer electrons on Coulombic scattering. Screening is actually a many-body phenomenon since it involves the collective motion of the electron gas. Using the Coulomb screening we get,

$$v(q,z_i) = \frac{e^2}{2\tilde{k}\varepsilon_0(q+q_s)}\iint \xi_i(z)\xi_j(z)e^{-q|z-z_i|}dz \tag{4}$$

Where $q_s = \frac{e^2}{2\tilde{k}\varepsilon_0}\iint \xi_i(z)\xi_j(z)e^{-q|z-z_i|}dz$ one obtains the scattering rate using Fermi's golden rules,

$$S(q,z_i) = \frac{2\pi}{\hbar^2}$$
$$\cdot \left(\frac{e^2}{2\tilde{k}\varepsilon_0(q+q_s)}\iint \xi_i(z)\xi_j(z)e^{-q|z-z_i|}dz \right)^2 \delta(E_k - E_{k'}) \tag{5}$$

where $\hbar$ is Planck's constant. E_k and $E_{k'}$ denote the initial and final energies of the mobile charge being scattered. Scattering of inversion layer mobile charges takes place due to Coulombic interactions with occupied traps at the interface and also with fixed charges distributed in the oxide. Defining the 2D charge density $N_{2D}\delta(z_i)$ at depth z_i inside the oxide as the combination of the fixed charge N_f and trapped charge N_{it} as

$$N_{2D}(z_i) = \begin{cases} N_{it} + N_f(0), & z_i = 0 \\ N_f(z_i), & z_i < 0 \end{cases} \tag{6}$$

Using the above approximation, one obtains the total transition rate. Since, Coulombic scattering is an elastic scattering mechanism, the scattering rate or equivalently the inverse of the momentum relaxation time is then calculated as

$$\frac{1}{\tau_m} = \frac{1}{(2\pi)^2}\cdot\frac{2\pi}{\hbar^2}\left(\frac{e^2}{2\tilde{k}\varepsilon_0} \right)^2$$
$$\cdot \int \left(\frac{1}{(q+q_s)}\iint \xi_i(z)\xi_j(z)e^{-q|z-z_i|}dz \right)^2 \delta(E_k - E_{k'})(1-\cos\theta)\delta k \tag{7}$$

Using the above relaxation time, one obtains the mobility of the i-th subband as,

$$\mu_i = \frac{e}{m^*}\frac{\int \sum_i \tau_m \varepsilon \frac{\partial f_0(\varepsilon)}{\partial\varepsilon}d\varepsilon}{\int \varepsilon \frac{\partial f_0(\varepsilon)}{\partial\varepsilon}d\varepsilon} \tag{8}$$

The average mobility, $\overline{\mu}$, is then given by [5]

$$\overline{\mu} = \frac{\sum_i p_i \mu_i^2}{\sum_i p_i \mu_i} \tag{9}$$

where p_i is the hole concentration in the i-th subband. Taking into the different scattering mechanism and using the Matthiessen's rule one obtains the total mobility μ.

IV. IMPLEMENTATION

We implemented the Coulomb scattering mobility model in Synopsys Sentaurus Device simulator. To activate the mobility model appropriate mobility values were defined in the fields of the parameter file. Our simulation data for the drain current (I_{ds}) versus gate voltage (V_{gs}) curves match the experimentally measured results very well [6]. Fig. 2 shows the I_{ds}-V_{gs} characteristics of the simulated p-type nanowire Ω-FinFET with a 10 nm-thick, and 100 nm-long Si-fin as the channel body. At room temperature, the devices show high ON-current (I_{ds} at V_{ds} = V_{gs} = 1.1 V) of ~0.68mA/μm, V_{th} ~ 0.2 V, and subthreshold swing (SS) of ~68 mV/dec. Low drain-induced barrier lowering

(DIBL) of ~10 mV/V is obtained, with $I_{ON}/I_{OFF} > 10^7$ at room temperature. These results are similar to those reported for nanowire Ω-FinFETs by Yang et al. [6].

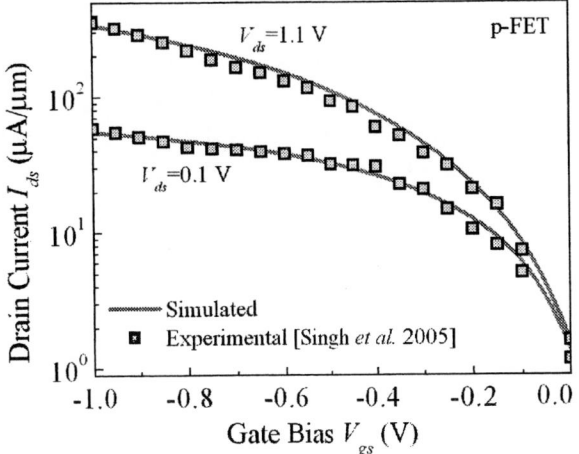

Fig. 2. Gate bias dependence of drain current for nanowire Ω-FinFETs (both simulated and experimental).

V. RESULTS AND DISCUSSIONS

Fig. 3 shows a lower drain current for Ω-FinFETs which underwent hot carrier stressing (compared to unstressed devices). Degradation in drain current indicates that hot-carrier induced positive charges are localized near the drain end.

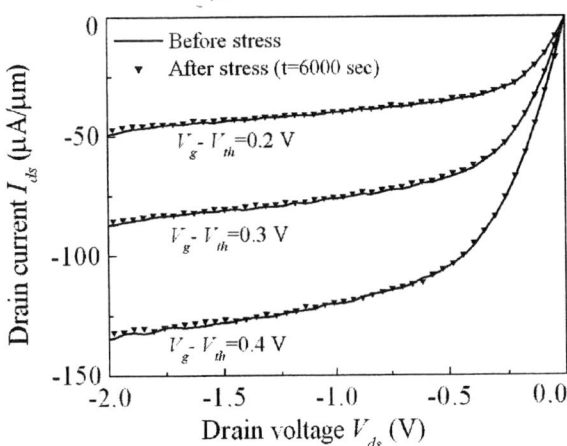

Fig. 3. Degradation of drain current under DC stress.

Fig. 4 shows the threshold voltage V_{th} shift with increasing stress time. The threshold voltage V_{th} shift indicates that net positive charges exist at the gate dielectric interface as a result of hole trapping. As the lateral electric field near the drain increases in short channel devices, electron-hole pairs are generated by impact ionization. These generated holes have energies far greater than the thermal-equilibrium value and are the hot holes. In surface-channel of Ω-FinFETs, hot holes are injected into the gate oxide via hot-carrier injection (HCI),

resulting in the formation of dangling silicon bonds due to the breaking of silicon-hydrogen bonds and lead to the interface traps generation [7]. The charge trapping in interface states causes a shift in threshold voltage and the decrease of transconductance, which degrades the device properties over a period of time.

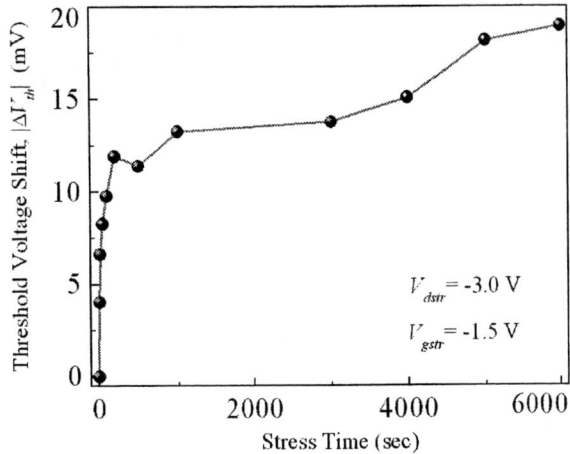

Fig. 4. Threshold voltage V_{th} shift with increasing stress time indicating an accumulation of negative charges due to electron trapping at the Si/SiO$_2$ interface.

The hot-carrier lifetime measurements were performed and the typical I_{dsat} degradation as a function of stress time is plotted in Fig. 5. The I_{dsat} degradation is consistent with V_{th} shift.

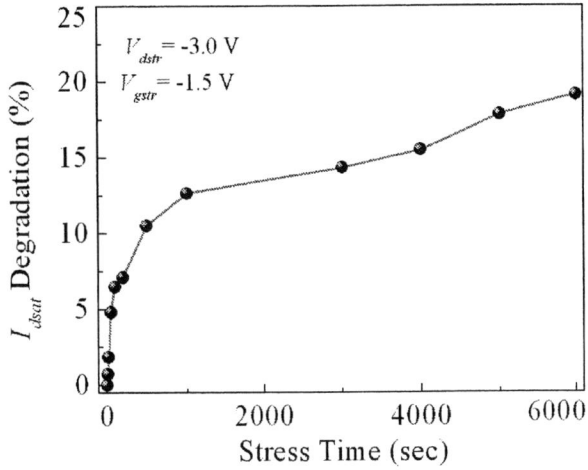

Fig. 5. I_{dsat} degradation as a function of stress time. Hot carrier lifetime in nanowire Ω-FinFETs after stressing for a given I_{sub}/I_d.

CONCLUSION

Hot-carrier induced degradation behavior in novel nanowire (NW) Ω-FinFETs is reported for the first time. A physics based Coulomb scattering mobility model for the scattering of inversion layer mobile carriers by occupied interface traps due to hot carrier injection has been developed and implemented in Synopsys Sentaurus Device simulator. Influence of the hot-carriers on the threshold voltage and saturation drain current in surface-channel of Ω-FinFETs has been studied in detail.

ACKNOWLEDGMENT

The work has been supported by DST, N. Delhi.

REFERENCES

[1] T. K. Maiti, S. S. Mahato, S. K. Sarkar, and C. K. Maiti, "Impact of Negative Bias Temperature Instability on Strain-engineered p-MOSFETs", in International Conference on Materials for Advanced Technologies (ICMAT 2007), Singapore, 2007, E-13-OR52.

[2] Y. Pan, "A physical-based analytical model for the hot carrier induced saturation current degradation of p-MOSFETs", IEEE Trans. Electron Devices, Vol. 41, No.1, 84-89, 1994.

[3] P. Heramans, R. Bellens, G. Groeseneken, and H. E. Meas, "Consistent model for the hot carrier degradation in n-channel and p-channel MOSFETs", IEEE Trans. Electron Devices, 2194-2209, 1998.

[4] N. Singh, A. Agarwal, L. K. Bera, R. Kumar, G. Q. Lo, N. Balasubramanian, and D. L. Kwong, "Gate-all-around MOSFETs: lateral ultra-narrow ($\leq$10 nm) fin as channel body", IEEE Electronics Lett., pp.1353-1354, 2005.

[5] K. Inoue and T. Matsuno, Phys. Rev. B47, 3771 (1993).

[6] N. Singh, F. Y. Lim, W. W. Fang, S. C. Rustagi, L. K. Bera, A. Agarwal, C. H. Tung, K. M. Hoe, S. R. Omampuliyur, D. Tripathi, A. O. Adeyeye, G. Q. Lo, N. Balasubramanian, and D. L. Kwong, "Ultra-Narrow Silicon Nanowire Gate-All-Around CMOS Devices: Impact of Diameter, Channel-Orientation and Low Temperature on Device Performance," IEDM Tech. Digest, pp. 1-4, 2006.

[7] C. Hu, C. S. C. Tam, F-C. Hsu, P. K. Ko, T. Y. Chan, and K. W. Terril, "Hot-electron-induced MOSFET degradation-Model, monitor, and improvement," IEEE Trans. Electron Device, vol. ED-32, no. 2, pp.375-385, 1985.

Degradation of Metal-Induced Laterally Crystallized n-type Poly-Si Thin-Film Transistors Under Dynamic Voltage Stress

Meng ZHANG, *Mingxiang WANG, Huaisheng WANG

Dept. of Microelectronics Soochow University, No. 178 Gan-jiang East Road, Suzhou 215021, China
*Phone: +86-512-6750 7765. Fax: +86-512-6724 8370. Email: mingxiang_wang@suda.edu.cn

Abstract-Degradation of metal-induced laterally crystallized n-type poly-Si TFTs is systematically investigated under synchronized V_g and V_d stresses. In low frequencies, degradation independent of frequency is the same as that in DC self-heating stress, while in high frequencies, frequency dependent degradation is attributed to an additional mechanism associated with pulse rising/falling edges.

I. INTRODUCTION

Metal-induced lateral crystallization (MILC) is a promising low temperature poly-Si technology to fabricate high performance thin film transistors (TFTs) for system-on-panel applications [1-2]. TFTs in driver circuits, unlike those in the pixels, are subjected to high-frequency voltage pulses [3]. Therefore, device reliability under dynamic operation should be understood. Recently, there were some studies [3-4] on device degradation of poly-Si TFTs under dynamic gate voltages (V_g), or under synchronized V_g and V_d voltage pulses [5]. While for MILC TFTs, no study on their degradation under dynamic operation has been reported.

In this work, degradation of n-type unilaterally crystallized MILC TFTs under synchronized V_g and V_d stresses (source grounded, see Fig.1) was systematically investigated. Pulse height, frequency and duty ratio are stress parameters. Different degradation phenomena for high and low frequency stresses are demonstrated. For low frequency stress, the degradation mechanism is the same as DC self-heating (SH) condition; while for high frequency stress, a hot carrier (HC) like mechanism is found to control the device degradation.

II. DEVICE FABRICATION AND STRESS CONDITIONS

TFTs used in this study were in conventional self-aligned top-gate structure. First, a 50nm amorphous-Si (a-Si) was deposited on an oxidized silicon wafer by LPCVD. After patterning of a-Si active islands and LPCVD of an 80nm low-temperature oxide (LTO), crystallization inducing window was opened through the LTO and 5nm Ni was evaporated by electron-beam at room temperature. The wafers were subsequently annealed for metal-induced unilateral crystallization of active islands at 550°C for 24 hours in N_2 ambient. After Ni removal, another extended anneal was done at 550°C for 48 hours. Then LTO was removed and another 100nm LTO layer was deposited as gate oxide, followed by LPCVD of

300nm poly-Si as gate. After gate patterning, a self-aligned phosphorous implantation with a dose of $4\times10^{15}cm^{-2}$ was done to form the source and the drain, and subsequently activated at 620°C for 3 hours. Contact holes were then opened before aluminum layer sputtering and patterning. Finally, wafers were sintered in Forming gas.

N-channel MILC TFTs were subjected to various synchronized V_g and V_d dynamic stresses with a fixed peak stress power density of 0.24mW/μm^2. Under DC SH condition at the chosen peak power density, a severe SH degradation would occur [2]. With peak power density fixed, such SH effect should be comparable for all devices investigated under various AC stresses. The devices were characterized before and after stress by using Agilent 4156C semiconductor parameter analyzer and Vector MX-1100B prober. Degradation mechanisms were studied by analyzing the evolution of device parameters including on-current (I_{on}), threshold voltage (V_{th}), minimum leakage current (I_{off_min}) and the gate-induced drain leakage (GIDL) current (I_{off}, defined at V_d=5V and V_g=−12V).

III. RESULTS AND DISCUSSIONS

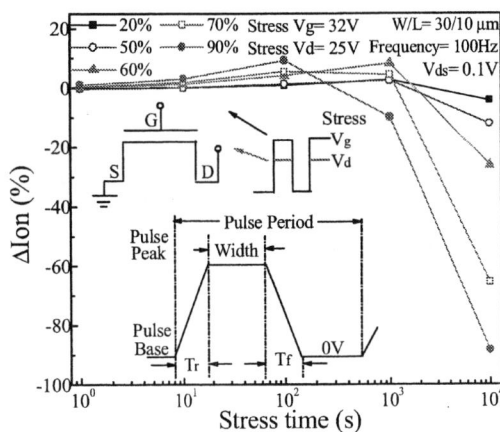

Fig. 1 Dependence of I_{on} degradation on stress time under various pulse duty ratios which are defined by Width/Pulse Period as shown in the inset. Stress condition and the pulse waveform is shown in the inset. Rising and falling time T_r=T_f=0.1μs.

Shown in Fig.1 is the dependence of I_{on} degradation on stress time under various pulse duty ratios. Devices under higher duty

978-1-4244-2039-1/08/$25.00 ©2008 IEEE

ratios exhibit more ΔI_{on}. This is not surprising because under the same stress time, higher duty ratio means longer accumulated time periods (duty time) during which device is subjected to SH stress. While in Fig. 2, when ΔI_{on} is plotted against duty time rather than the overall stress time, all curves with different pulse duty ratios almost overlap. As shown in the inset, the degradation becomes independent of duty ratio only if the duty time is the same. Although not given here, similar behavior is also observed for the V_{th} degradation. Therefore, in the low frequency range (<50 kHz) investigated, device degradation is related to the duty time, not to the apparent stress time. It has been reported that time constant characterizing the channel temperature rising upon SH stress is ~1μs for W/L=30/10μm [6], therefore under low frequency stresses the device channel temperature has enough time to follow the voltage pulse variation so that device is subjected to SH during the time intervals with high V_g and V_d level [2]. Such correlation between the degradation and the duty time strongly implies that the controlling degradation mechanism under the low frequency stress is SH.

Stress time dependences of ΔI_{on} under varied pulse frequencies, in logarithmic and linear scale are shown in Fig.3 and its inset, respectively. Pulse duty ratios are fixed at 50% for all curves. Apparently, I_{on} degradations under high (>50 kHz) or low frequencies are distinctly different. At low frequencies, the two-stage degradation [2] is typical for SH degradation of MILC TFTs, while a straightforward degradation without a turnaround phenomenon is observed under high frequencies. The degradation slope of ~0.93 for low frequencies is the same as in the DC SH degradation [2], which suggests an underlying degradation mechanism of defects generation either by dissolution of Si-H bones or breaking/distortion of Si-Si bonds at grain boundaries and/or the gate oxide/channel interfaces [2]. But for high frequencies, the slope is significantly lowered to ~0.25. The striking differences of the degradation under high frequency dynamic stress strongly imply that a different degradation mechanism other than SH is involved.

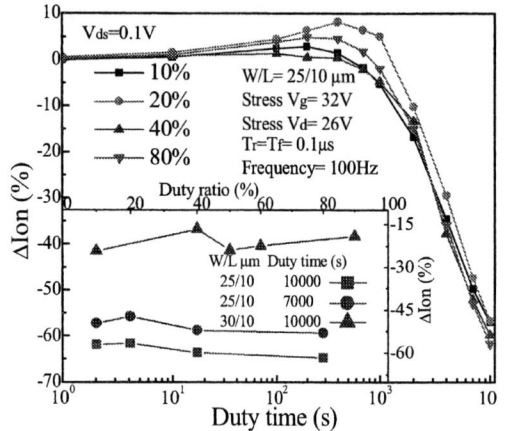

Fig. 2 Dependence of the I_{on} degradation on duty time under various pulse duty ratios. The inset is the ΔI_{on} plotted against the duty ratio with the same stress duty time of 7000 and 1000s.

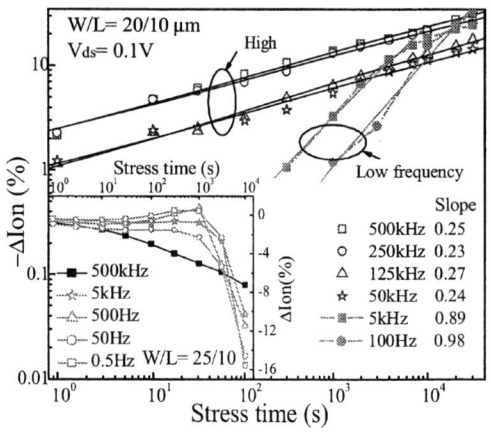

Fig. 3 Stress time dependent ΔI_{on} plotted in logarithmic scale for various pulse frequencies (V_g=32V, V_d=27V). The inset is an I_{on} degradation in linear scale (V_g=32V, V_d=26V)

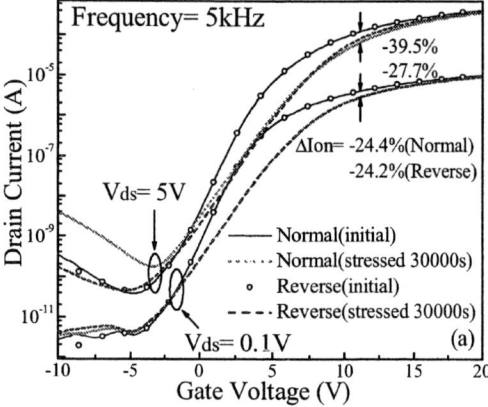

Fig. 4a Degradation of transfer characteristics under low frequency stress at V_{ds} =0.1 and 5V.

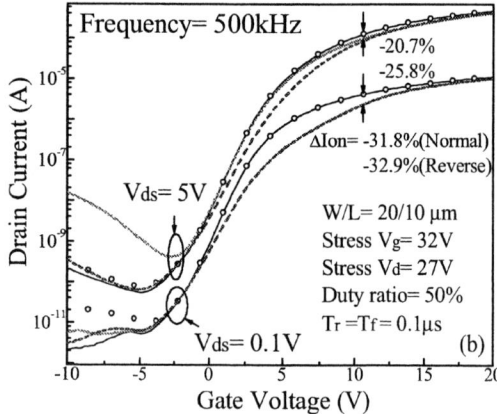

Fig. 4b Degradation of transfer characteristics under high frequency stress at V_{ds} =0.1 and 5V.

Fig. 4 is a comparison of the transfer curve degradation at V_{ds}=0.1 and 5V, between low and high frequency stresses. Under low frequency stress (Fig.4a), curves at ON-state region almost overlap when measured at V_{ds}=0.1V in normal or reverse mode. At V_{ds}=5V, only slight discrepancy is observed, reflecting the asymmetry of the SH within the channel. While the difference in I_{off} clearly indicates that leakage generating deep states are created near the drain side. No recovery phenomenon of the I_{on} degradation [7] is observed. As indicated by the arrows, in the normal mode, a degradation of 24.4% at V_{ds}=0.1V is largely aggravated to 39.5% at V_{ds}=5V. The same characteristics are also observed for DC SH stress. For high frequency stress (Fig.4b), recovery of the I_{on} degradation is apparent, especially in the normal mode, which can be explained by the drain-induced barrier lowering of trap state potential barrier near the drain side [7]. Interestingly, here one notes that at ON-state region the recovery phenomenon is similar to that of hot-carrier (HC) degradation [4, 7], while at OFF-state region the degradation is still the same as the SH degradation which is indicated by the I_{off} increase at high V_{ds} in the normal mode [1]. Therefore, it is considered that localized deep traps are generated near the drain side, while no HC injection occurs. Furthermore, we assume that traps caused by the high frequency stress should be localized closer to the drain edge compared to those by the low frequency stress, to account for the recovery phenomenon at high V_{ds} (Fig.4b).

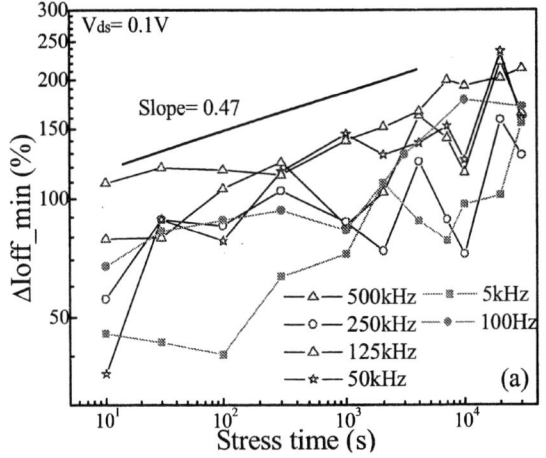

Fig. 5a Dependence of ΔI_{off_min} on stress time plotted in logarithmic scale for various pulse frequencies.

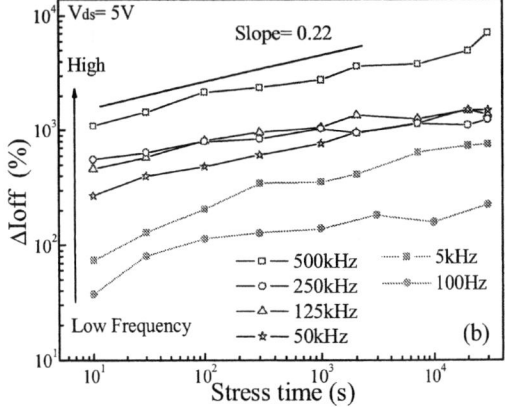

Fig. 5b Dependence of ΔI_{off} on stress time plotted in logarithmic scale for various pulse frequencies.

For OFF-state characteristics, stress time dependent ΔI_{off_min} and ΔI_{off} under various pulse frequencies are shown in Fig.5a and 5b, respectively. For I_{off_min}, degradation behaviors under different frequencies are indistinguishable and the average degradation slope is 0.47. While for the I_{off}, although the degradation slope is about the same as 0.22 for all frequencies, increasingly larger amount of I_{off} degradation occurs under higher frequency stresses. The correlation between the GIDL current and the frequency implies that trap generation here is related to pulse rising/falling edges because higher frequency stress brings more pulse rising/falling edges. After long stress time, it is noted that the degradation seems to saturate to some extent.

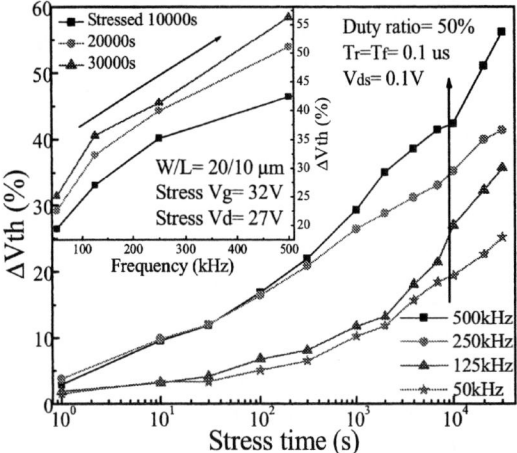

Fig. 6 Dependence of the V_{th} degradation on stress time under various high frequencies. The inset is the dependence of ΔV_{th} on frequency after being stressed for varied period of times.

In Fig.6, the dependence of the V_{th} degradation on stress time under various high frequencies is given. Devices under higher frequency stress degrade more severely. It's a different observation from that reported by K. Takechi [5], where the degradation is small at high frequencies due to insufficient SH time intervals during pulsed operation. Here, time constant characterizing the channel temperature rise is slightly less than 1μs [6], which is comparable to the time period at the highest frequency (500 kHz). Therefore, similar insufficient SH at high frequency should be expected. The opposite results presented here also suggest a different degradation mechanism other than SH for high frequency stresses which should occur during switching transients [8], agreeing with the above-mentioned argument.

It was previously suggested that when device being forced between ON and OFF state, HC mechanism occurs to create defects near the gate edge [3-4]. Under high frequency synchronized V_g and V_d stresses, our results suggest that some "HC-like" defect generation should also occur at the drain edge. Our most recent experiment shows that device degradation under the high frequency stress significantly depends on the rising edge of the voltage pulses and is little related to the falling edge. This lends further support that a new mechanism should be involved for high frequency stress. However, more investigations are still in progress to clarify the details of the mechanism.

IV. CONCLUSION

Degradation of n-type MILC TFTs is systematically investigated under synchronized V_g and V_d stresses. For the low frequency stress, device degradation independent of frequency is explained by the typical SH mechanism related to the duty time. While for the high frequency stress, besides the SH mechanism, an additional "HC-like" mechanism dependent on pulse rising/falling edges also contributes to the device degradation.

ACKNOWLEDGMENT

This work is supported by the National Natural Science Foundation of China under Contract No. 60406001.

REFERENCES

[1] Min Xue, Mingxiang Wang, Zhen Zhu, Dongli Zhang, Man Wong, "Degradation behaviors of metal-induced laterally crystallized n-type polycrystalline silicon thin-film transistors under DC bias stresses," *IEEE Transactions on Electron devices*, 2007,Vol.54, pp.225-232.

[2] Huaisheng Wang, Mingxiang Wang, Zhenyu Yang, Han Hao, Man Wong, "Stress power dependent self-heating degradation of metal-Induced laterally crystallized n-type polycrystalline silicon thin-film transistors," *IEEE Transactions on Electron devices*, 2007, Vol. 54, pp.3276-3284.

[3] Yukiharu Uraoka, Noboyuki Hirai, Hiroshi Yano, Tomoaki Hatayama, Takashi Fuyuki, "Hot carrier analysis in low-temperature poly-Si TFTs using picosecond emission microscope," *IEEE Transactions on Electron devices*, 2004, Vol. 51, pp.28-34.

[4] Kow Ming Chang, Yuan Hung Chung, Gin Ming Lin, "Hot carrier induced degradation in the low temperature processed polycrystalline silicon thin film transistors using the dynamic stress," *Japanese Journal of applied physics*, 2002, Vol. 41, pp.1941-1946.

[5] Kazushige Takechi, Mitsuru Nakata, Hiroshi Kanoh, Shigeyoshi Otsuki, Setsuo Kaneko, "Dependence of self-heating effects on operation conditions and device structures for polycrystalline silicon TFTs," *IEEE Transactions on Electron devices*, 2006, Vol. 53, pp.251-257.

[6] Yang Zhenyu, Wang Mingxiang, Wang Huaisheng, "Finite element analysis of temperature distribution of polysilicon TFTs under Self-Heating stress," *Chinese Journal of Semiconductors*, 2008, Vol.29(5), pp.954-959.

[7] Mutsumi Kimura, Satoshi Inoue, Tatsuya Shimoda, "Dependence of polycrystalline silicon thin-film transistor characteristics on the grain-boundary location," *Journal of applied physics*, 2001, Vol.89, pp.596-600.

[8] Alan G. Lewis, I-Wei Wu, Michael Hack, Anne Chiang, Richard H.Bruce, "Degradation of polysilicon TFTs during dynamic Stress," *IEDM*, 1991, pp.575-578.

Failure Analysis Matrix of Light Emitting Diodes for General Lighting Applications

Nam Hwang
Korea Photonics Technology Institute
971-35 Wolchul-dong Buk-gu, Gwangju 500-779, Korea
Phone: (82)-62-605-9270, Fax: (82)-62-605-9289, Email: nhwang@kopti.re.kr

Abstract- **Failure mechanisms of light emitting diodes (LEDs) have been investigated. Various failure patterns of LEDs for general lighting applications were simulated under high electrical and thermal stress test conditions. Using scanning electron microscopy related with I-V characterization before and after the stress simulation, a representative failure pattern of LEDs matrixed in terms of the electrical parameter shifts and physical conditions observed after decapsulation. By applying photo-emission microscopy it was found that local damage of LED surface was caused by excessive high temperature in the area of the pn junction of LEDs.**

I. INTRODUCTION

Recent technology development of light emitting diodes (LEDs) is very much promising to be applied to many applications such as automotive, streetlights, big out-door displays, and general lighting. The optical power and efficacy of LEDs are now compatible and superior to the conventional lighting systems. Furthermore, LEDs in solid state lighting systems are providing economical energy savings, robust reliability, and easy system integration [1-3]. However, due to the relatively simple structure and well-known good reliability of LEDs, the failure investigations of LEDs are not considered seriously. Specifically, the degradation of lighting parameters by employing LEDs as lighting sources has become very important for various solid state lighting applications, where the high electrical power is required to produce enough brightness for general lighting purpose. Consequently, the inevitable high thermal energy dissipated from LED chips is considered to be the key degradation factor. The failure analysis of LEDs is a quite sophisticated procedure since the LED failure is a combinational mechanism from electrical over stress to thermal dissipation that is critical to sustain a good reliability performance since continuous operation at high temperatures degrades LED characteristics with decrease of light output as well as the color shift of light emission [4-5]. In typical field failure samples, LEDs are being open due to its diode nature. Hence the failed LED is almost impossible to investigate the electrical character-istics that can show its internal electrical parameter shifts. Instead, the failure analysis has to rely on to its physical inspection leading to a speculative conclusion on its failure mechanism [6-7]. The purpose of this paper is, then, to investigate the failure mechanism of LEDs through the predefined intentional failure mode simulation. Typical failure modes of general semiconductor devices in operational environments are highlighted by the specific test conditions such as excessive current or voltage in forward or reverse bias.

By observing various individual failure patterns after applying failure mode simulation tests, the failure analysis of LEDs can become a failure matrix mosaicking with the identified failure patterns.

II. EXPERIMENT

The LEDs used in this study were packaged in a SMD PLCC type 3528 size including GaN/InGaN small size chips, by a domestic company. In this work, over 150 devices were studied. A semiconductor test and analyzer (ELECS EL-421C) was used to evaluate the operational electrical characteristics of LEDs. A series of tests were run on a total of 62 devices in the as-received condition to evaluate their conformity to the datasheet specifications:

i.   Input forward voltage   ($V_F@I_F = 20$ mA); < 3.5 V
ii. Reverse leakage current ($I_R@V_R = 5$ V); < 10 A
iii. Breakdown voltage   ($V_R@I_R = 100$ A); > 5 V

A total of 6 test conditions were predetermined to identify the effect on the failure patterns by individual stress factors as follows:

1. Forward over-current @ I_F=100 mA → 200 mA → 300 mA for 3 min. per level till to fail: Excessive for-ward current density injection.
2. Reverse over-current @ I_R=1 mA → 5 mA → 10 mA for 3 min. per level till to fail: Excessive re-verse bias stress.
3. HBM ESD stress @ 8 kV, 3 times, 1 sec: High volt-age impact.
4. MM ESD stress @ 1 kV, 3 times, 1 sec: High current impact.
5. Soft HBM ESD stress @ 0.5 kV, 3 times, 1 sec: Intermediate voltage shock to produce a latent dam-age.
6. Soldering over heat @ 300°C, 1 min: Thermal stress on leadframe or thermal dissipation on a chip

Table 1. Failure simulation matrix

Case	Failure Mode	Failure Simulation
1	Forward over current	Normal operation
2	Reverse over current	Power supply shock
3	HBM ESD stress	Over voltage shock
4	MM ESD stress	Over current shock
5	Soft HBM ESD stress	Latent damage
6	Soldering over heat	Thermal damage

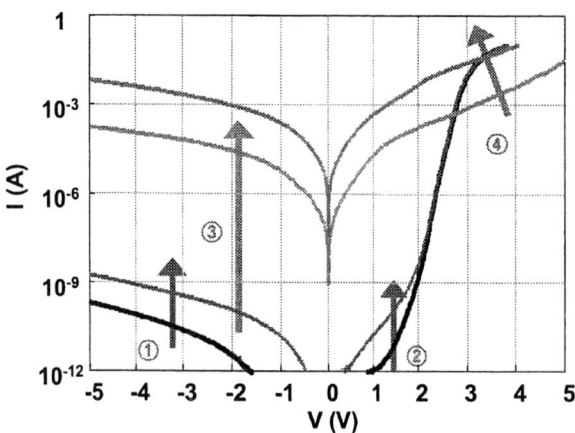

Fig. 1. Typical IV characteristics of LEDs before and after the forward over current test

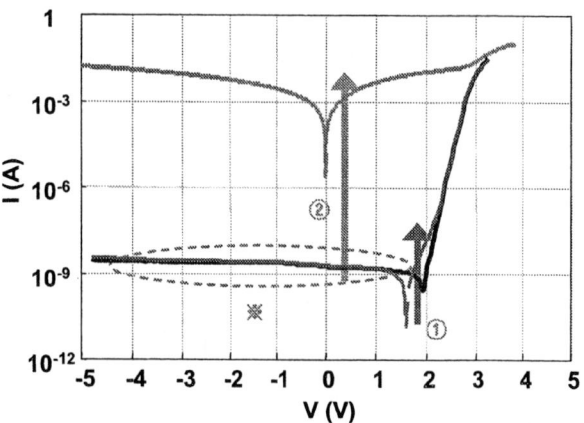

Fig. 3. Typical IV characteristics of LEDs before and after the reverse over current test.

III. RESULTS & DISCUSSIONS

Case 1: Forward over current

Fig. 1 shows the typical IV characteristics of LEDs before and after the forward over current stress test. The failure procedure was monitored as follows:
1) increase of the leakage current in the reverse region by a shunt component through the defects within pn junction,
2) increase of the recombination-generation current at a low voltage region by depletion region effect,
3) increase of the leakage current at full operational region by generation of defects within pn junction,
4) acceleration of leakage current generation from non-radiative recombination and a non-emission region,
5) electrical open failure.

Case 2: Reverse over current

Fig. 3 shows the typical IV characteristics of LEDs before and after the reverse over current test. The failure procedure was monitored as follows:
1) generation of diode parasitic effects with lower barrier height causing leakage current through a defective region within pn junction,
2) abrupt increase of the leakage current at full operational region by generation of defects within pn junction makes the acceleration of non-radiative recombination effect and large optical power reduction.
* photocurrent generation was found under room lighted conditions.

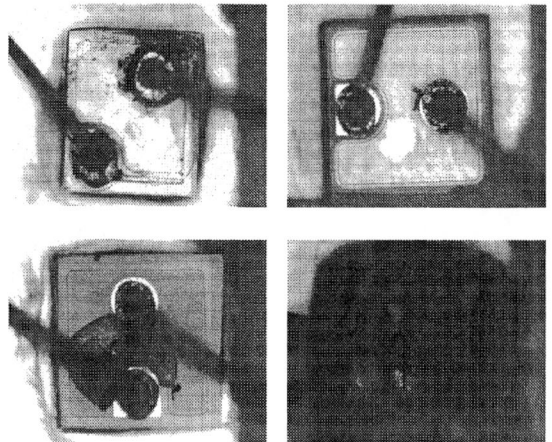

Fig. 2. Failure pattern of LEDs after the forward over current test.

Fig. 2 shows the failure patterns of LEDs after the forward over current stress test. The degenerated epoxy is found to be hard to remove under the normal depackaging process. Excessive thermal energy deformed the epoxy on the top surface between two electrodes.

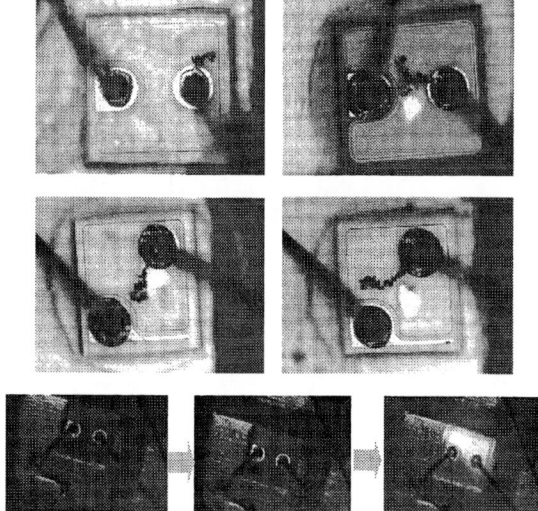

Fig. 4. Failure pattern of LEDs after the reverse over current test.

Fig. 4 shows the failure patterns of LEDs after the reverse over current stress test. The surface damage was found to be partitionable with a non-emitting region.

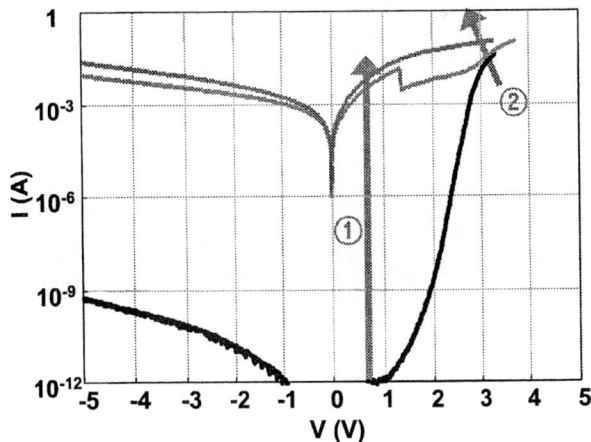

Fig. 5. Typical IV characteristics of LEDs before and after the HBM ESD stress test.

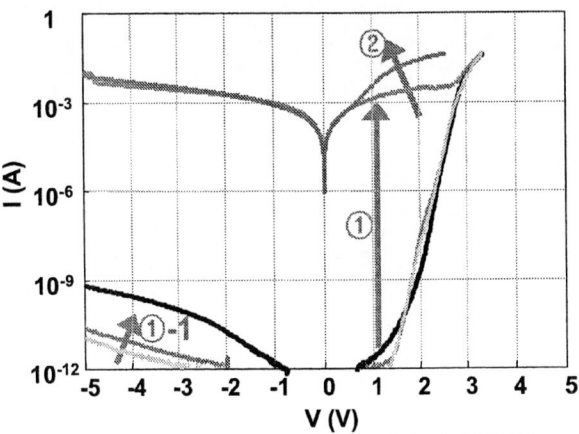

Fig. 7. Typical IV characteristics of LEDs before and after the MM ESD stress test.

Case 3: HBM ESD Stress

Fig. 5 shows the typical IV characteristics of LEDs before and after the HBM ESD stress test. The failure procedure was monitored as follows:

1) same as the failure pattern of the reverse over current stress, the high voltage impact shock seems induced high surface charge current leaving point burnt marks,

2) leakage current generation from a non-emitting region.

Case 4: MM ESD Stress

Fig. 7 shows the typical IV characteristics of LEDs before and after the MM ESD stress test. The failure procedure was monitored as follows:

1) same as the failure pattern of the HBM ESD stress, the high current impact shock seems accumulated high thermal energy leaving epoxy degeneration around the electrode,

2) leakage current generation from non-radiative recombination and a non-emitting region,

3) indiscernible damage with a minute optical power reduction at only green LED chips.

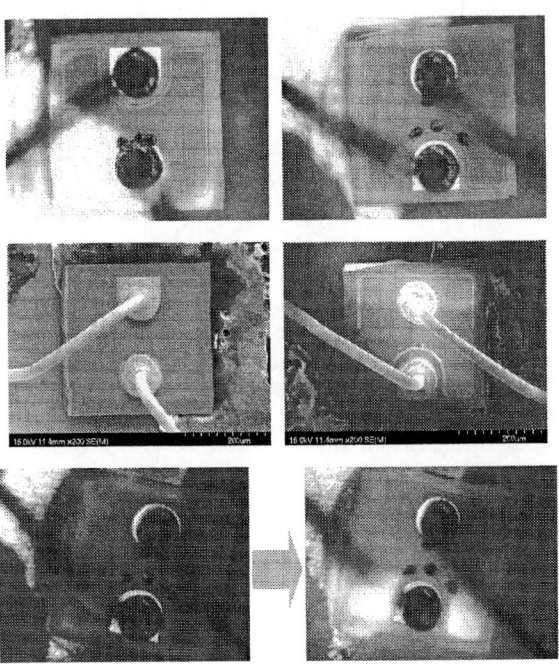

Fig. 6 Failure pattern of LEDs after the HBM ESD stress test.

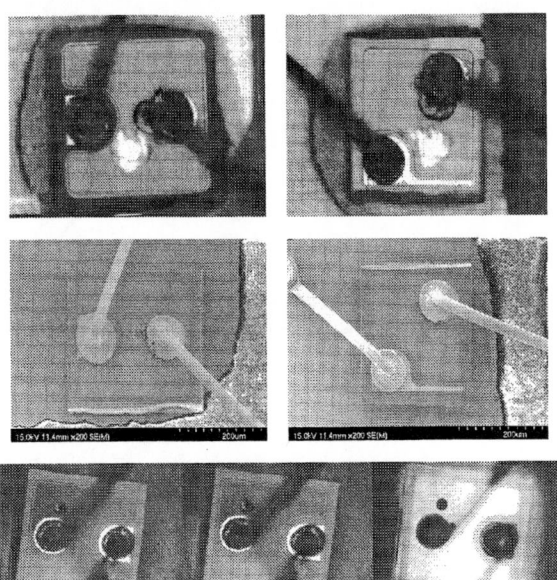

Fig. 8 Failure pattern of LEDs after the MM ESD stress test.

Fig. 6 shows the failure patterns of LEDs after the HBM ESD stress test. The severe burnt mark points were found as well as a non-emitting region.

Fig. 8 shows the failure patterns of LEDs after the MM ESD stress test. The surface damage is much like the combination of forward and reverse over current failures.

978-1-4244-2039-1/08/$25.00 ©2008 IEEE

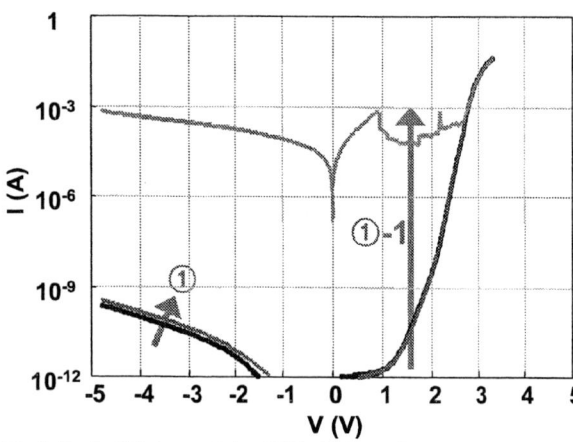

Fig. 9. Typical IV characteristics of LEDs before and after the soft ESD stress test.

Case 5: Soft HBM ESD Stress

Fig. 9 shows the typical IV characteristics of LEDs before and after the soft HBM ESD stress test. The failure procedure was monitored as follows:

1) increase of the leakage current and the small optical power reduction,
2) indiscernible damage with minute optical power reduction.

No noticeable surface damage was found as shown in Fig. 10. Intermediate thermal damage of epoxy was observed, but not enough to be evidentiary.

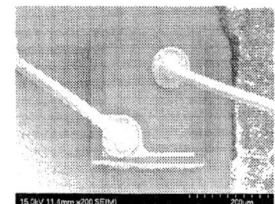

Fig. 10. Failure pattern of LEDs after the soft HBM ESD stress test.

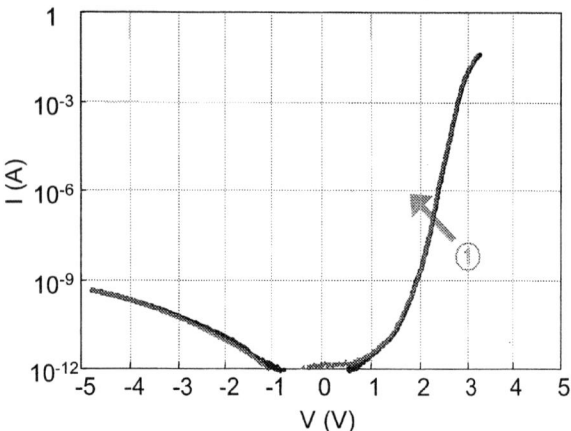

Fig. 11. Typical IV characteristics of LEDs before and after the MM ESD stress test.

Case 6: Soldering over heat

Fig. 11 shows the typical IV characteristics of LEDs before and after the soldering over heat stress test. As shown in Fig. 12, it should be pointed out that indiscernible chip damage and the optical power reduction by epoxy degradation through thermal damage were found.

Fig. 12. Failure pattern of LEDs after the soldering over heat stress test.

IV. CONCLUSION

In conclusion, failure analysis of LEDs has been studied under simulated failure tests. With the predesigned failure mode and its failure pattern, the investigation of failure mechanism is self-instructional by employing the results of failure mode simulation. The major failure mode of LEDs is found and confirmed to be related with internal thermal energy accumulation.

REFERENCES

[1] D. L. Barton, M. Osinski, P. Perlin, C. J. Helms, N. H. Berg, "Life tests and failure mechanisms of GaN/ AlGaN/InGaN light emittingdiodes," IRPS 1997, 8-10 Apr. 1997, pp. 276-281.

[2] G. Meneghesso, et. al., "Degradation mechanisms of GaN-based LEDs after accelerated DC current ag-ing," IEDM 2002, 8-11 Dec. 2002, pp. 103-106.

[3] S. Levada, M. Meneghini, G. Meneghesso, E. Zanoni, "Analysis of DC current accelerated life tests of GaN LEDs using a Weibull-based statistical model," IEEE Trans. on Device and Materials Reliability, vol. 5, issue 4, Dec. 2005, pp. 688-693.

[4] F. Wu, Y. Wu, B. An, F. Wu, "Analysis of Dark Stain on Chip Surface of High-Power LED," ICEPT '06, 26-29 Aug. 2006, pp. 1-4.

[5] M. Meneghini, L. R. Trevisanello, S. Levada, G. Me-neghesso, G. Tamiazzo, E.; Zanoni, T. Zahner, U. Zehnder, V. Harle, U. Straus, "Failure mechanisms of gallium nitride LEDs related with passivation," IEDM 2005, 5-7 Dec. 2005, pp. 4-7.

[6] D. L. Barton, M. Osinski, P. Perlin, P. G. Eliseev, J. Lee, "Degradation of single-quantum well InGaN green light emitting diodes under high electrical stress," IRPS 1998, 31 Mar. 1998, pp. 119-123.

[7] N. Hwang, P. S. R. Naidu, A. Trigg, "Failure analysis of plastic packaged optocoupler light emitting di-odes," EPTC 2003, 10-12 Dec. 2003, pp. 346-349.

978-1-4244-2039-1/08/$25.00 ©2008 IEEE

Failure Site Isolation on Passive RFID Tags

Bhanu Sood[1], Diganta Das[1], Michael Azarian[1], and Michael Pecht[1], Brian Bolton[2], Tingyu Lin[3]

1: Center for Advanced Life Cycle Engineering, University of Maryland, College Park, MD 20742 USA
Tel: (301) 405 5323 Fax: (301) 314 9269 Email: pecht@calce.umd.edu
2: Motorola Inc., 1340 Charwood Dr. Ste F, Hanover, MD 21075
3: Motorola Electronics Pte Ltd, Ang Mo Kio Ind. Park 3, 12 Ang Mo Kio St. 64, Singapore 569086

Abstract- **Passive RFID tag typically consists of an inert carrier substrate, an antenna and a semiconductor chip. Qualification process of the passive RFID tags includes temperature cycling in high humidity conditions and damp heat storage tests. In order to pass qualification tests, a passive RFID tags needs to respond to queries by a tag reader. There was an unacceptable level of failures of passive RFID tags after exposure to cyclic testing (-40 to 70°C, 95 %RH) and damp heat storage (85°C, 85 %RH) tests.**

We performed an evaluation of the materials, manufacturing and assembly processes and qualification testing environment to create a set of possible failure sites with associated failure modes, mechanisms and causes. A physical analysis plan was created and executed to investigate these sites of interest on as-manufactured, exposed and failed passive RFID tags. The analysis included electrical resistance measurements, chemical decapsulation, cross-sectioning, focused ion beam etching (FIB), scanning electron microscopy (SEM) and energy dispersive spectroscopy (EDS) analysis on cross-sections.

Based on the analysis, we found damaged passivation layer and delamination between bump and die metallization. We also found instances of poor physical contact between the semiconductor bump and antenna. The analysis of these results led to implementation of several design and assembly process changes, which resulted in high degree of improvements in reliability and performance.

I. INTRODUCTION

Radio frequency identification (RFID) is a rapidly developing technology which uses radio frequency signals for automatic identification of objects. RFID technology is very popular in both field and research applications. RFID provides an automatic identification method, relying on storing and remotely retrieving data using devices called RFID tags or transponders. An RFID tag is an object that can be attached to or incorporated into a product, animal, or person for the purpose of identification using radio frequency waves.

Due to their distinct advantages of read/write capabilities without a restriction of line of sight, RFID technology is gaining widespread acceptance [1], [2]. A typical RFID tag consists of an antenna printed on a substrate (usually flex substrate) and die (integrated circuit) mounted on the antenna and substrate. Many manufacturers of RFID tags use antennas made of silver, aluminum or copper. Recently, other materials have also become options because of low materials costs and compatibility with volume manufacturing [3].

RFID tag systems consist of the tags and a sensing or reading element. The system uses sensing elements to detect and record parameters such as temperature, humidity, movement and environmental conditions such as even radiation

data. For example, the same tags used to track items moving through the supply chain may also alert staff if they are not stored at the right temperature, or if meat has gone bad or if someone has injected a biological agent into food.

RFID die are attached to a substrate carrying the antenna with conductive adhesive materials known as anisotropically conducting adhesives, meaning that they conduct electrically in a selected direction only. In a RFID tag application, such adhesives usually conduct in the direction perpendicular to the plane of the tag (or Z) and are hence also known as Z-axis adhesives. Such adhesives prevent the terminals of the die to short with each other and also from lateral conduction of the radio frequency signals that would make the tag unusable. In the tag fabrication process, conductive materials, usually silver or aluminum, are printed to form the antenna, then, a controlled amount of anisotropically conducting adhesive is automatically dispensed on the region where the die will be attached. Later, the die with four bumps is placed on top of the adhesive, and the adhesive is then cured by thermocompression using opposing thermodes. Depending on the chemistry and nature of the anisotropically conducting adhesive, specific mechanical pressure levels and temperatures are selected to activate the bonds and create an electrical connection between the silver antenna and the die terminals. Special care is taken to control the mechanical pressure and temperature during the attach process so as not to damage any of the components of the die or silver antenna substrate.

Since RFID tags cannot usually be replaced after they are placed on a product, long term reliability and yield are major concerns for manufacturers. One of the key mechanical reliability issues for RFID tags is how well the die components and antenna adhere to the surface of the substrate onto which they are attached. Furthermore, delaminations are possible in addition to chemical interactions between the different components of the tag. These failure mechanisms are a result of manufacturing processes and application-specific environmental and mechanical loads that subject the tag to temperature variations and substrate deformations. Testing RFID for susceptibility to these failure mechanisms can be achieved through temperature cycling, damp heat storage tests and various bending and torsion loading tests.

A   Failure modes, mechanisms and effects analysis (FMMEA)

Failure modes, mechanisms and effects analysis (FMMEA) is a systematic methodology to identify potential failure mechanisms and models for all potential failures modes, and to prioritize failure mechanisms [3]. FMMEA enhances the value of the failure modes ands effects analysis (FMEA) and failure modes, effects and criticality analysis (FMECA) methods by identifying high-priority failure mechanisms in order to create an action plan to mitigate their effects. High priority failure mechanisms determine the operational stresses and the environmental and operational parameters that need to be controlled. Models for the failure mechanisms help in the design and development of the product. FMMEA is based on understanding the relationships between product requirements and the physical characteristics of the product (and their variation in the production process), the interactions of product materials with loads (stresses at application conditions) and their influence on the product susceptibility to failure with respect to the use conditions. This involves identifying the failure mechanisms and reliability models to quantitatively evaluate the susceptibility to failure.

In addition to the information gathered and used for FMEA, FMMEA uses life cycle environmental and operating conditions and the duration of the intended application with knowledge of the active stresses and potential failure mechanisms. An overview of the FMMEA process is shown in Fig. 1.

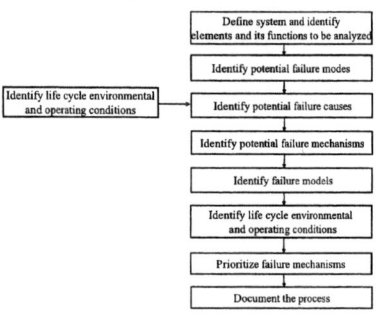

Fig. 1. FMMEA methodology.

For the analysis of RFID tags, FMMEA process begins by defining the system to be analyzed, which is viewed as a composite of subsystems or levels that are integrated to achieve a specific objective. The system is divided into various sub-systems or levels, continuing to the lowest possible level, which is referred to as component or element. The system breakdown can be either be performed by function (i.e., according to what the system elements "do"), by location (i.e., according to where the system elements "are"), or a combination of both (i.e., functional breakdown by location, or vice versa).

In a RFID tag, for example, location breakdown includes the die, antenna metallization, and the substrate. For each element, all of the associated functions are listed. Further the stresses acting upon the RFID tag are summarized and then the failure modes and mechanisms that would be responsible for a failure are collected. Table I summarizes the stress conditions, failure modes and associated failure mechanisms for RFID tags under study.

II. EXPERIMENTAL ANALYSIS

RFID samples that were exposed to different types of accelerated test conditions were received for failure analysis. The accelerated tests included damp heat storage at 85°C and 85%RH and cyclic testing from -40°C to +70°C (95%RH). Some unstressed samples were also provided.

A   Test materials

An overview of the RFID tags is shown in Fig. 2. Test points where electrical resistance is measured are marked Fig. 2.

B   Tests

Electrical resistance measurements were conducted on the various test points indicated in Fig. 3. Measurements were taken with a handheld Fluke 187 Multimeter. Resistance was measured between the points RF1-RF2, RF2-GND1, RF1-GND1 and also by changing the polarity of the test leads at

Table I
STRESSES, FAILURE MECHANISMS AND FAILURE MODES FOR RFID TAGS

Stresses	Failure Mechanisms	Failure Modes
Age	Overcured Adhesive	Short Between Bumps
High Temp	Undercured Adhesive	Short Between Antenna Pads
Low Temp	Corrosion of Antenna	Insufficient Impedance Between Bumps
Humidity	Corrosion of Bumps	Insufficient Impedance Between Antenna Pads
Temp & Humidity	Corrosion of Filler Particle	Open Between Bump and Antenna Pad
Temp Cycling	Adhesive Swelling	Excessive Impedance Between Antenna Pad and Bump at Bondline
Humidity Cycling	Die Lift	Excessive Impedance in Antenna (eg Partially Fractured Antenna Lead)
Temp & Humidity Cycling	Die Separation from Adhesive	Open In Antenna (eg Partially Fractured Antenna Lead)
ESD	Adhesive Separation from Antenna	Intermittent Contact Between Bump and Antenna Pad
Excessive Bond Force	Adhesive Void	Short In IC
Insufficient Bond Force	Insufficient Compression of Filler	Open In IC
Excessive Bond Temperature	Damaged IC	
Insufficient Bond Temperature	Insufficient Gap Between Die and Antenna	
Excessive Bond Time		
Insufficient Bond Time		
Mechanical Shear Stress		
Mechanical Bendng Stress		
Mechanical Compression		

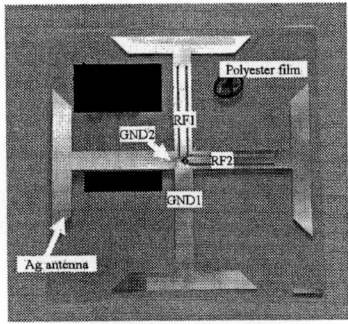

Fig. 2. Overview of RFID tag.

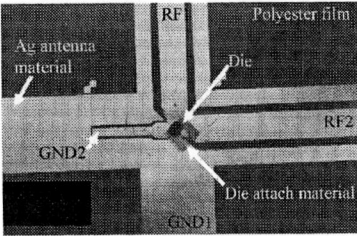

Fig. 3. Closeup of the die mounted on RFID tag. The four components of the antenna are marked.

RF1 and GND1 (RF1-GND1+ and RF1+GND1-). Sample electrical measurements for 3 RFID tags that were exposed to the different conditions are shown in Table II. The results show high electrical resistance on some samples that were exposed to damp heat storage and cyclic testing. Electrical resistivity testing on the stressed samples and comparison with unstressed samples provides convenient failure site identification. Appropriate failure analysis techniques can be selected based on the results of these measurements.

A possible reason for electrical open between different terminals of the RFID tag can be a complete separation between die bump and antenna metallization or a separated interface that leads to high resistance. Cross-sectional analysis is conducted on selected tags that exhibit high electrical resistance or open. Cross-sections of the RFID tags at the exposed plane facilitate close inspection with environmental scanning electron microscopy (ESEM) and energy dispersive spectroscopy (EDS) tools. The plane of cross-section is shown in Fig. 4.

For cross-sectioning, the area around the die on selected RFID tags is first cut out of the polyester film carrier and then potted in a transparent epoxy which is originally liquid. After the epoxy has cured and solidified, the sample is taken through the different steps of cross-sectioning starting from coarse and fine grinding, to final polishing. An environmental scanning electron microscope (ESEM) image of the die after cross-section steps is shown in Fig. 5. At this plane, the RF1 and GND2 bumps are visible along with edge of the silver antenna metallization and polyester film. Fig. 6 shows a close up of the GND2 bump. Cross-section analysis reveals a lower amount of

the silver antenna metallization material under the bump as compared to the sides and also reveals poor contact between bump and antenna material. There also exists a region of separation between the GND2 bump and silver flakes that comprise the antenna material. Based on energy dispersive spectroscopy (EDS) analysis at this cross-section plane, no intermetallic growth was observed in the gold bump.

Table II
SAMPLING OF RESULTS OF ELECTRICAL MEASUREMENTS ON RFID TAGS.
MEASUREMENTS SHOW HIGH ELECTRICAL RESISTANCE OR OPENS AFTER
EXPOSURE TO DAMP HEAT STORAGE AND CYCLIC TESTING.

Test Samples	Resistance measurements (Ω)				
	RF1-RF2	RF2-GND1	RF1-GND1	RF1+ GND2-	RF1- GND2+
Sample from Damp Heat Storage (85°C / 85% RH)	1.27	0.86	0.86	open	open
Sample from Cyclic Testing (-40°C to 70°C 95% RH)	1.27	0.89	0.92	0.58K	0.58K
Unstressed Sample	1.58	1.03	1.07	1.65	1.65

To confirm separation between the silver metallization and the gold bump at GND2 that was observed, a "bottom-up" cross-sectional analysis is conducted. Using this approach, it is possible to closely examine the quality of contact between the gold bump at GND2 and surrounding silver metallization. Fig. 7 shows an ESEM image of the cross-section plane. It can be seen from this image that there are certain areas around the bump that are not in complete contact with the bump. Note that this cross-section was performed on a separate RFID sample than one used for cross-section in Fig. 5 and Fig. 6.

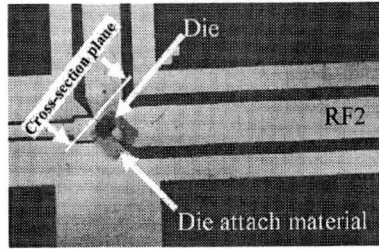

Fig. 4. Cross-section approach to reveal the joints between antenna and RF1 and GND2.

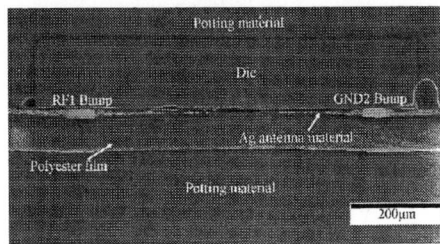

Fig. 5 RFID die after cross-sectioning steps.

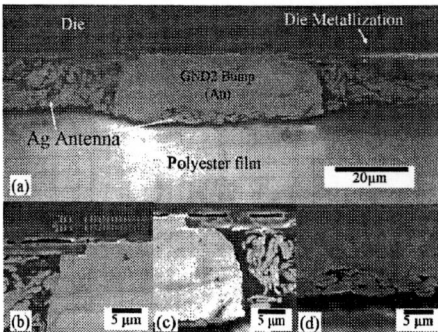

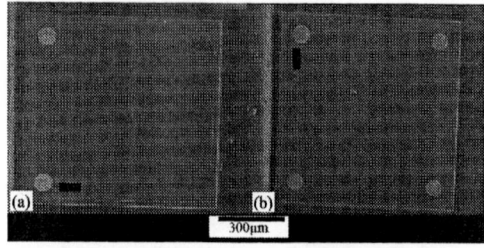

Fig. 8 ESEM images of decapsulated die. Die with two missing bumps (a) was exposed to cyclic testing (-40°C to +70°C, 95%RH) and (b) is a die from an unstressed RFID tag.

Fig. 6 (a) ESEM image of cross-section of GND2 bump shows poor contact between bump and Ag antenna material. Closeup ESEM inspection of the bump at (b), (c) and (d) also confirms the poor contact.

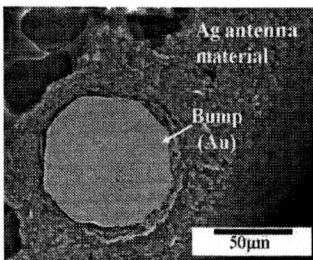

Fig. 7 ESEM image of cross-section plane revealed after top down cross-section of bump at GND2.

After cross-section analysis, chemical decapsulation is conducted on selected RFID tags to inspect the interface between the bumps and die and also the die surface. Chemical decapsulation involves dissolution of the polyester film substrate by means of nitric acid to expose the regions of interest. Concentration and temperature of the nitric acid are controlled so that the decapsulation process does not damage the die passivation layers to enable electrical resistance measurements on the die. Tags that show high electrical resistance are selected for the decapsulation and the dies from these tags are compared to decapsulated dies obtained from unstressed tags. Polyester film is cut around the die and immersed for 5 seconds in 90% red fuming nitric acid at 70°C. After removal from the acid the die is washed with distilled water and dried before observing under microscope and ESEM.

ESEM images in Fig. 8 show the die after decapsulation. Die decapsulated from the tag that was exposed to cyclic testing (-40°C to +70°C, 95%RH) shows two missing bump, whereas image of the die from unstressed tag shows that all four bumps are present. After conducting chemical decapsulation of additional stressed tags and comparing the die to unstressed samples, it was determined that the tags subjected to cyclic testing produced die with missing bumps. The missing bumps in most of the stressed samples are from RF1 and GND2 terminals.

To further investigate the cause of missing bumps in stressed tags, some additional samples were prepared for cross-section and for further analysis using focused ion beam (FIB) etching. FIB is selected because it enables selecting removal of material from the surface to expose the interface of interest. In the case of stressed samples, the interface of interest is between the bump and die pad metallization. The equipment used to make the final cross section cut is an early model FEI dual beam system. The system consists of a SEM column and a FIB column configured on the same chamber with an angle of about 53 degrees between the two beam lines. A stage with motorized translation and rotation and manual tilt and height control allows the features of interest on the sample to be positioned precisely at the intersection of the two beams. The advantage of this sample preparation is that the wall of a box cut being made by the focused ion beam is typically in clear view of the SEM allowing a visual guide.

This sample having already been cross-sectioned with conventional mechanical methods is mounted so the ion beam impinges at a glancing angle (<15 degrees) with the face of the existing cut. The first ion beam cut uses the largest current (11 nanoamps) in a thin and wide box (~10 by 100 microns). This removes the material residual from the earlier mechanical cut which is generally distorting or obscuring the view of the underlying structure. Beams with successively lower currents are then used to trim the face of the cut until the interfaces of interest are clearly imaged in the SEM. Slight adjustments in the position of each trim cut are made based on observation of the progress as indicated in SEM images. The shallow angle of the cut makes it difficult to follow the more common method of using ion beam images for positioning the trim cuts. However, the shallow angle made it possible to create a larger exposed face in a shorter time.

Fig. 9 shows SEM images of a die from a stressed tag. This sample was first cross-sectioned using techniques described earlier. The cross-sectioned sample was then to subjected to FIB etching at the interface of the die pad metallization and bump. Fig. 9 shows poor interface between the bump and die. The passivation layer that lies under the bump is damaged. A damaged interface between the bump and die pad metallization can potentially lead to a high electrical resistance path or a complete electrical open between the silver antenna and die. This is verified by conducting a similar cross-section on an unstressed tag and then subjecting to FIB etching.

Fig. 10 shows SEM image of an unstressed tag after the die and bump interface was subjected to FIB etching. It can be seen from the overview SEM image in (a) and closeup images of the sides in (b) and (c) that the passivation layer is not damaged.

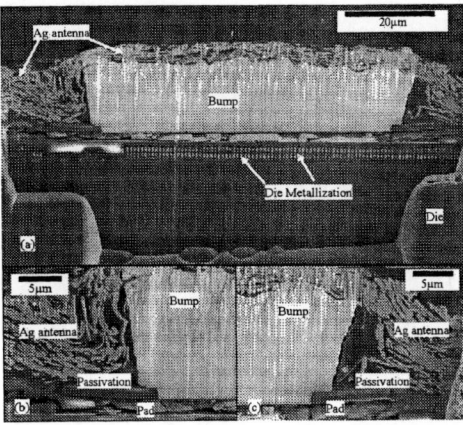

Fig. 9 SEM image of a die-bump interface after FIB etching. Overview of the interface in (a) shows the bump, die and silver antenna; (b) and (c) show closeup of the bump at two sides.

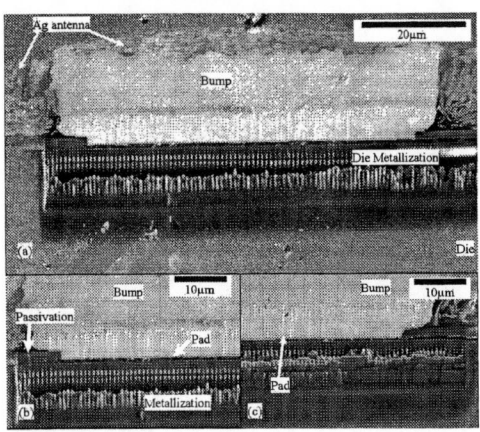

Fig. 10 SEM image of a die bump interface after FIB etching of an unstressed tag.

III. SUMMARY AND CONCLUSIONS

FMMEA analysis was performed on RFID tags to identify a set of possible failure sites with associated failure modes, mechanisms and causes. A systematic failure analysis plan was executed to investigate those possible sites on unstressed and stressed tags. Failure analysis steps include electrical resistance measurements for failure site isolation, cross-sectioning of stressed and unstressed tags to inspect interface between silver antenna and bumps, SEM and EDS analysis on cross-sections, chemical decapsulation to remove die from tag for further inspection and, focused ion beam (FIB) etching.

Based on results of SEM inspection of cross-sections, we have shown instances of inadequate physical contact between the silver antenna material and bumps of the die. Chemical decapsulation of the die from the tag and inspection revealed that certain bumps of die from stressed tags were not attached to the die, this was potentially caused due to poor adhesion of the bump to the die metallization or failure of the materials during processing or accelerated tests. Based on FIB etching, we found fractures in the passivation layer and delamination between semiconductor bump and internal metallization of the die. Recommendations provided to the manufacturer based on these findings have led to implementation of several design and assembly process changes, which resulted in significant improvements in reliability and performance of RFID tags.

ACKNOWLEDGEMENT

The authors would like to thank John Barry of IREAP, University of Maryland for FIB analysis.

REFERENCES

[1] K. Finkenzeller, "RFID handbook: Radio-frequency identification fundamentals and applications," John Wiley and Sons, 1999.

[2] K. V. S. Rao, "An overview of backscattered radio frequency identification system (RFID)," Asia Pacific Microwave Conference, vol. 3, pp. 746-749, November-December 1999.

[3] S. Cichos, J. Haberland, and H. Reichl, "Performance analysis of polymer based antenna-coils for RFID," International IEEE Conference on Polymers and Adhesives in Microelectronics and Photonics, pp. 120-124, June 2002.

[4] S. Ganesan, V. Eveloy, D. Das, and M.G. Pecht, "Identification and Utilization of Failure Mechanisms to Enhance FMEA and FMECA", Proceedings of the IEEE Workshop on Accelerated Stress Testing & Reliability (ASTR), October 3-5, 2005.

978-1-4244-2039-1/08/$25.00 ©2008 IEEE

Explore Further.

In 1989, FEI introduced the world's first ESEM™ Environmental SEM.

In 1993, FEI introduced the world's first DualBeam™ FIB/SEM.

In 2005, FEI introduced the Titan™ TEM, the world's most powerful commercially-available electron microscope.

Now, FEI introduces the Magellan™ family, the world's first Extreme High Resolution SEMs.

Prepare to see things you've never seen before. Visit FEI Company at IPFA 2008 in Singapore to learn how you can explore further with this new class of XHR SEM.

See more at fei.com/magellan

Background image: logic device, 1 kV, 1,200,000x. Logic device images courtesy of ST Microelectronics, Maila/Grenoble. Carbon nanotube image courtesy of Professor Raynald Gauvin and Camille Probst, Ph.D. student, McGill University.

©2008 FEI Company. All trademarks are the property of their respective owners.

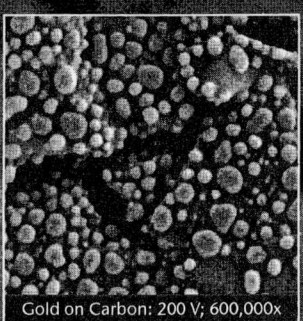

Gold on Carbon: 200 V; 600,000x

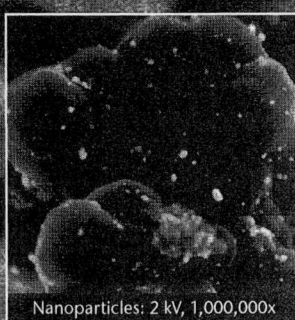

Nanoparticles: 2 kV, 1,000,000x

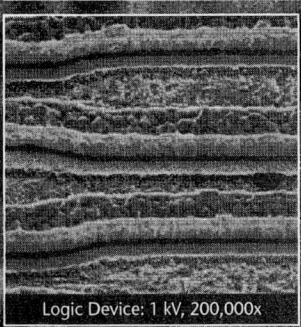

Logic Device: 1 kV, 200,000x

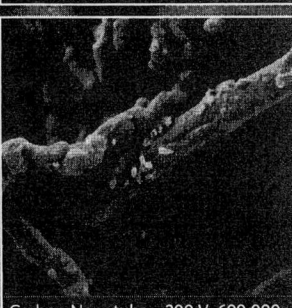

Carbon Nanotubes: 200 V, 600,000x

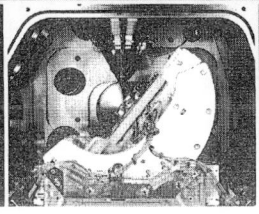

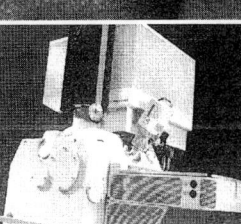

**FEI Company
Booth X20
IPFA 2008
Singapore**

Visit us at IPFA 2008 | booth X21

phoenix|x-ray
Part of GE's Sensing &
Inspection Technologies business

Improving product quality and reliability through high-resolution X-ray technology

3D Computed Tomography

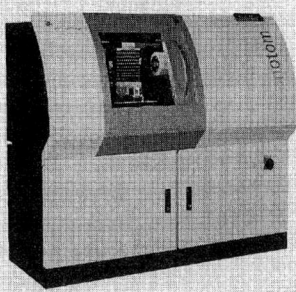

- ▶ nanotom® – first 180 kV nano-focus® CT system
- ▶ **nanoCT®** – a new dimension of 3D X-ray analysis with submicron voxel resolutions < 0.5 µm
- ▶ velo|CT – fast volume recon-struction

High-precision X-ray Inspection

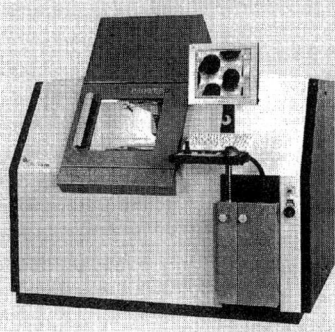

- ▶ nanome|x – total magnification above 24000x
- ▶ 180 kV / 15 W high power nanofocus® tube
- ▶ Detail detectability 0.2 - 0.3 microns
- ▶ Upgradeable to nanoCT®

Automated Solder Joint Inspection (µAXI)

- ▶ Highest magnification
- ▶ Maximum defect coverage
- ▶ Easy and fast CAD programming
- ▶ Live 3D CAD overlay

phoenix|x-ray offers a wide range of versatile application-oriented X-ray systems for failure analysis, process control and 3D CT.

GE imagination at work

phoenix|x-ray
Tel.: + 49.5031.172-0
Fax: + 49.5031.172-299
asia@phoenix-xray.com
www.phoenix-xray.com

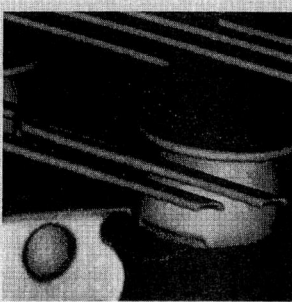
nanoCT® of CSP solder joints

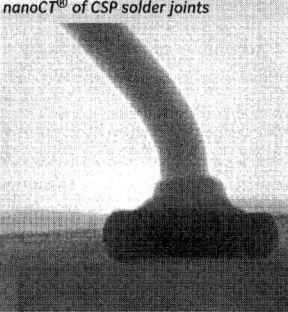
nanofocus® image of a ball bond

nanoCT® of a SMD inductor

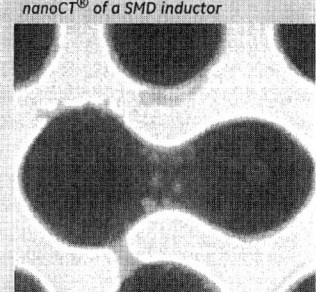
Flip Chip: micrometer sized voids

CT section of a die attach

Author's Index

AUTHOR'S NAME	PAGE
Ahmed K.	224, 255
Ahn D.S.	116
Albright G.	87
Ang G.B.	283
Antoniadis D.A.	1
Armstrong K.	214
Azarian M.	337
Balk L.J.	43
Bera M.K.	193, 196, 204, 325
Boit C.	9
Bolton B.	337
Bose P.K.	204
Boter D.	219
Ceric H.	78
Cervenka J.	78
Chakraborty A.	193, 196, 204, 325
Chakraborty P.	193, 196, 325
Chan D.S.H.	162
Chandrashekhar S.	224
Chang A.	170
Chang S.W.	245
Chao T.S.	260
Chapple-Sokol J.	308
Chauveau H.	219
Chen B.	128
Chen C.L	245
Chen C.Q.	150, 234
Chen J.F.	54
Chen K.S.	260
Chia V.K.F.	287
Chiang P.Y.	321
Chin J.M.	96, 104, 132, 279
Choub C.C.	321
Chu C.H.	170
Chua C.M.	20, 162
Chung K.	170
Conti G.	224
Craven A.J.	269
Crepel O.	30
Das D.	337
Deng Q.	82, 141
Deora S.	264
DeVries K.	308

Deyine-Barth A.	157
Dormans D.	219
Ee Y.C.	72
Eng Y.C.	185, 189
Er E.	26, 150, 234
Essely F.	15
Eum Y.H.	116
Fabris D.	312
Falk A.	87
Fernandez J.C.M.	178
Firiti A.	30
Foo E.G.	49
Fouillat P.	15
Gambino J.	308
Gan C.L.	72, 119
Ganguly U.	224
Gn F.H.	141
Goes W.	249
Goh S.H.	20
Goh Y.W.	145
Gong J.	321
Goubier V.	15
Grasser T.	249
Groeseneken G.	239
Gui D.	283, 304
Haller G.	15
He Z.H.	167
Hechtl M.	174
Heiderhoff R.	43
Hendarto E.	145
Hofmann R.	214
Hou Y.T.	260
Hua Y.N	283, 290, 304
Huang R.	170
Huang T.Y.	321
Huang Y.H.	283
Hwang N.	333
Isakov D.	162
Jacob P.	45
Jin L.J.	200
Johnson C.	308
Joman P.P.	132
Joo Y.C.	111
Joshi B.N.	181

Kang S.S.	185, 189
Karner M.	249
Ker M.D.	58
Khakifirooz A.	1
Kim D.H.	116
Kim B.J.	111
Kim J.	111
Kim S.A.	116
Kim Y.B.	116
Kitsuki H.	312
Koh L.S .	20, 162
Krishna N.	214, 224
Krishnan J.	124
Krueger B.	100
Kuan H.P.	200
Kuball M.	87
Lam D.	137
Lam J.	82, 141, 145
Lam T.F.	128
Lee K.	111
Lee M.L.	154
Lee C.K.	119
Lee J.H.	54
Lee Y.J.	260
Lewis D.	15, 30
Li K.	82, 234, 275
Li Y.	234
Li Y.G.	108
Lim V.	141
Lim G.T.	111
Lim M.K.	72
Lim S.H.	96
Lin C.Y.	58
Lin H.B.	145
Lin H.S.	92
Lin J.T.	185, 189
Lin K.C.	260
Lin P.H.	185, 189
Lin R.D.	154
Lin T.	337
Lin W.L.	260
Liu H.	304
Liu B.H.	290
Liu P.	234, 275
Lloyd J.	297
Lo W.C.	260
Loh S.K.	26, 150
Lok B.L.	316
Low F.	49
Lu W.	304
Luce S.	308
Lwin H.E.	104

Machouat A .	15
MacKenzie M.	269
Mahajan A.M.	181
Mahapatra S.	214, 224, 255, 264
Mahata C.	193, 196, 204, 325
Mahato S.S.	193, 196, 325
Maheta V.D.	255
Mai Z.H.	82, 141, 145
Maiti C.K.	193, 196, 204, 325
Maiti T.K.	193, 196, 325
Martin T.	87
Master R.N.	132
Master R.	7
McComb D.W.	269
McDevitt T.	308
Mertens R.	209
Mingo Liu	321
Mo Z.Q.	283, 290, 304
Mukhopadhya M.	200
Narang V.	96, 104, 279
Neo S.P.	26, 150
Ng E.	137
Nicoletti G.	45
Notte J.	36
Olsen C.	224, 255
On D.	124
Ong M.C.	132
Orio R.L.de	78
Ouyang T.	304
Pan L.Y.	231
Park Y.B.	111
Paul T.	167
Pecht M.	337
Perdu P.	15, 30, 157
Phang J.C.H.	20, 162
Phoa S.L.	279
Pohl H.	100
Pomeroy J.	87
Porth B.	308
Pouget V.	15
Pugatschow A.	43
Quah A.C.T	20
Raghavan N.	67
Rath D.	308

Ravikumar V.K.	279
Riko I.M.	119
Ritzenthaler R.	231
Robbins J.	308
Roth H.	167
Rutkowski T.	308
Saito T.	312
Salam B.	316
Sanchez K.	157
Sarkar S.K.	196
Sarua A.	87
Schoemann S.	100
Schumann F.	100
Scipioni L.	36
Selberherr S.	78
Sengupta M.	193, 196, 204, 325
Seutter S.	224
Sheppard C.J.R.	20
Shi G.J.	231
Shih J.R.	54
Sienkiewicz M.	30
Sim D.	49, 200
Singh K.K.	214, 224
Singh P.K.	214
Song X.	128
Sood B.	337
Stern L.	36
Su R.Y.	321
Sun L.	231
Sun M.	128
Sun W.R.	108
Suzuki M.	312
Sverdlov V.	249
Tan C.M.	67
Tan H.	141
Tan J.B.	72
Tan P.K.	82, 141, 145
Tan S.H.	108
Tan S.L.	162
Tan T.L.	72
Tan Y.C.	49
Tan Y.Y.	124
Tang W.T.	141
Tao G.Q.	219
Tee I.	275
Teo B.H.	96
Teo C.W.	96, 104
Teo J.K.J.	162
Teong J.	82, 150, 234, 275, 283, 290
Thompson C.V.	63
Thompson W.B.	36
Toh K.H.	162

Toh S.L.	82, 141, 145
Tsai J.L.	321
Tsai Y.C.	185
Tseng H.J.	185
Tseng Y.M.	185
Tsou Y.M.	92
Uren M. J.	87
Vasi J.	224
Vegt E.	219
Vereecken P.	308
Verhaar R.	219
Walton E.	308
Wang C.J.	245
Wang H.S.	329
Wang M.X.	329
Wang Q.	275
Wang Q.F.	82
Wang Q.X.	150
Wang X.	128
Wenner M,	308
Widodo J.	304
Wilhite P.	312
Wu C.	128
Wu K.	54, 245
Wu M.S.	92
Xing Z.X.	283, 304
Xu J.	231
Yamada T.	312
Yan L.L.	119
Yang C.Y.	312
Yang D.H.	54
Yeh B.	170
Yoon S.W.	119
Yu A.	119
Zhang B.C.	72
Zhang Z.G.	231
Zhang M.	329
Zhao X.L.	132
Zheng X.	137
Zheng X.H.	96
Zhu D.	26

9781424420391